Biotechnologies and Biomimetics for Civil Engineering

Fernando Pacheco Torgal · J.A. Labrincha
M.V. Diamanti · C.-P. Yu · H.K. Lee
Editors

Biotechnologies and Biomimetics for Civil Engineering

Springer

Editors
Fernando Pacheco Torgal
C-TAC Research Unit
University of Minho
Guimarães
Portugal

J.A. Labrincha
CICECO
University of Aveiro
Aveiro
Portugal

M.V. Diamanti
Politecnico di Milano
Milan
Italy

C.-P. Yu
Chinese Academy of Sciences
Institute of Urban Environment
Xiamen
China

H.K. Lee
Korea Advanced Institute of Science and Technology
Daejeon
Korea
Republic of South Korea

ISBN 978-3-319-35179-7 ISBN 978-3-319-09287-4 (eBook)
DOI 10.1007/978-3-319-09287-4

Springer Cham Heidelberg New York Dordrecht London

Printed on acid-free paper

Springer is part of Springer Science+Business Media (www.springer.com)

I am confident that humanity's survival depends on all of our willingness to comprehend feelingly the way nature works

Buckminster Fuller

I dedicated this book to my wife Adriana,
and to late Tico and Tucha, my companions
of writing, sources of all my drive,
inspiration and mental integrity, ever
present memories of our common
Earthling condition

Foreword

Although human ingenuity makes various inventions it will never discover inventions more beautiful, appropriate and more direct than in Nature because in her nothing is lacking and nothing is superfluous.

Leonardo da Vinci

In Nature there is an economic use of energy and materials. Water and air are vital for the plant and animal kingdoms to live and much of architecture is about how these are channelled in various climates in order to provide the best environment for the organism's survival. Much of our aesthetic is derived from the organic and fluid language that you find in Nature. It involves complex, three dimensional geometries but there is always a rigorous logic behind them. Animals, including humans, and plants have evolved various strategies for dealing with control to suit the local changing conditions such as thermal insulation, cooling via radiating surfaces, blood flow. In addition, plants are unique in being able to convert sunlight into integrated functionality in the process of photosynthesis.

The words *optimisation* and *integration* are often used by building design teams but often without any idea about how these can be achieved, even though there are methods in operational research such as dynamic, integer or linear programming available. Integration and optimisation in Nature appear as completely natural processes.

Now many researchers and designers believe in sustainable solutions for architecture using lessons from the natural world. The attraction of biomimetics for building designers is that it raises the prospect of closer integration of form and function. It promises to yield more interaction with the user by for example, learning from the sophisticated sensor systems in animals including the insect world. However, there are barriers including ever changing standards; the fragmentation of the construction industry at educational and professional levels; the persistent traditional culture with regard to matters like innovation and sacrificing value for cheap capital cost.

This book presents a true galaxy of ideas from biomimetics and how they maybe applied in engineering and architecture. The ideas here will have radical consequences for architecture. New materials can make not only low energy but also more beautiful facades that can produce healthier climates for people to work in. Energy systems using bacterial fuel cells, self-cleaning and self-healing materials and many other ideas are presented here by a distinguished group of international authors.

Not least biomimetics makes us think laterally. We can think the unthinkable because Nature is full of remarkable surprises and yet simplicity too. Our education in schools and universities needs to embrace all the creativity and wonder that Nature can show us. Biomimetics is at the interfaces of biology, engineering, material science, and chemistry and encourages an open dialogue, which can bring enlightenment about problems as displayed in this book.

Derek Clements-Croome

Contents

Chapter 1
Introduction to Biotechnologies and Biomimetics for Civil Engineering

F. Pacheco-Torgal

Abstract This chapter starts with an overview on the sustainable development crucial challenges. The ones directly or indirectly related to the field of civil engineering are highlighted. These include greenhouse gas emissions (GHG) related to the energy consumption of the built environment, aggravated by urbanization forecast expansion, and the recent increase in building cooling needs due to climate change. It also includes the depletion of nonrenewable raw materials and mining-related environmental risks in terms of biodiversity conservation, air pollution, and contamination of water reserves. Some shortcomings of engineering curriculum to address sustainable development challenges (especially civil engineering) are described. Possible contributions of biotechnologies and biomimetics to sustainable development and the rebirth of civil engineering curriculum are suggested. A book outline is also presented.

1.1 Sustainable Development Challenges

Four decades ago several investigators used a computer model based on the fixed-stock paradigm to study the interactions between population, food production, industrial production, pollution, and the consumption of nonrenewable resources. As a result, they predicted that during the twenty-first century the Earth's capacity would be exhausted resulting in the collapse of human civilization as we know it Meadows et al. (1972). Two decades after that an update of this study was published showing that some limits had already been crossed (Meadows et al. 1992).

Rockström et al. (2009) recently proposed a new approach to global sustainability defining nine interdependent planetary boundaries within which they expect that humanity can operate safely. This include:

F. Pacheco-Torgal (✉)
C-TAC Research Centre, University of Minho, Guimarães, Portugal
e-mail: torgal@civil.uminho.pt

F. Pacheco Torgal et al. (eds.), *Biotechnologies and Biomimetics for Civil Engineering*,
DOI 10.1007/978-3-319-09287-4_1

(1) climate change (CO_2 concentration in the atmosphere <350 ppm and/or a maximum change of +1 W m^{-2} in radiative forcing);
(2) ocean acidification (mean surface seawater saturation state with respect to aragonite ≥80 % of pre-industrial levels);
(3) stratospheric ozone (<5 % reduction in O_3 concentration from pre-industrial level of 290 Dobson Units);
(4) biogeochemical nitrogen (N) cycle (limit industrial and agricultural fixation of N_2 to 35 Tg N yr^{-1}) and phosphorus (P) cycle (annual P inflow to oceans not to exceed 10 times the natural background weathering of P);
(5) global freshwater use (<4,000 km^3 yr^{-1} of consumptive use of runoff resources);
(6) land system change (<15 % of the ice-free land surface under cropland);
(7) the rate at which biological diversity is lost (annual rate of <10 extinctions per million species).

Two additional planetary boundaries for which a boundary level was not yet determined are chemical pollution and atmospheric aerosol loading.

According to Rockström et al. (2009) "transgressing one or more planetary boundaries may be deleterious or even catastrophic due to the risk of crossing thresholds that will trigger nonlinear, abrupt environmental change within continental- to planetary-scale systems". These authors estimated that humanity has already transgressed three planetary boundaries for climate change, rate of biodiversity loss, and changes to the global nitrogen cycle. And a recent study (Garcia et al. 2014) confirms the devastating impacts of climate change on biodiversity loss. As a consequence of this worrying status, it remains crucial to act in order to address those problems in a context in which urban human population will almost double, increasing from approximately 3.4 billion in 2009 to 6.4 billion in 2050 (WHO 2014). Other authors also agree that this is the most vital challenge of the twenty-first century (Griggs et al. 2013; Gerst et al. 2014). As Spence et al. (2009) have showed this increase in urban population is economically motivated. The higher the urbanization rate of a country, the higher its GDP. Countries high a GDP per person over $10.000 have a urbanization rate over 60 % while countries with a GDP per person over $30.000 have a urbanization rate around 80 %. Internally the economic importance of working in cities can be assessed by the urban–rural income gap. In China the urban–rural residents' income ratio surged from 2.57:1 in 1978 to 3.13:1 in 2011 (Li et al. 2014a, b).

Climate change is one of the most important environmental problem faced by the Planet Earth (IPCC 2007; Schellnhuber 2008) being due to the increase of carbon dioxide (CO_{2eq}) in the atmosphere, for which the built environment is a significant contributor, with around one-third of global carbon dioxide emissions. In the early eighteenth century, the concentration level of atmospheric CO_{2eq} was 280 parts per million (ppm) at present it is already 450 ppm (Vijayavenkataraman et al. 2012).

Keeping the current level of emissions (which is unlikely given the high economic growth of less developed countries with consequent increases in emission rates) will imply a dramatic increase in CO_{2eq} concentration to as much as 731 ppm in the year 2130 leading to a 3.7 °C global warming above pre-industrial temperatures (Valero et al. 2011). Even if all the greenhouse gas emissions suddenly ceased, the amount already in the atmosphere would remain there for the next 100 years (Clayton 2001). Meaning the rise in the sea level, ocean acidification and the occurrence of extreme atmospheric events will continue. Hansen et al. (2013) are even more pessimistic believing that the climate has already been changed in an irreversible manner. A worrying sign that justifies Hansen's view comes from a recent study (McMillan et al. 2014) based on the measurements collected by the Cryosat-2 satellite which reported an annual loss of 159.000 million tons of the Antartic ice sheet. This represents a 200 % ice loss rate when compared to the 2005–2010 previous survey. This means that adaptation to climate change as well as mitigation of GHGs should be a priority to the built environment (Kwok and Rajkovich 2010; Varias 2013; Boucher et al. 2014; Reckien et al. 2014; Georgescu et al. 2014). Even because buildings are responsible for almost 40 % of energy consumption and energy efficiency improvements show the greatest potential of any single strategy to abate global GHG emissions from the energy sector (IEA 2012). And especially because as a consequence of climate change in the last two decades building cooling needs have increased in an exponential trend going from 6 TJ in 1990 to 160 TJ in 2010 (Balaras et al. 2007). According to Crawley (2008), "the impact of climate change will result in a reduction in building energy use of about 10 % for buildings in cold climates, an increase of energy use of up to 20 % for buildings in the tropics, and a shift from heating energy to cooling energy for buildings in temperate climates". Other authors mention that depending on the climate zone cooling loads are likely to increase by 50 to over 90 % until the end of the century (Roetzel and Tsangrassoulis 2012). Cooling needs will also be aggravated because of urban heat island (UHI) effect, which is one of the major problems in the twenty-first century posed to human beings as a result of urbanization and industrialization of human civilization (Rizwan 2008). And this scenario will get even worse due to the expected increase in urban population and also of predict number of deaths due to heat waves (and their synergic effects with air pollution) that may reach 89,000 deaths/year by the 2050s if no adaptation measures are taken (Pacheco-Torgal et al. 2015). This means that the energy efficiency of the built environment should and must constitute a priority in the field of civil engineering. However, only some parts of the world, like for instance Europe, are now start implementing ambitious building energy efficiency policies like for instance the "nearly zero-energy building" concept to be in effect beyond 2020 (Li et al. 2013; Pacheco-Torgal et al. 2013a, b). Since only several years ago, civil engineering curriculum starts giving this issue some attention. This means that the majority of civil engineering curriculum around the world are obsolete concerning building energy efficiency or the holistic and broader concept of green building (Zuo and Zhao 2014; Li et al. 2014a, b).

Another sustainable development serious problem which is directly related to the field of civil engineering concerns total resource inefficiency. Over the twentieth century, the world increased its fossil fuel use by a factor of 12, whilst extracting 34 times more material resources (COM 2011). Also during the last century, materials use increased eightfold and, as a result, Humanity currently uses almost 60 billion tons (Gt) of materials per year (Krausmann et al. 2009). The global construction industry alone consumes more raw materials (about 3,000 Mt/year, almost 50 % by weight) than any other economic activity, which emphasizes its unsustainable character. Also, in the next few years, the construction industry will keep on growing at a fast pace. China alone will need 40 billion square meters of combined residential and commercial floor space over the next 20 years—equivalent to adding one New York City every 2 years (Pacheco-Torgal and Jalali 2011). Recent estimates on urban expansion suggests that until 2030 a high probability exist (over 75 %) that urban land cover will increase by 1.2 million km^2 (Seto et al. 2012). This is equivalent to an area about the size of South Africa. The forecast urban expansion could lead to the loss of up to 40 % of the species and of 88 % of the global primary vegetation land cover had been destroyed in "biodiversity hotspots" (Pim and Raven 2000; Myers et al. 2000).

The most important environmental threat associated to materials production is not so much the depletion of nonrenewable raw materials (Allwood et al. 2011), but instead, the environmental impacts caused by its extraction, namely extensive deforestation and top-soil loss. In 2000, the mining activity worldwide generated 6,000 Mt of mine wastes to produce just 900 Mt of raw materials (Whitmore 2006).

This means an average use of only 0.15 %, resulting in vast quantities of waste, whose disposal represents an environmental risk in terms of biodiversity conservation, air pollution, and contamination of water reserves. It is worth mention that around 1.2 billion people live in areas of physical scarcity and 500 million people are approaching this situation. As a result, since the 1970s there were 30 serious environmental accidents in mines, 5 of which occurred in Europe (Pacheco-Torgal and Jalali 2011) like for instance the 2010 toxic red mud flood in the town of Kolontar (Hungary). This is rather disturbing because Europe has high environmental standards which mean that countries in which such high standards do not exist environmental disasters could happen much more frequently. Since materials demand will double in the next 40 years, the environmental impacts will therefore increase in a drastic manner (Allwood et al. 2011). Consequently, the World Business Council for Sustainable Development estimates that by 2050 a 4 to 10-fold increase in resource efficiency will be needed (COM, 571). Alwood et al. (2011) recognizes that part of the problem is related to the fact that so far researchers have paid too little attention to the crucial issue of materials efficiency. A possible explanation for that gap relates to the fact that sustainable development principles have not yet been apprehended by University curricula. In recent years, several authors theorized about the way to embed sustainable development in higher education and several institutions made some efforts on this issue (Lozano 2006; Pacheco-Torgal and Jalali 2007; Holmberg et al. 2008; De Vere et al. 2009; Lozano 2010; Waheed et al. 2011). Data from a recent survey completed by final year engineering students in three Irish Higher Education Institutions shows that the engineering students' knowledge on this subject is still deficient (Nicolao and Colon 2012).

Salcedo-Rahola and Mulder (2009) state that "If engineers are to contribute truly to sustainable development, then sustainability must become part of their everyday thinking. This, on the other hand, can only be achieved if sustainable development becomes an integral part of engineering education programs, not a mere "add-on" to the 'core' parts of the curriculum." As a result, the validation of any discipline in any engineering curriculum must be put to a test in which the one million dollar question is "How can your discipline contribute to sustainable development?" (Salcedo-Rahola and Mulder 2009). A more holistic approach is defended by Al-Rawahy (2013) who state that sustainable development has concentrated mainly on physical and tangible issues and assets and that that the most pressing ingredient and the most scarce resource facing the sustainability concept is the ethical and moral values that universities need to proactively and aggressively "infuse" into their respective curricula. This position was already defended by other authors. According to Dator (2005) "engineering is not more important than ethics... and science is not more important than policy and law" therefore a new kind of engineering education is therefore needed to address sustainable development principles. Grasso et al. (2010) mention that "a new kind of engineer is needed, one who can think broadly across disciplines and consider the human dimensions that are at the heart of every design challenge". This is especially important in the context of climate change, which raises many questions with ethical dimensions rooted in the human condition (Willis 2012; Kaklauskas et al. 2013).

1.2 Civil Engineering: The Rebirth of an Obsolete Curriculum Through Biotechnologies and Biomimetics

Recent studies show that students of civil and environmental engineering were reluctant to have sustainability integrated sustainability into existing classes (Watson et al. 2013). One of the latest trends concerning the update of civil engineering towards sustainable development is related to the inclusion of life-cycle assessment (LCA) skills in the education curriculum (Glass et al. 2013). Unfortunately, since almost all construction products are not environmentally friendly, this is the same as choosing between the less of two evils. Another drawback of LCA is the fact that it does not take into account the possible future environmental disasters associated with the extraction of raw materials. This means that, for instance, the LCA of the aluminum produced by the Magyar Aluminum factory, the one responsible for the toxic red mud flood in the town of Kolontar (Hungary), should account for this environmental disaster. Only then construction products will be associated with their true environmental impact. Since that it is almost impossible to put in practice this means that new and truly environmentally friendly construction materials are needed. However, not only is important that civil engineering curricula are updated so they may give future graduates appropriate skills to tackle the sustainable development challenges but it is also important that enough students are interested in following a career as civil engineers. Unfortunately, in the last decade, several Western countries have reported a severe applications reduction to civil engineering. A 50 % reduction was reported on undergraduate applications to civil engineering in UK (Byfield 2001; Edwards et al. 2004). In the UK, a shortfall of 9,000 civil engineers is predicted to occur until 2013 (Byfield 2003).

Nedhi (2002) also confirms that civil engineering is not traditionally viewed as "high tech" engineering and, as a result, student quality and enrollment have been declining across North America. The same also applies in the case of research funding in civil engineering programs. This also reduces the possibility of attracting high grade students. Also in my own country (Portugal) the reduction on the enrollment ratio exceeded 80 % in the last 5 years. To make matters worse, in the last 5 years, the grade of the last student to be admitted has fallen in all the top three Portuguese Universities meaning that civil engineering is less and less capable of attracting high grade students.

In the beginning of the twenty-first century, Yurtseven (2002) already mentioned that a general problem was common to all engineering professions thus affecting negatively the student recruitment. He stated that engineers were viewed as dull individuals by contrast "to the image of a true renaissance engineer, Leonardo da Vinci who was creative and literate… an accomplished painter, architect and scientist."

The explanation for that can be found in the words of Zielinski (2003) who states that "the traditional narrow technical formation produces graduates that are, using the German language expression "fachidiot." It is then of no surprise that engineers are often satirized as persons with zero social skills. For Hamill and Hodgkinson (2003), the responsibility lies in the "invisibility" of the civil engineering profession, the absence of positive role models, low starting salaries, and unattractive working conditions. Lawless (2005) mentions that South Africa faces the same recruitment problem. Adeli (2009) also mentions that the low enrollment ratio of students in civil and environmental engineering at many US universities constitutes a problem to be dealt with. This constitutes a strange fact in a country where civil engineering is viewed as a profession with high industry demand. India, a crucial worldwide player, is also facing a severe shortage of civil engineers to achieve its huge infrastructural development targets. Again, as it happens in the US, the demand is not the problem (construction industry in India needs civil engineers). This reason, however, however seems insufficient to motivate Indian students. Part of the explanation for the low attraction capability of civil engineering relates to the fact that, in India this course is viewed as "brick and mortar engineering" (Chakraborty et al. 2011). Even the "the word "civil" in "civil engineering" is anachronistic and does not represent the works of the so-called civil engineer." As a consequence, civil engineering is "the only engineering discipline to have a name that does not represent the works it undertakes" (Shings 2007). All of what was wrote can be seen as a proof that this curriculum is an obsolete one, which constitutes a worrying issue in the context of future of twenty-first century sustainable development challenges.

However, "recent" nanotechnology achievements regarding the replication of natural systems may provide a solution to solve some of the aforementioned sustainability challenges related to the field of civil engineering. Nanotechnology deals with an atom scale (1 nm = 1×10^{-9} m). A hydrogen atom has a diameter of about one tenth of a nanometer and it takes six bonded carbon atoms to reach a nanometer width. In Nature there are innumerous examples of the nanoscale but one of the most interesting in the "civil engineering context" is the 1–2 nm hydrophobic wax crystals that cover lotus leaves and are responsible for their self-clean ability (Varadan et al. 2010). This new field encompasses a holistic way of perceiving the potential of natural systems (Martin et al. 2010) in which traditional and predominant anthropocentric views are replaced by more eco-centrically approaches (Hofstra and Huisingh 2014) as prerequisite in order to build a sustainable future. It is worth mentioning that this ecological imperative is very far from the 1828 Royal Charter of the Institution of Civil Engineers main purpose, which defined civil engineering as the art of "directing the great sources of power in nature for the use and convenience of man…" (Muir-Wood 2012). Strangely as may seems most civil engineering curriculum and most civil engineering departments in the world still live by this two century outdated and unsustainable motto and some even went to the paradox extreme of try to marketing it as a curriculum forged in sustainable development principles.

The crucial importance of Nature's lessons relates to the fact that it always uses ambient conditions with minimum waste and no pollution, where the result is mostly biodegradable by the contrary man-made materials are processed by heating and pressurizing generating enormous hazardous wastes (Bar-Cohen 2006). On her inspired book Benyus (1997) quoted Mehmet Sarikaya, Professor of material's science and engineering at the University of Washington who wrote: "We are on the brink of a material's revolution that will be on par with the Iron Age and the Industrial Revolution. We are leaping forward into a new age of materials. Within the next century, I think biomimetics, will significantly alter the way in which we live. Learning from nature can become a great challenge for future management". And in fact some more or less recent papers on biological materials (Sarikaya et al. 2003; Sanchez et al. 2005; Chen et al. 2012; Yang et al. 2013; Amini and Miserez 2013) especially the highly cited papers of Markaya et al. (with 823 Scopus citations by May of 2014) and of Sanchez et al. (with 517 Scopus citations by May of 2014) and the extensively detailed paper of Chen et al. serve as a confirmation of the 1997 Saikaya's predictions.

The Biomimicry Institute, founded in 2006 by Janine Benyus, was precursor in this field providing the AskNature online library of research articles on biomimetic design indexed by function. The term biomimetics was used by the first time by Otto Schmitt during the 1950s and relates to the development of novel technologies through the distillation of principles from the study of biological systems. This author made a distinction between an engineering/physics approach to the biological sciences, which was termed "biophysics," and a biological approach to engineering, which he termed biomimetics (Vincent et al. 2006; Lepora et al. 2013). However, the study of biological systems as structures dates back to the early parts of the twentieth century with the work of D'Arcy W. Thompson, first published in 1917. In this work that some authors considered the first major one on this field D'Arcy W. Thompson looked at biological systems as engineering structures and obtained mathematical relationships that described their form (Chen et al. 2012).

According to Vincent (2001), biomimetics is the "technological outcome of the act of borrowing ideas from nature" and this concept would have also been termed as "biomimesis", "biognosis," and "bionics." For this author, the term "bionics" was coined in 1960 by Jack Steele of the US Air Force. In German-speaking countries, the term "Bionik" has become widely accepted for the corresponding field to "Biomimetics." "Bionik"—combining biology and technology (Gebeshuber et al. 2009).

Figure 1.1 gives an overview of the history of biomimetics research. Terms such as "biomimicry," "bioinspiration," and "bioinspired" are derived words from "biomimetic," and "bioinspired" is sometimes used to connote a presumed heir of the word biomimetic (Shimomura 2010).

The publications on the field of biomimetics have experienced an amazing increase from a few 10 papers per year in mid-1990s to the present, doubling every 2–3 years and reaching an annual production of 3,000 papers in 2011 (Lepora et al. 2013). A recent search on Elsevier's Scopus revealed that in 2013 the number of

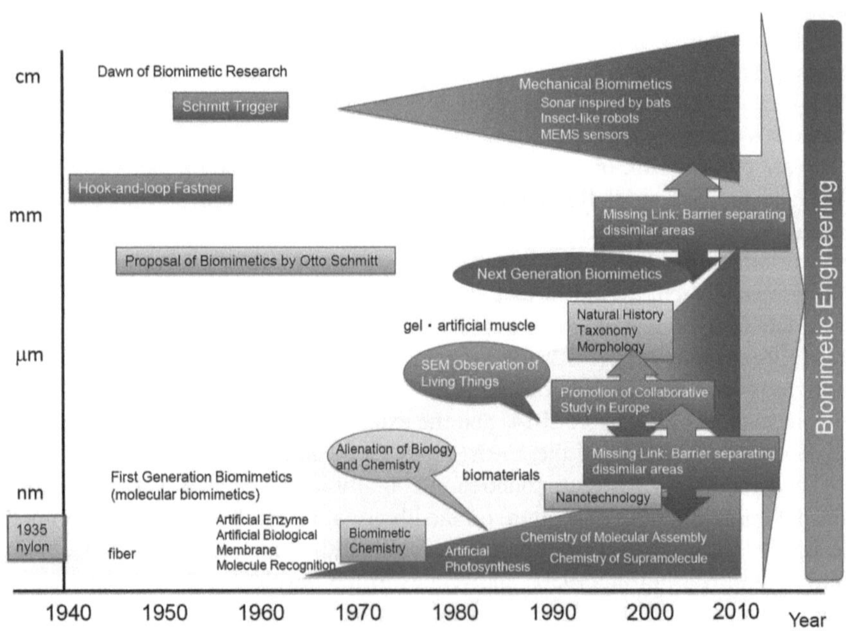

Fig. 1.1 History of biomimetics research (Simomura 2010)

journal papers containing the search term biomimetic reached 12,913, while the search term "bioinspiration" was found in 667 journal papers, and biomimicry in 380.

Analysis of bioinspired materials requires knowledge of both biological and engineering principles. As Vincent (2006) rightly put it "if engineers are going to be able to use ideas from biology, it cannot be stated too often that the biological system must be understood before allowing ideas to be transferred into the engineering environment". Bar-Cohen (2006) states that bridging between the fields of biology and engineering is crucial to harness the most from nature's capabilities. This remind us the words of Sir Isaac Newton about the need of more bridges and less walls, which is especially truth on scientific knowledge. It is important to remember that biologists themselves have recently started establishing bridges with physicists to investigate the weird field of "quantum biology" (the term of "quantum biology" was first mentioned in the beginning of the second half of the twentieth century by Lowdin (1963)). Some recent investigations suggest that plants use quantum "computing" to calculate how best to direct energy through their photosynthetic complexes (Engel et al. 2007; Sarovar et al. 2010; Vedral 2014). Other also suggest that some birds appear to use "quantum entanglement" to sense the Earth's magnetic field, helping to explain how they can migrate long distances (BBSRC 2012). This "weird" concept posits that entangled particles, once separated, can somehow "communicate" with each other instantly so that that a change in one automatically changes the other was famous for having been referred by Albert Einstein as spooky action at a distance (Kaku 2010).

Vincent and Mann (2002) compared solutions of some engineering problems such as cleaning and joining surfaces by natural organisms with those by using the Russian system of problem solving (TRIZ) and noted that TRIZ seemed to have the main qualifications of an effective bridge between biology and engineering. The use of TRIZ is suggested to be able to facilitate the transfer of ideas and analogues from biology for engineering (Vincent et al. 2006; Vincent 2007). Other authors (Denghai and Wuyi 2011) proposed a four-step systematic method of structural bionic design: selecting the most useful structural characteristic of natural organism; analyzing the structural characteristic finally chosen for engineering problem; completing the structural bionic design for engineering structure; and verifying the structural bionic design.

The allocation of biomimetics education to either natural science or engineering schools seems to be difficult to implement in both cases (Gebeshuber et al. 2009). The fact that biologists and engineers typically speak a very different language, may create communication challenges (Helms et al. 2009). In the words of Gebeshuber and Majlis (2010), "the biology papers are frequently inaccessible for engineers, since they are too descriptive and contain concepts and approaches such as taxonomy with its Latin names that are too far from any concept in engineering". These authors thus suggest the establishment of a tree of knowledge and the localisation of scientific articles on this tree.

An important difference between engineers and biologist concerns standardization. While the former are very familiar with standards this is hardly the case of the latter. That's MIT scientist Tom Knight once wrote that those differences could be illustrated by the following example "A biologist goes into the lab, studies a system and finds that it is far more complex than anyone suspected. He's delighted, he can spend a lot of time exploring that complexity and writing papers about it. An engineer goes into the lab and makes the same finding. His response is: 'How can I get rid of this?'" Meaning that contrary to biologists engineers excel at eliminating irrelevant complexity in order to build something that works and is fully understood (Brown 2004; Rai 2010).

Three examples below highlight the importance of biotechnologies and biomimetics for civil engineering.

The first one relates to Ordinary Portland cement (OPC) concrete, a typical civil engineering construction material, being the most used material on the Planet Earth. Its production reaches 10.000 million tons/year and in the next 40 years will increase around 100 % (Pacheco-Torgal et al. 2013a). Currently around 15 % of the total OPC production contains chemical admixtures to modify their properties, either in fresh or hardened state. Concrete super plasticizers based on synthetic polymers include melamine, naphthalene condensates or polycarboxylate copolymers. Environmental concerns justify a growing trend to the use of admixtures based on renewable bio-based feedstocks and or capable of biodegradation. Examples of biopolymers used in concrete include for instance lignosulfonate, pine root extract, protein hydrolysates or even vegetable oils. Biotechnological admixtures processes made in fermentation processes by using bacteria or fungi seem to receive an increase attention. This includes sodium gluconate, curdlan or Welan gum (Planck 2004, 2005).

An important biomimetic application for civil engineering concerns bio-inspired structural design. For instance, deployable structures can be mentioned among shape morphing structures that can change shape like the wing of the insects or the petals of the flowers or like the movable structure of human body (Friedman and Ibrahimbegovic 2013). These structures were born by the application of the basic ideas of tensegrities, as the foldable bridge realization, proposed by Rhode-Barbarigos et al. (2012). The tensegrity concept was born from the exceptional work of the inventor Buckminster Fuller (1962) aiming at maximal structural efficiency. He coined the word "tensegrity" from tensile integrity and defined it as "islands of compression inside an ocean of tension" (Kawaguchi 2002).

Snelson (1965) also worked on the tensegrity field termed as "floating compression" system and much later the cell biologist and bioengineer Ingber (1998) defined this concept as "the architecture of life."

Another important biomimetic civil engineering-related issue concerns biomimetic building "skins." The kinetics and adaptability implicit in this concept are quite the opposite of current trends on passive building design approach (Loonen et al. 2014; Schleicher et al. 2014; Reichert et al. 2014). Of course this concept would not make any sense in a heavy polluted city but only in a biophilic city. The concept of *biophilia*, popularized by Harvard myrmecologist and sociobiologist E.O. Wilson is defined as—the innately emotional affiliation of human beings to other living organisms. This author argues that humans have co-evolved with nature and that we carry with us our ancient brains and our need to connect with and affiliate with nature, to be happy and healthy (Beatley and Newman 2013). Recent findings even show that there is a strong correlation between the lack of green infrastructures in the urban environment and the increase of allergy-related health problems (Von Hertzen et al. 2011; Hanski et al. 2012). Besides since air pollution and higher concentrations of CO_2-induced increases in levels and allergenicity of allergenic pollens may contributing to increasing prevalence of allergic disease and asthma (Haahtela et al. 2013). This means that biomimetic building skins and biophilic cities will be crucial in the near future not only in terms of public health but also in the mitigation of UHI effects.

Since biotechnology is one of the world's fastest growing industries being one of the six Key Enabling Technologies-KETs that will be funded under the EU Framework Programme Horizon 2020 (Pacheco-Torgal 2014) this can also foster the development of start-ups in the field of bio materials and technologies for the construction industry. This is probably one of the most important advantages of the association between civil engineering and biotech areas just because civil engineering is one of the most notorious cases of a desert-like capacity concerning start-up development. This is a key issue because entrepreneurship is a key skill to the development of "start-up businesses, the motor propelling the development of the new economies" (Pacheco-Torgal 2004). Besides as some defend the second half of the twentieth century was the time for the scientific engineer while the twenty-first century will be the time for entrepreneurial engineer (Tryggvason and Apelian 2006; Shi and Vest 2014). This entrepreneurial-based civil engineering is very far from the old and traditional one (Muir-Wood 2012), which has persisted until the twenty-first century and this book intents to start changing. Besides, hot areas usually mean more investigation funds and high capability to attract bright students. The comparison between the impact factor of the journal "Nature nanotechnology" (IF = 31.17) or the journal "Nature biotechnology" (IF = 32.4) with the impact factors of most civil engineering-related journals (usually with an IF below 1.5) gives some insights about this issue. The importance of high impact factor journals in civil engineering can be seen in a recent study (Canas-Guerrero et al. 2013) about the research activity on this field that shows that the "high" impact factor of the Journal of Hazardous Materials (3.93) is one of reasons for its great influence over civil engineering researchers. This hardly constitutes a surprise because due to the existent very high number of journals and papers only the top 10 % will be get read, cited and have an impact (Hamilton 1990, 1991). However, this also shows the absurd of human actions associated to the production of enormous amounts of hazardous materials. The aforementioned study also shows that the average number of citations per paper in the field of civil engineering has been fallen steadily in the last 10 years.

Biotech and biomimetic liaisons can therefore constitute an opportunity to refresh the civil engineering curriculum in order to reverse its low career attractiveness and at the same time contribute to a more sustainable civil engineering industry. This will serve to fulfill ASCE's Vision for Civil Engineering in 2025 one in which civil engineers are entrusted to create a sustainable world (ASCE 2009).

Many books have been written on the field of biotechnology unfortunately the majority of them have absolutely nothing on civil engineering applications. And even the few that have something only have one or just two chapters on this issue. The literature on biomimetic civil engineering applications is even scarcer. It's easy to understand why that happens. Civil engineering is a very conservative field and its focus has saw little change in the last few decades. For instance, OPC was 50 years ago the most important construction material in this field and it remains still.

This book thus provides essential reading concerning biotechnologies and biomimetics for civil engineering. I hope that all of those involved in this field can benefit from the knowledge contained in the present book, which was kindly assembled by a team of international experts.

1.3 Book Outline

Basics of construction microbial biotechnology is the subject of Chap. 2. It includes considerations on the different biotechnological products and biotechnologies for civil engineering. Microorganisms in construction microbial biotechnology are analyzed. The application of microbial biopolymers in the construction industry and in geotechnical engineering is discussed. Biocements and biogrouts are reviewed. This chapter concludes with an analysis on the case of bioremediation of construction sites through biocementation.

Chapter 3 deals with general aspects of biomimetic materials. It includes a brief outline of the discipline and a discussion of general aspects related to the structure and synthesis of natural materials. It reviews the recent progress made in the development of biomimetic materials with improved mechanical resistance, optical, self-cleaning, adhesiveness, and anti-adhesion properties is reviewed with reference made to the most noteworthy examples.

Chapter 4 is concerned with the use of biomimicry as a tool for design for climate change adaptation and mitigation. Different biomimetic approaches to design are discussed and categorized, and a series of case study examples illustrate the benefits and drawbacks of each approach.

Chapter 5 reviews state-of-the-art examples of research concepts and design applications with bio-inspired adaptable solutions for the building envelope. The chapter concludes with an outlook of design support methodologies that can potentially incite the practical uptake of bio-inspired adaptive building skins in the future.

The importance of green building envelopes in promoting biophilic cities is the subject of Chap. 6. A discussion on the green building envelope strategy is included. Its contribution for air quality improvement, temperature regulation and insulating properties are reviewed. The chapter also includes an overview on green building details including its costs.

Chapter 7 concerns the use of microalgae photobioreactors (PhBR) as innovative construction systems for the production of bio-energy. An overview on photobioreactors is given. The concept of architectural photobioreactors (A-PhBR) is presented and discussed. Examples of PhBR in facades formed from blocks of translucent glass and of horizontal PhBR for roof and urban fountains are presented. Extensive graphical details are also presented.

The reduction of indoor air pollutants through biotechnology constitutes the subject of Chap. 8. The chapter starts with an analysis on the importance of indoor air pollution and its impact on health costs and human dead's followed by a review on current practices and the history of bioremediation on indoor air. The different air pollutants (volatile organic compounds, carbon dioxide, and others) are discussed. A comparison between physiochemical and biological methods are carried out. Hybrid physiochemical–biological systems, active biofiltration of indoor air, and phytoremediation and horticultural biotechnology are discussed. Considerations on the health benefits of indoor plants unrelated to air quality are included. Microbial systems as well as biological indoor air cleaning commercial systems are reviewed.

Chapter 9 deals with the mechanisms underlying bioinspired self-cleaning (hydrophilicity and hydrophobicity) and to the fields of application of these effects.

Common concepts on wettability are reminded. The different mechanisms of self-cleaning are reviewed and detailed. Examples of hydrophilic and superoleophobic plants and animals are given. The chapter concludes with a section on production techniques and applications which presents examples of biomimetic self-cleaning surfaces, and give details on how they were created.

Chapter 10 reviews the development of the bio-inspired concept on bridge design in the past two decades from two major forms: stationary forms and movable forms. The objective is to show how the inspiration from the biological world has influenced recent bridge designs and discusses how the bio-inspired idea could transform into a new language for the future bridge design industry. Four major challenges of the marriage between biology and engineering were discussed and latest endeavor on each aspect was presented.

Bioinspired sensors for structural health monitoring is the subject of Chap. 11.

Topics ranging from bio-inspired algorithms, creature-like robots, and skin-like sensors are presented.

Chapter 12 deals with bioinspired, flexible structures and materials. The potential of biomimetics in form finding and the development of structural systems based on constant or reversible elastic deformation are discussed. Elastic building materials and biomimetic abstraction techniques are introduced and two case studies are provided.

Bioinspired concrete is the subject of Chap. 13. An overview of Earths minerals is presented. Several bioinspired cements are covered. The environmental challenges with cements and concrete are revewed.

Chapter 14 describes the production of bacteria for structural concrete reviewing the mechanism of microbially induced calcium carbonate precipitation (MICP).

Chapter 15 deals with the use of bacteria for surface treatment reviewing the main mechanisms of the process and literature on biodeposition carbonates as surface treatment agents for the decrease of permeability of concrete materials and structures.

Chapter 16 describes a case study concerning the use of bacterial surface treatment for normal and lightweight concrete.

Chapter 17 identifies remediation techniques for contaminated soils including physical, chemical, biological, thermal, and other treatments.

Chapter 18 deals with the use of microbial fuel cells (MFCs) for wastewater treatment. The concept of MFCs is introduced, and the materials and design of MFCs are summarized. In-depth discussion of the microbiology of MFCs was also included.

References

Adeli H (2009) Vision for civil and environmental engineering departments in the 21st century. J Prof Issues Eng Educ Pract 135(1):1–3

Allwood J, Ashby M, Gutowski T, Worrell E (2011) Material efficiency: a white paper. Resour Conserv Recycl 55:362–381

Al-Rawahy KH (2013) Engineering education for sustainable development: the missing link. Proc Soc Behav Sci 102:392–401

Amini S, Miserez A (2013) Wear and abrasion resistance selection maps of biological materials. Acta Biomater 9:7895–7907

ASCE (2009) Achieving the vision for Civil Engineering in 2025. A road map for the profession. ASCE, Virginia

Balaras C, Grossman G, Henning H, Infante-Ferreira C, Podesser E, Wang L, Wiemken E (2007) Solar air conditioning in European overview. Renew Sustain Energy Rev 11:299–314

Bar-Cohen Y (2006) Biomimetics—using nature to inspire human innovation Bioinspir Biomim 1:1–12

BBSRC (2012) Biologists and physicists collaborate to investigate 'quantum biology'. http://www.bbsrc.ac.uk/news/research-technologies/2012/121203-f-quantum-biology.aspx. Accessed in 20 June 2014

Beatley T Peter, Newman P (2013) Biophilic cities are sustainable, resilient cities. Sustainability 5:3328–3345

Benyus J (1997) Biomimicry: innovation inspired by nature. Willeam Morrow Paperbacks

Boucher O, Forster P, Gruber N, Ha-Duong M, MLawrence M, Lenton T, aas A, Vaughan N (2014) Rethinking climate engineering categorization in the context of climate change mitigation and adaptation. Clim Change 5:23–35

Brown C (2004) BioBricks to help reverse-engineer life. EE Times. http://www.eetimes.com/document.asp?doc_id=1150423

Byfield MP (2001) Graduate shortage: the key to civil engineering's future? Proc Inst Civil Eng Civil Eng 144(4):161–165

Byfield MP (2003) British civil engineering skills: defusing the time bomb. Proc Inst Civil Eng Civil Eng 156(4):183–186

Cañas-Guerrero I, Mazarrón FR, Pou-Merina A, Calleja-Perucho C, Suárez-Tejero MF (2013) Analysis of research activity in the field "Engineering, Civil" through bibliometric methods. Eng Struct 56:2273–2286

Chakraborty S, Iyer N, Krishna P, Thakkar S (2011) Assessment of civil engineering inputs for infrastructure development. Indian Natl Acad Eng

Chen P-Y, McKittrick J, Meyers M (2012) Biological materials: functional adaptations and bioinspired designs. Prog Mater Sci 57:1492–1704

Clayton R (2001) Editorial: Is sustainable development an oxymoron? Trans I Chem E 79(Part B): 327–328

COM (2011) 571 final. Roadmap to a resource efficient Europe

Crawley DB (2008) Estimating the impacts of climate change and urbanization on building performance. J Build Perform Simul 1:91–115

Dator J (2005) Universities without quality and quality without universities. On Horiz 13(4):199–215. Q Emerald Group Publishing Limited, ISSN 1074-8121

Denghai X, Wuyi C (2011) Systematic method of applying structural characteristics of natural organisms to mechanical structures. Trans Tianjin Univ 17:293–297

De Vere I, Bissett Johnson K, Thong C (2009) Educating the responsible engineer: socially responsible design and sustainability in the curriculum. In: EPDE09/134, International conference on engineering and product design education, 10–11 September 2009. University of Brighton, UK

Edwards DJ, Dainty ARJ, Love PED (2004) A sustainable cohort of professional civil engineering graduates? Uncovering the United Kingdom graduate crisis. Int Educ J 5(3):374–384

Engel GS, Calhoun TR, Read EL et al (2007) Evidence for wavelike energy transfer through quantum coherence in photosynthetic systems. Nature 446:782–786

Friedman N, Ibrahimbegovic A (2013) Overview of higly flexible, deployable lattice structures used in architecture and civil engineering undergoing large displacements. J Built Environ 1:85–103

Fuller RB (1962) Tensile-integrity structures. US Pat 3063521

Garcia R, Cabeza M, Rahbek C, Araujo M (2014) Multiple dimensions of climate change and their implications on biodiversity. Science 2(3844):6183

Gebeshuber IC, Majlis BY (2010) New ways of scientific publishing and accessing human knowledge inspired by transdisciplinarity. Tribology 4:144–151

Gebeshuber IC, Gruber P, Drack M (2009) A gaze into the crystal ball: biomimetics in the year 2059. Proc IMechE Part C: J Mech Eng Sci 223:2899–2918

Georgescu M, Morefield P, Bierwagen B, Weaver C (2014) Urban adaptation can roll back warming of emerging megapolitan regions. Proc Natl Acad Sci USA PNAS 111:2909–2914

Girst MD, Raskin PD, Rockstrom J (2014) Contours of a resilient global future. Sustainability 6:123–125

Glass J, Dyer T, Georgopoulos C, Goodier C, Paine K, Parry T, Baumann H, Gluch P (2013) Future use of life-cycle assessment in civil engineering. Proc ICE Constr Mater 166:204–212

Grasso D, Burkins M, Helble J, Martinelli D (2010) Dispelling the myths of holistic engineering. In holistic engineering education: beyond technology. Springer, New York, pp 159–166

Griggs D, Stafford-Smith M, Gaffney O, Rockström J, Öhman MC, Shyamsundar P, Steffen W, Glaser G, Kanie N, Noble I et al (2013) Policy: sustainable development goals for people and planet. Nature 495:305–307

Haahtela T, Holgate S, Pawankar R, Akdis C, Benjaponpitak S, Caraballo L, Demain J, Portnoy J, Von Hertzen L, WAO Special Committee on Climate Change and Biodiversity (2013) The biodiversity hypothesis and allergic disease: world allergy organization position statement. World Allergy Organ J 6:3

Hamill L, Hodgkinson L (2003) Civil engineering's image in schools—and how to change it. Proc ICE Civil Eng 156(2):78–85

Hamilton DP (1991) Research papers: Who's uncited now? Science 251:25

Hamilton DP (1990) Publishing by and for the numbers? Science 250:1331–1332

Hansen J et al (2013) Assessing “Dangerous Climate Change”: required reduction of carbon emissions to protect young people. Future generations and nature. PLOS One. doi:10.1371/journal.pone.0081648

Hanski I, Von Hertzen L, Fyhrquist N, Koskinen K, Torppa K, Laatikainen T, Karisola P, Auvinen P, Paulin L, Mäkelä M, Vartiainen E, Kosunen T, Alenius H, Haahtela T (2012) Environmental biodiversity, human microbiota, and allergy are interrelated. Proc Natl Acad Sci USA. http://www.pnas.org/content/early/2012/05/01/1205624109.full.pdf

Helms M, Vattam S, Goel A (2009) Biologically inspired design: process and products. Des Stud 30:606–622

Hofstra N, Huisingh D (2014) Eco-innovation characterizaed: a taxonomic classification of relationships between humana and nature. J Clean Prod 66:459–468

Holmberg J, Svanström M, Peet DJ, Mulder K, Ferrer-Balas D, Segalàs J (2008) Embedding sustainability in higher education through interaction with lecturers: case studies from three European technical universities. Eur J Eng Educ 33(3):271–282

IEA (2012) World Energy Outlook 2012. OECD/IEA, Paris

Ingber DE (1998) The architecture of life. Sci Am Mag 278:48–57. http://pt.scribd.com/doc/35190367/Architecture-of-Life-Scientific-American-by-Ingber

Kaklauskas A, Pacheco-Torgal F, Grafakos S, Lapinskiene V (2013) Built environment life cycle process and climate change. In: Pacheco-Torgal F, Mistretta M, Cabeza L, Kalauskas A, Granqvist CG (eds) Nearly zero energy building refurbishment. A multidisciplinary approach. Springer, London, UK, pp 61–97

Kaku M (2010) The Bizarre and wonderful world of quantum theory—and how understanding it has ultimately changed our lives. http://bigthink.com/dr-kakus-universe/the-bizarre-and-wonderful-world-of-quantum-theory-and-how-understanding-it-has-ultimately-changed-our-lives. Accessed on 20 June 2014

Kawaguchi M (2002) Preface. In: Motro R (ed) Tensegrity: structural systems for the future. Butterworth-Heinemann, Oxford (2006)

Krausmann F, Gingrich S, Eisenmenger N, Erb K-H, Haberl H, Fischer-Kowalski M (2009) Growth in global materials use, GDP and population during the 20th century. Ecol Econ 68:2696–2705

Kwok A, Rajkovich N (2010) Addressing climate change in comfort standards. Build Environ 45:18–22

Lawless A (2005) A wake up call to address the capacity crisis in SA civil engineering. Civil Eng/Siviele Ingenieurswese 13(10):40–43

Lepora N, Verschure P, Prescott T (2013) The state of the art in biomimetics. Bioinspir Biomim 8:013001

Li Y, Yang l, He B, Zhao D (2014a) Green building in China: Needs great promotion. Suatain Cities Soc 11:1–6

Li Y, Wang X, Zhu Q, Zhao H (2014b) Assessing the spatial and temporal differences in the impacts of factor allocation and urbanization on urban–rural income disparity in China, 2004–2010. Hatitat Int 42:76–82

Li D, Yang L, Lam J (2013) Zero energy buildings and sustainable development implications—a review. Energy 54:1–10

Loonen RCGM (2014) Climate adaptive building shells. http://www.pinterest.com/CABSoverview/

Lowdin P-O (1963) Proton tunneling in DNA and its biological implications. Rev Mod Phys 35(3):724–732

Lozano R (2010) Diffusion of sustainable development in universities' curricula: an empirical example from Cardiff University. J Clean Prod 18(7):637–644

Lozano R (2006) Incorporation and institutionalization of SD into universities: breaking through barriers to change. J Clean Prod 14(9–11):787–796

Martin J, Roy E, Diemont S, Fergunson B (2010) Traditional ecological knowledge (TEK): ideas, inspiration, and designs for ecological engineering. Ecol Eng 36:839–849

McMillan Malcolm, Shepherd Andrew et al (2014) Increased ice losses from Antarctica detected by CryoSat-2'. Geophys Res Lett. doi:10.1002/2014GL060111
Meadows DL, Meadows DH, Behrene JRW (1972) The limits to growth. MIT Press, Boston
Meadows DL, Meadows DH, Randers J (1992) Beyond the limits: global collapse or a sustainable future. Earthscan, London
Muir-Wood D (2012) Civil engineering: a very short introduction. Oxford University Press, Oxford, UK
Myers N, Mittermeier RA, Mittermeier CG, da Fonseca GAB, Kent J (2000) Biodiversity hotspots for conservation priorities. Nature 403:853–858
Nehdi M (2002) Crisis of civil engineering education in information technology age: analysis and prospects. J Prof Issues Eng Educ Pract 128(3):131–137
Nicolaou I, Conlon E (2012) What do final year engineering students know about sustainable development? Eur J Eng Educ 37(3):267–277
IPCC (Intergovernmental Panel on Climate Change) (2007) Climate Change 2007: synthesis report. IPCC, Geneva
Pacheco-Torgal F (2015) Introduction to eco-efficient materials for mitigating building cooling needs. In: Pacheco-Torgal F, Labrincha JA, Cabeza l, Granqvist CG (eds) Eco-efficient materials for mitigating building cooling needs Woodhead Publishing Limited Abington Hall, Cambridge, UK (in press)
Pacheco-Torgal F (2014) Eco-efficient construction and building materials research under the EU framework programme Horizon 2020. Constr Build Mater 51:151–162
Pacheco-Torgal F, Cabeza L, Mistretta M, Kaklauskas A, Granqvist CG (2013a) Nearly zero energy building refurbishment. A multidisciplinary approach. Springer, London, UK
Pacheco-Torgal F, Labrincha JA, Jalali S, John VM (2013b) Eco-efficient concrete. Woodhead Publishing Limited Abington Hall, Cambridge, UK 592 p
Pacheco-Torgal F, Jalali S (2011) Eco-efficient construction and building materials. Springer, London, UK
Pacheco-Torgal F, Jalali S (2007) Does engineering students need sustainable development courses? In: ICEE 2007 international conference on engineering education—The moving frontiers of engineering, Coimbra
Pacheco-Torgal F (2004) EST-IPCB civil engineering curricula design. A proposal. Itinerary Mag (Institute of Educational Sciences) 157–180 (only in Portuguese)
Planck J (2005) Applications of biopolymers in construction engineering. Biopolymers Online 29–39. http://www.wiley-vch.de/books/biopoly/pdf_v10/vol10_17.pdf
Pimm SL, Raven P (2000) Biodiversity. Extinction by numbers. Nature 403:843–845
Plank J (2004) Application of biopolymers and other biotechnological products in building material. Appl Microbiol Biotechnol 66:1–9
Rai A (2010) Unstandard standardization: the case of biology. Commun ACM 53:37–39
Reckien D, Flacke J, Dawson R, Heidrich O, Olazabal M, Foley A, Hamann J-P, Orru H, Salvia M, De Gregorio Hurtado S, Geneletti D, Pietrapertosa F (2014) Climate change response in Europe: what's the reality ? Analysis of adaptation and mitigation plans from 200 urban areas in 11 countries. Clim Change 122:331–340
Reichert S, Menges A, Correa D (2014) Meteorosensitive architecture: biomimetic building skins based on materially embedded and hygroscopically enabled responsiveness. Comput Aided Des (in press)
Rhode-Barbarigos LGA, Bel Hadj ALI N, Motro R, Smith IFC (2012) Design aspects of a deployable tensegrity-hollow-rope footbridge. Int J Space Struct 27:81–96
Rizwan A, Dennis L, Liu C (2008) A review on the generation, determination and mitigation Urban Heat Island. J Environ Sci 20:120–128
Roetzel A, Tsangrassoulis A (2012) Impact of climate change on comfort and energy performance in offices. Build Environ 57:349–361
Rockström J, Steffen W, Noone K, Persson Å, Chapin FS, III, Lambin E, Lenton TM, Scheffer M, Folke C, Schellnhuber H, Nykvist B, De Wit CA, Hughes T, van der Leeuw S, Rodhe H, Sörlin S, Snyder PK, Costanza R, Svedin U, Falkenmark M, Karlberg L, Corell RW, Fabry

VJ, Hansen J, Walker B, Liverman D, Richardson K, Crutzen P, Foley J (2009) Planetary boundaries: exploring the safe operating space for humanity. Ecol Soc 14(2):32. http://www.ecologyandsociety.org/vol14/iss2/art32/
Salcedo-Rahola B, Mulder K (2009) European sustainable development master report. What has Europe got to offer? TU Delft
Sanchez C, Arribart H, Guille MMG (2005) Biomimetism and bioinspiration as tools for the design of innovative materials and systems. Nat Mater 4:277
Sarikaya M, Tamerler C, Jen AK-Y, Schulten K, Baneyx F (2003) Molecular biomimetics: nanotechnology through biology. Nat Mater 2(9):577–585
Sarovar M, Ishizaki A, Fleming G, Whaley B (2010) Quantum entanglement in photosynthetic light harvesting complexes. Nat Phys 6:462–467
Schellnhuber H (2008) Global warming. Stop worrying, start panicking? Proc Natl Acad Sci USA 105:14239–14240
Schleicher S, Lienhard J, Poppinga S, Speck T, Knippers J (2014) A methodology for transferring principles of plant movements to elastic systems in architecture. Comput Aided Des (in press)
Seto KC, Buneralp B, Hutyra LR (2012) Global forecasts of urban expansion to 2030 and impacts on biodiversity and carbon pools. PNAS 109(40):16083–16088
Shi H, Vest C (2014) Higher engineering education for the 21st century: American perspective and its implications for China. In: International conference on education reform and modern management. Atlantis Press, pp 139–143
Shimomura M (2010) The new trends in next generation biomimetics material technology: learning from biodiversity. Sci Technol Trends Q Rev 3(7):53–75
Singh A (2007) Civil engineering: anachronism and black sheep. J Prof Issues Eng Educ Pract 133(1):18–30
Snelson K (1965) Continuous tension, discontinuous compression structures. US Pat 3(169):611
Spence M, Annez PC, Buckley RM (2009) Urbanization and growth (Commission on growth and development). The International Bank for Reconstruction and Development, The World Bank
Tryggvason G, Apelian D (2006) Re-engineering engineering education for the challenges of the 21st Century. JOM 58(10):14–17
Valero A, Agudelo A, Valero A (2011) The crepuscular planet. A model for the exhausted atmosphere and hydrosphere. Energy 36:3745–3753
Varadan V, Pillai A, Mukherji D, Dwivedi M, Chen L (2010) Nanoscience and nanotechnology in engineering. World Scientific Publishing Company, Singapore
Varias N (2013) Brainstorming for a beautiful planet. ISBN 1484851331
Vedral V (2014) Quantum entanglement. Nat Phys 10:256–258
Vijayavenkataraman S, Iniyan S, Goic R (2012) A review of climate change, mitigation and adaptation. Renew Sustain Energy Rev 16(1):878–897
Vincent JFV (2006) Applications—influence of biology on engineering. J Bionic Eng 3(3):161–177
Vincent JFV, Bogatyreva O, Bogatyrev N, Bowyer A, Pahl A-K (2006) Biomimetics: its practice and theory. J R Soc Interface 22:9471–9482
Vincent JFV, Mann DL (2002) Systematic technology transfer from biology to engineering. Philos Trans R Soc Lon A 360:159–173
Vincent JFV (2001) Stealing ideas from nature. In: Pellegrino S (ed) Deployable structures. Springer, Vienna
Vincent J (2007) Re: Designing around existing patents through TRIZ, personal email communication, 5 May
Von Hertzen L, Hanski I, Haahtela T (2011) Natural immunity. Eur Mol Biol Organ EMBO Rep 12:1089–1093
Waheed B, Khan F, Veitch B, Hawboldt K (2011) An integrated decision-making framework for sustainability assessment: a case study of memorial University. High Educ Policy 24(4):481–498

Watson MK, Lozano R, Noyes C, Rodgers M (2013) Assessing curricula contribution to sustainability more holistically: experiences from the integration of curricula assessment and students' perceptions at the Georgia Institute of Technology. J Clean Prod 61(2013):106–116
Whitmore A (2006) The emperor's new clothes: sustainable mining. J Clean Prod 14:309–314
WHO (2014) Urban population growth. Global health observatory. http://www.who.int/gho/urban_health/situation_trends/urban_population_growth_text/en/
Willis A (2012) Constructing a story to live by: ethics, emotions and academic practice in the context of climate change. Emot Space Soc 5:52–59
Yang W, Chao C, McKittrick J (2013) Axial compression of a hollow cylinder filled with foam: a study. Acta Biomater 9:5297–5304
Yurtseven H (2002) How does the image of engineering affect student recruitment and retention? A perspective from the USA. Glob J Eng Educ 6(1):17–23
Zielinski W (2003) Vision of an engineer in the 21st century. In: Juan S (ed) 9th International conference on engineering education
Zuo J, Zhao Z-Y (2014) Green building research-current status and future agenda: a review. Renew Sustain Energy Rev 30:271–281

Chapter 2
Basics of Construction Microbial Biotechnology

V. Ivanov, J. Chu and V. Stabnikov

Abstract Construction Microbial Biotechnology is a new area of science and engineering that includes microbially-mediated construction processes and microbial production of construction materials. Low cost, sustainable, and environmentally-friendly microbial cements, grouts, polysaccharides, and bioplastics are useful in construction and geotechnical engineering. Construction-related biotechnologies are based on activity of different microorganisms: urease-producing, acidogenic, halophilic, alkaliphilic, denitrifying, iron- and sulfate-reducing bacteria, cyanobacteria, algae, microscopic fungi. The bio-related materials and processes can be used for the bioaggregation, soil biogrouting and bioclogging, biocementation, biodesaturation of water-satured soil, bioencapsulation of soft clay, biocoating, and biorepair of the concrete surface. Construction Microbial Biotechnology is progressing toward commercial products and large-scale applications. The biotechnologically produced materials and construction-related microbial biotechnologies have a lot of advantages over conventional construction materials and processes.

V. Ivanov (✉) · J. Chu
Department of Civil and Environmental Engineering, Iowa State University, Ames, USA
e-mail: vivanov@iastate.edu

J. Chu
e-mail: jchu@iastate.edu

V. Stabnikov
Department of Microbiology and Biotechnology, Ukrainian National University of Food Technologies, Kiev, Ukraine
e-mail: vstab@uduft.edu.ua

F. Pacheco Torgal et al. (eds.), *Biotechnologies and Biomimetics for Civil Engineering*,
DOI 10.1007/978-3-319-09287-4_2

2.1 Introduction

The discipline of Microbial Biotechnology includes a scientific and practical knowledge of using microorganisms and their products. We differentiate by the area of biotechnological applications such subdisciplines as Medical, Pharmaceutical, Industrial, Agricultural, and Environmental Biotechnology. Currently, a new subdiscipline of Microbial Biotechnology, Construction Microbial Biotechnology, can be differentiated. There are two major directions in Construction Microbial Biotechnology: (1) the microbial production of construction materials and (2) the applications of microorganisms in construction process. Many different biotechnological products and biotechnologies for civil engineering are developing within these directions (Fig. 2.1).

Production of cement, which is a major construction material, is energy-consuming and environmentally-unfriendly. Energy represents 20–40 % of the total cost of cement production because temperature above 950 °C is needed for transformation of limestone to cement clinker. New construction materials, microbial biocements, can be produced from limestone, dolomite, or iron ore at temperature 20–60 °C with less than 10 % of energy used for the manufacturing of conventional cement. Therefore, cost of biocements can be lower than that of conventional cement. There are also a lot of other advantages of microbially-based biocementing or bioclogging materials over conventional cements and grouts, for example sustainability due to production from organic matter, low viscosity, and low risk of negative environmental consequences. It is important that biocement can be produced from the same raw materials that are using for cement production.

Another type of biomaterials, which are used in construction industry are industrially produced or in situ synthesized microbial polysaccharides. Such industrially produced polysaccharides as xanthan, welan, succinoglucan, curdlan, chitosan are used in dry-mix mortars, wall plasters, self-leveling underlayers, injection grouts to improve viscosity, water retention, set retarding, flowability (Plank 2004). Other biopolymers for example, proteins or their hydrolysates as biosurfactant, can also be used for if the cost is acceptable. Sewage sludge of municipal wastewater treatment plants, which is a waste microbial biomass producing in quantities of several million tons a year, could be used also as a source of cheap microbial polymers.

Production of bacterial polysaccharides in soil after addition of bacterial cells and necessary nutrients in situ is used to modify soil properties. This approach could be used for such geotechnical applications as dam control, wind soil erosion control, earthquake liquefaction mitigation, construction of reactive barrier, and long-term stabilization of contaminated soils. Different kinds of organic wastes can be used as a source of organic matter for polysaccharide-producing microorganisms in large-scale geotechnical applications to diminish the cost of soil clogging. Initiated growth of exopolysaccharide-producing photosynthetic bacteria or algae on the surface of sand can be used for the wind erosion control. Surface growth of

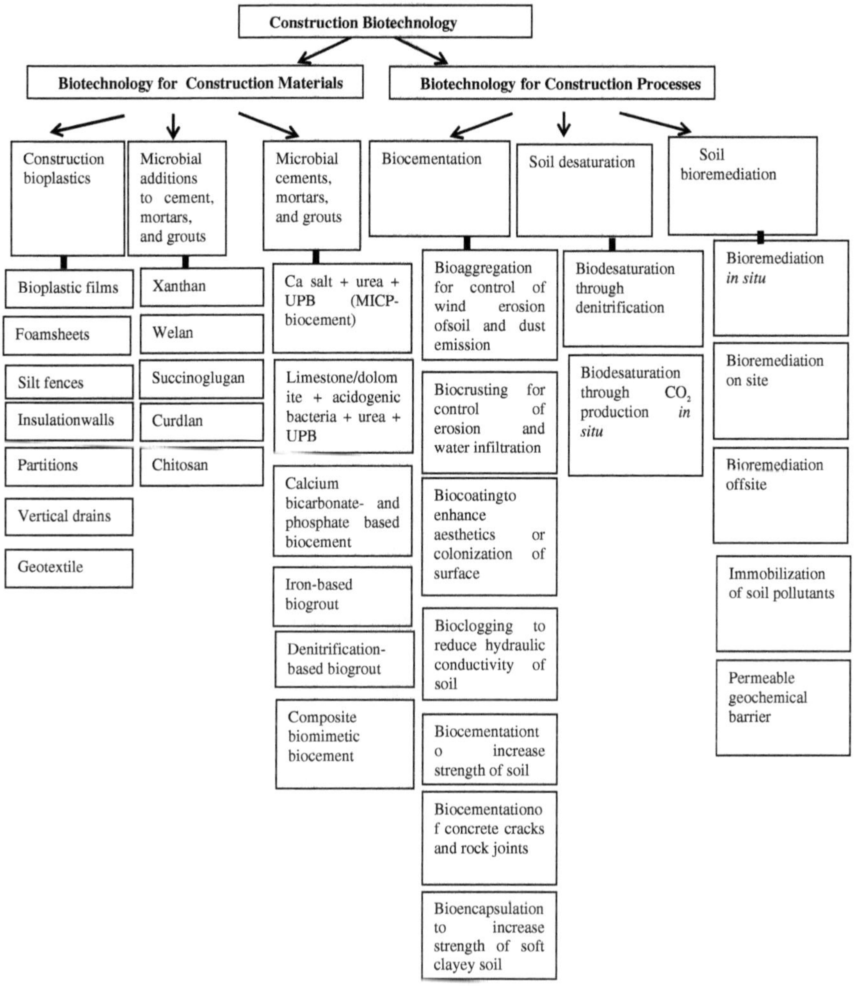

Fig. 2.1 The R&D directions of Construction Biotechnology

microbial photosynthetics in irrigation channel or aquaculture pond is an effective way for the seepage control.

The biotechnological production of construction biomaterials is sustainable process because renewable agricultural and biotechnological biomass residues are used as organic raw materials and as the components of composite biocement. One potentially important construction material, microbial bioplastic, can be produced from agricultural residuals and organic fraction of municipal solid wastes.

In some geotechnical processes, microorganisms themselves are performing useful function. There are at least eight types of construction-related biotechnological processes classified by the results of the microbial treatment:

- *bioaggregation* of soil or particles is a process to increase size of the fine particles so that water and wind soil erosions, sand movement, as well as dust emission will be reduced (Bang et al. 2011; Stabnikov et al. 2013a);
- *biocrusting* of soil surface is a process to form mineral or organic crust onto soil surface so that that erosion, dust emission, and water infiltration will be reduced (Stabnikov et al. 2011; Chu et al. 2012a);
- *biocoating* of solid surface is a process to form a layer on solid surface so that aesthetics or colonization of surface will be enhanced;
- *bioclogging* of soil or porous matrix is a process to fill in the pores and channels in soil/matrix so that hydraulic conductivity of soil or porous matrix will be significantly reduced (Ivanov et al. 2012);
- *biocementation* of soil or particles is a process to increase significantly strength of soil or particles (Ghosh et al. 2005; Mitchell and Santamarina 2005; Whiffin et al. 2007; Ivanov and Chu 2008; De Muynck et al. 2008a, b, 2010, 2012; Sarda et al. 2009; van der Ruyt and van der Zon 2009; Achal et al. 2010; Ivanov 2010; Van Tittelboom et al. 2010; Dosier 2013; Chu et al. 2012a; 2014;DeJong et al. 2010, 2013; van Paassen et al. 2010; Harkes et al. 2010; Dhami et al. 2012; Li and Qu 2012; Raut et al. 2014);
- *biodesaturation* of soil is a process to decrease saturation and liquefaction potential of soil through biogas production in situ (Chu et al. 2009a, 2013b; He et al. 2013; Rebata-Landa and Santamarina 2012);
- *bioencapsulation* of clay/soil/particles is a process to increase strength of soft clayey soil through the formation of strong shell around a piece of soft material (Ivanov et al. 2014);
- *bioremediation* of soil is a process to remove pollutants from soil or immobilize pollutant in soil before construction (Warren et al. 2001; Fujita et al. 2004; Mitchell and Ferris 2005).

Classification of construction biotechnologies by the results of their applications is illustrated in Fig. 2.2.

Any biotechnology for the production of construction materials includes three major stages:

(1) upstream processes such as preparation of medium, equipment, and microbial inoculum (seeds);
(2) core process such as cultivation of microorganisms;
(3) downstream processes such as concentration of biomass or microbial product, its drying, packing, washing of the equipment, treatment or disposal of wastes.

The example can be the biotechnology for the production of calcium- and urea-based biocement (Figs. 2.3, 2.4).

All these processes should be monitored and controlled to ensure efficiency of the processes.

Any biotechnological application of microorganisms in construction process also includes three major stages:

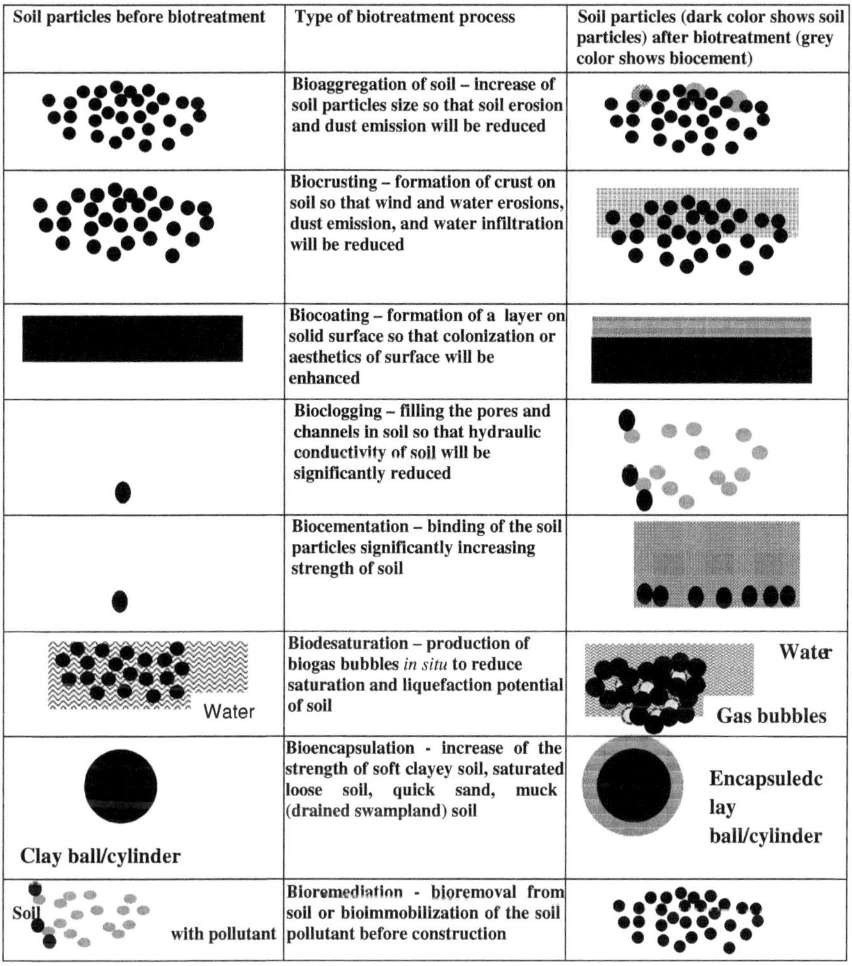

Fig. 2.2 The results of the construction biotechnologies

(1) upstream processes such as preparation of soil or construction material for the treatment, preparation of the reagents, equipment and microbial inoculum for the treatment;
(2) core process such as biotreatment of soil, construction material, or the construction objects;
(3) downstream processes such as disinfection (if needed) and solid, liquid, or gaseous wastes treatment or disposal.

All these processes should be monitored and controlled to ensure efficiency of the processes. The example can be the biotechnology for mitigation of saturated sand liquefaction using denitrifying and biosealing bacteria (Fig. 2.5).

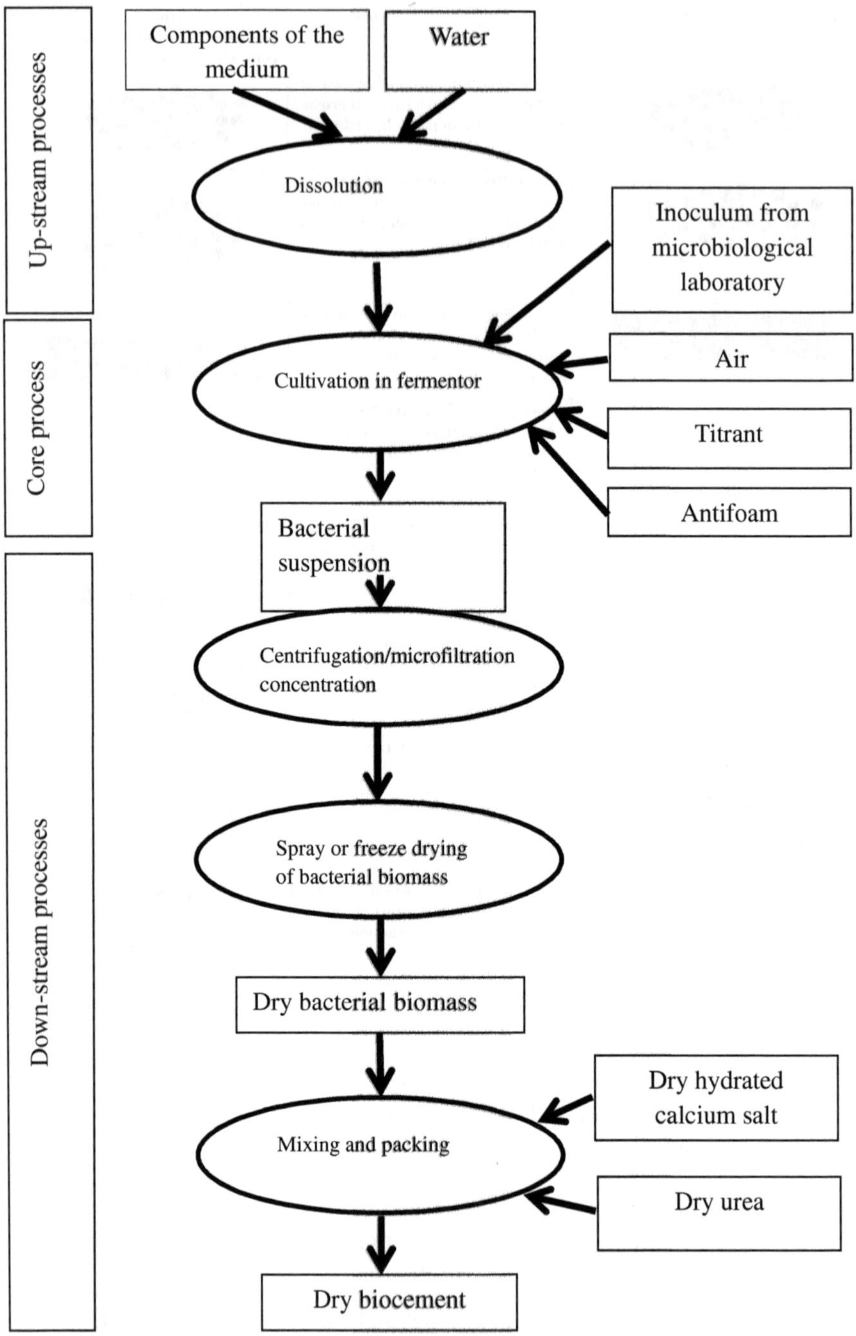

Fig. 2.3 Biotechnology of production of calcium- and urea-based biocement

Fig. 2.4 Cultivation of bacteria for biocement production in 50 L fermentor(photo taken at the Fermentation Facility, Iowa State University,USA)

Application of microorganisms for ground improvement can be performed by indigenous microorganisms of soil, without preparation and supply of the microbial inoculum, and without downstream processes. However, there could be such downstream step as detoxication of polluted air and water after biocementation.

The medium for microbial cultivation and the treatment medium for soil/particles can be mixed together or used separately in the form of solution, suspension, or slurry by the mixing of chemical reagents and agricultural fertilizers. Different kind of wastes or residuals such as mining and agricultural residuals, organic fraction of municipal solid wastes, sewage sludge, and reject water of municipal wastewater treatment plants can be used as a medium to reduce the cost of large-scale biogeotechnical work.

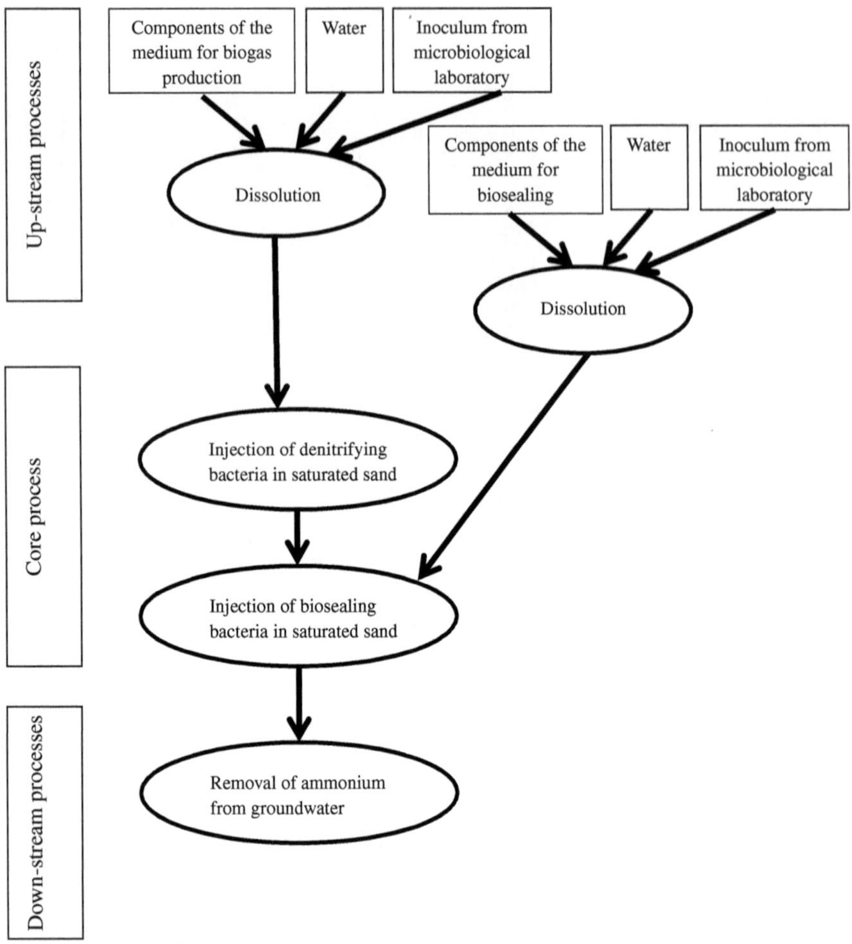

Fig. 2.5 Biotechnology for mitigation of saturated sand liquefaction using denitrifying and biosealing bacteria

2.2 Microorganisms in Construction Microbial Biotechnology

Most applicable microorganisms for Construction Microbial Biotechnology among kingdoms of *Archaea*, *Bacteria*, *Fungi*, *Plants*, and *Animals* are representatives of *Bacteria* because of their small cell size (0.5–10 μm), big physiological diversity (pH 2–10; temperature from −10 to +110 °C), big spectrum of biogeochemical reactions (oxidation-reduction of organics, oxygen, nitrate, ferric, sulfate), highest growth and metabolic rates. There are three evolutionary lines of chemotrophic (feeding by energy of chemical compounds) Gram-positive *Bacteria* (prokaryotes

Table 2.1 Physiological groups of chemotrophic prokaryotes

Evolutionary lines of prokaryotes	Physiological groups			
	Fermenting prokaryotes	Anoxic respiring prokaryotes	Microaerophilic and facultative anaerobic prokaryotes	Aerobic respiring prokaryotes
Gram-negative *Bacteria* (prokaryotes of aquatic evolutionary origin)	1	2	3	4
Gram-positive *Bacteria* (prokaryotes of terrestrial evolutionary origin)	5	6	7	8
Archaea (prokaryotes of extreme environments evolutionary origin)	9	10	11	12

of terrestrial evolutionary origin with thick and rigid cell wall), Gram-negative *Bacteria* (prokaryotes of aquatic evolutionary origin with thin and elastic cell wall) and *Archaea* (prokaryotes of environments with extreme temperature, pH, or strong anaerobic conditions). These three evolutionary lines contain four parallel physiological groups differentiated by the type of energy-yielding oxidation-reduction reactions: (1) fermenting, (2) anoxic respiring, (3) microaerophilic and facultative anaerobic, and (4) aerobic respiring prokaryotes. In total, there are 12 groups of chemotrophic prokaryotes differentiated by the type of energy-yielding oxidation-reduction reactions (Table 2.1).

Depending on the real conditions and requirements of the construction process, all these physiological groups can be involved in biotechnologies of construction materials or construction process biotechnologies. However, in majority cases anaerobic, anoxic, facultative anaerobic, and aerobic Gram-positive bacteria are most suitable for applications related to the soil improvement because of osmotic tolerance of these bacteria. Facultative anaerobic or aerobic Gram-negative bacteria are most suitable for biosynthesis of construction biomaterials. Applications of phototrophic (utilizing light energy) prokaryotes in civil engineering are rare. For example, Gram-negative phototrophic *Bacteria*, cyanobacteria can be used for the formation of soil crust to diminish water and wind erosion of soil.

The microorganisms that are used to start up the bioprocess are called inoculum by microbiologists or "seeds" by civil and environmental engineers. The inoculum could be a suspended, frozen, dried, or cooled microbial biomass. Cultivation after inoculation is performed in batch or continuous mode. Inoculum for construction materials production or biotreatment of soil/particles is selected using the following microbiological and molecular-biological methods:

(1) Obtaining and testing of the microbial strains from national collections of microorganisms, for example American Type Culture Collection (ATCC, USA) or German Collection of Microorganisms and Cell Cultures (DSMZ, Germany).

(2) Isolation, identification, and testing of wild strains from natural sites with environmental conditions close to the conditions that are needed for the biotreatment, for example, with high salinity, high or low temperature, aerobic or anaerobic conditions, alkaline, or acid pH. However, many bacteria are pathogenic (causing diseases) for human, animal, and plants. Therefore, biosafety of biotechnological process is always an important issue and only nonpathogenic isolated strains of bacteria can be used for civil engineering applications.
(3) Autoselection in continuous culture (Cheng and Cord-Ruwisch 2013), screening of the mutants (Li et al. 2011), and construction of the recombinant microbial strains from wild strains for the biotreatment. However, there are many restrictions on the applications of recombinant microbial strains so they can be used mainly for industrial production of such construction materials as polysaccharides or bioplastic.
(4) Selection and testing of suspended enrichment cultures using such selective conditions (selection pressure) as source of energy, carbon, nitrogen and phosphorus, temperature, pH, salinity (osmotic pressure), concentration of heavy metals, concentration of dissolved oxygen, spectrum and intensity of light (for photosynthetic microorganisms). Some autoselected features of the enrichment culture can be genetically unstable and could disappear after several generations when the selection pressure will be absent (Ivanov et al. 2012b).
(5) Selection and testing of aggregated enrichment cultures, such as flocs, biofilms, granules using such selective pressure as settling rate of microbial aggregates and adhesion of cells to solid surface. An example is formation of bacterial cells aggregates that cannot penetrate inside sand, settled onto the surface of sand and formed calcite crust (Stabnikov et al. 2011; Chu et al. 2012b).

In some cases, when soil is rich with indigenous microorganisms able to perform needed biogeochemical reaction, for example urease activity, soil biotreatment can be performed by indigenous microorganisms, without preparation and supply of microbial inoculum (Burbank et al. 2011, 2012a, b). To enhance the needed biogeochemical function of indigenous microorganisms, soil can be amended with the related reagent. For example, to enhance urease activity of indigenous microorganisms before the biotreatment, urea can be added to soil (Burbank et al. 2011). However, if microorganisms used in geotechnical bioprocess are indigenous it does not mean that they are safe for human, animals, and plants because nonselective conditions of the soil bioprocess, especially application of nutrients-rich medium, can enhance the proliferation of pathogens or opportunistic pathogens in soil.

Core process in the production of construction material, i.e., cultivation of microorganisms is performed in batch, semi-continuous (sequencing batch), complete mixing continuous, or plug-flow continuous mode in the bioreactor, where the components are mixing using stirring, upflow of liquid or gas through the reactor, or horizontal rotation of the bioreactor.

2.3 Application of Microbial Biopolymers in Construction Industry and Geotechnical Engineering

Animal biomaterials such as blood, urine, eggs, milk, lard, and plant biopolymers such as wood, straw, bark, cactus juice, flour have been used as admixtures from ancient times to improve properties of mortars and plasters. Straw and cattle dung were used and are used even at the present time in rural construction as the composite biomaterials to improve construction properties of clay. Probably, the Aztecs used fermented juice of nopal cactus (*Opuntia ficus indica*) to improve plasticity and water absorption capacity of lime mortar and earthen plasters due to the presence of cellulose fibers, gel polysaccharides, and fermentation products. Extracts of this nopal cactus and water hyacinth are proposed in our days to enhance viscosity of cement-based materials (León-Martínez et al. 2014; Sathya et al. 2013). The chemical derivates of plant biopolymers, for example carboxymethylcellulose, carboxymethylcellulose sulfate or such industrial waste as lignosulfonates are often used as cement and mortar admixtures for set retarding and increase of plasticity of self-consolidated concrete (Plank 2004; Yuan et al. 2013).

Chemical and biological admixtures in cement- and gypsum-based materials are using for dispersing/thikening effects, viscosity enhancement, water retention, set acceleration and retardation, air-entrainment, defoaming, hydrophobization, adhesion and film forming (Plank 2003) to improve such properties of the material as plasticity, water retention, adhesion, shrinkage reduction, flow ability, and stability. The global market of admixtures is estimated at the level of US$15 billions with the share of more than 500 different biological and biodegradable admixtures about 13 % (Plank 2004).

The advantage of microbial admixtures is that the biosynthesis rate of the microbial biopolymers is significantly higher, by 2–4 orders of magnitude than that of the plants and these substances can be produced in industrial scale on biotechnological factories. The major application of microbial biopolymers in construction industry is addition to concrete and dry-mix mortars. The examples of microbial admixtures that are used in concrete are protein hydrolysates and welan gum; and in case of dry-mix mortar these admixtures are succinoglycan and xanthan gum. The market share of microbial biopolymers is expected to increase because of technological advances and the growing trend to use naturally based or biodegradable products in building materials (Plank 2004; Ramesh et al. 2010). These microbial products of biotechnological industry are mainly viscosity-enhancing admixtures used to achieve high resistance to segregation of concrete. Major biotechnological admixtures are shown in Table 2.2.

Sewage sludge or dewatered dry sewage sludge of municipal wastewater treatment plants, are used in the cement and concrete industry (Mun 2007; Fytili and Zabaniotou 2008). It is mainly biomass of anaerobic Bacteria and Archaea performing acidogenic and methanogenic fermentations. This material contains various biopolymers, such as linear and branched polysaccharides, globular proteins and rRNA, and linear chains of DNA and mRNA. Our experiments with

Table 2.2 Major biotechnological admixtures used in building materials (based on Plank 2003, 2004; Mun 2007; Fytili and Zabaniotou 2008; Pacheco-Torgal and Jalali 2011; Pei et al. 2013; and other sources)

Admixture	Chemical type	Biotechnological process	Function, applications and dosage	Estimated price (US$)
Sodium gluconate	Organic salt	Biooxidation of glucose	Set retarder; superplasticizer; water reducer; corrosion inhibitor of the re-bar in concrete, used in gypsum plaster; concrete mix; mortars, grouts, at dosage 0.1–0.4 %	500–850
Xanthan gum	Polysaccharide	Biosynthesis by bacteria *Xanthmonas campestris*	Viscosifier (thickener), set retarder for self-consolidated concrete, floor screeds, paints, 0.2–0.5 %	2,000–5,000
Welan gum	Polysaccharide	Biosynthesis by bacteria *Alcaligenes sp.*	Viscosifier (thickener), set retarder for self-consolidated concrete, floor screeds, paints, 0.1–0.5 %	2,000–5,000
Scleroglucan	Polysaccharide	Biosynthesis by fungi from genus *Sclerotium, Corticium, Sclerotinia, Stromatinia*	Thermostable viscosifier (no loss of viscosity at 90 °C for 500 days), similar to xanthan gum	2,000–5,000
Succinoglycan	Polysaccharide	Biosynthesis by bacteria *Alcaligenes sp.*	High shear-thinning behavior with temperature-induced viscosity breakback for self-leveling compounds, soil stabilization, 1–15 g/L of water	2,000–5,000
Curdlan gum	Polysaccharide	Biosynthesis by bacteria from genera *Agrobacterium* or *Alcaligenes*	Viscosifier (thickener), set retarder for self-consolidated concrete, up to 10 g/L of water; can absorb water about 100 times	1,000–10,000
Polyaspartic acid	Polyanionic polyaminoacid	Chemical synthesis	Biodegradable dispersant, inhibitor of corrosion in concrete, air-entraining agent for concrete or mortar, set retarder for gypsum	1,000–10,000
Sodium alginate	Anionic polysaccharide	Extraction from brown seaweeds	Stabilizer, thickener, and emulsifier	2,000–10,000
Carrageenan	Linear sulfated polysaccharide	Extract of plants or red seaweeds	Foam for protecting freshly poured concrete from premature drying during highway construction	2,000–10,000
Dextran	Polysaccharide	Microbial synthesis		3,000–9,000

(continued)

Table 2.2 (continued)

Admixture	Chemical type	Biotechnological process	Function, applications and dosage	Estimated price (US$)
			Admixture to portland cement, self-leveling grouts, fresh or saltwater oil well cement slurries, microfine cements improving flow resistance (rheology-modified additive)	
Pullulan	Polysaccharide	Biosynthesis by fungi *Auerobasidium pullulans*	Viscosifier (thickener), set retarder for self-consolidated concrete	5000–20000
Sewage sludge	Mixture of biopolymers	Waste biomass of municipal wastewater treatment plants	Viscosifier (thickener), set retarder for self-consolidated concrete	0
			Production of sintered light-weight aggregated for nonstructural concrete (clay:sewage sludge ratio is from 1:1 to 1:3 by mass)	
			Co-combustion of sewage sludge in cement manufacturing (5 % of the clinker production capacity)	
Sewage sludge ash	Ash after incineration of sewage sludge	Waste of incineration plants	Increase of compressive strength of the mortars due to pozzolanic properties of sewage sludge ash	0
Bacterial cell walls	Structural polysaccharides and proteins	Aerobic cultivation of bacteria *Bacillus subtilis*; could be produced from sewage sludge	Microstructured filler for concrete increasing compressive strengths of concrete by 15 % and decreasing its porosity; 0.03–3.3 % by mass	2,000–10,000

addition of pure linear (xanthan, DNA), branched (amylopectin) or globular (albumin) biopolymers to Portland cement showed that even addition of 0.1 % of these hydrophilic biopolymers changed strength of concrete. It was higher with addition of biopolymers than in control after 3 days but was lower than in control after 7 days, probably because a thin layer of the hydrophilic biopolymer on the cement grain hindered its hydration (Wang et al. 2013, not published data).

Another important application of microbial biopolymers is production of bacterial polysacchartides in soil to modify its geotechnical properties (Stewart and Fogler 2001). The most suitable groups of microorganisms that produce insoluble extracellular polysaccharides to bind the soil particles and fill in the soil pores are as follows:

- oligotrophic bacteria from genus *Caulobacter* (Tsang et al. 2006);
- aerobic Gram-negative bacteria from genera *Acinetobacter*, *Agrobacterium*, *Alcaligenes*, *Arcobacter*, *Cytophaga*, *Flavobacterium*, *Pseudomonas*, and *Rhizobium* (Portilho et al. 2006; Ross et al. 2001);
- species of Gram-positive facultative anaerobic and aerobic bacteria, such as *Leuconostoc mesenteroides* producing water-insoluble exopolymer dextran (Stewart and Fogler 2001) and *Cellulomonas flavigena* producing a curdlan-type (beta-1,3-glucan) exopolysaccharide from cellulose (Kenyon et al. 2005).

It is well known that almost all bacteria produce exopolysaccharides under excess of carbohydrates or other water soluble sources of carbon over source of nitrogen. Therefore, such food-processing wastes or sub-products as corn glucose syrup, cassava glucose syrup and molasses with C: N ratio > 20 are used for industrial production of bacterial water-insoluble polysaccharides (Portilho et al. 2006). After growth of exopolysaccharide-producing bacteria in soil, its permeability for water is greatly reduced. Growth of exopolysaccharide-producing bacteria in soil can be used for different geotechnical applications such as selective zonal bioremediation, harbor and dam control, erosion potential minimization, earthquake liquefaction mitigation, construction of reactive barrier, and long-term stabilization of contaminated soils (Yang et al. 1993). Organic wastes such as organic fraction of municipal solid wastes, sewage sludge, composted poultry manure can be used as a source of organic matter for exopolysaccharide-producing microorganisms in large-scale applications to diminish the cost of soil clogging.

2.4 Construction Bioplastics

There is clear trend in construction industry for using of biodegradable materials and biopolymers (Plank 2004; Ramesh et al. 2010). There is considerable interest in the development of biodegradable bioplastics for construction industry. Its advantages are that use of this bioplastic will reduce the land for disposal of construction wastes and it is producing from renewable sources so their production will increase environmental and economic sustainability of construction industry.

However, the cost of bioplastics produced aseptically in fermenters is usually several times higher than the cost of petrochemical-based plastics, so the reduction of the bioplastic production costs using cheap raw materials and technological innovations is still essential for the bioplastic industry and applications. Most available types of bioplastics for construction industry are polyhydroxyalkanoates (PHAs), which are polyesters accumulated up to 80 % of dry bacterial biomass as a storage compound. Most important PHAs are poly-3-hydroxybutyrate (PHB) with monomer formula (-OCH(CH_3)-CH_2-C(O)-) and polyhydroxyvalerate (PHV) with monomer formula (-OCH(CH_2CH_3)-CH_2-C(O)-). Accumulated PHAs can be extracted from bacterial biomass and used in practice as bioplastic with melting temperature 160–180 °C, tensile strength 24–40 MPa, and elongation at break 3–142 %. Chemical and physical properties of PHAs can be found in numerous reviews (Lowell and Rohwedder 1974; Braunegg et al. 1998; Castilho et al. 2009; Sudesh et al. 2000; Sudesh and Abe 2010; Volova 2004; DeMarco 2005; Khanna and Srivastava 2005; Lenz and Marchessault 2005).

The following options for raw materials, biotechnology of production, and applications of bioplastic can help to diminish the cost of the bioplastic PHAs:

1. Use of cheap raw materials (Serafim et al. 2008): organic fraction of municipal solid wastes, liquid wastes of municipal wastewater treatment plants, food-processing waste, or agricultural wastes such as unbaled straw; corn cobs, stalks, and leaves (corn stover); silage effluent; horticulture residuals; farm yard manure; coconut fronds, husks, and shells; coffee hulls and husks; cotton (stalks), nut shells; rice hull, husk, straw, and stalks, sugarcane bagasse. Globally, 140 billion metric tons of biomass is generated every year from agriculture, which is equivalent to approximately 50 billion tons of oil. So, biomass wastes have attractive potentials for large-scale industries and community-level enterprises (UNEP 2009).
2. Batch or continuous nonaseptic cultivation for biosynthesis of bioplastic by mixed bacterial culture (Yu et al. 1999; Lu 2007);
3. Production of crude bioplastic for construction industry and agriculture avoiding its concentration and extraction of bioplastic using chemical treatment, filtration, centrifugation, or flotation.

For the biosynthesis of PHAs under nonaseptic conditions, organic wastes can be converted to organic acids through acidogenic fermentation of organics, then organic acids can be converted to PHAs (Yu 2006). Most typical material balance of acidogenic fermentation is as follows (Madigan et al. 2012):

$$C_6H_{12}O_6 + 0.82\ H_2O \rightarrow 1.13\ CH_3COOH + 0.35\ C_2H_5COOH + 0.26\ C_3H_7COOH + 1.67\ CO_2 + 2.47\ H_2, \quad (2.1)$$

where $C_6H_{12}O_6$ is a monomer of cellulose, CH_3COOH, C_2H_5COOH, C_3H_7COOH are acetic, propionic and butyric acids, respectively. The pH of organic fraction of municipal waste can be dropped below 5.5 during acidogenic fermentation (Barlaz et al. 2010), meanwhile optimal pH for acidogens is above 6.0 (Moosbrugger et al. 1993). To maintain optimal pH during acidogenic fermentation of organic wastes the fine powder of limestone ($CaCO_3$) or dolomite ($CaMgCO_3$) can be added. There will be two fractions at the end of the process: dissolved acetate salts of Ca and Mg that could be used for biosynthesis of bioplastic, and semi-solid residuals that can be used for biocementation (Ivanov and Chu 2008; Ivanov 2010) of sand and gravel in road, pond, or channel construction or for soil erosion control. Using this semi-solid residual all solid wastes from bioplastic production can be used as the components of biocement (see below).

Batch biosynthesis of bioplastic is simpler but less productive than continuous process, which productivity can be about 1 kg of PHAs/day/m^3 of bioreactor (Ben Rebah et al. 2009). Production of PHAs can be done as semi-continuous cultivation of a mixed culture using a feast-famine cycle comprising a feast phase and a famine phase in one bioreactor. This cycling process promotes not only accumulation of PHAs in biomass but also selection of PHAs-producing microorganisms (Beun et al. 2006; van Loosdrecht et al. 1997, 2008).

All known methods of PHAs extraction suffer from a high cost or environmental pollution and are difficult to be industrialized. Therefore, crude bioplastic, without extraction of PHAs, could be used for construction applications. Major advantage of PHAs for construction applications is biodegradability of bioplastic to carbon dioxide and water for about 1.5 months in anaerobic sewage, 1.5 years in soil, and 6.5 years in seawater (Mergaert et al. 1992; Reddy et al. 2003; Castilhio et al. 2009). Dead bacterial biomass with PHAs contains also polysaccharides of cell wall, proteins, polynucleotides, and phospholipids, which content is about 15, 50, 25, and 10 % of dry biomass without PHAs, respectively, and biodegradation rate is higher than that of PHAs. Therefore, from the point of view of biodegradability of bioplastic construction wastes there is no need to extract PHAs from biomass but to use dry biomass with PHAs as a crude nanocomposite material. Such nanocomposites should be more flexible and better biodegradable than extracted PHAs. Sustainability of this biodegradable construction materials is due to: (1) production of bioplastic from renewable sources or even from organic wastes; (2) fast biodegradability of this material under the conditions of landfill or composting so negative effect of construction waste on environment will be minimized.

One area of applications of nanocomposite bioplastic from bacterial biomass containing PHAs is the production and use of biodegradable construction materials, which do not require removal and incineration after temporary application. Biodegradable bioplastic foam can be used for insulation walls and partitions, construction of nonstructural (internal) elements such as separating walls and partitions, and for the temporarily constructions that can be landfilled for fast degradation. Other examples of potential application of crude nanocomposite from bacterial biomass and PHAs are construction silts and dust fences that can be landfilled for fast biodegradation or composted as biomass. Biodegradable plastics could be also useful for vertical drains, geotextile, geomembranes, soil stabilization mats. These materials are used temporarily for soft soil stabilization, filtration and drainage (Park et al. 2010; Ogbobe et al. 1998; Arunaye and Mwasha 2011; Chu et al. 2009b), so biodegradability of the material can eliminate the cost of extraction and disposal of the temporal objects. There could be a big market for biodegradable bioplastic construction material, which does not require removal and incineration after use.

2.5 Biocements and Biogrouts

There are possible and used different types of biocementation based on diverse biogeochemical reactions performed by microorganisms (Ivanov and Chu 2008; Ivanov 2010).

2.5.1 Calcium- and Urea-Dependent Biocementation

Most popular type of biocementation is based on so-called microbially-induced calcium carbonate precipitation (MICCP), which is formation of calcium carbonate minerals such as calcite, vaterite, or aragonite on the surface of soil particles due to: (1) adhesion of cells of urease-producing bacteria (UPB) on the surface of particle; (2) creating a microgradient of concentration of carbonate and pH in the site of cell attachment due to hydrolysis of urea by urease of UPB. The biogeochemical reactions of this biocementation process are as follows:

$$(NH_2)_2CO + 2\,H_2O \xrightarrow{\text{Urease}} CO_2\uparrow + 2NH_4OH, \quad (2.2)$$

$$CO_2 + H_2O \leftrightarrow H_2CO_3 \xrightarrow{\text{Carbonicanhydrase}} H^+ + HCO_3^- \leftrightarrow 2H^+ + CO_3^{2-}, \quad (2.3)$$

$$CaCl_2 + H_2CO_3 \rightarrow CaCO_3\downarrow + 2HCl, \quad (2.4)$$

$$2\,HCl + 2\,NH_4OH \rightarrow 2\,NH_4Cl + 2\,H_2O, \quad (2.5)$$

Total

$$(NH_2)_2CO + 2\,H_2O + CaCl_2 \xrightarrow{\text{Urease and carbonicanhydrase}} CaCO_3\downarrow + 2NH_4Cl. \quad (2.6)$$

Enzyme urease (EC 3.5.1.5) is produced by a wide range of microorganisms because urea is a final product of nitrogen metabolism of human and animals and plays a role of nitrogen source for many microorganisms in nature. Another enzyme with important role in MICP is carbonic anhydrase (EC4. 2.1.1) catalyzing the reversible hydration of CO_2 (Dhami et al. 2014).

MICP is developing and testing in the field mainly for numerous geotechnical applications (Seagren and Aydilek 2010; DeJong et al. 2010, 2013; Sarayu et al. 2014): to enhance stability of the slopes and dams (van Paassen et al. 2010; Harkes et al. 2010); for road construction and prevention of soil erosion (Mitchell and Santamarina 2005; Whiffin et al. 2007; Ivanov and Chu 2008; Ivanov 2010); for the construction of the channels, aquaculture ponds, or reservoirs in sandy soil (Chu et al. 2013a, b; Stabnikov et al. 2011); for sand immobilization and suppression of dust (Bang et al. 2011; Stabnikov et al. 2013a); to reinforce sand in near-shore areas (van der Ruyt and van der Zon 2009).

The applications of MICP in civil engineering can be as follows: the production of bricks (Sarda et al. 2009; Dhami et al. 2012; Raut et al. 2014); the remediation of cracks in concrete and rocks and increase of durability of concrete structures (De Muynck et al. 2008a, b, 2010, 2012; Achal et al. 2010; van Tittelboom et al. 2010; Ghosh et al. 2005; Li and Qu 2012); the concrete improvement (Pacheco-Torgal and Labrincha 2013a; Pacheco-Torgal and Jalali, 2014); the self-remediation of concrete (Jonkers 2007; Jonkers et al. 2010; De Muynck et al. 2008a, b; Wiktor and Jonkers 2011; Ghosh et al. 2006; Siddique and Chahal 2011; Wang et al. 2012); the modification of mortar (Ghosh et al. 2009; Vempada et al. 2011); consolidation of porous stone (Jimenez-Lopez et al. 2008); the bioremediation of weathered-building stone surfaces (Fernandes 2006; Webster and May 2006; Achal et al. 2011); the fractured rock permeability reduction (Cuthbert et al. 2013); dust suppression (Bang et al. 2011; Stabnikov et al. 2013a); the construction of ponds and channels (Chu et al. 2012b, 2013a; Stabnikov et al. 2011); the mitigation of earth quake-caused soil liquefaction (DeJong et al. 2006, 2013; Chu et al. 2009a; Weil et al. 2012; Montoya et al. 2012); the encapsulation of soft clay (Ivanov et al. 2014); the coating of surfaces with calcite for enhanced marine epibiota colonization (Ivanov et al. not published data).

In majority of the biocementation research, the Gram-positive bacterial species *Sporosarcina pasteurii* (former *Bacillus pasteurii*), especially the strain *S. pasteurii* ATCC 11859 (DSM 33), is used because of its high urease activity and ability to grow at pH above 8.5 and at high concentration of calcium, at least at 0.75 M Ca^{2+}. Last property is especially important for MICP. Other physiologically similar species using for biocementation are the representatives of the genus *Bacillus: B. cereus* (Castanier et al. 2000); *B. megaterium* (Bang et al. 2001; Dhami et al. 2014), *B. sphaericus* (Hammes et al. 2003; De Muynck et al. 2008a, b; Wang et al. 2012), *B. pseudofirmus* (Jonkers et al. 2010), *B. subtilis* (Reddy et al. 2010), *B. alkalinitrilicus* (Wiktor and Jonkers 2011), *B. licheniformis* (Vahabi et al. 2014), *B. lentus* (Sarda et al. 2009) and not identified species (Stabnikov et al. 2011, 2013b; Hammes et al. 2003; Lisdiyanti et al. 2011). It is well known that some halotolerant species of genus *Staphylococcus* exhibited high urease activity (Jin et al. 2004; Christians et al. 1991). Halotolerant urease-producing strain of Gram-positive bacteria of *Staphylococcus succinus* was isolated from water of the Dead Sea with salinity 34 % (Stabnikov et al. 2013b). However, the strains of *S. succinis* are often hemolytic and toxigenic ones (Zell et al. 2008) and

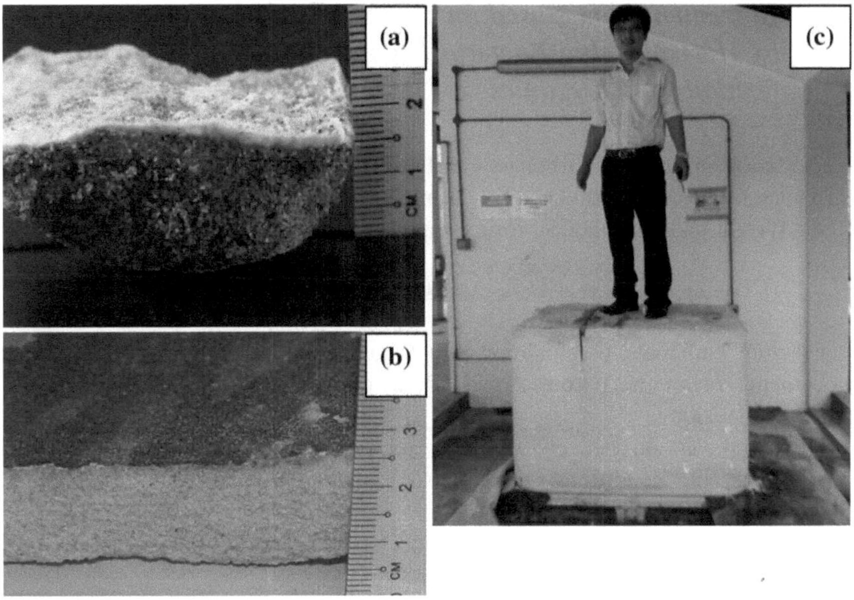

Fig. 2.6 Spatial types of biocementation: formation of the crust on surface of sand (**a**), formation of the biocemented layer of the defined thickness (**b**), biocementation of monolith (**c**)

were associated with some infectious diseases (Novakova et al. 2006; Taponen et al. 2008). Therefore, the isolated strain was not used for biocementation studies and applications.

Biocementation can be performed as bulk biocementation through the supply of bacterial suspension altogether or separately with solutions of calcium and urea by injection, and using surface percolation (Cheng and Cord-Ruwisch 2012; Stabnikov et al. 2011) or surface spraying (Stabnikov et al. 2011, 2013a; Chu et al. 2012a). Modifying types of the treatment it is possible to form the crust on surface of soil (Fig. 2.6a), biocemented layer of defined thickness (Fig. 2.6b) or biocemented monolith (Fig. 2.6c).

There are several drawbacks in the conventional MICP process: (1) by-product of urea hydrolysis is ammonium and ammonia that are toxic substances for workers, harmful for aquatic environment and atmosphere, and increases the risk corrosion because of high pH (Pacheco-Torgal and Labrincha 2013a); (2) the brittleness of calcite crystals bonding the soil particles; and (3) the cost of calcium reagent and urea are higher than the cost of conventional cement. Therefore, the improvements of MICP as well as new types of biocementation have to be developed to overcome these disadvantages of conventional MICP.

2.5.2 *Biocementation Based on Production of Carbonates by Heterotrophic Bacteria During Aerobic or Anoxic Oxidation of Organics*

Precipitation of calcium carbonate can be due to increase pH and production of carbonate by heterotrophic bacteria during aerobic oxidation of organics (Ehrlich 1999; Wright and Oren 2005), for example, in such biogeochemical reactions as

$$(CH_3COO)_2Ca + 4\ O_2 \rightarrow CaCO_3 \downarrow + 3\ CO_2 \uparrow + 3\ H_2O. \quad (2.7a)$$

Calcium carbonate precipitation due to oxidation of organics was used for biocementation of the porous stones (Rodriguez-Navarro et al. 2003; Jimenez-Lopez et al. 2008).

Dissolved salts or chelates of Fe(II) produced by iron-reducing bacteria under strong anaerobic conditions can be also transformed to ferrous carbonate by anaerobic heterotrophic bacteria:

$$(CH_3COO)_2Fe + 4\ O_2 \rightarrow FeCO_3 \downarrow + 3\ CO_2 \uparrow + 3\ H_2O. \quad (2.7b)$$

Precipitation of calcium carbonate due to increase pH and production of carbonate by heterotrophic bacteria during anoxic oxidation of organics, for example due to bioreduction of nitrate:

$$(CH_3COO)_2Ca + 8/5\ NO_3^- \rightarrow CaCO_{3\downarrow} + 4/5\ N_2 \uparrow + 3\ CO_2 \uparrow + 3\ H_2O + 8/5\ OH^- \quad (2.8)$$

Calcium carbonate precipitation due to nitrate bioreduction of organics is useful for the combination of biocementation with nitrogen gas production in situ during partial desaturation of sandy soil, which is an effective method for the mitigation of earthquake-caused soil liquefaction (Chu et al. 2009a; Rebata-Landa and Santamarina 2012; He et al. 2013). Bioreduction of nitrate (bacterial denitrification process) can also increase pH and initiate precipitation of $CaCO_3$ without pH buffering (Hamdan et al. 2011). For example, precipitation of $CaCO_3$ can be performed using bioreduction of calcium nitrate by ethanol:

$$12\ NO_3^- + 5\ C_2H_5OH \rightarrow 6\ N_2 \uparrow + 10\ CO_2 \uparrow + 9\ H_2O + 12\ OH^-, \quad (2.9)$$

$$6\ Ca^{2+} + 12\ OH^- + 6\ CO_2 \rightarrow 6\ CaCO_3 \downarrow + 6\ H_2O, \quad (2.10)$$

$$12\ NO_3^- + 5\ C_2H_5OH + 6\ Ca^{2+} \rightarrow 6\ CaCO_3 \downarrow + 6\ N_2 \uparrow + 4\ CO_2 \uparrow + 15\ H_2O. \quad (2.11)$$

The biogeochemical reactions in case of denitrification and iron-based bio-clogging are similar:

$$12\,NO_3^- + 5\,C_2H_5OH \rightarrow 6\,N_2\uparrow + 10\,CO_2\uparrow + 9\,H_2O + 12\,OH^-, \quad (2.12)$$

$$6\,Fe^{2+} + 12\,OH^- \rightarrow 6\,Fe(OH)_2\downarrow. \quad (2.13)$$

$$12\,NO_3^- + 5\,C_2H_5OH + 6\,Fe^{2+} \rightarrow 6\,Fe(OH)_2\downarrow + 6\,N_2\uparrow + 10\,CO_2\uparrow + 9\,H_2O. \quad (2.14)$$

2.5.3 Biogas Production in Situ for Mitigation of Soil Liquefaction

Earthquake is one of the most devastating types of geohazards on Earth causing great economic losses including damages to infrastructures and properties. Many of the damages were related to soil liquefaction—a phenomenon whereby soil substantially loses strength and stiffness. Conventional ground improvement for mitigating liquefaction-induced geotechnical hazards are vibroreplacement, compaction grouting, and deep dynamic compaction methods. However, these methods are energy-consuming and expensive. Furthermore, dynamic compaction cannot be used for retrofitting or in the city or built-up areas. Recent fundamental studies in soil mechanics showed that inclusion of gas bubbles in saturated sand can reduce its susceptibility for liquefaction substantially (Xia and Hu 1991; Yang et al. 2004; Yegian et al. 2007; Eseller-Bayat et al. 2012). This finding has paved the way for the development of one of the best solutions to the mitigation of liquefaction disasters. It has been demonstrated that the liquefaction resistance of saturated sand can be significantly increased when the sand is slightly de-saturated with some voids displaced by nitrogen gas produced by denitrifying bacteria (Chu et al. 2009a, ; Rebata-Landa and Santamarina 2012; He et al. 2013). The biogas production in situ (Eqs. 2.15–2.20) has three major advantages over the other methods: (1) the distribution of gas is uniform because bacteria and reagents are in the liquid form and can be distributed evenly in sand; (2) the gas bubbles generated by bacteria are tiny and thus the gas bubbles are relatively stable; and (3) nitrogen gas is inert and has low solubility.

Different organic and inorganic substances can be biooxidized by nitrate but ethanol (C_2H_5OH), acetic acid (CH_3COOH), or glucose ($C_6H_{12}O_6$) that can be as used as 75 % (w/v) syrup from corn, are most suitable electron donors because of their low cost, availability, and high solubility in water. Their biooxidation by nitrate (denitrification), is shown below:

$$1.67\,C_2H_5OH + 2\,NO_3^- \rightarrow N_2\uparrow + 3.33\,CO_2 + H_2O + 2\,OH^-, \quad (2.15)$$

$$1.25\,CH_3COOH + 2\,NO_3^- \rightarrow N_2\uparrow + 2.5\,CO_2 + 1.5\,H_2O + 2\,OH^-, \quad (2.16)$$

$$0.42\,C_6H_{12}O_6 + 2NO_3^- \rightarrow N_2\uparrow + 2.5\,CO_2 + 1.5\,H_2O + 2\,OH^-. \quad (2.17)$$

These electron donors are very similar from stoichiometrical and economical points of view (Table 2.3).

Table 2.3 Comparison of stoichiometrical and economical parameters of partial biodesaturation of saturated soil with 50 % porosity

Electron donor	Use of electron donor (kg/m^3 of N_2)	Use of electron acceptor (sodium nitrate) (kg/m^3 of N_2)	Cost of electron donor ($/kg)	Cost of electron acceptor ($/kg)	Estimated cost of electron donor and acceptor for 10 % desaturation of soil ($/m^3 of N_2)
Ethanol	3.4	7.6	$0.60–0.70	$0.4–0.5	$0.25–$0.31
Acetic acid	3.4	7.6	$0.60–0.70	$0.4–0.5	$0.25–$0.31
Glucose	3.4	7.6	$0.60–0.80	$0.4–0.5	$0.25–$0.31

The stoichiometrical parameters of these reactions are almost same: consumption of electron donor is 3.4 kg/m^3 of N_2 and consumption of electron acceptor (sodium nitrate) is 7.6 kg/m^3 of N_2. The consumption of electron donor and acceptor for 10 % (volume of gas/volume of water) desaturation of soil with porosity 50 % is 0.55 kg/m^3 of saturated soil. Production of carbon dioxide in reactions 2.1–2.3, which is from 120 to 159 g/m^3 of N_2 or from 12 to 16 g/m^3 of water in saturated soil with 50 % porosity, is not accounted for desaturation of soil because solubility of CO_2 in water at 10 °C is 2500 g/m^3.

There is almost no cost difference between these electron donors: the cost of electron donor is from $0.5 to $0.7/kg, the cost of electron acceptor (sodium nitrate) is from $0.4 to $0.5/kg, so the estimated cost of electron donor and acceptor for partial desaturation is from $5.1 to $6.2/m^3 of N_2. The estimated cost of electron donor and acceptor for 10 % (volume of gas/volume of water) desaturation of soil with porosity 50 % is from $0.25 to $0.31/m^3 of saturated soil. However, even stoichiometrical and economic parameters of the electron donors are similar, ethanol could be more preferable electron donor then acetic acid or glucose sirup for geotechnical applications because it is liquid with neutral pH and not corrosive substance with low viscosity.

Biocementation of loose sand using a MICP process to increase the liquefaction resistance of sand has also been reported by DeJong et al. (2006), Montoya et al. (2012). It was shown (Montoya et al. 2012) that the resistance of sand to liquefaction, as measured by a decrease in the excess pore water pressure ratio, was significantly increased after MICP. However, sufficiently strong biocementation of saturated sand, at the level of unconfined compressive strength 250–500 kPa, could be at the content of precipitated calcium carbonate of 75–100 g/kg of sand (Ivanov et al. 2012a; Cheng et al. 2013). Therefore, it could be material-consuming process requiring about 88 kg $CaCl_2$ and 96 kg of urea per 1 m^3 of sand, which will cost at least $41/m^3 of saturated soil. This value is about 140 times higher than 10 % desaturation of soil using biogas production in situ. So, biocementation of soil to mitigate liquefaction could be too expensive to be applicable for large-scale geotechnical practice.

2.5.4 Calcium- and Magnesium-Based Biocementation

The biocement for this biocementation can be produced through dissolution of dolomite, which is a common raw material for the production of cement, in hydrochloric acid:

$$Ca^{2+} + Mg^{2+} + CO(NH_2) + UPB \rightarrow CaMg(CO_3) \downarrow + 2\,NH_4^+. \quad (2.18)$$

Our experiments showed that this biocementation produced the unconfined compressive strength of the biocemented sand column 12.4 MPa at the content of precipitated Ca and Mg 6 % (w/w). However, the hydraulic permeability of the biocemented sand was high, $7 \times 10^{-4}\ ms^{-1}$. For calcium-and urea based biocementation the hydraulic conductivity and strength are correlated and high strength is accompanied with low hydraulic conductivity of the biocemented sand. Probably, high strength but at the same time high permeability of sand after magnesium-based biocementation is due to coating of whole surface of sand grains with crystals, while calcium- and urea-based biocementation produced crystal mainly in the contact areas of the sand grains (Fig. 2.7).

Combined calcite and struvite (NH_4MgPO_4) precipitation using triple superphosphate and magnesium salt to avoid formation of soluble ammonia and release of ammonia to atmosphere during biocementation:

$$Ca(H_2PO_4)_2 + 2\,Mg^{2+} + CO(NH_2)_2 + H_2O + \text{acid urease} \rightarrow 2\,NH_4MgPO_4 \downarrow + CaCO_3 \downarrow. \quad (2.19)$$

2.5.5 Calcium-Phosphate Biocementation

Calcium phosphate precipitation from calcium phytate (*myo*-inositol hexakisphosphate, calcium salt) solution, (the main storage form of phosphorus in the plant seeds) using phytase activity of microorganisms (Roeselers and van Loosdrecht 2010) producing a mixture of the crystal forms such as monetite ($CaHPO_4$), whitlockite [$Ca_9(Mg,Fe^{2+})(PO_4)6HPO_4$], and hydroxyapatite [$Ca_5(PO_4)_3OH$] with the Ca-to-P molar ratio 1.55. The problem of this type of biocementation is a low solubility of calcium phytate (in the described study the concentration was 5.6 mM), so big volumes of solution must be pumped through soil.

Triple superphosphate, which is a relatively cheap commodity, has also low solubility about 0.08 M, and can be used for calcium phosphate precipitation using acidotolerant urease-producing microorganisms as shown below.

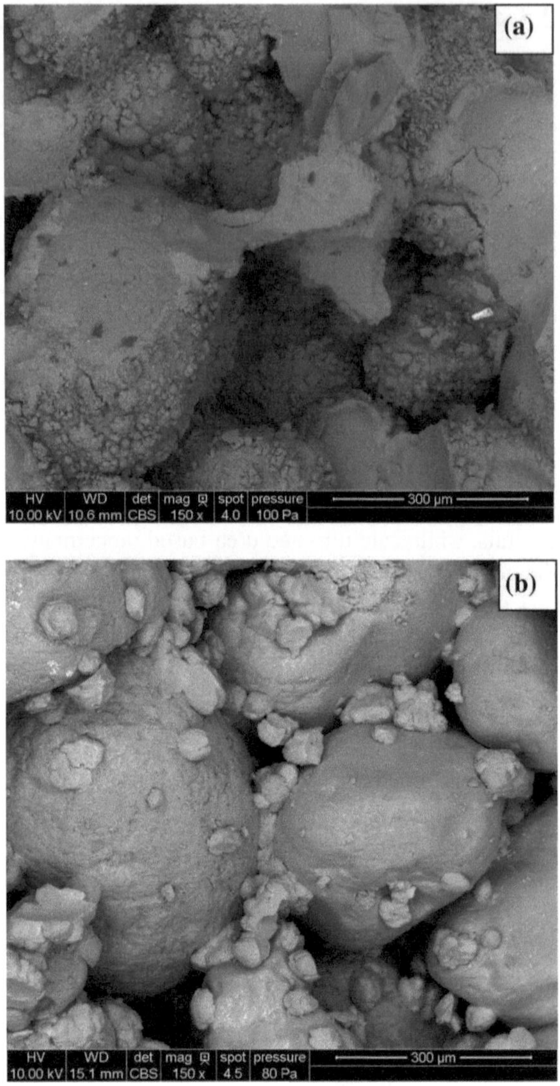

Fig. 2.7 Precipitates/crystals on sand surface after Ca–Mg based biocementation (**a**) and calcium-based biocementation (**b**)

Monetatite precipitation:

$$Ca(H_2PO_4)_2 + CO(NH_2)_2 + H_2O + \text{acid urease} \rightarrow CaHPO_4 \downarrow + CO_2 + (NH_4)_2HPO_4. \quad (2.20)$$

Hydroxyapatatite precipitation:

$$5Ca(H_2PO_4)_2 + 8\ CO(NH_2)_2 + 8\ H_2O\ +\ \text{acid urease} \rightarrow Ca_5(PO_4)_3(OH)\downarrow\ + 2NH_4HCO_3 + 6CO_2 +\ 7\ (NH_4)_2HPO_4. \quad (2.21)$$

2.5.6 Calcium Bicarbonate Biocementation

Important biocementation technology could be precipitation of calcite using removal of CO_2 from solution of calcium bicarbonate (Ehrlich 1999) because it releases low quantity of ammonia and can be performed without increase of pH to 8.5–9.0 as conventional MICP:

$$Ca(HCO_3)_2 + CO(NH_2)_2 + H_2O\ +\ \text{acid urease} \rightarrow CaCO_3\downarrow + CO_2 +\ (NH_4)_2CO_3. \quad (2.22)$$

Solubility of calcium bicarbonate is relatively high, about 1 M, to perform practically feasible biocementation. This method is a model of the naturally occuring dissolution-precipitation of calcium carbonate

$$CaCO_3 +\ CO_2 +\ H_2O\ \leftrightarrow Ca(HCO_3)_2. \quad (2.23)$$

The difference is that biocementation has to be performed at high concentration of calcium bicarbonate and with significantly higher rate than in nature. The rate of precipitation of calcium carbonate from calcium bicarbonate in nature is determined by the removal rate of CO_2 from the reaction. The rate of biocementation can be accelerated due to increase of pH during hydrolysis of urea by enzyme urease. The problem of bicarbonate biocement is its instability, so the solution must be produced and stored at elevated partial pressure of CO_2.

2.5.7 Iron-Based Bioclogging and Biocementation

Iron-based biocementation could be suitable for geotechnical applications if to combine three bioprocesses shown below.

- Acidogenic fermentation of cellulose-containing agricultural or food-processing residuals producing mainly acetic acid (see Eq. 2.1);
- bioreduction of cheap commodity, iron ore, using products of acidogenic fermentation or many organic electron donors (Ivanov et al. 2009; Guo et al. 2010), see equation for the reduction of ferric ions by iron-reducing bacteria using acetate:

$$4\,Fe_2O_3 + CH_3COOH + 16\,H^+ \rightarrow 8Fe^{2+} + 2\,CO_2 + 10\,H_2O, \quad (2.24)$$

$$8\,Fe^{2+} + 16\,CH_3COOH \rightarrow 8\,Fe(CH_3COOH)_2 + 16\,H^+, \quad (2.25)$$

$$4\,Fe_2O_3 + 17\,CH_3COOH \rightarrow 8\,Fe(CH_3COOH)_2 + 2\,CO_2 + 10\,H_2O. \quad (2.26)$$

- oxidation and bioprecipitation of ferrous chelates (Stabnikov and Ivanov 2006; Ivanov et al. 2012a):

$$Fe^{2+} + 1.5(NH_2)_2CO + 0.25\,O_2 + 5.5\,H_2O + UPB \rightarrow Fe(OH)_3\downarrow + 1.5\,(NH_4)_2CO_3 + 2H^+. \quad (2.27)$$

Urease-producing bacteria and urea are used to maintain the pH to be above the neutral value because oxidation of ferrous ions and hydrolysis of ferric ions is accompanied with acidification of solution. The advantages of using iron hydroxide as the clogging compound are that the soil treated by iron minerals is more ductile and able to resist low pH conditions. The soil treated using iron based biocement is not as strong as that treated using calcium based biocement (Ivanov et al. 2012a) but the clogging effect of precipitated iron hydroxide is higher than that of calcium carbonate (Fig. 2.8).

Precipitation of ferric and manganese hydroxide from chelates of Fe^{2+} and Mn^{2+} for soil bioclogging can be done using neutrophilic iron-oxidizing bacteria. For example, precipitation of iron/manganese minerals by iron-oxidizing bacteria *Leptothrix discophora* is a promising technology for modifying engineering soil properties and mitigating geologic hazards (Weaver et al. 2011).

2.5.8 Eco-Efficient Biocement

By the analogy with eco-efficient concrete (Pacheco-Torgal and Jalali 2011; Pacheco-Torgal et al. 2012; Pacheco-Torgal and Jalali, 2013), eco-efficient biocement can be produced using mining tails/residuals of limestone, dolomite, iron ore and organic agricultural, food-processing, or municipal wastes using acidogenic fermentation and bioreduction of iron with production of dissolved salts of calcium, magnesium, and iron.

The problem with the brittleness of biocementation could be solved using biomimetic approach (Sarikaya 1994; Mayer and Sarikaya 2002) using composite strengthening through combination of mineral and organic nano- and micro-particles. By the analogy with nanomaterials in cement (Pacheco-Torgal and Jalali

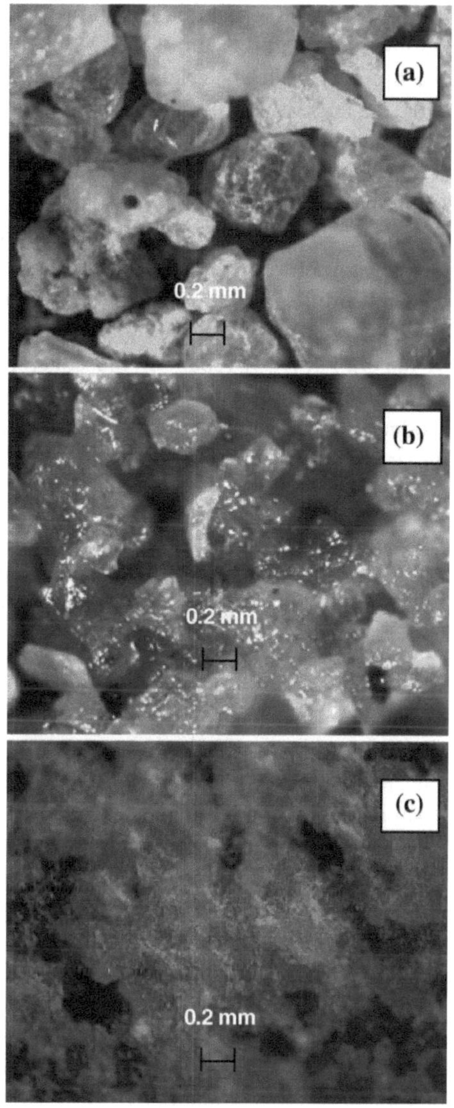

Fig. 2.8 Micrographs of untreated sand (**a**) and sand treated with iron-based biogrout (**b**) or calcium-based biogrout (**c**)

2011), applications of composite micro- and nano-materials can also be useful to increase strength and ductility of biocement. Theoretically, ductile biocement could be made as a bioinspired material (Pacheco-Torgal and Labrincha 2013b), with the 3D-composite structure of hierarchically arranged nano- and micrometric units (Imai and Oaki 2010), or just simply with the layers or inclusions, where inorganic crystals of calcium carbonate (calcite, aragonite, vaterate), calcium phosphate (hydroxyapatite), oxides of Si and Fe and others create the hardness and the organic components such as proteins and polysaccharides ensure flexibility of the biocemented structure. This property is well known from the structure of the natural biominerals such as bones, shells, and corals as well as artificial engineering composite materials (Yao et al. 2011; Mayer and Sarikaya 2002). However, the cost of micro- and nano-composites could be too high to be suitable for construction practice.

One prospective application of biocementation is repair of the cracks in concrete and self-healing concrete. The repair of the cracks in the surface layer of concrete is a major portion of multi-billion maintenance and repair cost of the concrete structures (Neville 1996; FHWA 2001). Self-healing concrete is based on the embedding into concrete the glass or plastic capsules with material, which could be released after simultaneous cracking of concrete and capsules. One type of material for self-repair of the concrete proposed to be based on MICP (Ramachandran et al. 2001; Jonkers 2007; Jonkers et al. 2010). However, volume of produced $CaCO_3$ will be always significantly smaller of the volume of capsule and the crack to be filled.

2.6 Bioremediation and Biodecontamination of Construction Site Through Biocementation

When site can be used for construction but is polluted with chemical substances above permitted levels, microbial remediation of this site could be the cheapest option in comparison with mechanical or chemical cleaning of the construction site. Bioremediation can be done using such biogeochemical reactions as oxidation, reduction, and transformation of pollutants to nontoxic or not dissolved substances. Bioremediation of polluted sites is now well developed area with hundreds of published research papers, reviews and books as well as a lot of commercially available biotechnologies and equipment.

Meanwhile, bioremediation of construction sites through biocementation is a relatively new area. Usually, the aim of these methods of construction biotechnology is to prevent the dispersion of hazardous substances from the accidentally polluted site to environment. It is performed through either biocementation of soil or formation of biogeochemical barrier. For example, MICP has the ability to co-precipitate toxic radionuclides ^{90}Sr, ^{60}Co and metal contaminants such as Cd and this can be used to prevent their dispersion in environment (Warren et al. 2001; Fujita et al. 2004; Mitchell and Ferris 2005). After MICP treatment of sand surface with the quantity just 15.6 g Ca/m^2 the release of the sand dust and its artificial pollutants to atmosphere decreased in comparison with control by 99.8 % for dust, 92.7 % for phenantherene, 94.4 % for led nitrate, and 99.8 % for bacterial cells of *Bacillus megaterium* due to bioaggregation of the fine sand particles. Bioaggregation treatment of the soil surface could be useful method to prevent the dispersion of dust and the dust-associated chemical and bacteriological pollutants in water, air and soil (Stabnikov et al. 2013a), so it could be useful in construction and probably to protect atmosphere in fixation response to a radiation dispersal devise attack (Cordesman 2002; Parra et al. 2009).

Another way of environmental protection from the polluted sites can be construction of permeable, reactive, biogeochemical barriers, which will precipitate heavy metals and degrade toxic chemicals due to microbial activity inside the barrier (Kavamura and Esposito 2010; Gibert et al. 2013).

2.7 Conclusions

Construction biotechnology includes microbial production of construction materials and microbially-mediated construction processes. Microbial cements, polysaccharides, and construction bioplastics can be made using biotechnologies. Microbial cement could be produced with lower cost than conventional cement. Biodegradability of the bioplastic constructions after their demolishing reduces the cost of construction wastes disposal. Microbial polysaccharides are used to modify the cement properties and soil bioclogging. Construction biotechnologies are based on the activity of urease-producing, acidogenic, halophilic, alkaliphilic, denitrifying, iron- and sulfate-reducing bacteria, cyanobacteria, algae, and microscopic fungi. The bio-related materials and processes can be used for the particles aggregation, soil grouting and clogging, cementation of the particles, desaturation of soil, encapsulation of soft clay, and coating of the solid surfaces. The biotechnologically produced materials and construction biotechnologies have a lot of advantages in comparison with the conventional construction materials and processes so the practical implementations of the construction biotechnologies could give significant economic and environmental benefits.

References

Achal V, Mukherjee A, Reddy MS (2010) Microbial concrete: way to enhance the durability of building structures. J Mater Civil Eng 23:730–734. doi:10.1061/(ASCE)MT.1943-5533.0000159

Achal V, Mukherjee A, Reddy MS (2011) Effect of calcifying bacteria on permeation properties of concrete structures. J Ind Microbiol Biotechnol 38:1229–1234. doi:10.1007/s10295-010-0901-8

Arunaye FI, Mwasha A (2011) On behavior of limited life geotextile materials for reinforcing embankment on soft ground. World J Eng 8:195–199. doi:10.1260/1708-5284.8.2.195

Bang SS, Galinat JK, Ramakrishnan V (2001) Calcite precipitation induced by polyurethane-immobilized *Bacillus pasteurii*. Enzyme Microb Technol 28:404–409

Bang S, Min SH, Bang SS (2011) Application of microbiologically induced soil stabilization technique for dust suppression. Int J Geo-Eng 3:27–37

Barlaz MA, Staley BF, De Los Reyes III FL (2010) Anaerobic biodegradation of solid waste. In: Mitchell R, Gu JD (eds) Environmental microbiology, 2nd edn. Wiley, Hoboken, NJ, USA, pp 281–299

Ben Rebah F, Prevost D, Tyagi RD, Belbahri L (2009) Poly-beta-hydroxybutyrate production by fast-growing rhizobia cultivated in sludge and in industrial wastewater. Appl Biochem Biotechnol 158:155–163

Beun JJ, Dircks K, van Loosdrecht MCM et al (2006) Poly-β-hydroxybutyrate metabolism in dynamically fed mixed microbial cultures. Water Res 36:1167–1180

Braunegg G, Lefebvre G, Genser KF (1998) Polyhydroxyalkanoates, biopolyesters from renewable resources: Physiological and engineering aspects. J Biotechnol 65:127–161

Burbank MB, Weaver TJ, Green TL, Williams BC, Crawford RL (2011) Precipitation of calcite by indigenous microorganisms to strengthen liquefiable soils. Geomicrobiol J 28:301–312. doi:10.1080/01490451.2010.499929

Burbank MB, Weaver TJ, Williams BC, Crawford RL (2012a) Urease activity of ureolytic bacteria isolated from six soils in which calcite was precipitated by indigenous bacteria. Geomicrobiol J 29:389–395. doi:10.1080/01490451.2011.575913

Burbank M, Weaver T, Lewis R, Williams T, Williams B, Crawford W (2012b) Geotechnical tests of sands following bioinduced calcite precipitation catalyzed by indigenous bacteria. J Geotech Geoenviron Eng 139:928–936. doi:10.1061/(ASCE)GT.1943-5606.0000781

Castanier S, Le Métayer-Levrel G, Orial G., Loubière JF, Perthuisot JP (2000) Bacterial carbonatogenesis and applications to preservation and restoration of historic property. In: Ciferri O, Tiano P, Mastromei G (eds) Of microbes and art. the role of microbial communities in the degradation and protection of cultural heritage. Kluwer Academic/Plenum Publisher, New York, pp 203–218

Castilho LR, Mitchell DA, Freire DMG (2009) Production of polyhydroxyalkanoates (PHAs) from waste materials and by-products by submerged and solid-state fermentation. Bioresour Technol 100:5996–6009

Cheng L, Cord-Ruwisch R (2012) In situ soil cementation with ureolytic bacteria by surface percolation. Ecol Eng 42:64–72. doi:10.1016/j.ecoleng.2012.01.013

Cheng L, Cord-Ruwisch R (2013) Selective enrichment and production of highly urease active bacteria by non-sterile (open) chemostat culture. J Ind Microbiol Biotechnol 40:1095–1104. doi:10.1007/s10295-013-1310-6

Cheng L, Cord-Ruwisch R, Shahin MA (2013) Cementation of sand soil by microbially induced calcite precipitation at various degrees of saturation. Can Geotech J 50:1–10

Christians S, Jose J, Schäfer U, Kaltwasser H (1991) Purification and subunit determination of the nickel-dependent *Staphylococcus xylosus* urease. FEMS Microbiol Lett 80:271–275

Chu J, Ivanov V, Jia H, Chenghong G, Naeimi M, Tkalich P (2009a) Microbial geotechnical engineering for disaster mitigation and coastal management. In: Proceedings of WCCE-ECCE-TCCE joint conference: earthquake & tsunami, Istanbul, Turkey

Chu J, Varaksin S, Klotz U, Mengé P (2009b) Construction processes. state-of-the-art-report. In: Proceedings of the 17th international conference on soil mechanics and geotechnical engineering, Alexandria, Egypt, 5–9 Oct 2009 vol 4, pp 3006–3135

Chu J, Stabnikov V, Ivanov V (2012a) Microbially induced calcium carbonate precipitation on surface or in the bulk of soil. Geomicrobiol J 29:1–6

Chu J, Indraratna B, Shuwang Yan S, Rujikiatkamjorn C (2012b) Soft soil improvement through consolidation: an overview. In: Proceedings of the international conference on ground improvement and ground control, pp. 251–280. doi:10.3850/978-981-07-3559-3103-0007

Chu J, Ivanov V, Stabnikov V (2013a) Microbial method for construction of aquaculture pond in sand. Géotechnique 63:871–875. doi:10.1680/geot.SIP13.P.007

Chu J, He J, Ivanov V (2013b) Mitigation of liquefaction of saturated sand using biogas. Geotechnique 63:267–275. doi:10.1680/geot.SIP13.P.004

Chu J, Ivanov V, Naeimi M, Stabnikov V, Liu HL (2014) Optimization of calcium-based bioclogging and biocementation of sand. Acta Geotech 9:277–285. doi:10.1007/s11440-013-0278-8

Cordesman AH (2002) Terrorism, asymmetric warfare, and weapons of mass destruction: defending the U.S. homeland. Praeger Publishers, Westport

Cuthbert MO, McMillan LA, Handley-Sidhu S, Riley MS, Tobler DJ, Phoenix VR (2013) A field and modeling study of fractured rock permeability reduction using microbially induced calcite precipitation. Env Sci Technol 47:13637–13643. doi:10.1021/es402601g
DeJong J, Fritzges M, Nusstein K (2006) Microbially induced cementation to control sand response to undrained shear. J Geotech Geoenv Eng 32:1381–1392
DeJong JT, Mortensen BM, Martinez BC, Nelson DC (2010) Bio-mediated soil improvement. Ecol Eng 36:197–210
DeJong JT, Soga K, Kavazanjian E et al (2013) Biogeochemical processes and geotechnical applications: progress, opportunities and challenges. Geotechnique 63:287–301
DeMarco S (2005) Advances in polyhydroxyalkanoate production in bacteria for biodegradable plastics. Basic Biotechnol eJ 1:1–4
De Muynck W, Cox K, Verstraete W, De Belie N (2008a) Bacterial carbonate precipitation as an alternative surface treatment for concrete. Constr Build Mater 22:875–885
De Muynck WD, Debrouwer D, De Belie ND, Verstraete W (2008b) Bacterial carbonate precipitation improves the durability of cementitious materials. Cem Concr Res 38:1005–1014
De Muynck W, De Belie N, Verstraete W (2010) Microbial carbonate precipitation in construction materials: a review. Ecol Eng 36:118–136
De Muynck W, Verbeken K, De Belie N, Verstraete W (2012) Influence of temperature on the effectiveness of a biogenic carbonate surface treatment for limestone conservation. Appl Microbiol Biotechnol 97(3):1335–1347 doi:10.1007/s00253-012-3997-0
Dhami NK, Reddy MS, Mukherjee A (2012) Improvement in strength properties of ash bricks by bacterial calcite. Ecol Eng 39:31–35. doi:10.1016/j.ecoleng.2011.11.011
Dhami NK, Reddy MS, Mukherjee A (2014) Synergistic role of bacterial urease and carbonic anhydrase in carbonate mineralization. Appl Biochem Biotechnol 172:2552–2561 doi:10.1007/s12010-013-0694-0
Dosier GK (2013) Methods for making construction material using enzyme producing bacteria. US Patent App. 13/093,335, 2011
Ehrlich HL (1999) Microbes as geologic agents: their role in mineral formation. Geomicrobiol J 16:135–153
Eseller-Bayat E, Yegian MK, Alshawabkeh A, Gokyer S (2012) Prevention of liquefaction during earthquakes through induced partial saturation in sands. Geotechnical engineering: new horizons. IOS Press, Amsterdam, pp 188–194
Fernandes P (2006) Applied microbiology and biotechnology in the conservation of stone culture heritage materials. Appl Microbiol Biotechnol 73:291–296
FHWA-RD-01-156 (2001) Corrosion cost and preventive strategies in the United States, Report by CC Technologies Laboratories, Inc. to Federal Highway Administration (FHWA), Office of Infrastructure Research and Development
Fujita F, Redden GD, Ingram JC, Cortez MM, Ferris FG, Smith RW (2004) Strontium incorporation into calcite generated by bacterial ureolysis. Geochim Cosmochim Acta 68:3261–3270. doi:10.1016/j.gca.2003.12.018
Fytili D, Zabaniotou A (2008) Utilization of sewage sludge in EU application of old and new methods—a review. Renew Sustain Energy Rev 12:116–140. doi:10.1016/j.rser.2006.05.014
Ghosh P, Mandal S, Chattopadhyay BD, Pal S (2005) Use of microorganism to improve the strength of cement mortar. Cem Concr Res 35:1980–1983. doi:10.1016/j.cemconres.2005.03.005
Ghosh P, Mandal S, Pal S, Bandyopadhyaya G, Chattopadhyay BD (2006) Development of bioconcrete material using an enrichment culture of novel thermophilic anaerobic bacteria. Indian J Exp Biol 44:336–339
Ghosh S, Biswas M, Chattopadhyay BD, Mandal S (2009) Microbial activity on the microstructure of bacteria modified mortar. Cem Concr Compos 31:93–98
Gibert O, Cortina JL, de Pablo J, Ayora C (2013) Performance of a field-scale permeable reactive barrier based on organic substrate and zero-valent iron for in situ remediation of acid mine drainage. Env Sci Pollut Res 20:7854–7862. doi:10.1007/s11356-013-1507-2

Guo CH, Stabnikov V, Ivanov V (2010) The removal of nitrogen and phosphorus from reject water of municipal wastewater treatment plant using ferric and nitrate bioreductions. Bioresour Technol 101:3992–3999

Hamdan N, Kavazanjian E, Rittman BE, Karatas I (2011) Carbonate mineral precipitation for soil improvement through microbial denitrification. In: Han J, Alzamora DA (eds) Geo-frontiers 2011: advances in geotechnical engineering. American Society of Civil Engineers, Dallas

Hammes F, Boon N, de Villiers J, Verstraete W, Siciliano SD (2003) Strain-specific ureolytic microbial calcium carbonate precipitation. Appl Env Microbiol 69:4901–4909

Harkes MP, van Paassen LA, Booster JL, Whiffin VS, van Loosdrecht MCM (2010) Fixation and distribution of bacterial activity in sand to induce carbonate precipitation for ground reinforcement. Ecol Eng 36:112–117

He J, Chu J, Ivanov V (2013) Mitigation of liquefaction of saturated sand using biogas. Geotechnique 63:267–275

Imai H, Oaki Y (2010) Bioinspired hierarchical crystals. MRS Bull 35(2):138–144. doi:10.1557/mrs2010.634

Ivanov V, Chu J (2008) Applications of microorganisms to geotechnical engineering for bioclogging and biocementation of soil *in situ*. Rev Env Sci Biotechnol 7:139–153

Ivanov V, Kuang SL, Guo CH, Stabnikov V (2009) The removal of phosphorus from reject water in a municipal wastewater treatment plant using iron ore. J Chem Technol Biotechnol 84:78–82

Ivanov V (2010) Environmental microbiology for engineers. CRC Press, Taylor & Francis Group, Boca Raton, p 402

Ivanov V, Stabnikov V, Hung YT (2012) Screening and selection of microorganisms for the environmental biotechnology process. In: Yung-Tse H, Lawrence KW, Nazih KS (eds) Handbook of environment and waste management: air and water pollution control. World Scientific Publishing Co., Inc., Singapore, pp 1137–1149

Ivanov V, Chu J, Stabnikov V, Li B (2014) Strengthening of soft marine clay using biocementation. Marine Georesour Geotechnol. doi:10.1080/1064119X.2013.877107

Jimenez-Lopez C, Jroundi F, Pascolini C, Rodriguez-Navarro C, Piñar-Larrubia G, Rodriguez-Gallego M, González-Muñoz MT (2008) Consolidation of quarry calcarenite by calcium carbonate precipitation induced by bacteria activated among the microbiota inhabiting the stone. Int Biodeterior Biodegrad 62:352–363

Jin M, Rosario W, Walter E, Calhoun DH (2004) Development of a large-scale HPLC-based purification for the urease from *Staphylococcus leei* and determination of subunit structure. Protein Expr Purif 34:111–117

Jonkers HM (2007) Self healing concrete: a biological approach. In: van der Zwaag S (ed) Self healing materials: an alternative approach to 20 centuries of materials science. Springer, Berlin, pp. 195–204

Jonkers HM, Thijssen A, Muyzer G, Copuroglu O, Schlangen E (2010) Application of bacteria as self-healing agent for the development of sustainable concrete. Ecol Eng 36:230–235

Kavamura VN, Esposito E (2010) Biotechnological strategies applied to the decontamination of soils polluted with heavy metals. Biotechnol Adv 28:61–69. doi:10.1016/j.biotechadv.2009.09.002

Kenyon WJ, Esch SW, Buller CS (2005) The curdlan-type exopolysaccharide produced by *Cellulomonas flavigena* KU forms part of an extracellular glycocalyx involved in cellulose degradation. Antonie Van Leeuwenhoek 87:143–148

Khanna S, Srivastava AK (2005) Recent advances in microbial polyhydroxyalkanoates. Proc Biochem 40:607–619

León-Martínez FM, de Cano-Barrita PFJ, Lagunez-Rivera L, Medina-Torres L (2014) Study of nopal mucilage and marine brown algae extract as viscosity-enhancing admixtures for cement based materials. Constr Build Mater 53:190–202. doi:10.1016/j.conbuildmat.2013.11.068

Lenz RW, Marchessault RH (2005) Bacterial polyesters: biosynthesis, biodegradable plastics and biotechnology. Biomacromolecules 6:1–8

Li M, Guo H, Cheng X (2011) Application of response surface methodology for carbonate precipitation production induced by a mutant strain of *Sporosarcina pasteurii*. In: Geo-Frontiers, ASCE, pp 4079–4088

Li P, Qu W (2012) Microbial carbonate mineralization as an improvement method for durability of concrete structures. Adv Mater Res 365:280–286

Lisdiyanti P, Suyanto E, Ratnakomala S, Fahrurrozi, Sari MN, Gusmawati NF (2011) Bacterial carbonate precipitation for biogrouting. In: Proceedings of the national symposium on ecohydrology, Jakarta, pp 204–211

Lowell WL, Rohwedder WK (1974) Poly-beta-hydroxyalkanoate from activated sludge. Env Sci Technol 8:576–579

Lu Y (2007) Advance on the production of polyhydroxyalkanoates by mixed cultures. Front Biol China 2:1673–3509

Madigan MT, Martinko JM, Stahl D, Clark DP (2012) Brock biology of microorganisms (13th Edn). Pearson, San Francisco

Mayer G, Sarikaya M (2002) Rigid biological composite materials: structural examples for biomimetic design. Exp Mech 42:395–403

Mergaert J, Anderson C, Wouters A, Swings J, Kerster K (1992) Biodegradation of polyhydroxyalkanoates. FEMS Microbiol Rev 103:317–322

Mitchell AC, Ferris FG (2005) The coprecipitation of Sr into calcite precipitates induced by bacterial ureolysis in artificial groundwater: temperature and kinetic dependence. Geochim Cosmochim Acta 69:4199–4210

Mitchell JK, Santamarina JC (2005) Biological considerations in geotechnical engineering. J Geotech Geoenv Eng ASCE 131:1222–1233

Montoya BM, DeJong JT, Boulanger RW, Wilson D W, Gerhard R, Ganchenko A, Chou JC (2012) Liquefaction mitigation using microbial induced calcite precipitation. In: GeoCongress, pp 1918–1927

Moosbrugger RE, Wentezel MC, Ekama GA, Marais GV (1993) Weak acid/bases and pH control in anaerobic systems: a review. Water South Africa 19:1–10

Mun KJ (2007) Development and tests of lightweight aggregate using sewage sludge for nonstructural concrete. Constr Build Mater 21:1583–1588. doi:10.1016/j.conbuildmat.2005.09.009

Neville AM (1996) Properties of concrete, 4th edn. Pearson Higher Education, Prentice Hall, New Jersey

Novakova D, Sedlacek I, Pantucek R, Stetina V, Svec P, Petras P (2006) *Staphylococcus equorum* and *Staphylococcus succinus* isolated from human clinical specimens. J Med Microbiol 55:523–528

Ogbobe O, Essien KS, Adebayo A (1998) A study of biodegradable geotextiles used for erosion control. Geosynth Int 5:545–553

van der Ruyt M, van der Zon W (2009) Biological *in situ* reinforcement of sand in near-shore areas. Geotech Eng 162:81–83

van Loosdrecht MCM, Pot MA, Heijnen JJ (1997) Importance of bacterial storage polymers in bioprocesses. Water Sci Technol 33:41–47

van Loosdrecht MCM, Kleerebezem R, Muyzer G, Jian Y, Johnson K (2008) Process for selecting polyhydroxyalkanoate (PHA) producing micro-organisms. WO/2009/153303. International application No.: PCT/EP2009/057571, 18 June 2008

van Paassen L, Ghose R, van der Linden T, van der Star W, van Loosdrecht M (2010) Quantifying biomediated ground improvement by ureolysis: large-scale biogrout experiment. J Geotech Geoenv Eng 136:1721–1728

van Tittelboom K, De Belie N, De Muynck W, Verstraete W (2010) Use of bacteria to repair cracks in concrete. Cem Concr Res 40:157–166. doi:10.1016/j.cemconres.2009.08.025

Pacheco-Torgal F, Jalali S (2011) Nanotechnology: advantages and drawbacks in the field of construction and building materials. Constr Build Mater 25:582–590. doi:10.1016/j.conbuildmat.2010.07.009

Pacheco-Torgal F, Jalali S, Labrincha J, John VM (2012) Eco-efficient concrete using industrial wastes: a review. Mater Sci Forum 730–732, 581–586. doi:58110.4028/www.scientific.net/MSF.730-732.581

Pacheco-Torgal F, Labrincha JA (2013a) Biotech cementitious materials: Some aspects of an innovative approach for concrete with enhanced durability. Constr Build Mater 40:1136–1141. doi:10.1016/j.conbuildmat.2012.09.080

Pacheco-Torgal F, Labrincha JA (2013b) Biotechnologies and bioinspired materials for the construction industry: an overview. Int J Sustain Eng. doi:10.1080/19397038.2013.844741

Pacheco-Torgal F, Jalali S (2013) Eco-efficient construction and building materials. Springer, London, p 247 doi:10.1007/978-0-85729-892-8

Park JH, Yuu J, Jeon HY (2010) Green geosynthetics applications to sustainable environmental fields from the viewpoint of degradability. In: Proceedings of the international symposium and exhibition on geotechnical and geosynthetics engineering: challenges and opportunities on climate change, Bangkok, pp. 43–50, 7–8 Dec (2010)

Parra RR, Medina VF, Conca JL (2009) The use of fixatives for response to a radiation dispersal devise attack – a review of the current (2009) state-of-the-art. J Env Radioact 100:923–934

Pei R, Liu J, Wang S, Yang M (2013) Use of bacterial cell walls to improve the mechanical performance of concrete. Cem Concr Compos 39:122–130. doi:10.1016/j.cemconcomp.2013.03.024

Plank J (2003) Applications of biopolymers in construction engineering. In: Steinbüchel A (ed) Biopolymers, V.10. General aspects and special applications. Wiley, Weinheim

Plank J (2004) Application of biopolymers and other biotechnological products in building material. Appl Microbiol Biotechnol 66:1–9

Portilho M, Matioli G, Zanin GM, de Moraes FF, Scamparini AR (2006) Production of insoluble exopolysaccharide *Agrobacterium sp.* (ATCC 31749 and IFO 13140). Appl Biochem Biotechnol 129–132:864–869

Ramachandran SK, Ramakrishnan V, Bang SS (2001) Remediation of concrete using microorganisms. ACI Mater J 98:3–9

Ramesh BNG, Anitha N, Rani HKR (2010) Recent trends in biodegradable products from biopolymers. Adv Biotechnol 9:30–34

Raut SH, Sarode DD, Lele SS (2014) Biocalcification using *B. pasteurii* for strengthening brick masonry civil engineering structures. World J Microbiol Biotechnol 30:191–200. doi:10.1007/s11274-013-1439-5

Rebata-Landa V, Santamarina JC (2012) Mechanical effects of biogenic nitrogen gas bubbles in soils. J Geotech Geoenv Eng 138:128–137

Reddy CS, Ghai R, Rashmi C, Kalia VC (2003) Polyhydroxyalkanoates: an overview. Bioresour Technol 87:137–146

Reddy S, Rao M, Aparna P, Sasikala C (2010) Performance of standard grade bacterial (*Bacillus subtilis*) concrete. Asian J Civ Eng (Build Hous) 11:43–55

Rodriguez-Navarro C, Rodriguez-Gallego M, Ben Chekroun K, Gonzalez-Muñoz MT (2003) Conservation of ornamental stone by *Myxococcus xanthus*-induced carbonate biomineralization. Appl Env Microbiol 69:2182–2193. doi:10.1128/AEM.69.4.2182-2193.2003

Roeselers G, van Loosdrecht MCM (2010) Microbial phytase-induced calcium-phosphate precipitation– a potential soil stabilization method. Folia Microbiol 55:621–624

Ross N, Villemur R, Deschenes L, Samson R (2001) Clogging of limestone fracture by stimulating groundwater microbes. Water Res 35:2029–2037

Sarda D, Choonia HS, Sarode DD, Lele SS (2009) Biocalcification by Bacillus *pasteurii urease*: a novel application. J Ind Microbiol Biotechnol 36:1111–1115

Sarayu K, Iyer NR, Murthy AR (2014) Exploration on the biotechnological aspect of the ureolytic bacteria for the production of the cementitious materials—a review. Appl Biochem Biotechnol. doi:10.1007/s12010-013-0686-0

Sarikaya M (1994) An introduction to biomimetics: a structural viewpoint. Microsc Res Tech 37:360–375

Sathya A, Bhuvaneshwari P, Niranjan G, Vishveswaran M (2013) Influence of bio admixture on mechanical properties of cement and concrete. Asian J Appl Sci. doi:10.3923/ajaps.2013
Seagren EA, Aydilek AH (2010) Bioremediated geomechanical processes. In: Mitchell R, Gu J-D (eds) Environmental Microbiology, 2nd edn. Wiley, Hoboken, pp 319–348
Serafim LS, Lemos PC, Albuquerque MGE, Reis MAM (2008) Strategies for PHA production by mixed cultures and renewable waste materials. Appl Microbiol Biotechnol 81:615–628
Siddique R, Chahal NK (2011) Effect of ureolytic bacteria on concrete properties. Constr Build Mater 25:3791–3801. doi:10.1016/j.conbuildmat.2011.04.010
Stabnikov VP, Ivanov VN (2006) The effect of various iron hydroxide concentrations on the anaerobic fermentation of sulfate-containing model wastewater. Appl Biochem Microbiol 42:284–288
Stabnikov V, Chu J, Naeimi M, Ivanov V (2011) Formation of water-impermeable crust on sand surface using biocement. Cem Concr Res 41:1143–1149
Stabnikov V, Chu J, Myo AN, Ivanov V (2013a) Immobilization of sand dust and associated pollutants using bioaggregation. Water Air Soil Pollut 224:1631. doi:10.1007/s11270-013-1631-0
Stabnikov V, Chu J, Ivanov V, Li Y (2013b) Halotolerant, alkaliphilic urease-producing bacteria from different climate zones and their application for biocementation of sand. World J Microbiol Biotechnol 29:1453–1460. doi:10.1007/s11274-013-1309-1
Stewart TL, Fogler HS (2001) Biomass plug development and propagation in porous media. Biotechnol Bioeng 72:353–363
Sudesh K, Abe H, Doi Y (2000) Synthesis, structure and properties of polyhydroxyalkanoates: biological polyesters. Prog Polym Sci 25:1503–1555
Sudesh K, Abe H (2010) Practical guide to microbial polyhydroxyalkanoates. Smithers Rapra Technology, UK, p 160
Taponen S, Björkroth J, Pyörälä S (2008) Coagulase-negative staphylococci isolated from bovine extramammary sites and intramammary infections in a single dairy herd. J Dairy Res 75:422–429
Tsang PH, Li G, Brun YV, Freund LB, Tang JX (2006) Adhesion of single bacterial cells in the micronewton range. Proc Natl Acad Sci USA 103:11435–11436
UNEP (2009) Converting waste agricultural biomass into a resource. Compendium of technologies, United Nations Environment Programme, p 441
Vahabi A, Ramezanianpour AA, Sharafi H, Shahbani HZ, Vali H, Noghabil KA (2014) Calcium carbonate precipitation by strain *Bacillus licheniformis* AK01, newly isolated from loamy soil: a promising alternative for sealing cement-based materials. J Basic Microbiol 53:1–7. doi:10.1002/jobm.201300560
Vempada SR, Reddy SSP, Rao MVS, Sasikala C (2011) Strength enhancement of cement mortar using microorganisms-an experimental study. Int J Earth Sci Eng 4:933–936
Volova TG (2004) Polyhydroxyalkanoates - plastic materials of the 21st Century. Nova Publishers, Hauppauge, p 282
Warren LA, Maurice PA, Parmar N, Ferris FG (2001) Microbially mediated calcium carbonate precipitation: implications for interpreting calcite precipitation and for solid-phase capture of inorganic contaminants. Geomicrobiol J 18:93–115. doi:10.1080/01490450151079833
Wang JY, De Belie N, Verstraete W (2012) Diatomaceous earth as a protective vehicle for bacteria applied for self-healing concrete. J Ind Microbiol Biotechnol 39:567–577. doi:10.1007/s10295-011-1037-1
Webster A, May E (2006) Bioremediation of weathered-building stone surfaces. Trends Biotechnol 24:255–260. doi:10.1016/j.tibtech.2006.04.005
Weil MH, DeJong JT, Martinez BC, Mortensen BM, Waller JT (2012) Seismic and resistivity measurements for real-time monitoring of microbially induced calcite precipitation in sand. ASTM Geotech Test J 35:330–341
Whiffin VS, van Paassen LA, Harkes MP (2007) Microbial carbonate precipitation as a soil improvement technique. Geomicrobiol J 24:417–423

Weaver TJ, Burbank M, Lewis A, Lewis R, Williams B (2011) Bio-induced calcite, iron, and manganese precipitation for geotechnical engineering applications. In: Geo-Frontiers: Adv Geotech Eng ASCE, 211:3975–3983. doi:10.1061/41165(397)406)

Wiktor V, Jonkers H (2011) Quantification of crack-healing in novel bacteria-based self-healing concrete. Cem Concr Compos 33:763–770

Wright DT, Oren A (2005) Nonphotosynthetic bacteria and the formation of carbonates and evaporites through time. Geomicrobiol J 22:27–53

Xia H, Hu T (1991) Effects of saturation and back pressure on sand liquefaction. J Geotech Eng 117:1347–1362

Yang IC, Li Y, Park JK, Yen TF (1993) Subsurface application of slime - forming bacteria in soil matrices. In: Proceedings of the 2nd international symposium *in situ* and on site bioreclamation, April 1993, San Diego, CA. Lewis Publishers, Boca Raton

Yang J, Savidis S, Roemer M (2004) Evaluating liquefaction strength of partially saturated sand. J Geotech Geoenv Eng 130:975–979

Yao L, Yang J, Sun J, Cai L, He L, Huang H, Song R, Hao Y (2011) Hard and transparent hybrid polyurethane coatings using in situ incorporation of calcium carbonate nanoparticles. Mater Chem Phys 129:523–528

Yegian M, Eseller-Bayat E, Alshawabkeh A, Ali S (2007) Induced-partial saturation for liquefaction mitigation: experimental investigation. J Geotech Geoenv Eng 133:372–380

Yuan F, Xiao H, Wu J, Ying WF, Yu YP, Chi HX (2013) Effect of carboxylmethyl cellulose sulfate (CMC-S) on the hydration process of cement paste. Adv Mater Res 838–841:123–126

Yu PH, Chua H, Huang A, Ho KP (1999) Conversion of food industrial wastes into bioplastics with municipal activated sludge. Macromol Symp 148:415–424

Yu J (2006) Production of biodegradable thermoplastic materials from organic wastes. US Patent 7,141,400. Accessed 28 Nov 2006

Zell C, Resch M, Rosenstein R, Albrecht T, Hertel C, Götz F (2008) Characterization of toxin production of coagulase-negative staphylococci isolated from food and starter cultures. Int J Food Microbiol 127:246–251

Chapter 3
General Aspects of Biomimetic Materials

P.M.M. Pereira, G.A. Monteiro and D.M.F. Prazeres

Abstract Natural materials like bone, ligaments, wood, shells, and scales are remarkably efficient in terms of fulfilling complex and multiple functional requirements with minimal amounts of matter. Mimicking design features found in these biomaterials like hierarchical structure and composite nature, and resorting to bio-inspired manufacturing processes like biomineralization and self-assembly could yield man-made materials that are multifunctional, lightweight, benign, and recyclable. More specifically, the incorporation of many of the characteristics and properties found in natural materials into paints, coatings, films, concrete, glass, ceramics, fibers, and insulation has the potential to revolutionize the way infrastructures and buildings are constructed. This chapter provides a concise coverage of the area of biomimetic materials. A brief outline of the discipline is followed by a discussion of general aspects related to the structure and synthesis of natural materials. Next, the recent progress made in the development of biomimetic materials with improved mechanical resistance, optical, self-cleaning, adhesiveness, and anti-adhesion properties is reviewed with reference made to the most noteworthy examples.

3.1 Introduction

Biomimetics or bio-inspired design relies on an understanding of biological functions, structures, and principles of various entities found in nature with the goal of designing products, processes, and systems of commercial and/or industrial value, which mimic certain features of the biological counterparts (Benyus 1997). The rationale behind the biomimetic approach is the realization that throughout

P.M.M. Pereira · G.A. Monteiro · D.M.F. Prazeres (✉)
Department of Bioengineering, Institute for Biotechnology and Bioengineering, Instituto Superior Técnico, University of Lisbon, Lisbon, Portugal
e-mail: miguelprazeres@tecnico.ulisboa.pt

F. Pacheco Torgal et al. (eds.), *Biotechnologies and Biomimetics for Civil Engineering*,
DOI 10.1007/978-3-319-09287-4_3

millions of years of evolution, Nature has most often selected the solutions best adapted to the prevailing environmental conditions. The environmental sustainability, which is inherent to natural organisms, is also another allure of the field of biomimetics.

Bio-inspired design combines a deep understanding of the physical, chemical, and biological phenomena and mechanisms behind a given shape, process, or natural system with a rigorous scientific and technical development of a functional application (Vincent and Mann 2002). While in many occasions in the past scientists, engineers, and artists have looked into biological systems for inspiration (e.g., Leonardo da Vinci, the Wright brothers, Antoni Gaudí), the explosion of biomimetics as a research field occurred essentially in the last 10–15 years, fueled by developments of many engineering and scientific disciplines and by the availability of ever more powerful scientific instrumentation and manufacturing technologies. A recent survey of the literature has revealed that the area has expanded from less than 100 papers per year in the 1990s to around 300 papers per year in 2013 (Lepora et al. 2013). The range of bio-inspired applications under development or which have already seen commercial success cover disparate disciplines, from materials science, nanotechnology and architecture to robotics, computer science, and biomedical engineering (Lepora et al. 2013).

The field of materials science is one of the areas where biomimetics is likely to have more impact (Whitesides and Grzybowski 2002). One of the reasons for this has to do with the fact that when compared with synthetic, human-made materials, biological materials are remarkably efficient in terms of fulfilling complex requirements with minimal amounts of matter (Wegst and Ashby 2004). The transposition of the ability of many biological systems to sense, regulate, react, grow, regenerate, and heal into new materials could yield unprecedented properties and performance (Youngblood and Sottos 2008). Furthermore, the vast majority of biomaterials are biodegradable and their basic components can be recyclable. Mimicking and incorporating this aspect in man-made materials would have a critical impact in our society both from an environmental and an economic point of view.

From a biomimetic view, the most interesting properties of biomaterials which we would like to replicate are mechanical resistance (Barthelat 2007), optical properties (Vukusic and Sambles 2003), self-cleaning (Bhushan et al. 2009), adhesiveness and anti-adhesion (Hasan et al. 2013), self-healing (Trask et al. 2007), drag reduction (Fish 2006), and thermal storage (Nikolić et al. 2003; Sharma et al. 2009). The incorporation of many of these properties into materials like paints, coatings, and films (e.g., Hasan et al. 2013; Zhao et al. 2011), concrete (e.g., Kumar et al. 2011), glass (e.g., Mirkhalaf et al. 2014), ceramics (e.g., Munch et al. 2008), fibers (e.g., Li et al. 1995) and insulation (e.g., Wang et al. 2012) which are lightweight, benign, and recyclable could revolutionize the way infrastructures and buildings are constructed (Fig. 3.1). Examples of biological materials which have attracted scientific curiosity and technological ingenuity, and which are worth looking at in the context of civil engineering include teeth, horn, bone, skin, tendon, ligament, silk, shells, wood, bamboo, leaves, seeds, scales (fish, butterfly, reptile), feathers, fur, and wool, to name a few (Wegst and Ashby 2004).

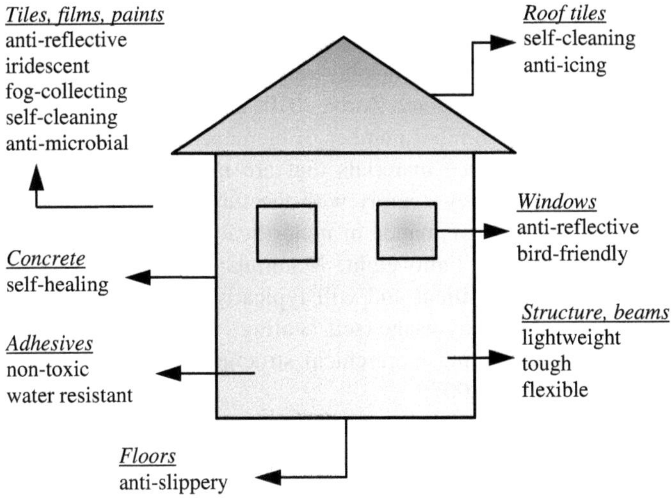

Fig. 3.1 Some potential applications of biomimetic materials in the construction of buildings and infrastructures

3.2 General Aspects of Biomaterials

The transfer of a concept or strategy observed in Nature into a new material is not trivial, requiring first a careful analysis and study of the natural model, and then a certain degree of creativity, interpretation and abstraction in order to identify the underlying principles and mechanisms. Only then will it be possible to use that fundamental information to solve a specific technological problem or to create something new. In many cases, the final practical result can be very far from the inspiring biological model. As is the case with any material, the properties of materials of natural origin are intimately determined by the way their different components are structured and bio-manufactured. Thus, in order to create new materials that mimic the behavior and performance of natural biological materials, it is important to understand their key structural features and the processes by which such structures are formed.

3.2.1 Key Structural Features of Biomaterials

While the variety of biomaterials found in Nature is tremendous, one can identify a number of structural and functional features which are recurrently found across different groups of materials of biological origin. Examples of thematic features widely disseminated across biomaterials include (i) hierarchy, (ii) composite nature, (iii) multifunctionality, and (iv) self-healing.

Hierarchy. Hierarchical structures are complex systems made up by a combination of interconnected parts across the nano-, micro- and millimeter scales (and beyond), which have evolved to display physical properties that solve specific functional challenges (Stratakis and Zorba 2010; Bae et al. 2014). The majority of these structures contain a limited number of elements arranged and oriented in a controlled way as to produce materials that are most often heterogeneous and anisotropic. Furthermore, they comply with the fundamental biological principle of minimizing the amount of matter in noncritical areas, a characteristic which leads to materials that are lightweight. Materials displaying these hierarchical structures are extremely resilient and will typically display ability to self-repair cracks and fractures caused by usage (self-healing). Two well-known examples of biological materials displaying hierarchical structure are bone (Oyen 2008) and hair (Popescu and Höcker 2007).

Composite nature. Nearly all biomaterials are composites made up of a relatively small number of polymeric and mineral building blocks (Wegst and Ashby 2004). These composites will typically arrange themselves by a process of self-assembly into complex hierarchical structures (see above). Natural composites contain a considerable proportion of mineral material without becoming brittle. The combination with biological polymers confers strength and toughness values superior to man-made ceramics. In comparison with the thousands of compounds used by humankind, living organisms rely on a smaller range of molecules to produce materials with disparate properties. Examples of minerals widely present across living organisms include (i) calcium carbonate (e.g., in mollusk shells, eggs, crustaceous exoskeletons), (ii) calcium phosphate (e.g., in teeth, bones) and (iii) amorphous silica (e.g., in diatoms and sponges). On the other hand, proteins like collagen (tendons, muscle, bones), keratin (hair, horns, beaks), and elastin (skin) and carbohydrates like cellulose (wood) and chitin (arthropods exoskeleton) provide a soft matrix for elastic behavior and create the support required for the self-assembly and stacking up of the mineral building blocks in biomaterials.

Multifunctionality. Multifunctionality is a wide spread characteristic of biological materials and structures. This means that a given architecture and composition typically originates a material that displays a variety of functions. For example, the microstructure of the scales that cover butterfly wings is responsible for functionalities which range from super-hydrophobicity and directional self-cleaning to structural color and chemical sensing (Liu and Jiang 2011).

Self-healing. Self-healing is one of the secrets of biological materials. The toughness behind many materials is translated, at the molecular level, by the fracture of sacrificial bonds (Bhushan 2009). Whenever the load stops, the material has the ability to regenerate. Most biomaterials have this ability, which, besides allowing its recovery, opens the possibility for adaptation (Meyers et al. 2008). Some biomaterials, like bone, are dynamic, continuously restructuring themselves in response to the applied load. The increment in life-time attributed by self-healing is a valuable characteristic, which many researchers are trying to replicate (Wool 2008).

Summing up, materials of biological origin use a few components (like keratin) to make a wide array of different structures in controlled orientations with durable interfaces between different materials. They are dependent or sensitive to water and produced with benign chemistry. Their properties and performances can vary in response to the environment. These complex, controlled shapes are resilient and often able to repair themselves.

3.2.2 Synthesis of Biomaterials

The processes used by living organisms to synthesize biomaterials are substantially different from those used by us to produce the range of materials that support modern societies. The most impressive features of biosynthesis are self-assembly, biomineralization, water-based chemistry, and resource efficiency.

Self-assembly. Biological materials and structures are typically assembled from the bottom up into an organized structure or pattern as a consequence of the autonomous organization of their components (Whitesides and Grzybowski 2002). The self-assembly of biomaterials involves a range of noncovalent and reversible interactions, which may include metal coordination, hydrogen bonding, van der Waals forces, hydrophobic effects, π–π interactions, metal coordination, and electrostatic interactions. These relatively weak forces act in concert to bring and hold pre-existing building blocks (e.g., biomacromolecules, ions) together (Aili and Liedberg 2010). Many times the interactions that take place between individual components are highly specific, spontaneously giving rise to architectures as complex as viruses capsules, even in vitro (Pampaloni and Masotti 2010).

Biomineralization. The processes used by organisms to incorporate mineral compounds into biomaterials are carried out at moderate conditions of pressure, temperature, and pH. Mineral crystals are formed at the nanoscale and put together from the bottom up with the help of proteins and carbohydrates, which provide the optimal conditions for this assemblage (Kröger et al. 1999). The same process allows the formation of metal inclusions by some bacteria and diatoms (Klaus-Joerger et al. 2001). The resulting materials present an extraordinary homogeneity along their small-scale structures. The composites are tailor-made according to the organisms requirements in a highly controlled fashion.

Water-based chemistry. When reviewing the thousands of diverse materials existing on the living world—strong shells, tough skeletons, elastic silk, metal granules in the magnetosomes of *Magnetospirillum gryphiswaldense*, polyhydroxyalkanoates (bioplastic) granules in several microorganisms or the feathers and fur that act as thermal and moist insulators—it is astonishing to realize that they were produced resorting exclusively to water-based chemistry. Since the major component of organisms is water, molecular systems have evolved to perform highly specific and efficient reactions in aqueous solution (Swiegers 2012). Biological processes confirm that most materials can be produced without the need for extreme conditions or toxic reagents. Even at mild conditions, biomaterials are assembled with an impressive precision, from the nano and micro to the macroscale.

Resource efficiency. In Nature, resources are limited and transportation is expensive. For this reason, living organisms live in close balance with the surrounding environment. The local environment is the source of elements, not only for the organisms that live there, but also for their offspring. For this reason, bioprocesses have been perfected by natural selection to maximize efficiency and reduce vital losses along the metabolic routes. Moreover, Nature uses cyclic processes, ensuring that each element is returned to its original form, ready to be used again. Nature found a way to recycle every biomaterial, reduce the waste generation and the impact of organisms on the environment, and constantly feed the cycle of the elements.

3.3 Development of Biomimetic Materials

3.3.1 Mechanical Properties

Despite their attractive properties, the application of highly mineralized materials such as ceramics and glasses in man-made products used in construction is often limited by their characteristic brittleness. In nature, however, the incorporation of a small percentage of biomolecules and a controlled nano-, micro- and meso-hierarchical architecture of the building blocks originates materials that are both flexible and robust, providing organisms with an ability to manage impact and shock in the most varied circumstances (Mirkhalaf et al. 2014). In these materials, structural, elastomeric biomolecules such as proteins and polysaccharides contribute to improve the robustness of mineral components which are characteristically harder, but more fragile. This combination yields natural composites that present both strength and toughness several orders of magnitude above the ones displayed by pure minerals (Barthelat 2007). Examples of natural ceramic composites that have been the subject of study range from nacre (Corni et al. 2012) and bone (Launey et al. 2010), to bird eggshells (Freeman et al. 2010), sponge spicules (Meyers 2008) and the armored skins of crocodiles, armadillo, turtles, and fish (Zimmermann et al. 2013; Chintapalli et al. 2014). Natural biopolymers (e.g., silk) and cellular matrices (e.g., wood, bamboo) also present properties, which result from nano-, micro- and meso-structures that are worth replicating into materials used to withstand mechanical loads. A few examples are briefly described next that are especially relevant for the development of new construction materials (e.g., glasses, protective coatings, tiles, etc).

Nacre. In order to protect mollusks from predators and water-carried debris, the shells of organisms like the abalone or oysters have evolved into a stiff and impact resistant material named nacre or mother of pearl (Corni et al. 2012). The structuring of alternating aragonite (an orthorhombic crystal form of calcium carbonate)

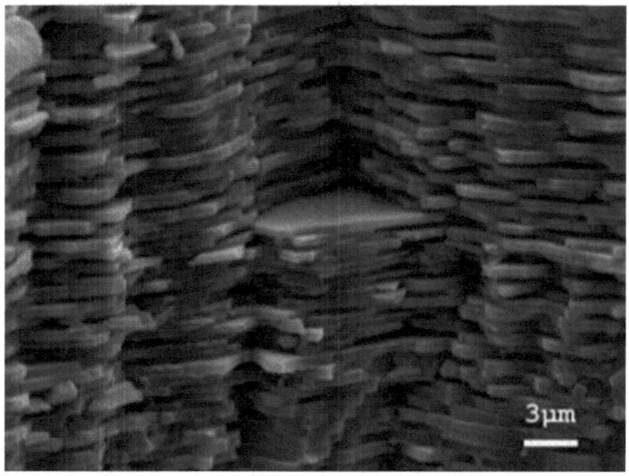

Fig. 3.2 Electron microscopy image showing the characteristic arrangement of layers of aragonite "bricks" in nacre. *Image source* Wikimedia commons, Fabian Heinemann, public domain

tablets (95 %wt) and protein layers (5 %wt) into a brick and mortar fashion (Fig. 3.2) provides a bending strength one order of magnitude higher than pure aragonite and a 3,000 times higher work of fracture (Meyers et al. 2008).

At high tensile stress, the crystal tablets will slide on one another and the organic layer will suffer viscoplastic deformation, allowing nacre to withstand stretch and dissipate energy (Barthelat 2007). Fabrication techniques such as freeze casting, hot-press assisted slip casting, extrusion and roll compaction, layer-by-layer self-assembly, and paper-making method have successfully yielded materials that mimic the mechanical properties of nacre (Corni et al. 2012).

Enamel. The hardest material in vertebrates is enamel, a specialized tissue present in the outer layer of teeth. The microscopic arrangement of 5 µm rods of hydroxyapatite (the biologically common form of calcium phosphate crystals) and a high mineralization confer a hardness of 4 GPa to enamel (Meyers et al. 2008). The low toughness of enamel is balanced by the dentin that composes the inner layer of teeth. It comprises 30 % vol. of collagen, the structural protein present in most body tissues, and 25 % vol. of water, along with hydroxyapatite crystals, which assemble themselves around the collagen tubules to form a porous, though tough and light, material (Corni et al. 2012). The development of hard and resistant teeth by animals allowed them to explore a wider range of food sources.

Bone. The mimicking of the porous and branched structure of bone is ubiquitous in the work of some renowned architects, originating structures that are lightweight and tough [e.g., Casa Batlló and La Sagrada Famílía (Barcelona, Spain) by Antoni Gaudí or Gare do Oriente (Lisbon, Portugal) by Santiago Calatrava]. Bone has evolved to provide efficient structural support to vertebrates and protect their vital organs and is thus an interesting material to emulate for construction and safety applications. It is composed of hydroxyapatite mineralized collagen fibrils arranged across seven hierarchical levels into lamellar-structured osteons in a porous matrix. The strength of bone is controlled by its constituents at two levels: an intrinsic toughening based on plastic deformation of organic molecules at the submicron scale and an extrinsic toughening based on crack-tip shielding of osteons at macro scale. The tensile strain of human cortical bone is 2 % while for elk antler it can reach 12 % (Launey et al. 2010). In contrast to nacre and enamel, bone is a dynamic material with anisotropic distribution of mass, where the regions subjected to higher loads are reinforced in comparison to the overall bone. Besides reducing the density, bone porosity grants the elasticity necessary to competently deflect impacts (Young's modulus of bone is between 8 and 24 GPa) (Meyers et al. 2008). This property has been explored by Wilfredo Méndez Vázquez (from Escuela de Arquitectura de la Pontificia Universidad Católica de Puerto Rico) to create Stick.S, a more sustainable concrete associated with a design less vulnerable to natural disasters (earthquakes, hurricanes, etc.).[1]

Silk. In addition to mineralization, another strategy used in the natural world to produce strong tissues like silk, arthropods' exoskeleton, hoofs, and horns is to increase the cross-linking of polymeric or fibrous materials. Spider silk possesses a remarkably high tensile strength and extensibility (the strain at failure can exceed 1,600 %), which evolved in order to quickly absorb the kinetic energy of insects trapped in webs (Meyers et al. 2008). The viscoelastic behavior of silk fibers is attributed to the entropic unfolding of its constitutive proteins and to the transfer of load to protein nanocrystals embedded in the disordered matrix (Meyers et al. 2013). The properties of each silk thread are determined by the proportion of each component, the flexible glycine-rich matrix and the strong anti-parallel β-sheets crystals, as well as by the entanglement of the fibers. The published literature on the production of artificial silk suggests that the spinning process is crucial to obtain fibers with similar characteristics (Eadie and Ghosh 2011). Another remarkable property of silk is its ability to retain 10,000 times its weight in water, which can be used to create humidity-responsive fibers. Clearly, silk is an outstanding material that can serve as an inspiration to produce stronger and lighter cables, textiles, or protective clothing.

Wood. The mechanical properties of cellular matrixes deserve some credit as well. Wood, for example, has been used by humankind since its rise in applications ranging from buildings to tools. Its success as a structural material is explained by the high specific strength (comparable to steel) provided by its hierarchical structure and optimized orientation of fibrils reinforcement (Meyers et al. 2008). Wood is mainly composed of cellulose, hemi-cellulose, and lignin polymers arranged in parallel hollow tubular cells (Eadie and Ghosh 2011). The optimization of the number of layers and biopolymer orientation in some specific types of woods, like bamboo, can double their tensile strengths up to 520 MPa (Li et al. 1995). The mimicking of these reinforcement patterns may be applied when designing the structural support of buildings and in the manufacturing of composite materials (Li et al. 1995; Banthia et al. 2012).

Pummelo. Using the hierarchical structure of cellular matrices, nature has found an ingenious solution to resist impact in the form of the *Citrus maxima* fruit (also

[1] STICK.S lightweight structural system, Ask Nature, Biomimicry 3.8. Accessed on the 14th March 2014, http://www.asknature.org/product/8779abb44e4bcd7a9e8f3bf18a2c0a89.

known as pummelo). This Southeast Asian native fruit grows 10–15 m above the ground and is the largest fruit of the *Citrus* genus (15–25 cm diameter). Once ripened, it falls to the ground, quickly dissipating the kinetic energy ($\sim$90 J) to resist outstanding deceleration forces without being visibly damaged. The 2 cm thick pummelo's peel, with its alveolar structure, is acknowledged for this ability. Metal foams have already been produced by emulating the peel's microstructure, which reasonably reproduce the impact resistance properties of the pummelo fruit (Fischer et al. 2010). Applications of pummelo-like foams could range from low weight package filling to protective coating for electronic devices or safety coating (helmets).

3.3.2 Optical Properties

The ability to manage light (visible and nonvisible) is important for many organisms in the most varied contexts. For example, by adequately absorbing and reflecting light, organisms can send visual signals, which may serve to attract individuals from the same species or to keep potentially harmful species at bay. The handling of infrared, visible, and ultraviolet radiation may also be critical to optimize thermoregulation, enhance photosynthesis or aid in navigation. A number of examples are described next that are especially relevant to develop new materials with impact in the aesthetics of facades and building envelopes.

Structural color. Living organisms, from flowers and fruits to birds, fishes and butterflies, display a diversity of colors that has long marveled people and artists, who thrive to capture them in their works. In many cases, color is imparted by the presence of pigments, i.e., molecules capable of selectively absorbing visible wavelengths, inducing the perception of color. However, research on the microstructure of hair-like setae of polychaete worms (Parker et al. 2001), butterfly wing scales (Kinoshita et al. 2002), beetle cuticle (Seago et al. 2009), bird feathers (Stavenga et al. 2011) and fruit peals (Vignolini et al. 2012) has revealed that in several situations bright and vivid colors have a structural component (Prum et al. 2006; Ingram and Parker 2008). This so-called structural color (Fig. 3.3) is generated as a result of the interaction of the micro- and nano-structure of scales, feathers, cuticles, petals or fruit peels with the incident light (Vukusic and Sambles 2003; Eadie and Ghosh 2011). This effect is based on the principle that physical features with a size comparable to the wavelength of incident light can interfere with it via multilayer interference, diffraction or scattering mechanisms (Prum et al. 2006).

Thus, structural color allows a precise control over the way light is reflected, propagated or transmitted from a surface. This feature may present a significant advantage in vital processes such as camouflage, long distance signaling and thermoregulation (Vukusic 2006). Structural coloring is more brilliant and intense than pigment-induced color, it may possess iridescence properties or only be visible at certain angles, and is less prone to bleaching. These characteristics have

Fig. 3.3 Examples of structural colors displayed by **a** the wasp *Hedychrum rutilans* and **b** the butterfly *Morpho peleides*. *Image source* Wikimedia commons, public domain

attracted the interest of many researchers who have been striving to come up with technologies capable of producing materials with structural color (Eadie and Ghosh 2011; Li et al. 2012). So far, the principles behind the structural color displayed by butterfly scales and peacock feathers has inspired the development of color-shifting paints (ChromaFlair, JDSU) and fabrics (Diphorl, Kuraray Corporation and Morphotex, Teijin Fibers), anti-counterfeiting technologies (NOtES, NanoTech Security), high efficiency solar collectors (Zhao et al. 2011), electronic screens without environmental light interference (Mirasol, Qualcomm), radiant light films (3 M^{TM} Radiant Films, 3 M) for home décor and architectural glass, among others.

Anti-reflection. The outstanding optical properties displayed by living organisms do not end with their color techniques. The eyes of the elephant hawk-moth (*Deilephila elpenor*) are coated with a regular pattern of conical nano protuberances (200–300 nm height), which significantly minimize light reflection. This ability improves night vision and is critical for moths to escape predators. The anti-reflective properties of these structures is superior to any man-made cover and can be replicated in anti-reflective films for electronic devices (Motheye and Marag, MacDermid Autotype[2]) and solar panels (Yang et al. 2013) or to improve X-ray imaging instruments (Pignalosa et al. 2012). The production of such films for construction applications would enable architects to construct or retrofit buildings with anti-reflective facades. This would be especially valuable in the case of metal- and glass-clad skyscrapers implanted in densely populated areas that can potentially create hazardous reflectivity issues for neighboring buildings and passing traffic.

Bird-friendly glass. The latest architectural trends include an increase in the incorporation of glass windows in buildings, to better harness sunlight, and improve energy efficiency. Nonetheless, glass facades of city skyscrapers came with a high cost to non-human inhabitants, namely bird populations. As it happens with humans, birds are not capable of seeing glass, and thus often collide with

[2] MacDermid Autotype, Wantage, UK.

windows in buildings during flight, most of the times with fatal consequences. Collision with glass is estimated to cause hundreds of millions of bird casualties every year in the United States alone, adding up to the man-derived causes for the decline of bird populations, from hummingbirds to falcons (Sheppard 2011). Conventional strategies to prevent collisions are based on the imprinting of patterns visible to birds on the glass. However, this strategy reduces the passage of sunlight and may interfere with the building's aesthetics. Searching for a better solution, the European manufacturer Arnold Glas designed and produced Ornilux,[3] a bird-friendly glass with a UV-reflective cover invisible to human eye. This solution was inspired by spider signaling through web decorations. In order to prevent their webs from being damaged by flying birds, spiders include special silk structures in their pattern that reflect UV light (Bruce et al. 2005). This reflex allows birds to locate and avoid a spider's web without warning their prey. The coating of glass with a similar pattern of UV reflectors has been claimed to reduce bird collisions by 76 % (Ornilux 2014).

3.3.3 Superhydrophobicity and Self-Cleaning Properties

The external and internal surfaces of buildings are recurrently exposed to liquid, gaseous, and solid water, whether as a result of rain hitting facades, water vapor condensing in the ceramic tiles of bathrooms or snow crystals falling down on roofs. The creation of new materials capable of handling water under these different circumstances in a more efficient way can be foreseen as a way of improving durability, comfort, and performance of a vast array of surfaces (e.g., tiles, floors, roofs, walls, etc.) found in buildings. Nature is a source of inspiration in this regards, since strategies to manage water at surfaces and control wettability are widely found across organisms due to their relevance in the context of properties like self-cleaning, drag reduction, adhesion, water capture, condensation and evaporation kinetics or biofilm formation (Koch and Barthlott 2009).

Self-cleaning. In some cases, the accumulation of water layers over the surface of living organisms is detrimental to vital processes. This is especially critical for plants, since the deposition of water over their leaves may impede gas exchange, hinder respiration and interfere with thermoregulation. Moreover, water films and droplets can partially block the sunlight that is essential for photosynthesis and allow potentially pathogenic microorganisms to proliferate (Koch and Barthlott 2009). Thus, some plants have engineered ways to render their surfaces hydrophobic. On hydrophobic surfaces, the energy necessary for liquids to spread is larger than the cohesive energy between the liquid molecules. As an outcome, the contact angle, i.e., the angle between the surface and the liquid–air interface, is high (above 150° for superhydrophobic surfaces) inducing the formation of drops

[3] ORNILUX Bird Protection Glass. Accessed on the 14th March 2014, http://www.ornilux.com/.

instead of a film. Most of the times, superhydrophobicity is coupled with a low tilting angle (i.e., the angle at which a surface has to be tilted in order for the drop to slide), so that drops easily roll off from the surface. The strategies used to achieve hydrophobicity span from hydrophobic waxes (e.g., grapes, Koch and Ensikat 2008), air-retaining microstructures (e.g., *Salvinia* fern, Barthlott et al. 2010) and high hair density (e.g., *Alchemilla monticola*, Otten and Herminghaus 2004), to hierarchical topographies (e.g., lotus, Bhushan et al. 2009).

One of the best known examples of the impact of hierarchical topography on hydrophobicity is given by the lotus (*Nelumbo nucifera*), a plant whose leaves have one of the highest known contact angles (164°). This very low wettability of the lotus results from a combination of a regular pattern of micropapilae (convex structures ~20 μm) with superimposed wax crystals, (Bhushan and Jung 2011). Other combinations of micro and nano-patterns with impact on surface wettability have been observed in organisms like butterflies (Yan et al. 2007), english weeds (Pereira et al. unpublished results, Fig. 3.4), rice leaves (Feng et al. 2002), and rose petals (Feng et al. 2008).

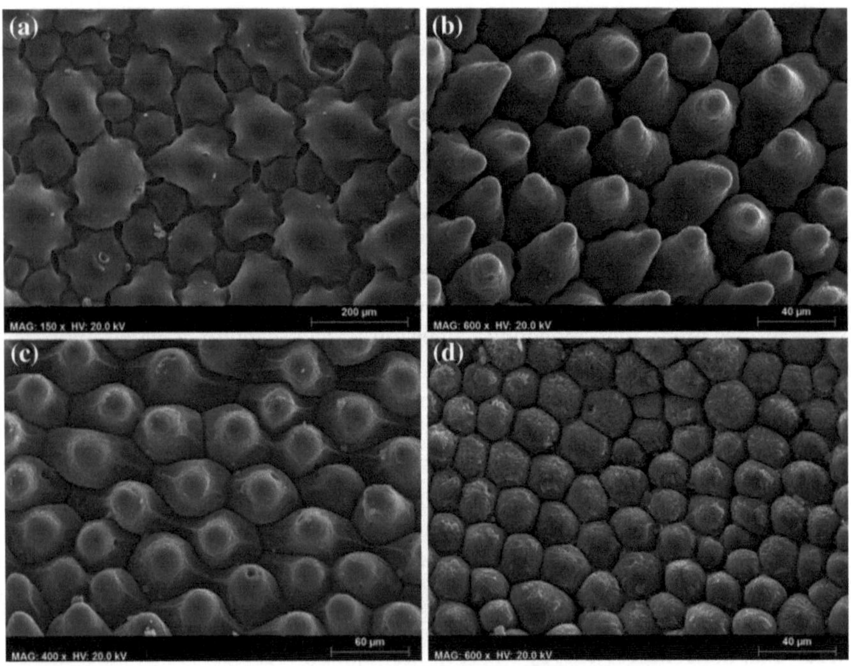

Fig. 3.4 Scanning electron microscope image of epoxy replicas of **a** the underside of the English weed leaf (scale bar = 200 μm), **b** the petals of the flowers of English weed (scale bar = 40 μm), **c** the petals of a Cape Daisy (scale bar = 60 μm) and **d** the petals of a rose (scale bar = 40 μm). (Pereira et al. unpublished)

The low wetting is explained by the Cassie-Baxter principle, which predicts the formation of air bubbles between the microstructures, avoiding the spread of the liquid. In addition to superhydrophobicity, these hierarchical architectures decrease the contact area between the surface and any particles settled over it. Therefore, water drops rolling off the surface carry sediments away, providing the leaves with self-cleaning ability.

The superhydrophobicity and self-cleaning properties conferred by microtopographic patterns have been explored by the construction industry in the form of self-cleaning paints (e.g., StoCoat® Lotusan®[4]) or clay roof tiles (Erlus Lotus®[5]) designed to reduce building maintenance. By mimicking the microstructure of lotus leaves, these materials display a high level of water and dirt-repellency, a pronounced self-cleaning effect and a high resistance to soiling and growth of mold, mildew, and algae. As a result, water (e.g., from rain) and dirt flow off immediately and facades and roofs remains dry and attractive. The textile industry has also been allured by the lotus effect and as a result a number of impervious, stain-free fabrics have been launched in the market (e.g., Greenshield®[6], Nanosphere®,[7] and Mincor[8]) which could be used for example in sunshades, awnings, or overhangs.

Anti-fogging. On cold days, with high humidity weather, driving can become a very dangerous and unpleasant activity due to the condensation of water over the windshield and mirrors of one's car. Likewise, condensation of water over tiles and walls in bathrooms after a warm bath contributes to the development of microbial growth. Superhydrophobic surfaces inspired by plants like the lotus cannot be used as anti-fogging since very small drops can get trapped between the underlying microstructures (Cheng and Rodak 2005). Moreover, such structures interfere with light, reducing the visibility. Inspiration to solve these problems came from insect eyes (e.g., mosquito, moth, flies), which are maintained clean and dry, even in moisture rich environments. The hexagonal structures that form the mosquito eyes (ommatidia) are covered with a pattern of smaller ($\sim$100 nm) hexagonal features evenly spaced ($\sim$47 nm), which effectively prevent condensation, at normal conditions (Fig. 3.5; Gao et al. 2007). Using mosquito eyes as an inspiration and resorting to soft lithography, researchers created a superhydrophobic compound eye-like microstructure consisting of arrays of micro-hemispheres covered with nipple-like SiO_2 nanospheres surface with a contact angle of 155° (Gao et al. 2007). The expectation is that further developments will lead to the development of multifunctional coatings with antifogging and easy-cleaning properties tuned to match the requirements of applications in the realm of civil engineering (Gao et al. 2007).

---

[4] Sto Corp., Atlanta, Georgia.

[5] Erlus AG, Neufharn, Germany.

[6] BigSky Technologies LLC

[7] Schoeller Textile AG, Sevelen, Switzerland.

[8] BASF AG, Germany.

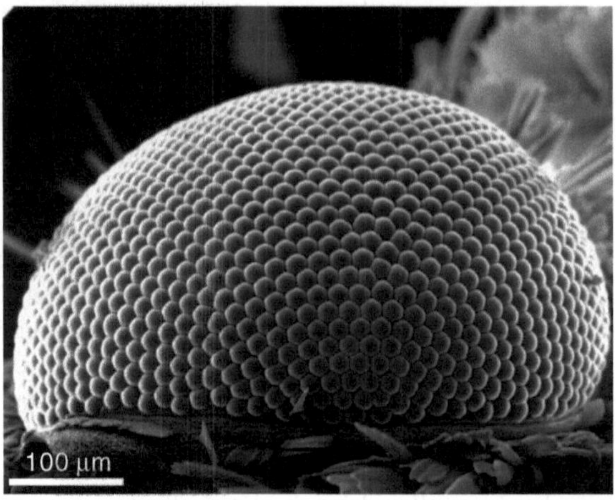

Fig. 3.5 Scanning electron microscope image of the eye on a leaf miner moth. *Image source* Louisa Howard, Dartmouth college, Electron Microscopy Facility, public domain

Fog-collection. More than 450 million people currently live under water scarcity in arid regions and two-thirds of the world population are expected to live in water stressed areas by 2025 as a result of the exponential demographic growth and the ever increasing rate of water exploitation (Diop and Rekacewicz 2008). Although an inconvenience to drivers, fog is a valuable resource that could be tapped to alleviate this problem. For example, in arid regions of coastal Africa (Shanyengana et al. 2002) and in the coastline of the Atacama desert in Chile (Cereceda et al. 2008) oceanic fog represents an inexpensive water source, most of the times more reliable than precipitation or surface water. The integration of fog collection materials, technologies and structures into buildings, urban environments and agricultural landscapes could potentially provide fresh water for human consumption in poor dry countries, provide water to the top of skyscrapers without pumping, and reduce the exploitation of groundwater for farming purposes. A wide variety of species, ranging from cactuses to ferns and insects can be found in arid regions which rely on elaborate fog-collecting strategies that are worth mimicking in this context. For example, the Namibian desert beetles (from the genus *Stenocara*) use a strategy to collect water from the morning fog, which enables them to survive under the harsh weather conditions characteristic of their habitat. The water-collecting mechanism used by the Namibian beetle relies on the bumpy surface of its wings, which combine hydrophilic regions that induce coalescence of water drops, with hydrophobic wax-covered paths that transport water to the insect's mouth (Parker and Lawrence 2001). Several researchers have directed their efforts towards the development of fog-catching surfaces inspired in the topography of the wings of *Stenocara* beetles (Zhai et al. 2006; Garrod et al. 2007; Dorrer and Rühe 2008; Thickett et al. 2011). The mechanisms behind the fog-harvesting ability of cactuses spines (Ju et al. 2012; Cao et al. 2014) and spider silk (Zheng et al. 2010) have also be studied in detail, constituting excellent biomimetic models for the development of water collection strategies.

Anti-icing. Icing is another climatic event capable of affecting vital infrastructures of modern society. Ice may block roads, take down power cables, cause the rupture of pipelines and promote pedestrian slip events. The aeronautics industry has invested considerable resources on this problem, since the accumulation of ice over the fuselage of aircrafts increases its weight, unbalances the flight and increases drag, consequently reducing fuel efficiency and raising safety concerns (Bahadur et al. 2011). The ice formed when supercooled suspended drops aggregate on airplane surfaces has been appointed as culprit on many plane crashes over the years. Conventional solutions to create anti-icing surfaces are not sustainable or reliable. On one hand, such surfaces depend on an energy intensive heating system, while on the other, they depend on anti-cryogenic chemicals, such as glycerol, which are toxic and require permanent replenishment. Researchers have looked up to nature for inspiration and were rewarded with two potential ideas. For example, superhydrophobic surfaces have been created using topographic features inspired on the defogging mosquito eyes, which also prevent ice crystals nucleation, and on the water repellent hairs of buoyant insects (from order *Hemiptera*). Assays on the behavior of droplets impacting on these surfaces at supercooled conditions revealed that they can avoid ice formation down to −25 °C (Mishchenko et al. 2010). The second strategy uses slippery liquid-infused porous surfaces (SLIPS) inspired in carnivorous plants from the *Nepenthes* genus that lower the nucleation temperature of supercooled water (Wilson et al. 2012). These plants attract insects to their pitcher cups, hampering their escape as a result of the slippery inner surface (Kim et al. 2012). The pitfall traps have a porous lining permeated with an aqueous solution, which is incompatible with the oils sheathing the insect's feet. The SLIPS systems consist of a microporous scaffold infused with a lubricating fluid with which the liquid to repel (in this case water) is immiscible (Wilson et al. 2012). In nature, a structure is often multifunctional, and SLIPS are no exception. Besides preventing ice formation, they are self-cleaning, anti-corrosive, and anti-biofouling (i.e., they prevent the adhesion of organisms).

3.3.4 Anti-biocolonization Properties

The biological colonization of walls, tiles, flooring, and other building surfaces (external and internal) by organisms like algae, bacteria and fungi can cause direct and indirect damage (e.g., blackening, greening, fissuration) to materials and elements of construction (Macedo et al. 2009), even if some species (e.g., lichens, mosses) can enhance the appearance and character of buildings and structural monuments. Such biological and microbiological growths are usually dependent on both the bioreceptivity of the underlying materials and on the environmental conditions (e.g., humidity, temperature, solar exposure). While this type of material decay and soiling can be partially alleviated by resorting to a planned preventive maintenance, the development of anti-biocolonization materials could constitute a valuable alternative to the construction industry.

The attachment of microorganisms to surfaces also presents a major concern in public health control. With the recent outbreaks of flu epidemics and the persistent incidence of hospital-acquired infections, the potential of the inanimate environment (door and stair handles, hospital surfaces, public transports, etc.) for spreading diseases has been seriously discussed (Page et al. 2009). In hospital facilities, for example, there is a confluence of multiple pathogenic organisms, often drug-resistant or contagious strains, which are easily transported in, as well as out, unless effective anti-microbial measures are taken. Again, the availability of anti-microbial surfaces could offer architects and engineers a valuable option to minimize microbial spreading in buildings like hospitals, day-care centers and food-preparation facilities (Hasan et al. 2013; Page et al. 2009). A successful example of a bio-inspired anti-biocolonization surface is described next.

Shark-skin. The search for materials to which organisms struggle to adhere was primarily motivated by the naval industry. From barnacles, mussels, tubeworms and algae to microalgae, fungi and bacteria, all sorts of marine organisms get attached to the hull of ships (Chapman et al. 2013). As a consequence, the hydrodynamic drag increases, fuel consumption soars, the ship's framework corrodes faster and regular maintenance is required to remove the biological layer (i.e., biofouling) and replenish the cover (Schultz et al. 2011). One of the most interesting developments in the search for new, long-lasting, nontoxic alternatives to minimize fouling came out from the realization that dermal denticles that cover the skin of sharks had an unprecedented capacity to prevent microorganisms from settling. Further studies attributed this capacity to the distinct pattern of tiny riblets, arranged in diamond shape, on the shark's denticles (Chung et al. 2007). When microorganisms settle on a surface, they signal other individuals of the same species to induce the formation of biofilms, which help the colonization besides avoiding detachment. Over the shark skin though, the signalling and biofilm establishment is inhibited by the topography (Chung et al. 2007). This discovery led to the development of Sharklet^{TM}, "the only nontoxic, biocide-free and no-kill surface pattern used to inhibit bacterial attachment, survival and touch transference" (Sharklet 2011). In Sharklet^{TM} textured films, the characteristic topography of shark's dermal denticles was re-interpreted, simplified and optimized, originating a diamond pattern of ribbons which is easy to produce at large-scale. As a result, in surfaces covered with Sharklet^{TM} films the settlement of green algae is reduced by 85 % and bacterial attachment by more than 76 % (Sharklet 2011). The films are currently marketed as being ideal to cover frequently touched bacteria-prone surfaces in healthcare buildings or in other facilities where bacterial absence is desired.

Other natural examples of anti-biocolonization surfaces are the skin of pilot whale (*Globicephala melas*), which secretes an enzymatic gel that breaks down the components of natural glues (Baum et al. 2003), the eggs of the white dogwhelk sea snail (*Dicathais orbita*), which combine a micro-roughness with oily secretions (Lim et al. 2007) and the wings of cicadas (*Psaltoda claripennis*), which display a set of pillars with diameters around 100 nm that mechanically disrupts bacterial cells (Ivanova et al. 2012).

3.3.5 Adhesive Properties

The use of adhesives, i.e., materials that can join surfaces (similar or dissimilar) together is widely spread across the construction industry, both in structural and nonstructural situations. Adhesives are used for example to join and attach internal building panels and elements, to attach brick slips, ceramic tiles, and flooring or to produce plywood. A large proportion of these adhesives are based on organic solvents and polymeric resins like polyesters, polyurethanes, and acrylics. Replacing these petroleum-based materials for more benign, water-based adhesives would definitely contribute to environmental sustainability. Nature offers a vast array of solutions to promote adhesion that are worth exploring in this context, as described next with a few examples.

Marine-inspired adhesives. Marine organisms like the sandcastle worm (*Phragmatopoma californica*, Shao et al. 2008), limpets (*Diodora aspera*, Ellem et al. 2002), sea cucumbers (Holothuroidea, Flammang et al. 2009), barnacles (*Megabalanus spp.*, Knight 2009), mussels (*Mytilus edulis*, Lee et al. 2006, 2007), and sea urchins (Santos and Flammang 2012) secrete protein-based glues unrivalled by any man-made adhesive. Moreover, unlike common adhesives, they resist humidity and are produced using water-based chemistry. For example, mussels secrete a cocktail of proteins rich in lysine and 3,4-dihydroxyphenylalanine (DOPA) amino acids, which solidify at the pH of sea water (around 8.2) (Flammang et al. 2009). DOPA residues are able to establish strong coordinate bounds with metal ions of inorganic substrate and covalent bounds with the molecules of organic substrates (Lee et al. 2006). This versatility provides outstanding permanent adhesion for defying the power of ocean waves.

If technically and economically feasible, the substitution of traditional glues like the ones based on the recognized carcinogen formaldehyde by water-based and water-resistant glues formulated with DOPA mimetic molecules could revolutionize the adhesive market. Excellent results have already been obtained by manipulating the abundant and inexpensive soy proteins to resemble marine adhesive cocktails. The plywood manufacturer Columbia Forest Products[9] reduced the use of carcinogenic components and the emission of hazardous air pollutants by successfully introducing these soy-based, formaldehyde-free adhesives in the manufacturing of hardwood plywood (PureBond® technology[10]).

Velcro. One of the most famous biomimetic products, Velcro, was conceived back in 1948 by George de Mestral following the observation that the seeds of the burdock plant (*Arctium lappa*) relied on hook-like structures to stick to clothes or animal fur as a strategy for dispersion (Bhushan 2009). Mestral ultimately patented a process for producing the well-known fabric fastener and named it Velcro. Since then, the production of Velcro has grown, its applications ranging from textiles,

---

[9] Columbia Forest Products, Greensboro, North Carolina.

[10] PureBond Technology. Ask Nature, Biomimicry 3.8. Accessed on the 20th March 2014, http://www.asknature.org/product/22aa5601fcdb68b5d2dc9e3d3a22f7f1.

agriculture and transportation to industry, healthcare, and space exploration (Eadie and Ghosh 2011).

Gecko. Gecko lizards are specialists on temporary adhesion. As a result of the branching of feet hairs in multiple setae, which in turn gives raise to multiple spatulae, there is a strong adhesion to a variety of dry surfaces (Greiner 2010). The contact splitting in billions of nano hair-like structures provides a high number of van der Waals interaction points between the feet and the surface, allowing the gecko to walk on glass or even on ceilings (Bhushan 2012). Since the adhesion does not rely on chemical components, but rather on a physical effect, Gecko feet can attach and detach from a surface numerous times without loosing the adhesive effect. Detachment is possible because, although as a bulk the billions of setae on the gecko feet provide a strong interaction with the substrate ($\sim$10 N/cm^2), individually the adhesive force is easy to overcome ($\sim$10 nN per spatula) (Geim, et al. 2003; Huber et al. 2005). The same strategy is used by several insects, with a smaller mass to contact area ratio though. Replicating the array of nano-hairs in the gecko feet, a glue free adhesive tape (TacTiles™) was created by the carpet manufacturer Interface.[11] While originally developed for the installation of carpets without glue, applications are wider. The environmental impact associated to the use of TacTiles™ is 90 % lower when compared with regular carpet glue, since no toxic chemicals are used and no volatile organic compounds are released.[12] As a way of overcoming the limited adhesion of gecko-inspired tape in wet surfaces, researchers covered a nanostructured polymeric material with a synthetic DOPA mimic (Lee et al. 2007). The adhesive, patented under the name Geckel, presents a wet adhesion 15 times stronger than the nanostructured polymer alone and is able to maintain the adhesion over 100 cycles.

3.4 Conclusions

The array of materials produced by living organisms is incredible, considering the reduced number of constituents used. Moreover, the fact that all biomaterials are produced at mild-conditions using highly specific interactions that take place in aqueous solutions is impressive. As a contrast, humankind resorts to thousands of different compounds to create the essential products used every day. In addition, those products generally require extreme temperatures or pH and incorporate organic solvents or toxic compounds. Our impact on the planet is widely exceeding its capacity. Despite all this, Nature's materials still hold properties unraveled by human cutting-edge technology. By studying and understanding the principles behind the way materials are produced in Nature, we may innovate in

[11] InterfaceFloor, LaGrange, Georgia.

[12] InterfaceFlor, Learning from Nature. Accessed on the 20th March 2014, http://issuu.com/interfaceflor/docs/biomimetic_brochure_can.

the field of materials sciences. These new bioinspired materials could find wide application in civil engineering, increasing the comfort, resilience and functionality of buildings and infrastructures, and setting them in tune with the local environment.

References

Aili D, Liedberg B (2010) Biomimetic approaches to self-assembly of nanomaterials. In: Kumar C (ed) Biomimetic and bioinspired nanomaterials. Wiley-VCH Verlag GmbH & Co. KGaA, Weinheim, pp 343–377

Bae W-G, Kim HN, Kim D, Park S-H, Jeong HE, Suh K-Y (2014) 25th Anniversary article: scalable multiscale patterned structures inspired by nature: the role of hierarchy. Adv Mater 26:675–700. doi:10.1002/adma.201303412

Bahadur V, Mishchenko L, Hatton B, Taylor JA, Aizenberg J, Krupenkin T (2011) Predictive model for ice formation on superhydrophobic surfaces. Langmuir 27:14143–14150. doi:10.1021/la200816f

Banthia N, Sappakittipakorn M, Jiang Z (2012) On permeable porosity in bio-inspired fibre reinforced cementitious composites. Int J Sustain Mater Struct Syst 1:29–41. doi:10.1504/IJSMSS.2012.050451

Barthelat F (2007) Biomimetics for next generation materials. Philos Trans R Soc A 365:2907–2919. doi:10.1098/rsta.2007.0006

Barthlott W, Schimmel T, Wiersch S, Koch K, Brede M, Barczewski M, Walheim S, Weis A, Kaltenmaier A, Leder A, Bohn HF (2010) The *Salvinia* paradox: superhydrophobic surfaces with hydrophilic pins for air retention under water. Adv Mater 22:2325–2328. doi:10.1002/adma.200904411

Baum C, Simon F, Meyer W, Fleischer LG, Siebers D, Kacza J, Seeger J (2003) Surface properties of the skin of the pilot whale *Globicephala melas*. Biofouling 19:181–186

Benyus JM (1997) Echoing nature. In: Benyus JM (ed) Biomimicry: innovation inspired by nature. Harper Collins, New York, pp 1–10

Bhushan B (2009) Biomimetics: lessons from nature—an overview. Philos Trans R Soc A 367:1445–1486. doi:10.1098/rsta.2009.0011

Bhushan B (2012) Bioinspired structured surfaces. Langmuir 28(3):1698–1714. doi:10.1021/la2043729

Bhushan B, Jung Y, Koch K (2009) Micro-, nano- and hierarchical structures for superhydrophobicity, self-cleaning and low adhesion. Philos Trans R Soc A 367:1631–1672. doi:10.1098/rsta.2009.0014

Bhushan B, Jung YC (2011) Natural and biomimetic surfaces for superhydrophobicity, self-cleaning low adhesion and drag reduction. Prog Mater Sci 56:1–108. doi:10.1016/j.pmatsci.2010.04.003

Bruce MJ, Heiling AM, Herberstein ME (2005) Spider signals: are web decorations visible to birds and bees? Biol Lett 1:299–305. doi:10.1098/rsbl.2005.0307

Cao M, Ju J, Li K, Dou S, Liu K, Jiang L (2014) Facile and large-scale fabrication of a cactus-inspired continuous fog collector. Adv Funct Mater 24:3235–3240. doi:10.1002/adfm.201303661

Cereceda P, Larrain H, Osses P, Farías M, Egaña I (2008) The spatial and temporal variability of fog and its relation to fog oases in the Atacama Desert, Chile. Atmos Res 87:312–323. doi:10.1016/j.atmosres.2007.11.012

Chapman J, Hellio C, Sullivan T, Brown R, Russel S, Kiterringham E, Nor L, Regan F (2013) Bioinspired synthetic macroalgae: examples from nature for antifouling applications. Int Biodeteor Biodegrad 86:6–13. doi:10.1016/j.ibiod.2013.03.036

Cheng YT, Rodak D (2005) Is the lotus leaf superhydrophobic? Appl Phys Lett 86:144101. doi:10.1063/1.1895487

Chintapalli RK, Mirkhalaf M, Dastjerdi AK, Barthelat F (2014) Fabrication, testing and modeling of a new flexible armor inspired from natural fish scales and osteoderms. Bioinspir Biomimet 9:036005. doi:10.1088/1748-3182/9/3/036005

Chung KK, Schumacher JF, Sampson EM, Burne RA, Antonelli PJ, Brennan AB (2007) Impact of engineered surface microtopography on biofilm formation of *Staphylococcus aureus*. Biointerphases 2:89–94. doi:10.1116/1.2751405

Corni I, Harvey TJ, Wharton JA, Stokes KR, Walsh FC, Wood RJ (2012) A review of experimental techniques to produce a nacre-like structure. Bioinspir Biomimet 7:031001. doi:10.1088/1748-3182/7/3/031001

Diop S, Rekacewicz P (ed) (2008) Vital water graphics—an overview of the state of the world's fresh and marine waters. UNEP

Dorrer C, Rühe J (2008) Mimicking the Stenocara beetle—dewetting of drops from a patterned superhydrophobic surface. Langmuir 24:6154–6158. doi:10.1021/la800226e

Eadie L, Ghosh TK (2011) Biomimicry in textiles: past, present and potential. An overview. J R Soc Interface 8:761–775. doi:10.1098/rsif.2010.0487

Ellem GK, Furst JE, Zimmerman KD (2002) Shell clamping behaviour in the limpet Cellana tramoserica. J Exp Biol 205(Pt 4):539–547

Feng L, Li S, Li Y, Li H, Zhang L, Zhai J, Song Y, Liu B, Jiang L, Zhu D (2002) Super-hydrophobic surfaces: from natural to artificial. Adv Mater 14:1857–1860. doi:10.1002/adma.200290020

Feng L, Zhang Y, Xi J, Zhu Y, Wang N, Xia F, Jiang L (2008) Petal effect: a superhydrophobic state with high adhesive force. Langmuir 24:4114–4119. doi:10.1021/la703821h

Fischer SF, Thielem M, Loprang RR, Seidel R, Fleck C, Speck T, Bührig-Polaczek A (2010) Pummelos as concept generators for biomimetically inspired low weight structures with excellent damping properties. Adv Eng Mater 12:B658–B663. doi:10.1002/adem.201080065

Fish FE (2006) The myth and reality of Gray's paradox: implication of dolphin drag reduction for technology. Bioinspir Biomimet 1:R17–R25. doi:10.1088/1748-3182/1/2/R01

Flammang P, Lambert A, Bailly P, Hennebert E (2009) Polyphosphoprotein—containing marine adhesives. J Adhesion 85:447–464. doi:10.1080/00218460902996358

Freeman CL, Harding JH, Quigley D, Rodger PM (2010) Structural control of crystal nuclei by an eggshell protein. Angew Chem Int Ed 49:5135–5137. doi:10.1002/anie.201000679

Garrod RP, Harris LG, Schofield WCE, McGettrick J, Ward LJ, Teare DOH, Badyal JPS (2007) Mimicking a Stenocara beetle's back for microcondensation using plasmachemical patterned superhydrophobic–superhydrophilic surfaces. Langmuir 23:689–693. doi:10.1021/la0610856

Gao X, Yan X, Yao X, Xu L, Zhang K, Zhang J, Yang B, Jiang L (2007) The dry-style antifogging properties of mosquito compound eyes and artificial analogues prepared by soft lithography. Adv Mater 19:2213–2217. doi:10.1002/adma.200601946

Geim AK, Dubonos SV, Grigorieva IV, Novoselov KS, Zhukov AA, Shapoval SY (2003) Microfabricated adhesive mimicking gecko foot-hair. Nat Mater 2:461–463. doi:10.1038/nmat917

Greiner C (2010) Gecko-inspired materials. In: Kumar CSSR (ed) Biomimetic and bioinspired nanomaterials. Nanomaterials for the life sciences, vol 7. Willey-VCH Verlag GmbH & Co. KGaA, Weinheim

Hasan J, Crawford RJ, Ivanova EP (2013) Antibacterial surfaces: the quest for a new generation of biomaterials. Trends Biotechnol 31:295–304. doi:10.1016/j.tibtech.2013.01.017

Huber G, Gorb SN, Spolenak R, Arzt E (2005) Resolving the nanoscale adhesion of individual gecko spatulae by atomic force microscopy. Biol Lett 1:2–4. doi:10.1098/rsbl.2004.0254

Ingram AL, Parker AR (2008) A review of the diversity and evolution of photonic structures in butterflies, incorporating the work of John Huxley (The Natural History Museum, London from 1961 to 1990). Philos Trans R Soc B 363:2465–2480. doi:10.1098/rstb.2007.2258

Ivanova EP, Hasan J, Webb HK, Truong VK, Watson GS, Watson JA, Baulin VA, Pogodin S, Wang JY, Tobin MJ, Löbbe C, Crawford RJ (2012) Natural bactericidal surfaces: mechanical rupture of *Pseudomonas aeruginosa* cells by cicada wings. Small 8:2489–2494. doi:10.1002/smll.201200528

Ju J, Bai H, Zheng Y, Zhao T, Fang R, Jiang L (2012) A multi-structural and multi-functional integrated fog collection system in cactus. Nat Commun 3:1247. doi:10.1038/ncomms2253

Kim P, Wong TS, Alvarenga J, Kreder MJ, Adorno-Martinez WE, Aizenberg J (2012) Liquid-infused nanostructured surfaces with extreme anti-ice and anti-frost performance. ACS Nano 6:6569–6577. doi:10.1021/nn302310q

Kinoshita S, Yoshioka S, Yasuhiro F, Okamoto N (2002) Photophysics of structural color in the Morpho butterflies. Forma 17:103–121

Klaus-Joerger T, Joerger R, Olsson E, Granqvist CG (2001) Bacteria as workers in the living factory: metal accumulating bacteria and their potential for materials science. Trends Biotechnol 19:15–22. doi:10.1016/S0167-7799(00)01514-6

Knight K (2009) Barnacle cures like blood clots. J Exp Biol 212. doi: 10.1242/jeb.038927

Koch K, Barthlott W (2009) Superhydrophobic and superhydrophilic plant surfaces: an inspiration for biomimetic materials. Philos Trans R Soc A 367:1487–1509. doi:10.1098/rsta.2009.0022

Koch K, Ensikat H-J (2008) The hydrophobic coatings of plant surfaces: epicuticular wax crystals and their morphologies, crystallinity and molecular self-assembly. Micron 39:759–772. doi: 10.1016/j.micron.2007.11.010

Kröger N, Deutzmann R, Sumper M (1999) Polycationic peptides from diatom biosilica that direct silica nanospheres formation. Science 286:1129–1132

Kumar VR, Bhuvaneshwari B, Maheswaran S, Palani GS, Ravisankar K, Iyer NR (2011) An overview of techniques based on biomimetics for sustainable development of concrete. Curr Sci 101:741–747

Launey ME, Buehler MJ, Ritchie RO (2010) On the mechanistic origins of toughness in bone. Ann Rev Mater Res 40:25–53. doi:10.1146/annurev-matsci-070909-104427

Lee H, Lee B, Messersmith PB (2007) A reversible wet/dry adhesive inspired by mussels and geckos. Nature 448:338–341. doi:10.1038/nature05968

Lee H, Scherer NF, Messersmith PB (2006) Single-molecule mechanics of mussel adhesion. Proc Natl Acad Sci 103:12999–13003. doi:10.1073/pnas.0605552103

Lepora NF, Verschure P, Prescott TJ (2013) The state of the art in biomimetics. Bioinspir Biomimet 8:013001. doi:10.1088/1748-3182/8/1/013001

Li J, Liang G, Zhu X, Yang S (2012) Exploiting nanoroughness on holographically patterned three-dimensional photonic crystals. Adv Funct Mater 22:2980–2986. doi:10.1002/adfm.201200013

Li SH, Zeng QY, Xiao YL, Fu SY, Zhou BL (1995) Biomimicry of bamboo bast fiber with engineering composite materials. Mater Sci Eng C 3:125–130. doi:10.1016/0928-4931(95)00115-8

Lim NSH, Everuss KJ, Goodman AE, Benkendorff K (2007) Comparison of surface microfouling and bacterial attachment on the egg capsules of two molluscan species representing Cephalopoda and Neogastropoda. Aquat Microb Ecol 47:275–287. doi:10.3354/ame047275

Liu K, Jiang L (2011) Multifunctional integration: from biological to bio-inspired materials. ACS Nano 5:6786–6790. doi:10.1021/nn203250y

Macedo MF, Miller AZ, Dionísio A, Saiz-Jimenez C (2009) Biodiversity of cyanobacteria and green algae on monuments in the Mediterranean Basin: an overview. Microbiology 155:3476–3490. doi:10.1099/mic.0.032508-0

Meyers MA, Chen P-Y, Lin AY-M, Seki Y (2008) Biological materials: structure and mechanical properties. Prog Mater Sci 53:1–206. doi:10.1016/j.pmatsci.2007.05.002

Meyers MA, McKittrick J, Chen P-Y (2013) Structural biological materials: critical mechanics-materials connections. Science 339:773–779. doi:10.1126/science.1220854

Mirkhalaf M, Dastjerdi AK, Barthelat F (2014) Overcomming the brittleness of glass through bioinspiration and micro-architecture. Nat Commun 5:3166. doi:10.1038/ncomms4166

Mishchenko L, Hatton B, Bahadur V, Taylor JA, Krupenkin T, Aizenberg J (2010) Design of ice-free nanostructured surfaces based on repulsion of impacting water droplets. ACS Nano 4:7699–7707. doi:10.1021/nn102557p

Munch E, Launey ME, Alsem DH, Saiz E, Tomsia AP, Ritchie RO (2008) Tough, bio-inspired hybrid materials. Science 322:1516–1520. doi:10.1126/science.1164865

Nikolić R, Marinović-Cincović M, Gadžurić S, Zsigrai IJ (2003) New materials for solar thermal storage—solid/liquid transitions in fatty acid esters. Solar Energy Mater Solar Cells 79:285–292. doi:10.1016/S0927-0248(02)00412-9

Ornilux Bird Protection Glass, Arnold Glas (2014) http://www.ornilux.com/. Accessed from March 2014

Otten A, Herminghaus S (2004) How plants keep dry: a physicist's point of view. Langmuir 20:2405–2408. doi:10.1021/la034961d

Oyen ML (2008) The materials science of bone: lessons from nature for biomimetic materials synthesis. MRS Bull 33:49–55. doi:10.1557/mrs2008.14

Page K, Wilson M, Parkin IP (2009) Antimicrobial surfaces and their potential in reducing the role of inanimate environment in the incidence of hospital-acquired infections. J Mater Chem 19:3819–3831. doi:10.1039/b818698g

Pampaloni F, Masotti A (2010) Biomimetic and bioinspired self-assembled peptide structures. In: Kumar CSSR (ed) Biomimetic and bioinspired nanomaterials. Nanomaterials for the life sciences, vol 7. Willey-VCH Verlag GmbH & Co. KGaA, Weinheim

Parker AR, Lawrence CR (2001) Water capture by a desert beetle. Nature 414:33–34. doi:10.1038/35102108

Parker AR, McPhedran RC, McKenzie DR, Botten LC, Nicorovici NA (2001) Photonic engineering: Aphrodite's iridescence. Nature 409:36–37. doi:10.1038/35051168

Pignalosa P, Liu B, Chen H, Smith H, Yi YS (2012) Giant light extraction enhancement of medical imaging scintillation materials using biologically inspired integrated nanostructures. Opt Lett 37:2808–2810. doi:10.1364/OL.37.002808

Popescu C, Höcker H (2007) Hair—the most sophisticated biological composite material. Chem Soc Rev 36:1282–1291. doi:10.1039/B604537P

Prum RO, Quinn T, Torres RH (2006) Anatomically diverse butterfly scales all produce structural colors by coherent scattering. J Exp Biol 209:748–765. doi:10.1242/jeb.02051

Santos R, Flammang P (2012) Is the adhesive material secreted by sea urchin tube feet species-specific? J Morphol 273:40–48. doi:10.1002/jmor.11004

Schultz MP, Bendick JA, Holm ER, Hertel WM (2011) Economic impact of biofouling on a naval surface ship. Biofouling 27:87–98. doi:10.1080/08927014.2010.542809

Seago AE, Brady P, Vigneron J-P, Schultz TD (2009) Gold bugs and beyond: a review of iridescence and structural colour mechanisms in beetles (Coleoptera). J R Soc Interface 6:S165–S184. doi:10.1098/rsif.2008.0354.focus

Shanyengana ES, Henschel JR, Seely MK, Sanderson RD (2002) Exploring fog as a supplementary water source in Namibia. Atmos Res 64:251–259. doi:10.1016/S0169-8095(02)00096-0

Shao H, Bachus KN, Stewart RJ (2008) A water-borne adhesive modeled after the sandcastle glue of *P. californica*. Macromol Biosci 9(5):464–471. doi:10.1002/mabi.200800252

Sharklet Technologies (2011) Understanding Sharklet surface protection products—a novel approach to improving cleanliness of high-touch environmental surfaces. Sharklet Technologies Inc., Aurora

Sharma A, Tyagi VV, Chen CR, Buddhi D (2009) Review on thermal energy storage with phase change materials and applications. Renew Sustain Energy Rev 13:318–345. doi:10.1016/j.rser.2007.10.005

Sheppard C (2011) Bird-friendly building design. American Bird Conservancy, 58 p. http://www.abcbirds.org/newsandreports/BirdFriendlyBuildingDesign.pdf. Accessed March 2014

Stavenga DG, Leertouwer HL, Marshall NJ, Osorio D (2011) Dramatic colour changes in a bird of paradise caused by uniquely structured breast feather barbules. Proc R Soc B 278:2098–2104. doi:10.1098/rspb.2010.2293

Stratakis E, Zorba V (2010) Biomimetic artificial nanostructured surfaces. In: Kumar CSSR (ed) Biomimetic and bioinspired nanomaterials. Nanomaterials for the life sciences, vol 7. Wiley-VCH Verlag GmbH & Co. KGaA, Weinheim

Swiegers G (ed) (2012) Bioinspiration and biomimicry in chemistry: reverse-engineering nature. Wiley, New York. doi: 10.1002/9781118310083

Thickett SC, Neto C, Harris AT (2011) Biomimetic surface coatings for atmospheric water capture prepared by dewetting of polymer films. Adv Mater 23:3718–3722. doi:10.1002/adma.201100290

Trask RS, Williams HR, Bond IP (2007) Self-healing polymer composites: mimicking nature to enhance performance. Bioinspir Biomimet 2:P1. doi:10.1088/1748-3182/2/1/P01

Vignolini S, Rudall PJ, Rowland AV, Reed A, Moyroud E, Faden RB, Baumberg JJ, Glover BJ, Steiner U (2012) Pointillist structural color in *Pollia* fruit. Proc Natl Acad Sci 109:15712–16715. doi:10.1073/pnas.1210105109

Vincent JFV, Mann DL (2002) Systematic technology transfer from biology to engineering. Philos Trans R Soc Lond A 360:159–173. doi:10.1098/rsta.2001.0923

Vukusic P (2006) Structural color in Lepidoptera. Curr Biol 16(16):R621–R623. doi:10.1016/j.cub.2006.07.040

Vukusic P, Sambles JR (2003) Photonic structures in biology. Nature 424:852–855. doi:10.1038/nature01941

Wang T, Konga S, Changa L, Wong C (2012) Preparation and heat-insulating property of the bio-inspired ZrO2 fibers based on the silk template. Ceram Int 38:6783–6788. doi:10.1016/j.ceramint.2012.05.074

Wegst UGK, Ashby MF (2004) The mechanical efficiency of natural materials. Philos Mag 84:2167–2181. doi:10.1080/14786430410001680935

Whitesides GM, Grzybowski B (2002) Self-assembly at all scales. Science 295:2418–2421. doi:10.1126/science.1070821

Wilson PW, Lu W, Xu H, Kim P, Kreder MJ, Alvarenga J, Aizenberg J (2012) Inhibition of Ice nucleation by slippery liquid-infused porous surfaces (SLIPS). Phys Chem Chem Phys 15:581–585. doi:10.1039/C2CP43586A

Wool RP (2008) Self-healing materials. Soft Mater 4:400–418. doi:10.1039/b711716g

Yan F, Gang S, TongQing W, Qian C, LuQuan R (2007) Hydrophobicity mechanism of non-smooth pattern on surface of butterfly wing. Chin Sci Bull 52:711–716. doi:10.1007/s11434-007-0120-5

Yang Q, Zhang XA, Bagal A, Guo W, Chang C-H (2013) Antireflection effects at nanostructured material interfaces and the suppression of thin-film interference. Nanotechnology 24:235202. doi:10.1088/0957-4484/24/23/235202

Youngblood JP, Sottos NR (2008) Bioinspired materials for self-cleaning and self-healing. MRS Bull 33:732–741. doi:10.1557/mrs2008.158

Zhai L, Berg MC, Cebeci FC, Kim Y, Milwid JM, Rubner MF, Cohen RE (2006) Patterned superhydrophobic surfaces: toward a synthetic mimic of the Namib desert beetle. Nano Lett 6:1213–1217. doi:10.1021/nl060644q

Zhao Q, Fan T, Ding J, Zhang D, Guo Q, Kamada M (2011) Super black and ultrathin amorphous carbon film inspired by anti-reflection architecture in butterfly wing. Carbon 49:877–883. doi:10.1016/j.carbon.2010.10.048

Zheng Y, Bai H, Huang Z, Tian X, Nie F-Q, Zhao Y, Zhai J, Jiang L (2010) Directional water collection on wetted spider silk. Nature 463:640–643. doi:10.1038/nature08729

Zimmermann EA, Gludovatz B, Schaible E, Dave NKN, Yang W, Meyers MA, Ritchie RO (2013) Mechanical adaptability of the Bouligand-type structure in natural dermal armour. Nat Commun 4:2634. doi:10.1038/ncomms3634

Chapter 4
Can Biomimicry Be a Useful Tool for Design for Climate Change Adaptation and Mitigation?

Maibritt Pedersen Zari

Abstract As professionals of the built environment need to solve more urgent and difficult problems related to mitigating and adapting to climate change, it may be useful to examine examples of how the same problems have been solved by other living organisms or ecosystems. Looking to plants or animals that are highly adaptable or ones that survive in extreme climates or through climatic changes may provide insights into how buildings could or should function. Examining the qualities of ecosystems that enable them to be adaptable and resilient may also offer potential avenues to follow. This chapter examines whether biomimicry, where organisms or ecosystems are mimicked in human design, can be an effective means to either mitigate the causes of climate change the built environment is responsible for, or to adapt to the impacts of climate change. Different biomimetic approaches to design are discussed and categorised, and a series of case study examples illustrate the benefits and drawbacks of each approach. In light of the conclusions reached during the course of the research, it is argued that design that mimics ecosystems and utilises synergies between mitigation and adaptation strategies in relation to climate change could be a beneficial long-term biomimetic built environment response to climate change. The foundations of the theory to support this are also presented.

4.1 Introduction: Climate Change and the Built Environment

Changes to the climate, and therefore related impacts on the built environment are expected to increase in intensity in the future, suggesting that re-evaluation of the built environment and rapid expansion of policies and actions to mitigate

M.P. Zari (✉)
School of Architecture, Victoria University, PO Box 600, Wellington 6011, New Zealand
e-mail: maibritt.pedersen@vuw.ac.nz

F. Pacheco Torgal et al. (eds.), *Biotechnologies and Biomimetics for Civil Engineering*,
DOI 10.1007/978-3-319-09287-4_4

greenhouse gas (GHG) emissions are urgently required (IPCC 2007a). The built environment is responsible for approximately a third of global anthropogenic GHG emissions, leading to climate change (de la Rue du Can and Price 2008). Despite international climate change protocols and initiatives, global GHG emissions are still increasing. Building sector carbon emissions including those from energy generation used to power buildings have increased annually by 2 % since 1970 for example, while emissions from commercial buildings have increased by 3 % annually since 2002 (IPCC 2007a).

The built environment will also have to adapt to climate change impacts, as the main site of human economic, social and cultural life. More than half of all humans now live in urban built environments, and people spend more than 70 % of their time indoors (Chan and Cheng 2006; Grimm et al. 2008). Even if all GHG emissions were immediately halted, the climatic change caused by past emissions would still be experienced due to the slow response of the planet's atmosphere, oceans and other carbon sinks (IPCC 2007b). It is important therefore that built environment professionals are not only able to work towards mitigating the causes of climate change, but are also able to devise strategies to adapt to the impacts concurrently.

It is appropriate then that established responses to climate change in the built environment broadly fall into two categories. The first is mitigating the causes of climate change by reducing GHG emissions. The second is adapting the existing and future built environment to predicted climate change impacts. Changes in the climate that will affect the built environment are numerous and although difficult to quantify, have been explored by several researchers (Koeppel and Ürge-Vorsatz 2007; Wilby 2007). The impacts that climate change will have on the built environment are both direct and indirect. Direct impacts will affect the actual physical fabric of the built environment. Indirect impacts incorporate economic and social changes that will also affect the built environment. The results of a review of international research examining the main direct impacts of climate change on the built environment have been summarised in Table 4.1 (see: Pedersen Zari 2012 for methodology, full results and sources). Impacts will vary greatly depending on the location and local quality and density of the existing built environment.

4.2 Biomimicry and Climate Change

By looking to the living world, there may be organisms or systems that can be mimicked to create and maintain a resilient and adaptable built environment, and improve its capacity for regeneration of the health of ecosystems (Pedersen Zari and Storey 2007). Biomimicry is the emulation of strategies seen in the living world as a basis for design and is a source of innovation, with potential to contribute to the creation of more sustainable architecture. It is the mimicry of an organism, organism behaviour or an entire ecosystem, in terms of its form,

Table 4.1 Direct climate change impacts on the built environment

Potential direct climate change impacts	Consequences for the built environment	Possible scale of the negative impact
Changes in temperatures (Likely to increase in most areas)	Increased overheating and air conditioning load	High
	Intensified urban heat island effect	High
	Decreased winter space heating	Low (potentially positive)
	Decreased water heating energy	Low (potentially positive)
Increased intense weather events	Damage to buildings and infrastructure	High
Changes in precipitation patterns	Increased inland flooding	High
	Increased erosion, landslips, rock falls	High
	Changes in aquifers and urban water supply and quality	High
	Heavier snow or ice loads	Medium
	Increased fire risk associated with more frequent droughts	Medium
	Damage to foundations, underground pipes /cables, etc.	Medium
	Increased subsidence (clay soils)	Medium
	Increased pressure on urban drainage systems	Medium
	Increased storm water run-off and leaching of pollutants into water ways or aquifers	Low
	Damage to facades and internal structure due to rain penetration	Low
Thermal expansion of oceans and changes in the cryosphere (ice systems) such as retreating snow lines and ice packs, and melting glaciers	Increased coastal flooding	High
	Increased erosion and loss of land	High
	Relocation or displacement from coastal areas	High
	Changes in water tables and possible increased salinity of aquifers and estuaries	High
	Loss of inter tidal areas acting as buffer zones	High
	Impeded drainage	Medium
Changes in wind patterns and intensities	Changes in wind loading on buildings	Medium

(continued)

Table 4.1 (continued)

Potential direct climate change impacts	Consequences for the built environment	Possible scale of the negative impact
Increased air pollution	Impacts on interior air quality management	Medium
	Damage to building facades	Low
Impacts on urban biodiversity	Changes in cooling, shading and evapotranspiration benefits from urban biodiversity	Medium
	Changes to storm water management	Low

material, construction method, process strategies or function (Table 4.2). Mimicking living organisms or ecosystems involves a process of translation into suitable solutions for the human context. This process of translation often results in designs that are not immediately similar to the organism or ecosystem that inspired them, but utilise the same functional concepts. It is important to acknowledge that not all solutions arrived at through evolutionary processes found in organisms will be perfect, or suitable for a human context, or result in better ecological performance outcomes. Several noteworthy contemporary examples of biomimetic architecture or technologies that can assist the built environment in adapting to climate change are examined in the following sections. Additional historic examples are detailed by Vincent et al. (2006), Vogel (1998), Benyus (1997).

4.3 Biomimicry for Mitigating GHG Emissions from the Built Environment

Several contemporary examples of biomimetic architecture or technologies that can assist the built environment in climate change issues will be examined to ascertain whether biomimicry can be used to address climate change, and if so in what way. Common mitigation strategies in a built environment context include:

1. Increasing the density and limiting sprawl of urban form to reduce building energy use and emissions from vehicles[1];

[1] Although this is a common mitigation strategy in the context of the built environment (Reisinger et al. 2011) it is not necessarily advocated as a suitable solution in all contexts. There is growing evidence that increasing the density of cities may not actually contribute to mitigating the causes of climate change and that it negatively affects several other environmental performance aspects of cities, such as flood risk, storm water issues, habitat provision, building energy efficiency, and energy generation potentials (see: Pedersen Zari 2012, Sect. 1.3.2.1).

2. Creating or maintaining urban forest and green space;
3. Design for energy conservation;
4. Provision of renewable energy sources; and
5. Carbon storage or sequestration.

The first two strategies relate to urban planning and represent long-term climate change adaptation strategies. This section therefore will deal with the latter three, and most common (Steemers 2003), categories. The first category examined in this chapter is biomimetic examples that mimic the energy efficiency or effectiveness of living organisms and systems. The impetus is that by being more energy efficient, less fossil fuel is burnt and therefore fewer GHGs are emitted. The second approach is to devise new ways of producing energy to reduce human dependence on fossil fuels and their associated GHGs. Various biomimetic technologies and products have been developed for the purposes of improving energy efficiency and exploring new energy generation possibilities. A third biomimetic approach to mitigating GHG emissions is investigating organisms or ecosystems for examples of processes within them that can sequester and store carbon. This is to prevent emission of GHGs through human activities causing additional climate change. Brief case studies are presented in the following sections.

4.3.1 Biomimicry for Energy Efficiency

There are numerous examples of living organisms and systems that are highly energy effective and that may yield an understanding of how humans could carry out their activities without a dependence on fossil fuels (Allen 2010, p. 169, Pawlyn, pp. 91–101). Some well-known examples of biomimicry fit into this category, such as DaimlerChrysler's prototype Bionic car[2] (2005) (Fig. 4.1) and Mick Pearce's Eastgate building in Harare, Zimbabwe (1996) (Fig. 4.2).

The large volume, small wheel base concept Bionic Car, was based on the hydrodynamic and strength characteristics of the boxfish (*Ostracion meleagris*). This resulted in the design of a more fuel efficient car with the low drag coefficient of 0.19, and panels with 40 % more rigidity than a standard car (Anon 2005). The chassis and structure of the car were also biomimetic, having been designed using a computer modelling method designed by Claus Mattheck that mimics how trees are able to grow in a way that minimises stress concentrations (Pawlyn 2011, p. 21, Mattheck 1998, Mattheck et al. 1996). The resulting car structure looks almost skeletal (Fig. 4.1). Total car weight was reduced by at least a third because material was allocated only to the places where it is most needed (Vincent et al. 2006). This process was abandoned because according to the car manufacturers it was too time intensive (Vincent 2010).

[2] The Bionic Car is also referred to as the 'Mercedes Bionic Car' (Pawlyn 2011, p. 5). Mercedes Benz is part of the DaimlerChrysler group.

Table 4.2 Different kinds of biomimicry (adapted from Pedersen Zari 2007)

Level of biomimicry		Example—A building that mimics termites:
Organism level (mimicry of a specific organism)	*Form*	The building looks like a termite
	Material	The building is made from the same material as a termite; a material that mimics termite exoskeleton /skin for example
	Construction	The building is made in the same way as a termite; it goes through various growth cycles for example
	Process	The building works in the same way as an individual termite; it produces hydrogen efficiently through metagenomics for example
	Function	The building functions like a termite in a larger context; it recycles cellulose waste and creates soil for example
Behaviour level (mimicry of how an organism behaves or relates to its larger context)	*Form*	The building looks like it was made by a termite; a replica of a termite mound for example
	Material	The building is made from the same materials that a termite builds with; using digested fine soil as the primary material for example
	Construction	The building is made in the same way that a termite would build; piling earth in certain places at certain times for example
	Process	The building works in the same way as a termite mound would; by careful orientation, shape, materials selection and natural ventilation for example, or it mimics how termites work together
	Function	The building functions in the same way that it would if made by termites; internal conditions are regulated to be optimal and thermally stable for example (Fig. 4.6). It may also function in the same way that a termite mound does in a larger context
Ecosystem level (Mimicry of an ecosystem)	*Form*	The building looks like an ecosystem (a termite would live in)
	Material	The building is made from the same kind of materials that an (termite) ecosystem is made of; it uses naturally occurring common compounds, and water as the primary chemical medium for example

(continued)

Table 4.2 (continued)

Level of biomimicry		Example—A building that mimics termites:
		The building is assembled in the same way as an (termite) ecosystem; principles of succession and increasing complexity over time are used for example
	Process	The building works in the same way as an (termite) ecosystem; it captures and converts energy from the sun, and stores water for example
	Function	The building is able to function in the same way that an (termite) ecosystem would and forms part of a complex system by utilising the relationships between processes; it is able to participate in the hydrological, carbon, nitrogen cycles, etc. in a similar way to an ecosystem for example

When biological analogues are matched with human identified design problems, the fundamental approach to solving a given problem omits examining the issue of how buildings relate to each other and the ecosystems they are within. The underlying causes of a non-sustainable built environment are not necessarily addressed when mechanisms and structures found in biology are incorporated into an existing inherently unsustainable engineering or architecture paradigm (Vincent 2010). This is typical of most current attempts at architectural or engineering biomimicry[3] (Vincent 2010) and could be a reason for the lack of measurable improved sustainability outcomes attributable to biomimicry either architecturally or generally.

The Bionic Car (Fig. 4.1) again illustrates the point. Compared to a standard non-biomimetic car, it is efficient in terms of fuel use because the body is more aerodynamic due to the mimicking of the box fish. It is also more materially efficient through mimicking tree growth patterns to identify the minimum amount of material needed in the structure of the car. The car itself, however, is not a new approach to transport. Instead, small improvements have been made to existing technology without a re-examination of the idea of the car as an answer to personal transport. It is also no more fuel or materials efficient than ultra-efficient cars designed without employing biomimicry. Because the impression is given that it looks fish-like, it is somehow more sustainable; the car has been criticised for being an example of biomimicry green-wash (Gebeshuber et al. 2009).

3 For examples of this see Pawlyn (2011).

Fig. 4.1 Daimlercrysler's bionic car inspired by the boxfish and tree growth patterns (*car* photo by R. Somma. *Boxfish insert* photo by R. Ling)

A related issue with this kind of biomimicry is that designers are able to research potential biomimetic solutions without an in-depth scientific understanding or collaboration with a biologist or ecologist. With a limited scientific understanding, translation of such biological knowledge to a human design setting has the potential to remain at a shallow level. It is in some cases easy to mimic forms and certain mechanical aspects of organisms but difficult to mimic other aspects, such as chemical processes, without scientific collaboration (Bensaude-Vincent et al. 2002).

An architectural example of this is Mick Pearce's Eastgate building. This is perhaps the most often cited example of architectural biomimicry (see for example: Koelman 2004; Zhao and Xu 2010; Pronk et al. 2008). Eastgate was designed to have a relatively thermally stable interior environment with minimal mechanical cooling. This resulted in reductions in energy use of between 17 and 52 % compared to similar buildings in Harare (Baird 2001; Smith 1997), and therefore also reduced GHG emissions. Pearce based the design in part on principles of induced flow and the use of thermal capacity to regulate temperature observed in termite (Macrotermes michaelseni) mounds of southern Africa, although the correctness of the science has been disputed (Turner and Soar 2008). The temperature regulation in the mounds is achieved through careful orientation, spatial organisation and techniques of passive ventilation. It is not therefore the termite that is mimicked, but the results of its mound building behaviour. Biomimicry at the behaviour level is much less common than biomimicry at the organism level.

In-depth scientific analysis since Eastgate was designed has revealed that termite mounds interact with the environment in a much more sophisticated way to

Fig. 4.2 Eastgate building in Harare, Zimbabwe (Photo By: G. Bembridge) and termite mound in Africa (Photo By: D. Siu)

regulate temperature than was previously thought. This has led to suggestions that a new generation of 'living' biomimetic buildings could utilise a more accurate understanding of the termite mound as a multi-purpose extension of termite physiology more analogous to the function of lungs (Turner and Soar 2008). Although Eastgate could be considered to be a successful building in terms of energy efficiency and GHG emissions reduction, it is an example of how architects and designers are able to use a metaphor of how organisms work without a full understanding of the science behind the mechanism. This demonstrates the importance of working with ecologists and biologists to avoid 'bio-mythologically-inspired' design (Pawlyn 2011, p. 78).

An example of a building which attempts to take advantage of this new understanding of termite mounds (but perhaps does not fully exploit it) is the 2006 Davis Alpine House by Wilkinson Eyre, Dewhurst MacFarlane and Atelier Ten located in Kew Gardens, London, England (Fig. 4.3). The building was designed to avoid energy intensive refrigeration typically needed for the display of alpine plants, and instead uses a stack effect to cool the interior passively, while essentially remaining a glass house with high rates of air circulation (Pawlyn 2011, pp. 87–88). A removable shading sail is included in the design to prevent too much sunlight reaching the plants. The stack effect is enhanced through the high internal space created by the double arches, sequential apex venting as temperature increases, by vents at the bottom of the glass structure, and through a Barossa termite inspired decoupled thermal mass labyrinth below the building. The concrete block labyrinth is set between a double concrete slab that also acts to resist the forces exerted by the

Fig. 4.3 Davis alpine house, London, England (photo by P. Galletly)

tension rods that support the glass ceiling (Bellew 2006). The air that is cooled within the labyrinth is recirculated so it cools the low level plants. The labyrinth is vented at night to take advantage of cooler temperatures. The mass remains at a temperature usually cooler than that required for the space itself.

> The system is arranged so that a small fan can drive the air through the labyrinth tunnels and expel it outside at night, so that the structure is cool in the morning. As the building starts to heat up the outdoor air is directed straight into the glasshouse, via vertical pipes that terminate with directional outlets within the plant beds. As the outdoor temperature continues to rise above 18 °C the air is diverted through the labyrinth where it is cooled before passing through the same pipes to cool the plants. This is cooling without refrigeration (Bellew 2006).

This form of cooling allows more effective control, and in the way termites open and close vents within mounds to regulate temperature, the 'coolth' can be drawn upon or rejected. The economic payback period calculated for the labyrinth cooling system was 9 years (Pawlyn 2011, p. 88).

A similar system has been installed by Atelier Ten in the atrium-like Federation Square space in Melbourne, Australia (Fig. 4.4). Mechanical cooling was eliminated and the energy needed to power the system's fan is only 10 % of a conventional overhead cooling system. If the atrium is at a comfortable temperature excess 'coolth' is diverted to adjoining museum galleries (Bellew 2006).

Though both the Bionic car and Eastgate make contributions to energy efficiency it is doubtful that they are more energy efficient than comparable

Fig. 4.4 Federation square, Melbourne, Australia (photo by: R. Flett)

counterparts with similar energy efficiency results achieved without the use of biomimicry. This is also the case with the 2006 Council House 2 (CH2) in Melbourne with which Pearce was also involved (Fig. 4.5).

In CH2 100,000 L of water is extracted and cleaned from the sewers beneath the building and used to condition the air. This is reminiscent of how certain termite species use the proximity of aquifer water as an evaporative cooling mechanism. African Barossa termites make tunnels tens of metres deep to reach the water table, so that its cooling effect can be used in extreme heat to keep the mound within a 1° temperature fluctuation range. In CH2 some of the cleaned water is passed through shower towers on the outside of the building. This cools the water, particularly at night. The water passing through the shower towers also cools surrounding air which is then used to ventilate the commercial premises on the ground floor. The water continues to the basement where it passes through a system that stores the 'coolth' by using phase change materials. The water is then used in a closed loop in chilled beams that provide cooling to the building interior. In combination with the effects of additional sustainability features,[4] CH2 is estimated to use 85 % less energy, reduce GHG emissions by 87 % and use 70 %

[4] The building generates much of its own energy through wind turbines, photovoltaic panels and co-generation (Tan 2007).

Fig. 4.5 CH2 building, Melbourne, Australia (photo by: W. Gurak)

less water than typical comparable buildings that are built without any sustainability features (Jones 2008; Tan 2007). CH2 is a 6-star Australian Green Star building. It was built at a cost premium of 22.1 % (Aranda-Mena et al. 2008), but due to productivity increases of 10.9 % from staff attributed to the new building (Paevere and Brown 2008), as well as substantial energy savings, pay back was between 5 and 7 years (Bond 2010). These results suggest that the building is successful and performs better than conventional buildings.

Other Australian buildings achieve similar or better results and are not biomimetic, however, such as the Gauge by Land Lease (by architect Darren Kindrachuk) built in 2009, also a 6 Green Star building in Melbourne. It is claimed that the building recycles 92 % of its water on-site, and produces 25 % of its own energy. The 2008 Workplace 6 (Sydney, also 6 Green Star rated by architects Nettleton Tribe) reports 90 % water savings, 70 % reduction in carbon footprint, and 45 % energy savings (Bond 2010).[5] This serves to illustrate that this kind of biomimicry (process biomimicry at the behaviour level) should be considered as one aspect of a larger and more holistic response to climate change than incorporates an understanding of human behaviour elements in design rather than a sort of quick technological fix.

Improving general energy efficiency is an important part of addressing climate change, but should be regarded as an intermediate step (McDonough and Braungart 2002, p. 62). Improving efficiencies helps to reduce the intensity of GHG emissions but does not challenge assumptions about how and why technologies are made and used. Nor does it address the underlying causes of climate change such as dependence on fossil fuels. Other researchers also detail increased rather than decreased energy use as a result of energy efficiency initiatives. This is referred to as 'Jevons Paradox' (Jevons 1865). Simplistically, as efficiencies increase, the price of a technology goes down often resulting in a net increase in consumption. Improving energy efficiencies does allow positive incremental changes to be made to existing technologies and buildings rather than assuming the only way forward is a complete rebuild of the built environment (which is not possible). Energy efficiency could, therefore, be important in the short term for the built environment becoming better able to address climate change.

4.3.2 Biomimetic Energy Generation

Several biomimetic technologies or systems aim to replace the use of fossil fuels as the primary human energy source. Looking to the living world for inspiration is appropriate in this regard, because almost all organisms source energy from renewable sources, which predominantly is directly or indirectly from contemporary sunlight. Examples of biomimetic systems for development of alternative energy sources include mimicking the process of artificial photosynthesis in solar energy cell technology (Moore et al. 2004; Collis et al. 2005; Davidson 2003),[6] microbial fuel cells generated from electron donors in wastewater, biomass conversion systems, radiosynthesis-based systems[7] (Gebeschuber et al. 2009) and the development of ocean energy technologies that mimic how sea kelp or certain fish move efficiently in water.

[5] Both The Gauge and Workplace6 hold two 6 Green Star ratings, one based on the design and one 'as built'. This means estimations of performance given above are based on the building after construction, not just on the design. Values, however, are still based on simulations rather than measured performance. Bond (2010) discusses some of the issues between estimations given by simulations and actual measured performance that can lead to quite different results. Performance estimates given above should therefore be treated with caution.

[6] See also the work of the Australian Artificial Photosynthesis Network (AAPN).

[7] Radiosynthesis refers to the process of melanin in micro-organisms capturing high-energy electromagnetic radiation to generate metabolic energy (Gebeschuber et al. 2009).

Fig. 4.6 Image of biowave unit (*Source* Biopower systems)

The Australian company BioPower has, for example, developed underwater power generators called BioWAVE that oscillate in ocean waves and currents rather than rotate like turbines (Fig. 4.6). Baker (2011) says of the technology: 'The BioWave system sits on the floor of the ocean with three buoyant blades extending from the base. These buoyant blades, which can reach heights of 25 and 19.5 m wide (for all three blades combined), sway back and forth with the motion of the waves. These blades reach almost to the top of the water where the flow of the water is most powerful. The motion of the blades drives an on-board generator to produce electricity, which is then transported from the system to the shore through cables on the seabed.'

Through the use of permanent magnet motors, the low-speed high-torque oscillation is converted into high-speed low-torque rotation. Units are anchored to the ocean floor by the use of a series of small root-like devices to avoid extensive, habitat damaging and complicated drilling and installation. The generators rotate freely to orient themselves towards currents and in the same way can lie flat in storm events to avoid damage (BioPower Systems 2011; Finnigan and Caska 2006). The units are made from lightweight, non-corrosive, durable composite materials. They are more lightweight than comparable wave power systems and possibly more material and cost-effective (Baker 2011). A 250 kW 4-year grid connected pilot study of the technology, funded in part by the Victorian Government in Australia, began in 2012 near Port Fairy. Installation of the pilot BioWAVE is estimated to be completed in 2015. BioPower state they have independent verification of the viability of their systems but as yet there are no results available (see however: Gonzalez et al. 2009 for a technical report of BioWave testing). Once in operation the energy generated is claimed to be cost competitive with that from typical wind farms (BioPower Systems 2011). Bio-Power has also developed tidal power conversion systems based on the efficient propulsion mechanisms (termed 'thunniform motion') of species such as shark, tuna and mackerel called BioSTREAM (Gebeshuber et al. 2009).

Finding methods to replace fossil fuels with renewable energy sources is a long-term solution to climate change (Atkinson 2007a). This is potentially problematic given the time available to find alternatives to the use of fossil fuel before more irreversible damage is done to the climate however (Turner 2008). Due to the economic and social impacts of climate change (Stern 2006; Howden-Chapman et al. 2010) it is possible that resources to research and develop alternatives to fossil fuels will become more scarce (Mitchell 2012). It should also be remembered that despite new developments in energy generation technologies, the uptake of such technologies is hindered because industrialised countries still tend to subsidise fossil fuels, hampering research into renewable energy alternatives. One study estimated that the subsidy is as high as $US200 billion per year (Gebeschuber et al. 2009).

While reducing or ceasing to use fossil fuels as an energy source would prevent creation and release of additional GHG emissions, biomimicry has also been investigated to find ways to remove excess CO_2 already in the atmosphere.

4.3.3 Biomimetic Sequestering and Storage of Carbon

There are several organisms and processes in nature that are able to store, sequester or recycle carbon. Understanding how they do this could be used in the development of technologies for industrial processes and the built environment. In Quebec, CO_2 Solutions is developing carbon sequestration technology which mimics certain chemical processes that occur in the bodies of mammals (Geers and Gros 2000). The technology mimics the enzyme carbonic anhydrase which is able to convert CO_2 into bicarbonates. This enzyme enables mammals to manage CO_2 during respiration. The process works at atmospheric pressure and ambient temperatures (Fradette 2007). It generates bicarbonate which can be used to neutralise certain industrial wastes, store CO_2 or can be transformed into carbonate compounds such as limestone to be used in processes in cement works or paper mills. The aqueous solution, where the conversion of CO_2 to bicarbonate occurs, is reused in a closed loop. The technology can be retrofitted onto existing facilities such as power plants, cement works, aluminium smelters and oil sands operations, or integrated into new ones (Atkinson 2007b; CO_2 Solution 2008). The process is more energy effective, and therefore cost-effective, than conventional carbon capture technology that use monoethanolamine, with energy savings of the order of 30 % (Carley 2012). The enzymes used allow the use of different solvents (such as tertiary amines, carbonates and amino acids) that are also less toxic, corrosive and volatile than those used in similar non-biomimetic systems. For example, this system does not result in the emission of the known carcinogen nitrosamine which is the case in systems that use piperazine (a chemical promoter used to accelerate CO_2 absorption) (Carley 2012). Initial testing on an Alcoa aluminium smelter in Deschambault, Quebec, indicated removal of 80 % of CO_2 that would otherwise have been emitted into the atmosphere (Hamilton 2007). The CO_2 emissions are combined with waste bauxite residue, which is difficult to manage and potentially dangerous, in order to neutralise it and create a product that has secondary uses and revenue generating potential (Carley 2012). A larger pilot project on a power plant (for a company wishing to keep the project confidential) was conducted in 2011. CO_2 Solutions reported positive but as yet unquantified and unpublished results. Discussions of pilot projects in cement works and with major oil producers in Alberta are underway to tailor the technology to these sectors (Carley 2012). In 2014 the 'largest-ever scale test of a biocatalytic process for carbon capture' was completed by CO_2 solutions in relation to its Alberta oil sands project. Results showed an improvement of 33 % in energy consumption compared to the existing carbon capture technologies for the capture of 90 % of the CO_2 emissions (CO_2 Solutions 2014).

A similar example of biomimetic carbon sequestration technology can be found in Stanford University research. Brent Constantz investigated how coral creates its hard aragonite exoskeleton of calcium and magnesium carbonate and bicarbonate minerals by using minerals, sea water and CO_2. Based on his discovery of how calcite and aragonite (polymorphs of calcium carbonate) are nucleated by the marine organisms and subsequent replication of that in a laboratory, in 2007 he formed the company Calera. Calera is developing technology that sequesters carbon from industrial flue gas emissions, adds it to brine wastewater, sea water, manufactured alkaline solutions, or brines extracted from geological deposits, and from this process converts the gas first to carbonic acid and then to stable solid minerals. These materials are used to produce high reactive cements that do not require the calcining of the carbonate typically required to produce conventional Portland cement. The resulting Calera cement process results in fewer CO_2 emissions than conventional cement production (Calera 2012).

The technology has been applied at a demonstration level to California's Moss Landing gas fired electricity power plant in the Monterey Bay area. Here sea water is used, along with 92 % of the plant's 3.5 million tonnes of annual CO_2 emissions and some of the waste heat and fly ash from the flue to create cement (Lovins and Cohen 2011, p. 285). Further independent analysis of the Calera process confirms that up to 90 % of carbon emissions could be captured if the system was applied to other suitable plants (Andersen et al. 2011). The same report from the Institute for Governance and Sustainable Development states: 'The advantage of the Calera process relative to other proposed CCS (carbon capture and storage systems) is that: it is available near-term at a lower estimated cost, it is [a] modular retrofit to existing power plants making it scalable, and it has the potential to capture carbon while producing a useful product'.

Every tonne of cement made in the process sequesters between 0.5 and 1.0 tonnes of CO_2. This supposedly eliminates additional carbon emissions caused by producing and transporting cement (Lovins and Cohen 2011, p. 285; Nidumolu et al. 2009; Andersen et al. 2011). The sea water used in the process is not polluted and can be returned to the ocean or used as pre-treated water to make drinking water through desalinisation because the hardness has been removed (McKeag 2010). Trace metals in emissions are also removed during the process and mercury is captured and converted into a non-leachable form (Calera 2012). Calera has developed several new processes to enable their system to work, one of which is a process to create the alkaline solution needed for the electrochemical part of the system. This uses one-third to one-fifth of the energy of comparable state-of-the-art practices. Testing began in 2011 on the suitability of the products of the Calera process for building and construction processes[8] and cement-based products have been used in several demonstration projects. Non-structural applications (foot paths, tiles) are likely to be end uses for the cement (Monteiro et al. 2013).

A final example of efforts to increase the sequestration of carbon using biomimicry is illustrated by research conducted by Dr Jeffrey Brinker at the Sandia National Laboratories in the United States, investigating how the abalone or paua is able to grow a crack resistant shell approximately 200 % harder than human ceramics using only sea water and a series of proteins (Brinker et al. 1999). The research could lead to lightweight, extremely strong, optically clear building materials (Sellinger et al. 1998) or to alternatives to concrete. Cement production accounts for approximately 5 % of the world's anthropogenic CO_2 emissions (Vanderley 2003). This process of biomineralisation stores carbon much like the growing of forests locks carbon into the structure of the trees and soil until released. The concept of a new material able to grow through self-assembly over a structure, with the simple additive of sea water, by activating proteins on the structure imitated from the abalone has been investigated, but results are not available (Koelman 2004). Biomimetic biomineralistion is also discussed by Vincent (2010) and Armstrong (2009).

The utilisation of detritus, or waste, is an important part of the process of cycling nutrients and is a fundamental part of maintaining the health of an ecosystem. In using biomimicry to address excess carbon in the atmosphere it may be possible to use carbon as a resource rather than it being a source of pollution or waste. The obvious example from biology is how plants utilise CO_2 during the photosynthesis process, converting it into the products needed for plant growth and development, such as cellulose. For plants, CO_2 in the correct quantities is a necessary resource, rather than a pollutant. A company formed out of research by Dr Geoff Coates at Cornell University called Novomer is mimicking this aspect of carbon sequestration in plants, by using CO_2, mostly captured as factory emissions, as a resource for new carbon-based polymers (McKeough 2009). The resulting plastics are 40 %

8 Compressive strength data is available on the Calera website: http://calera.com/ (accessed May 2014).

transformed CO_2 by weight and could be used as coatings for building materials or foam insulation (Fister Gale 2008; McKeough 2009). The catalyst that is needed for the process works at ambient temperature and low pressure (150psi), and the process is therefore less energy intensive and expensive than conventional bioplastics production (Greenemeier 2007). The use of CO_2 and carbon monoxide (CO) as feedstocks, rather than corn or starch as in bioplastics, means the carbon-based plastic does not compete with food production, and both captures and stores carbon, while reducing demand on oil reserves (if the result is less production of conventional oil based plastics). Ongoing research suggests that the stored carbon in the carbon-based plastic is released upon decomposition, although depending on the exact nature of the feedstock and catalyst, the biodegradability of the carbon-based polymers can be varied to enable longer term carbon storage (Patel-Predd 2007; Gerngross and Slater 2003). In recent advances within Novomer, acrylic acid, acrylate esters, butanediol and succinic anhydride have been synthesised cost competitively from bio-based feedstocks using existing technology. These materials can be combined with Novomer's catalyst to make materials and chemicals with a potentially negative carbon footprint, with the suggestion that such a process could lead to plastics which sequester CO_2 over the product lifecycle, while being 30 % cheaper than conventional plastics to manufacture (Novomer 2010). The first large manufacturing run of Novomer polypropylene carbonate (PPC) polyol was completed in 2013 through Albemarle at their Orangeburg, S.C. factory to enable large-scale commercial testing. Novomer chemicals may also be useful in the production of paint, coatings, textiles and nappies with an estimated increased energy productivity of chemical manufacturing by 30–70 %, reduced CO_2 footprint of 40–110 % depending on the target chemical, energy savings of 20 trillion BTUs per year within 10 years with complete sequestration of waste gases (Novomer 2013).

An issue with this approach to addressing climate change impacts is that sequestering carbon does not examine or solve the problem of excessive burning of fossil fuels. Nor does it take into account the depletion of oil reserves. Rather, sequestration is another interim step in the development of a more sustainable human society and economy, possibly creating time to develop technology which does not just pollute less, but instead does not pollute at all. There are several additional logistical, economic, technological and environmental problems with current attempts at carbon sequestration (Schiermeier 2006). Technologies that allow polluting practices to continue for longer, even if at decreased rates, may distract people from the necessary task of reorganising human industry, and with it consumption at a fundamental level, and may instead perpetuate a 'business as usual' paradigm. This could make the eventual highly probable collapse of such systems more difficult for people. It is likely that transitions to non-polluting and non-fossil fuel-based ways of making energy and products, including buildings, would be easier and less disruptive ecologically and economically if done before there was no other option due to ecosystem collapse or the end of cheap fossil fuels. If such a scenario is allowed to occur, it is possible that the impacts would be so great that transition may not be possible (Turner 2008). The benefit of using biomimicry as a means to capture or sequester carbon, however, is that such techniques may help to retrofit and adapt existing building infrastructure, while addressing GHG emissions in the short to medium term. The existing built environment will need to be part of a long-term solution to climate change because of the relatively long life of buildings and slow rate of renewal compared to consumer items such as clothing or electronic equipment.

Several of the examples of a biomimetic approach to carbon sequestration or storage discussed here reveal that useful secondary products related to a built environment context can be made from wastes without toxic by-products and without using high amounts of energy. There may also be important restorative capacity in lowering the amount of atmospheric carbon by using CO_2 as a feedstock for new materials. The built environment uses approximately 40 % of the materials consumed by the global economy (Rees 1999). Building materials that store carbon long term, or that are made from CO_2 and do not release all this upon biodegradation, and are durable and safe, could make a contribution to mitigating climate change or even lowering levels of atmospheric carbon over the long term, if combined with other initiatives. It should be remembered that most of the built environment that will exist for the next 50–90 years (in Western countries) has already been built, so there is a limit to how much of an impact new kinds of materials can make within a time frame that would allow the built environment to address GHG emissions before the impacts of climate change prevent large amounts of additional building.

4.4 Biomimetic Strategies for Adaptation to Climate Change in the Built Environment

Work that discusses how to strategically adapt new or existing buildings or urban environments to climate change, tends to be limited to discussion on how to make buildings more adaptable in general through design techniques such as design for deconstruction, materials recycling and reuse and lightly treading foundations (Fernandez 2004; Steemers 2003). This raises the question of whether new techniques or technologies that could contribute to adaptation to climate change could be revealed by study of how certain organisms and the ecosystems they create or are part of are already able to do this. There are several examples of biomimicry being employed as a strategy that professionals of the built environment can harness to adapt buildings to climate change. The first and most common category of examples are those that respond to anticipated direct impacts of climate change on the built environment (see Table 4.1). The second is a more comprehensive approach to altering the built environment so it becomes more adaptable and resilient as a whole system.

4.4.1 Responding to Direct Impacts of Climate Change

The living world is made of numerous organisms that effectively solve the same problems that the built environment will face as climate change continues. While the potential impacts of climate change are numerous and dependent on local conditions, the list of organisms and ecosystems that effectively manage similar issues is also long. There are approximately 1.8 million species which have been described and categorised. Estimates of the total number of species range, however, from 2 to 100 million, with a 'best guest' at 14 million (Purvis and Hector 2000). There are organisms and ecosystems that manage overheating, high winds and erosion. Organisms may be specifically tailored to these conditions because they are part of their habitat niche. Other organisms demonstrate strategies to adapt to changes on a temporary basis that could be useful for humans to study, while others adapt over the longer term or over generations, through the processes of evolution. Several architectural biomimicry projects respond to direct impacts of climate change. The architecture discussed here may form a suitable response to changes in precipitation patterns and projected water shortages for example.

Grimshaw Architects in collaboration with Charles Paton of Seawater Greenhouse have taken an understanding of the Namib desert beetle and proposed a unique desalinisation process that will form part of a large outdoor theatre called Teatro del Agua on a shore of the Canary Islands. The stenocara beetle lives in desert with little rainfall but with short infrequent morning fogs. It is able to capture moisture from the swift moving fog by tilting its body into the wind. Water condenses on the surface of the beetle's back because its shell is cooler than the surrounding air. Droplets form on the shell, and the alternating hydrophilic, hydrophobic surface of the beetle's back enables the drops to roll down into its mouth (Garrod et al. 2007; Parker and Lawrence 2001). Research conducted in the United Kingdom has also shown that surfaces based on the beetle's shell are several times more effective at harvesting fog than typical methods using nets (Trivedi 2001), and could be useful in improving the design of de-humidification and distillation equipment (Knight 2001).

The Teatro del Agua mimics aspects of the beetle by passing sea water over a series of evaporative grills. As the sea breeze moves through these grills, some of the water evaporates leaving salt behind. The moist air then continues until it hits pipes holding cool sea water, pumped up from the nearby ocean. As the warm moist air touches the cool pipes, condensation forms and clean fresh water trickles down the outside of the pipes to be collected. The sea water pumps are powered by wind turbines using the same uni-directional sea breeze. The building is projected to be self-sufficient in water with surplus being transferred to neighbouring buildings and landscapes (Pawlyn 2011, p. 70).

Another established example of biomineralisation that could relate to climate change adaptation and may have application to the building industry is Biorock (Fig. 4.7). Biorock was developed by marine biologist Thomas Goreau and engineer Wolf Hilbertz in the 1970s to restore coral reefs (Pawlyn 2011, p. 50).

Fig. 4.7 Biorock structure at Pemuteran Bay, Bali, Indonesia (photo by: R. Morrow-Euigk)

Frames of steel are placed onto ocean floors and low voltage current that is not harmful to marine life is passed through the frames. This encourages minerals dissolved in the sea water to crystallise. A layer of minerals appears on the frames within a few days upon which coral can be attached. Accretion rates are up to 50 mm per year with a load bearing strength of between 24 and 80 MPa (concrete's compressive strength is typically between 17 and 28 MPa) (Goreau 2010; Pawlyn 2011, p. 50). The process continues for as long as the current moves through the metal substrate. The resulting material has self-repairing characteristics. The exact make-up of the material is dependent on the type of sea water it grows from.

The original intent of the technology was to develop low cost structures on land (Pawlyn 2011, p. 50). Suggestions have been made that integrating Biorock with offshore wind turbines or ocean energy power generators by diverting small currents to foundations could be used to regenerate surrounding marine ecosystems (Pawlyn 2011, p. 50). The physical presence of the frames, the low voltage currents and an environment suitable for coral attract other marine life, and over time diverse ecosystems mature. Coral and oysters grow three to five times faster on Biorock and have a survival rate in higher ocean temperatures caused by climate change of 16–50 times higher than background rates. Biorock has been used since 1975 in at least 20 countries (Goreau 2010). This is an example of a biomimetic technology focused on adapting to climate change and reducing biodiversity loss that is regenerative of ecosystem health.

Responses to the direct impacts of climate change have a number of benefits and associated difficulties. They are helpful for a gradual response to the impacts of climate change, particularly if the financial resources needed to research, develop and test technologies continue to be available. They require accurate knowledge of what the impacts of climate change will be for a given site over time, which is difficult to predict accurately in many cases. A benefit of this approach is that technologies and architectural responses to direct impacts may be transferable to other places with similar issues. The biomimetic system based on the Namib beetle described above may be useful for small island communities or coastal areas that have difficulties sourcing fresh water, exacerbated by climate change for example. An additional benefit of the process of developing technological solutions for individual buildings is that it fits into the current method of extending and renewing the built environment, which is typically a building-by-building or addition-by-addition process over time. This means it may be suitable for a gradual retrofitting of the existing built environment.

Developing individual technologies or even whole buildings to deal with the myriad of direct climate change impacts on the built environment does not however ready the built environment for unpredicted changes or indirect climate change impacts. Focusing solely on adapting the built environment to the direct impacts of climate change also does not address multiple concurrent impacts. The challenge of other related drivers of change is also not tackled, which may in the long run be of greater consequence to the built environment and to humanity as a whole (Atkinson 2007a). Understanding local built environments as whole systems in terms of their strengths and weaknesses and utilising these to create greater resilience may be a more effective way to plan for unpredictable future climatic changes.

4.4.2 Systemic Improvement of the Built Environment

Ecosystems are typically resilient and many are able to move though infrequent abrupt changes while still supporting the survival of organisms (Gunderson and Holling 2002). The ability of ecosystems to adapt to the rapid changes that may come about due to climate change is difficult to predict however (Walther et al. 2002). Despite this, mimicking ecosystems can offer insights into how the built environment could function more like a system than as a set of unrelated object-like buildings, and thus become better able to adjust to change. An aspect of ecosystems which enables them to successfully adapt to constant change comes from the fact they are made up of organisms and processes that are in close relationships. High diversity in terms of these relationships between organisms typically leads to increased system redundancy, and results in a greater ability for the whole system to adapt to change.

4.4.2.1 Mimicking How Ecosystems Work: Process Strategies

Systems-based climate change adaptation challenges conventional architectural design and procurement thinking, particularly the typical boundaries of a building site and design timescales. By mimicking process strategies in ecosystems, designers may have successful models to follow in devising how systems in buildings or urban environments should be put together and how they should work. Research into intensive aquaculture systems stretching back several decades demonstrates in the advantages of mimicking ecosystem processes to create resilient and effective systems. Typically, such systems mimic the process in ecosystems where waste becomes a resource for another component of the system, or where energy is shared ensuring the system eliminates or reduces duplication of effort. The Happy Shrimp Farm in the Netherlands (2007–2009) was strategically located next to E.ON Benlux's pulverised coal power station in Rotterdam's Maasvlakte (a part of the city's harbour and industrial area). Greenhouse enclosed basins for raising shrimp were kept at a steady 30 °C by a 2.5 km pipe which transferred heat from the power plant's waste coolant water (60 °C) through a heat exchange system to the farm. The 5,000 m^2 farm also supported the growth of algae. Algae fed on the nitrogen-rich waste produced in the shrimp farming process and became in turn feed for the shrimps, while helping to keep the basins clear. This resulted in productivity up to 200 times higher than if traditional feeds were used (Braungart et al. 2008).

Well-known examples of successful industrial ecology, such as Denmark's Kalundborg industrial region, also illustrate how the process of cycling materials in ecosystems can be mimicked, even between diverse companies. In Kalundborg this sharing of waste as resource results in a reduction of 30 million m^3 of groundwater used, and a reduction of 154,000 tonnes of CO_2 and 389 tonnes of mono-nitrogen oxides (NO_x) emitted. Five companies and one local municipality make up the industrial park where 20 different bi-product exchanges occur (Jacobsen 2006). The UK Cardboard to Caviar (or ABLE) Project created by Graham Wiles of the Green Business Network in Kirklees and Calderdale and the design of a zero emissions beer brewery near Tsumeb, Namibia demonstrate similar concepts of mimicking the waste cycling of ecosystems and both projects report significant beneficial social outcomes (Mathews 2011). The elimination of toxins and pollutants that lead to the degradation of ecosystems is also addressed with such an approach. Analysis of further ecosystem processes other than cycling of wastes or sharing of energy, suggests additional strategies for the built environment to mimic (Korhonen 2001).

In aiming to capture cross-disciplinary understanding of how ecosystems work, a comparative analysis of related knowledge in the disciplines of ecology, biology, industrial ecology, ecological design and biomimicry was conducted to formulate a group of ecosystem process strategies for biomimetic design (Table 4.3). For methodology, justifications and sources see: Pedersen Zari (2012). Although many aspects of ecosystems are suitable for designers to investigate in the creation of sustainable built environments, features of ecosystems that make them resilient are

Table 4.3 Ecosystem processes for biomimetic design

Tier one. Ecosystem context
1.1 The context that life exists in is constantly changing 1.2 Living entities that make up ecosystems generally work to remain alive
Tier two. Therefore
2.1 Ecosystems adapt and evolve within limits at different levels and at different rates 2.2 Ecosystems are resilient. They can persist through time even as components within them change 2.3 Ecosystems enhance the capacity of the biosphere to support life, and functioning and processes in ecosystems and within organisms tend to be benign 2.4 Ecosystems are diverse in species, relationships and information
Tier Three. The implications of this are that
3.1 Ecosystems are self-organising decentralised and distributed 3.2 Ecosystems function through the use of complex feedback loops or cascades of information 3.3 Organisms within ecosystems operate in an interdependent framework 3.4 Ecosystems and organisms are dependent upon and responsive to local conditions 3.5 Ecosystems and the organisms within them optimise the whole system rather than maximise components 3.6 Organisms within ecosystems are resourceful and opportunistic. Abundances or excesses are used as a resource
Tier Four. This is supported by the fact that
4.1 Ecosystems have the capacity to learn from and respond to information and self-assemble 4.2 Ecosystems and the organisms within them have the capacity to heal within limits 4.3 Variety can occur through emergent effects (rapid change) 4.4 Variety can occur by recombination of information and mutation (gradual change) 4.5 Ecosystems are organised in different hierarchies and scales 4.6 Ecosystems and organisms use cyclic process in the utilisation of materials 4.7 Ecosystems often have in-built redundancies 4.8 Parts of ecosystems and organisms are often multifunctional 4.9 Local energy /resources become spatial and temporal organisational devices 4.10 Ecosystems and the organisms within them gather, use and distribute energy effectively 4.11 The form of ecosystems and organisms is often a result of functional need 4.12 Organisms that make up ecosystems are typically made from commonly occurring elements

useful in the context of climate change. These were identified as: use of abundances or excess (i.e. renewable energy sources); optimisation of the whole system rather than single parts; a responsiveness and dependence on local conditions; diversity in types of organisms and relationships; a capacity for decentralised organisation; interdependence of organisms; benign functioning; complex information feedback loops; and a capacity for self-healing or self-correction.

4.4.2.2 Mimicking What Ecosystems Do: Ecosystem Functions

Mimicking the functions of ecosystems (what they do) is different from mimicking their processes (how they do it). For the purposes of applying ecosystem biomimicry to the built environment is it useful to understand the concept of ecosystem services from the study of ecology (Alcamo et al. 2003; Daily et al. 2000; de Groot et al. 2002). Ecosystem services are fundamental to human survival. Mimicking the functions of ecosystems enables design teams to know what the quantifiable ecological goals should be for a development in a given location and climate if it is to integrate with existing ecosystems and contribute to health rather than deplete it. Previous research by the author focused on ascertaining which ecosystem services were most applicable to a built environment context. For the sake of brevity, only the results are reproduced here (Table 4.4), however methodology, sources and justifications can be found in: Pedersen Zari (2012). Daily (1997) provides scientific details of each function. It is commonly known that the built environment has a large negative effect on these ecosystem services (Graham 2003). One way to reduce this is to create built environments that mimic or provide these ecosystem services and therefore reduce pressure on ecosystems, as the urban environment grows and as the climate continues to change.

This list given in Table 4.4 suggests that in a similar way to the functioning of an ecosystem, a building or development could be designed to form a system, or be part of a system that: produces food; produces renewable energy; produces raw materials for the future built environment; collects and purifies water; purifies air and soil; regulates climate through mitigating GHG emissions and the heat island effect; contributes to soil formation and fertility through careful cycling of biodegradable wastes and recycling of non-biodegradable wastes; and deliberately provides habitat for species suitable for co-inhabitation with humans in the urban built environment. Such new ecologically regenerative developments in turn could act as filters (mechanisms that purify air and water), producers (of food and materials) and generators (of energy) for the rest of the built environment which is still degrading ecosystems and is likely to persist for at least another 50–90 years (O'Connell and Hargreaves 2004). If these regenerative nodes became part of the built environment and start to perform even small aspects of ecosystem functions, it is possible that built environment caused climate change would be to some extent mitigated, and at the same time the built environment could become more adaptable to climate change, while creating beneficial biodiversity outcomes (Pedersen Zari 2012).

Similar ideas are discussed by proponents of eco-effectiveness (McDonough and Braungart 2002). The difference between those concepts and the ecosystem biomimicry proposed here is that the measurable targets such as emissions levels, carbon storage, water catchment, energy production and resource production, are determined through an understanding of suitable ecosystems or the pre-development ecosystem of the site, and thus are based on ecological reality, rather than human political needs or trends. A benefit of mimicking the functions of ecosystems is that, through careful urban planning and an integrated and multidisciplinary design method, buildings as part of a larger system, able to mimic ecosystem processes and/or functions in their creation, use, and eventual end of life, may have the potential to adapt more readily to climate change. Whole system adaptation of built environments using an understanding of ecosystems may have the potential to address climate change impacts and biodiversity changes

Table 4.4 Ecosystem services for urban/architectural design

Ecosystem services for the built environment to mimic		Applicability to the built environment	Ecological significance	Negative environmental impact caused by the built environment		Climate change / biodiversity implications
Supporting Services	1. Habitat provision (including: provision of genetic information; biological control; fixation of solar energy; and species maintenance)	Medium	High	High at a local scale	Plan for habitat for non-human. Consider reducing fragmentation of habitat.	Increased/maintained biodiversity may have links to increased ecosystem resilience, and allow for better adaptation to climate change; reduction of the urban heat island effect; sequestration of carbon; increased air, water and soil quality; and more fertile soil meaning greater potential for the growth of biomass and food production and therefore increased human health and resilience. Remediation of some forms of water, air and soil pollution through increased urban vegetation. Possible protection from wind or wave surges; reduced erosion; more adaptable ecosystems as the climate changes; reduction of stormwater peak flows; and cooler urban temperatures (due to increased urban vegetation.
	2. Nutrient cycling (including: decomposition; soil building; and provision of raw materials)	Medium	High	High at a regional/global scale	Nutrients (materials) should be able to be biodegraded or recycled in closed loops and retained in the system. Development should contribute actively to soil formation and the renewal of fertility. Development should be considered a potential source of future building materials.	Reduction of waste and ecosystem pollution caused by materials production and transportation. Reduced need for mining / growing / production / transportation of materials and energy leading to reduction in GHG emissions, waste and ecosystem disturbance. Decreased use of energy and therefore GHG emissions. Increased health of ecosystems and humans and reduced biodiversity loss.
Regulating Services	3. Purification	High	High	High at a local/ regional scale	Water, air and soil should be purified on site before return to non-human ecosystems. Water and air should be cleaner leaving the development than when it entered. Surrounding soil should become more fertile over time.	Increased health of ecosystems and living organisms within them (including humans). Increased terrestrial, riparian and marine productivity. Reduction of air and water pollution. Eutrophication reduction. Remediation of polluted sites. Reduced ozone damaging gas emissions and reduced GHG emissions Reduced biodiversity loss.

(continued)

Table 4.4 (continued)

Ecosystem services for the built environment to mimic		Applicability to the built environment	Ecological significance	Negative environmental impact caused by the built environment		Climate change / biodiversity implications
	4. Climate regulation	High	High	High at a global scale	Development should contribute to regulating climate by: sequestering carbon; providing protection from decreased ozone; and remediating the heat island effect.	Mitigation of the causes of climate change. Mitigation of the urban heat island effect. More adaptable communities. Improved health of living organisms Improved ability of ecosystems and the organisms within them to adapt to climate change.
Provisioning Services	5. Provision of fuel / energy for human consumption	High	Medium	High at a global scale	Development should capture rainwater and conserve water to meet its own needs are and preferably more so that excess can be distributed to neighbours. Design should encourage effective energy use.	Reduced transport and energy generation related GHG emissions. More self-reliant and therefore robust urban environments. Reduction of air, water and soil pollution. Reduction of fossil fuel mining, drilling and transportation impacts on biodiversity.
	6. Provision of fresh water	High	High	High at a regional scale	Development should capture rainwater and conserve water to meet its own needs are and preferably more so that excess can be distributed to neighbours. Design should encourage the conservation of water.	Reduction of water pollution. Increased health of riparian systems. Reduction of the urban heat island effect. Increased quality of water. Increased health of living organisms. Reduced wastage of water through pipes.
	7. Provision of food *(including: provision of biochemicals)* Medium	Medium	Medium	High at a global scale	Development should produce food.	Reduced transport and energy generation food related GHG emissions. More self-reliant and therefore robust urban environments. Reduction of air, water and soil pollution. Reduction of fossil fuels needed to produce the products of large scale agriculture. Healthier humans Reduced land use transformations (i.e. from natural ecosystems to agricultural systems).

concurrently, but the concept needs to be better understood and further developed. Further research of ecosystem biomimicry is ongoing, however a comprehensive analysis of how ecosystem services could specifically be supplemented by the built environment forms the basis of a comprehensive case study of Wellington, New Zealand and is available in Pedersen Zari (2012).

That a greater understanding of ecology and systems design is required on the part of design teams is implicit with such an approach. Increased collaboration between fields that traditionally seldom work together such as architecture or urban design, and biology or ecology would also be required. Challenges with such an approach include the current competitive economic context of the built environment in many places in the world as discussed by Hunt (2004). Encouraging greater interdependence and the sharing or exchange of resources or knowledge between buildings or neighbourhoods requires for example a different economic, legal and attitudinal framework. Because this approach requires communities to become interdependent systems, coordination and cooperation from landowners and authorities is needed to enable progress at a large scale. The built environment varies greatly between different climatic, economic and cultural contexts, and systems that are appropriate to specific places will therefore also vary greatly. Although each differing geographic region will have to evolve its own unique system over time, knowledge of how to create or evolve such systems can be transferred.

Although there are some draw backs to a whole system adaptation of the built environment, it is a suitable solution for a longer term response to climate change impacts, because it addresses many of the underlying issues with current urban environments that are in need of re-evaluation (Grimm et al. 2008; Wahl and Baxter 2008). The difference can be likened to a long-term treatment of the underlying cause of an illness in an individual, rather than a short-term treatment of symptoms which may in fact aggravate the underlying condition. In this case, this is the fact that the majority of human urban settlements are dependent on fossil fuels to heat, feed and transport people in a linear system which creates pollution leading in part to climate change. This system also causes the degradation of waterways, air quality, soil and human health while at the same time consumes non-renewable resources in such a way that they cannot be reused. A whole-systems approach to built environment design acknowledges that human developments and therefore humans are not in any way separate from the ecosystems they exist in.

4.5 Conclusions

Commonly given examples of successful biomimicry are typically products or materials rather than buildings or building systems, and tend to mimic a single organism. Mimicking aspects of organisms can produce innovations that address sustainability issues in some cases, but without an understanding of the ecological context of these organisms, such innovations can too easily become simple technological add-ons to conventional buildings. These could fall into the category of 'green-wash', if buildings remain average in terms of overall sustainability performance over whole lifecycles. Such solutions also miss an opportunity to examine the possibility of systemic change in the built environment and to re-evaluate the nature of the relationship between people, their built environment the ecosystems they exist in.

While existing technologies and techniques will be crucial in the short and medium terms, biomimicry could form an important part of long-term solutions to climate change and biodiversity loss. Biomimicry could be useful as part of strategies designed to replace the use of fossil fuels, to develop technologies or techniques to address direct climate change impacts on the built environment and in the systemic improvement of the built environment using ecosystem biomimicry. Technologies that increase energy efficiencies and can sequester or store carbon may form part of an important short to medium-term approach, but should be seen as intermediate steps. As well as a reduced or potentially negative carbon footprint for the built environment, examples of biomimetic technologies reveal approaches that use current excess CO_2 as a resource for new materials. Biomimetic technologies that address direct climate change impacts and biomimetic technologies or systems that prevent further GHG emissions have also been examined and could be implemented alongside wider systemic change in the built environment (including a consideration of people's consumption behaviour and lifestyle expectations).

The case studies examined in this chapter suggest that ecosystem biomimicry may be the most effective kind of biomimicry to respond to climate change impacts and utilise synergies between mitigation and adaptation strategies. This is also the least explored aspect of biomimicry in-built form. Positive integration with ecosystems leading to a regenerative rather than damaging effect on them may contribute to maintaining biodiversity and the ecosystem services that humans are dependent upon for survival, particularly as the climate continues to change. Such a concept goes beyond encouraging a basic understanding of ecological processes over time, as is increasingly advocated in recent publications and in educational institutions. Instead it is the thorough integration of ecological knowledge into architecture and urban design to alter how buildings function fundamentally in relation to ecosystems and to each other. Buildings should be expected to become active contributors to ecosystems and social systems, rather than remaining unresponsive agents of ecosystem degeneration.

Acknowledgments This research is based in part on earlier published paper: M. Pedersen Zari (2010), Biomimetic design for climate change adaptation and mitigation, *Architectural Science Review*, 53(2), pp. 172–183. It is revised, expanded and updated.

References

Alcamo J, Ash N, Butler C, Baird Callicott J, Capistrano D, Carpenter S et al (2003) Ecosystems and Human Wellbeing. Island Press, Washington

Allen R (2010) Bulletproof feathers: how science uses nature's secrets to design cutting edge technology. University of Chicago Press, Chicago

Andersen SO, Zaelke D, Young O, Ahmadzai H, Anderson F, Atkinson M, Carson E, Carson RJ, Christensen S, Deventer JSV, Hanford S, Hoenig V, Miller A, Molina M, Price L, Ramanathan V, Tope H, Wilkinson J, Yamabe M (2011) Scientific synthesis of calera carbon sequestration and carbonaceous by-product applications. Consensus Findings of the Scientific Synthesis Team. Washington DC, Institute for Governance and Sustainable Development

Anon (2005) Natural innovation: the growing discipline of biomimetics. Strateg Dir 21(10):35–37

Aranda-Mena G, Sperling J, Wakefield R (2008) E-business adoption: case study on Melbourne city council ch2 building. In: Paper presented at the World sustainable building conference (SB08)

Armstrong R (2009) Living buildings: plectic systems architecture. Technoetic Arts 7(2):79–94

Atkinson A (2007a) Cities after oil—1: 'sustainable development' and energy futures. City 11(2):201–213

Atkinson WI (2007) Mouthwash for a smokestack. Globe and Mail

Baird G (2001) The architectural expression of environmental control systems. Spon Press, New York

Baker J (2011) Prospectus of wave energy. http://sustainus.org/docs/citsci/winners/2011/J%20Baker%202011.pdf. Accessed May 2014

Bellew P (2006) Going underground. Ingenia 28, September

Bensaude-Vincent B, Arribart H, Bouligand Y, Sanchez C (2002) Chemists and the school of nature. New J Chem 26:1

Benyus J (1997) Biomimicry—innovation inspired by nature. Harper Collins Publishers, New York

BioPower Systems (2011) BioPower systems press releases 8 August and 1 December 2011. Sydney, BioPower Systems. http://www.biopowersystems.com. Accessed May 2014

Bond S (2010) Lessons from the leaders of green designed commercial buildings in Australia. Pac Rim Property Res J 16(3):314–338

Braungart M, Darlington A, Garland J, Lucey WP, Peck S et al (2008) Ecological design and engineering for urban environments. Ind Biotechnol 4(3):211–223

Brinker J, Lu Y, Sellinger A (1999) Evaporation-induced self-assembly: nanostructures made easy. Adv Mater 11(7):579–585

Calera (2012) Calera Website. Calera. http://calera.com/index.php/. Accessed May 2012

Carley J (2012) Enzyme enabled carbon capture. Lowering the CCS Cost Barrier. In: Presentation at the 15th annual energy, utility, and environment conference (EUEC). Phoenix, Arizona

Chan M, Cheng BN (2006) Performance evaluation of domestic ionizer type air cleaners. Architectural Sci Rev 49(4):357(356)

Collis GE, Campbell WM, Officer DL, Burrell AK (2005) The design and synthesis of porphyrin/oligiothiophene hybrid monomers. Org Biomol Chem 3:2075–2084

CO_2 Solutions (2008) Innovation in carbon capture—2008 Annual Report. http://www.co2solutions.com/uploads/file/ebedaa3590ce4b94804822d19b09a19c66df992b.pdf. Accessed May 2014

CO_2 Solutions (2014) CO_2 solutions successfully completes second oil sands project milestones. Press Release

Daily GC (1997) Nature's services societal dependence on natural ecosystems. Island Press, Washington D. C

Daily GC, Soderqvist TSA, Arrow K, Dasgupta P, Ehrlich PR et al (2000) The value of nature and the nature of value. Science 289:395–396

Davidson S (2003) Light factories. ECOS 117:10–12
de Groot R, Wilson MA, Boumans RMJ (2002) A typology for the classification, description and valuation of ecosystem function, goods and services. Ecol Econ 41:393–408
de Ia Rue du Can S, Price L (2008) Sectoral trends in global energy use and greenhouse gas emissions. Energy Policy 36(4):1386
Fernandez JE (2004) Design diverse lifetimes for evolving buildings. In: Steemers K, Stean MA (eds) Environmental diversity in architecture. Routledge, New York, pp 65–82
Finnigan T, Caska A (2006) Simulation of a biomimetic ocean wave energy device using blade-element theory. In: Paper presented at the sixteenth international offshore and polar engineering conference, San Francisco, California
Fister Gale S (2008) Carbon dioxide turns useful. PC Magazine, November
Fradette DS (2007) CO_2 solution and climate change. BioInspired! 5(2)
Garrod RP, Harris LG, Schofield WCE, McGettrick J, Ward LJ, Teare DOH et al (2007) Mimicking a Stenocara Beetle's Back for Microcondensation Using Plasmachemical Patterned Superhydrophobic-Superhydrophilic Surfaces. Langmuir 23(2):689–693
Gebeshuber I, Gruber P, Drack M (2009) A gaze into the crystal ball: biomimetics in the year 2059. J Mech Eng Sci 223(12):2899–2918
Geers C, Gros G (2000) Carbon dioxide transport and carbonic anhydrase in blood and muscle. Physiol Rev 80(2):681–715
Gerngross T, Slater S (2003) Biopolymers and the environment. Science 299(5608):822–825
Goreau TJ (2010) Reef technology to rescue Venice. Nature 468(7322):377
Graham P (2003) Building Ecology—First Principles for a sustainable built environment. Blackwell Publishing, Oxford
Greenemeier L (2007) Making plastic out of pollution. Scientific American, November
Grimm NB, Faeth SH, Golubiewski NE, Redman CL, Wu J, Bai X et al (2008) Global change and the ecology of cities. Science 319(5864):756–760
Gunderson LH, Holling CS (2002) Panarcy. Understanding transformations in human and natural systems. Island Press, Washington D.C
Hamilton T (2007) Capturing carbon with enzymes. A new process turns the greenhouse gas into useful materials. MIT Technology Review, February 22
Howden-Chapman P, Chapman R, Hales S, Britton E, Wilson N(2010) Climate change and human health: Impact and adaptation issues for New Zealand. In: Nottage RWDBJ, Jones K (eds) Climate change adaptation in New Zealand. Future scenarios and some sectorial perspectives. Wellington, New Zealand Climate Change Centre
Hunt J (2004) How can cities mitigate and adapt to climate change? Building Res Inf 32(1):55–57
IPCC (Intergovernmental Panel on Climate Change) (2007a) Climate change 2007: mitigation contribution of working group iii to the fourth assessment report of the IPCC. Cambridge University Press, Cambridge
IPCC (Intergovernmental Panel on Climate Change) (2007b) Climate change 2007: the physical science basis. contribution of working group I to the fourth assessment report of the IPCC. Cambridge, Cambridge University Press
Jacobsen NB (2006) Industrial symbiosis in Kalundborg, Denmark: a quantitative assessment of economic and environmental aspects. J Ind Ecol 10(1–2):239.
Jevons, W (1865) The Coal Question. An Inquiry Concerning the Progess of the Nation and the Probable Exhaustion of our Coal Mines. London and Cambridge. Macmillan and Co
Jones T (2008) Distributed energy systems. In: Newton P (ed) Transitions. Pathways towards sustainable urban development in Australia. Collingwood, CSIRO
Knight W (2001) Beetle fog-catcher inspires engineers. New Scientist 13:38
Koelman O (2004) Biomimetic buildings: understanding and applying the lessons of nature. BioInspire 21
Koeppel S, Ürge-Vorsatz D (2007) Assessment of policy instruments for reducing greenhouse gas emissions from buildings. Report for the UNEP SBCI (United Nations Environmental Programme Sustainable Buildings and Construction Initiative), Central

Korhonen J (2001) Four ecosystem principles for an industrial ecosystem. J Clean Prod 9:253–259
Lovins LH, Cohen B (2011) Climate Capitalism. Capitalism in the age of climate change, New York
Mathews F (2011) Towards a deeper philosophy of biomimicry. Organ Environ 24(4):364–387
Mattheck C (1998) Design in nature. Learning From Trees, Berlin
Mattheck C, Baumgartner A, Gräbe D, Teschner M (1996) Design in nature. Struct Eng Int 6(3):177–180
McDonough W, Braungart M (2002) Cradle to cradle—Remaking the way we make things. North Point Press, New York
McKeag T (2010) Bringing biomimicry to cement, offices and daily life GreenBiz.com. http://www.greenbiz.com/blog/2010/07/15/bringing-biomimicry-cement-offices-and-daily-life?page=0%2C0. Accessed May 2014
McKeough T (2009) Novomer's plastic reduces greenhouse gas-but will it biodegrade? Fast Company Newsletter
Mitchell RB (2012) Technology is not enough. J Environ Develop 21(1):24–27
Monteiro PJM, Clodic L, Battocchio F, Kanitpanyacharoen W, Chae SR, Ha J et al (2013) Incorporating carbon sequestration materials in civil infrastructure: A micro and nano-structural analysis. Cement Concr Compos 40:14–20
Moore T, Moore A, Gust D, Hambourger M, Brune A (2004) Artificial photosynthesis and hydrogen production: strategies for sustainable energy production. In: Paper presented at the 13th international congress on photosynthesis, Montreal, Canada
Nidumolu R, Prahalad CK, Rangaswami MR (2009) Why sustainability is now the key driver of innovation. Harvard Business Review
Novomer (2010) Novomer. Catalyzing green chemistry novomer. http://www.novomer.com/. Accessed March 2012
Novomer (2013) Novomer catalytic process using waste co2 and shale gas targets $20 Billion market and up to 110 % carbon footprint reduction content. Press Release
O'Connell M, Hargreaves R (2004) Climate change adaptation study Report No. 130, Branz
Paevere P, Brown S (2008) Indoor environment quality and occupant productivity in the ch2 building: post-occupancy summary. CSIRO, Melbourne
Parker AR, Lawrence CR (2001) Water capture by a desert beetle. Nature 414(6859):33
Patel-Predd P (2007) Carbon-dioxide plastic gets funding. A startup is moving ahead with an efficient method to make biodegradable plastic. Technol Rev (November 14)
Pawlyn M (2011) Biomimicry in architecture. RIBA Publishing, London
Pedersen Zari M (2007) Biomimetic approaches to architectural design for increased sustainability. In: Paper presented at the sustainable building conference, Auckland
Pedersen Zari M (2010) Biomimetic design for climate change adaptation and mitigation. Archit Sci Rev (ASR) 53(2):172−183
Pedersen Zari M (2012) Ecosystem services analysis for the design of regenerative urban built environments. Victoria University of Wellington, Wellington
Pedersen Zari M, Storey JB (2007) An ecosystem based biomimetic theory for a regenerative built environment In: Paper presented at the Lisbon sustainable building conference 07, Lisbon, Portugal
Pronk A, Blacha M, Bots A (2008) Nature's experiences for building technology. In: 6th international seminar of the international association for shell and spatial structures (IASS) working group 15 'structural morphology', Acapulco
Purvis A, Hector A (2000) Getting the measure of biodiversity. Nature 405(6783):212–219
Rees W (1999) The built environment and the ecosphere: a global perspective. Build Res Inform 27(4/5):206–220
Reisinger A, Wratt D, Allan S, Larsen H (2011) The role of local government in adapting to climate change: lessons from New Zealand. In: Ford JD, Berrang-Ford L (eds) Climate change adaptation in developed nations, vol 42, pp 303–319. Springer Netherlands

Schiermeier Q (2006) Putting the carbon back. The hundred billion tonne challenge. Nature, 442:620–623

Sellinger A, Weiss P, Nguyen A, Lu Y, Assink R, Gong W and Brinker C (1998) Continuous self-assembly of organic-inorganic nanocomposite coatings that mimic nacre. Nature 394(6690):256−260

Smith F (1997) Eastgate, Harare, Zimbabwe. Arup J 32(1):3–8

Steemers K (2003) Towards a research agenda for adapting to climate change. Build Res Inform 31(3/4):291–301

Stern N (2006) Stern review: the economics of climate change. Independent Review for the Government of the United Kingdom

Tan S (2007) CH2 6 stars, but how does it work? Architecture Australia, 96(1):101–104

Trivedi BP (2001) Beetle's shell offers clues to harvesting water in the desert. National Geographic Today

Turner G (2008) A comparison of the limits to growth with 30 years of reality. CSIRO, Canberra

Turner, J and Soar R (2008) Beyond Biomimicry: What Termites can tell us about Realizing the Living Building. First International Conference on Industrialized, Intelligent Construction (I3CON). Loughborough University

Vanderley MJ (2003) On the sustainability of concrete. UNEP J Ind Environ

Vincent J (2010) New materials and natural design. In: Allen R (ed) Bulletproof feathers. University of Chicago Press, Chicago

Vincent JFV, Bogatyreva OA, Bogatyrev NR, Bowyer A, Pahl A-K (2006) Biomimetics—its practice and theory. J R Soc Interface 3(9):471–482

Vogel S (1998) Cat's paws and catapults. Norton and Company, New York

Wahl DC, Baxter S (2008) The designer's role in facilitating sustainable solutions. Design Issues 24(2):72–83

Walther G-R, Post E, Convey P, Menzel A, Parmesan C, Beebee TJC et al (2002) Ecological responses to recent climate change. Nature 416(6879):389

Wilby RL (2007) A review of climate change impacts on the built environment. Built Environ 33(1):31–45

Zhao J, Xu Y (2010) Ecological wisdom inspired from termite mounds. Analysis on biomimetric design of Zimbabwe Eastgate Center. Build Sci 2

Chapter 5
Bio-inspired Adaptive Building Skins

R.C.G.M. Loonen

Abstract How do living organisms capture, convert, store and process energy, water and sunlight? How does nature cool down, heat up, provide shade, and control light? Adaptability, the ability of a system to act in response to variations in environmental conditions often plays a key role in this context. Unlike living organisms, buildings are typically conceived as static, inanimate objects. Because a building's surroundings and internal conditions are constantly changing, there is a lot to learn about how inspiration from nature can foster more adaptability of the façade for enhanced building performance. After highlighting the need for more adaptability in the built environment, this chapter reviews state-of-the-art examples of research concepts and design applications with bio-inspired adaptable solutions for the building envelope. All examples are in the scope of building physics and energy efficiency with a focus on improving indoor environmental quality. The chapter concludes with an outlook of design support methodologies that can potentially incite the practical uptake of bio-inspired adaptive building skins in the future.

5.1 Introduction

Building envelopes separate the ambient environment from indoor spaces and their occupants. On the one hand, the building envelope has to offer protection against inclement influences from outside, such as wind, rain, excessive radiation and extreme temperatures. On the other hand, the façade functions as the connecting element between occupants and the outside world, by regulating the exchange of energy and admitting access to, e.g. views, daylight and fresh air.

R.C.G.M. Loonen (✉)
Unit Building Physics and Services, Eindhoven University of Technology, Eindhoven, The Netherlands
e-mail: r.c.g.m.loonen@tue.nl

F. Pacheco Torgal et al. (eds.), *Biotechnologies and Biomimetics for Civil Engineering*,
DOI 10.1007/978-3-319-09287-4_5

The design of a building envelope is of crucial importance for the quality of the indoor environment it encloses (Knaack et al. 2007). Designing comfortable spaces for working and living has always been a primary objective for architects and engineers alike. Accruing evidence now shows that indoor environmental quality is not only important for comfort perception, but also has a significant, direct effect on office worker productivity, health, reduced sick leave, and associated economic implications (Fisk and Rosenfeld 1997; Singh et al. 2010; Aries et al. 2014). From the viewpoint of design for whole life-cycle value, these insights put even more emphasis on the need to develop building envelope constructions, which can support the realization of high-performance indoor environments (Jin and Overend 2014).

In many climatic zones, exterior conditions throughout most of the year are too extreme to be considered comfortable. The building envelope can usually moderate the negative effects to only a limited extent. To ensure a healthy and comfortable indoor climate all year long, intensive use of artificial lighting, and mechanical systems for heating, ventilation and air-conditioning (HVAC) is often required. In the current fossil fuel economy, this leads to large pressures for sustainable development, to the extent that one-third of all end-use CO_2 emissions comes from the building sector (International Energy Agency 2012). This present situation represents a barrier for sustainable growth of the current generation, but at the same time also forms an immense opportunity for innovative solutions. The Intergovernmental Panel on Climate Change has identified the building sector as the sector with the highest potential for cost-effective mitigation of carbon dioxide emissions (Ürge-Vorsatz and Novikova 2008).

The need for increasing energy efficiency levels in the built environment is currently being enforced through energy performance regulations for new buildings and renovation projects, such as the *Directive on the energy performance of buildings* (EPBD) in the European Union. These new regulations have a focus on demand reductions, which, in many instances, is manifested through the development of low-energy buildings with highly insulated, airtight, sealed envelopes (Meijer et al. 2009). Such designs, however, leave little connection between occupants and the outside world, thereby conducing to the detachment of occupants from the exterior environment.

The here described features of the *passive* design approach are difficult to reconcile with the biophilic design movement, which advocates the necessity of maintaining, enhancing and restoring the connection between nature, human beings and the built environment (Kellert et al. 2011). Moreover, at times, the *passive* strategy seems to be at odds with the requirements for high-quality indoor environments, which could take advantage of the positive contributions from natural energy sinks and sources in a more active way. Indeed, several studies show the associated risk for unintended comfort issues in buildings where the primary focus was on low-energy building design (Mlecnik et al. 2012; McLeod et al. 2013).

An alternative direction in the efforts to reconcile energy efficiency with the need for high-quality indoor climates integrates *adaptability* into the building envelope, as inspired by principles found in nature. Modern construction techniques, developments in material sciences, and availability of controllable kinetic façade components now offer ample opportunities for innovative building envelope solutions that respond better to the environmental context, thereby allowing the façade to "behave" as a living organism (Loonen et al. 2013; Loonen 2014). Natural envelope-like structures such as shells, skins and coverings have evolved over time to limit energy expenditure while maintaining conditions to thrive and reproduce. In many instances, adaptability, either on the short- or longer-term (Kasinalis et al. 2014), forms an essential mechanism for resilience and survival. Mimicking these strategies, by finding functional analogies, and bringing them to the domain of architecture, can form a key ingredient for environmentally conscious sustainable development in the building sector. On a wider scale, bio-inspired adaptable building skin design can play an important role in creating a better mutual relationship between buildings and their surrounding local micro-climate, through careful integration of principles from urban physics (Moonen et al. 2012), for enhancing, e.g. natural ventilation potential (Ramponi et al. 2014), pollutant transport in street canyons (Xie et al. 2007), acoustic properties (Van Renterghem et al. 2013) and wind comfort (Montazeri et al. 2013). In the longer term, such design considerations will help in the transition from biological principles on the building scale towards the more holistic notion of "biophilic cities" (Beatley 2011).

This chapter presents an overview of building envelope surfaces, materials, structures, functions, constructions, mechanisms, principles and processes, all inspired by nature. We approach the building envelope design question from the viewpoint of visual, acoustic and thermal aspects, and do this with a specific focus on adaptability, the ability of a system to change in response to varying environmental conditions. The next section continues with a discussion of systems-based terminology to be able to better analyse the many ways of how adaptability can be embedded in the building envelope. This is followed by a state-of-the-art structured overview with examples of bio-inspired adaptive building skins described in literature. Finally, this chapter concludes with discussing strengths and challenges of design methods that can support a more systematic design methodology for bio-inspired adaptive building skins.

5.2 Adopting Bio-inspired Adaptation Principles in Building Envelope Design

The existence and manifestation of *adaptation* is an important condition for sustaining life in natural (eco-)systems. In nature, adaptation can be found at multiple physical scales and time resolutions. It is not always straightforward to convey the complexity and subtlety of these biological concepts in a form that is directly applicable to the domain of architecture and building envelope design. The systems engineering literature provides a well-described taxonomy that can help make this connection (Chalupnik et al. 2013). The same terminology can be used to highlight the need for adaptation in the built environment (Loonen et al. 2013). With respect to building envelopes, the underlying principles of adaptation can offer advantages on the following three levels: adaptability, multi-ability and evolvability.

5.2.1 *Adaptability*

Adaptability is defined by Ferguson et al. (2007) as the ability of a system to deliver intended functionality considering multiple criteria under variable conditions through the design variables changing their physical values over time. Building skins having this attribute can seize the opportunity to deliberately act in response to changes in ambient conditions, such solar radiation, wind speed and direction, temperature, rainfall, etc. Doing this offers a potential for energy savings compared to conventional, static buildings because the valuable energy resources in our environment can be actively exploited, but only at times when these effects are deemed favourable (Loonen et al. 2011). Adaptable facades can thus act as climate mediator, negotiating between comfort needs and what is available in the ambient environment (Wigginton and Harris 2002). With embedded adaptability, façades no longer have to be a compromise solution for the whole year, performing acceptable under a wide range of conditions, but never optimal regarding a specific situation (Ochoa et al. 2012). Moreover, it gives opportunities to adjust to the individual user (Bakker et al. 2014), rather than a best average for all, as usually prescribed in comfort standards. In addition to the immediate effect, adaptability of building envelopes also helps in achieving gains through smart utilization of building constructions' thermal storage capacity. The dual effects of shifting peak demands can help in mitigating comfort problems, and also limit redundancy in installed heating and cooling capacity (Hoes et al. 2011).

5.2.2 *Multi-ability*

The concept of multi-ability originates from the existence of non-simultaneous performance requirements, or the need to fulfil new roles over time. The 'balcony that can be folded' (Hofman and Dujardin 2008) is an illustrative example of a responsive building envelope that features multi-ability: depending on ambient conditions and users' preference, it changes function from window to a balcony-on-demand (Weaver et al. 2008). Multi-ability differs from adaptability in the sense that multiple requirements are fulfilled successively, not at the same time (Ferguson et al. 2007). Unlike conventional systems, designed to satisfy a single set of conditions, it allows for addressing change via a plurality of individually-optimized states. In this way, multi-ability promotes more efficient use of resources, which adds to the list of benefits already mentioned in the previous paragraph. A second application of multi-ability is the potential for spatial versatility. At the same moment in time, the properties of the building envelope can be different for various positions of the building shell. In this way, different orientations of the building can independently react to the ambient conditions or to distinct comfort preferences requested by individual users in separate zones.

5.2.3 Evolvability

While adaptability and multi-ability mostly handle short-term changes, evolvability is a property of flexible systems that deals with variations over a longer time-horizon (Silver and de Weck 2007). Perhaps even more than uncertainty in day-to-day operation, future building requirements and boundary conditions are highly unpredictable, or cannot even be known in the design stage (Holmes and Hacker 2007; De Wilde and Tian 2010). Building shells that can evolve over time are a means of extracting value from the uncertainty of these unforeseen events. Evolvability is nevertheless considered more a positive side-effect, rather than a main design objective; the ability to keep options open preserves opportunities to react to changes as they unfold in the future or perhaps might not occur. Concerning the building and construction industry, evolvability, or sometimes called survivability, can be used to deal with changing conditions coming from the outside (e.g. climate change, changing urban environment, wearing of the façade) or from the inside (e.g. organizational function changes of the building, new space layout). In all these cases, having a façade that can react to changes increases the chances that the building can continue operation as intended, without suffering from the potential negative impacts of unforeseen future conditions.

5.3 Overview and Analysis of Bio-inspired Facades

In both research settings and design practice, a tendency for more bio-inspired adaptable façade projects can be observed. In addition, several universities recently started to teach courses on biomimetic building design as part of their architecture curricula (Badarnah 2012). Although Braun (2008) presents a wide overview of more than 45 potential role models for bio-inspired building envelope design, the skins of plants and humans tend to be seen as the most straightforward emulation model and inspiration source for multifunctional and truly sustainable enclosure systems (Wigginton and Harris 2002; Koch et al. 2009). In nature, skins perform a multitude of different functions, respond to a range of operating conditions and are self-regulating. Transferring these traits as metaphors to the building envelope is a commendable, but challenging task (Mazzoleni 2010; Drake 2007).

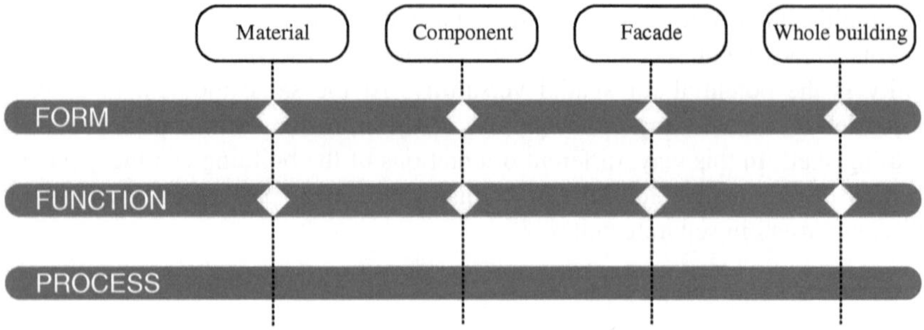

Fig. 5.1 Overview of bio-inspiration at different levels

Initial manifestations of biomimetic applications in the construction sector were mainly focused on individual buildings and experimental architecture. More and more, however, efforts now tend to shift to the level of materials and components, henceforth aiming at easier repeatability and wider applicability (Loonen et al. 2014). Figure 5.1 provides an overview of the different scales and function levels at which bio-inspired adaptable facade design is examined in this chapter.

From the level of the building envelope, bio-inspiration can take place at various different spatial scales, ranging from (i) material, (ii) component, (iii) façade, to (iv) whole-building. Following biomimetic literature, the types of bio-inspiration can be classified in one of the following three groups, (a) form, (b) function, and (c) process (Pedersen Zari 2010).

Form is the most straightforward type of bio-inspired design, because it directly relates to imitation of the morphological appearance of the biological system or organism. There is a clear visual resemblance between the building envelope and its biological equivalent, but it does not necessarily embody a functional feature that has a positive effect on performance of the building, and if it performs a function, it is not associated with the original application. Although out of scope and hence not presented in this chapter, many examples of non-adaptive bio-inspired structures, such as organism-shaped or zoomorphic architecture (Aldersey-Williams 2004) fit into this category.

Function, on the other hand, mimics the underlying biological mechanism, i.e. it tends to be more concerned with what the facade does rather than the way it looks (although not completely excluding similarity in terms of form). Functional bio-inspiration can either be direct or indirect. The first approach *directly* copies the observed functional principle into a building envelope technology which performs the same role. The *indirect* approach is loosely based on a selected biological principle, but requires an intermediate abstraction step in the transfer from biological principle to building envelope technology.

Process, the third type of bio-inspiration, is an important constituent in many features of biomimicry. However, it is not commonly found on the level of individual buildings, but more often in inter-building connections on the urban or district scale. It is therefore not part of the analysis presented in this chapter.

5.3.1 Form

5.3.1.1 Form: Material

Adaptability of form at the material scale implies that changes take place at the scale of nanometres. A myriad of responsive or smart building materials are capable of changing their configuration at this level of spatial resolution (Bastiaansen et al. 2013; Klooster et al. 2009; Leydecker et al. 2008). The change itself, however, can normally not be observed with the naked eye, but is manifested in the form of variations in aggregate optical (Baetens et al. 2010), thermophysical (Quesada et al. 2012), or surface properties (Chen et al. 2012). This then results in a change of function, rather than form, and for this reason, this class of materials is discussed in Sect. 5.3.2.1: Function—Material.

5.3.1.2 Form: Component

The efficiency of building-integrated photovoltaic (BIPV) façade systems can take advantage of natural design strategies. SMIT Solar Ivy is such an innovative BIPV system, inspired by the way foliage leafs are arranged to capture sunlight in an optimal way. Instead of cladding the envelope with flat surface-mounted panels, the system consists of smaller loose-hanging elements, the so-called solar leaves that freely move and partly overlap each other. This arrangement not only increases the usable surface for photovoltaic energy generation, but also induces a natural, wind-driven cooling effect, which further enhances the energy generation efficiency of the thin-film solar cells (Klooster et al. 2009; Brownell 2010).

One of the most well-known and widely studied (e.g. Dawson et al. 1997) examples of movement found in nature is the opening and closing behaviour of pine cones in response to hygroscopic conditions. Reichert et al. (2014) recently adopted this phenomenon, and transformed it into an innovative façade component. Their approach, named Meteorosensitive Architecture, harnesses the elastic deformation of veneer composite materials to make a tunable, humidity-responsive façade opening system. An interesting feature of this concept is that the material acts as sensor and actuator at the same time (Menges and Reicher 2012). The adaptive capacity is an inherent feature of the subsystems comprising the building shell. An advantage of this type of control, compared to conventional, external control hardware is that these types of systems can immediately change their configuration without expending any fuel or electricity to facilitate the state transition. In addition, the subtlety of the technology is a main asset: the number of components is limited since no add-on systems like control units, processors or wires are necessary (Loonen et al. 2013). The material can be "programmed" in such a way that it responds in different ways to prevailing environmental

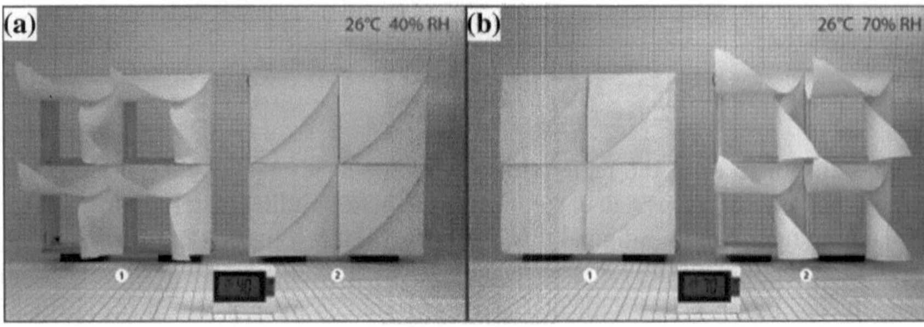

Fig. 5.2 Example of opening and closing behaviour of the Meteorosensitive Architecture prototypes. (*Source* Reichert et al. (2014), reprinted with permission)

conditions. Figure 5.2, for instance, shows two configurations, where (a) the façade closes and (b) opens with increasing relative humidity levels. Many types of architectural expressions can be achieved by adapting the existing design concept.

5.3.1.3 Form: Facade

Natural ventilation is a prominent design strategy for low-energy building operation. Controlling the opening and closing schedules of windows is, however, a challenging task because the resulting air flow can easily be too high or too low. "Living glass" tries to offer a solution to this problem by controlling the opening of a façade with gill-like slits, looking similar to the mechanism for gaseous exchange found in aquatic organisms (Geiger 2010; Linn 2014). Using strips of shape memory alloy that expand or contract in response to (man-made) CO_2 concentrations, a perfect balance between façade opening, pressure difference and momentary ventilation requirements can be achieved.

5.3.1.4 Form: Whole Building

Unmistakably inspired by the wings of a bird, the Burke Brise Soleil, covering the Quadracci Pavillion at the Milwaukee Art Museum in Wisconsin, is an iconic piece of architecture (Fig. 5.3). The eye-catcher consists of 72 steel fins that span the roof of a 27-m tall glass dome, and collectively open and close with the opening hours of the museum (Trame 2001). The morphological resemblance with a bird is not just cosmetic but also serves a functional purpose. The brise soleil dynamically protects the sunspace from excessive radiation, but in a form that is detached from the ornithological background where the façade system took its inspiration. Moreover, one can argue if this type of shading strategy is actually effective when compared to more conventional solutions.

Fig. 5.3 Burke Brise Soleil at the Milwaukee Art Museum. Architect: Calatrava (*Source* http://www.flickr.com/photos/crazyegg95/ under CC BY-ND 2.0 license)

5.3.2 Function

5.3.2.1 Function: Material

Building materials and natural systems have a long common history. In the first human settlements, building envelopes were made from locally available materials such as animal skins, timber or bamboo. This connection was basically lost with the advent of modern constructions made from concrete, steel and glass, but is now subject of a revival with the rediscovery of the sustainable virtues of bioclimatic or ancient vernacular architecture (Foruzanmehr and Vellinga 2011; Zhai and Previtali 2010), such as wind-catchers for natural ventilation in hot and arid climates (Montazeri and Azizian 2008), courtyard architecture (Taleghani et al. 2014) or the potential for building underground (Van Dronkelaar et al. 2014). Also in many augmented or composite construction materials, the connection between nature and building materials is re-established. A typical example of a nature inspired way of building are the green walls or roofs, which directly transform the use of greenery into a functional building material (Ottelé 2014). Vegetation as a seasonally adaptive solar shading strategy furthermore shows good potential for energy conservation as well as physiological and psychological performance benefits (Ip et al. 2010; Magnone and van der Linden 2014). Constructions enriched with biofibers and biocomposites such as hemp, flax or mycelium (Pacheco-Torgal 2014; Mayoral 2011) are other examples of construction materials with a direct link to nature. A pioneering new development is currently exploring the possibility of making building materials in a bottom-up way, using enzyme producing bacteria (Ednie-Brown 2013). In this biomanufacturing approach, the energy-intensive processes of making materials such as bricks or concrete are replaced by an environmentally friendly growing process. The developers argue that the only elements needed to produce the construction material, named bioMASON are: loose pieces of aggregate, enzyme producing bacteria, an amount of urea and an amount of calcium ions (Dosier 2011).

In addition to these materials, there are also applications of buildings which attain adaptability by directly incorporating living organisms in the building envelope construction, but do this while extending their functionality beyond what is found in the original application domain. An example is the bio-reactive façade of the BIQ house at the International Building Exhibition 2013 in Hamburg (Fig. 5.4). The BIQ house is equipped with façade-integrated bio-reactors—transparent containers with microalgae cultivated in the façade component. While growing, the algae in these bio-reactors serve multiple functions at the same time: they act as shading system, solar thermal collectors and absorb carbon dioxide. After the cultivation phase, the algae can be separated from the water, to be used as biomass with an expected energy gain of 30 kWh/m^2 per year (Wurm 2013).

Sweating polymers recently receive attention as a novel way for cooling buildings without energy consumption (Rotzetter et al. 2012). Perspiration, the way mammals release heat to their environment via moisture secretion is used as a guiding principle for this concept. The technology uses thermoresponsive hydrogels, which enable utilization of evaporative cooling. At a critical switching temperature of 32 °C, the material undergoes a phase transition and accompanying release of moisture in a sweating-like manner. Autonomous operation is ensured by regeneration of the system with rain water. Experiments in a reduced-scale set-up demonstrated an estimated energy-saving potential for cooling of up to 60 % (Rotzetter et al. 2012).

Worldwide, avian mortality due to collisions with building facades is estimated in the billions every year (Klem 2009), mainly because traditional windows appear invisible for birds. In the search for solutions to this problem, bio-inspired principles may form a valuable resource. The ultraviolet (UV) reflection pattern of spider webs makes them discernable for the birds' sensory system. It is proposed that window systems with similar UV reflecting properties can act as a significant deterrent for collisions, with little or no obstructed view for humans (Evans Ogden 2014). Commercial products based on this proposition are currently entering the market. The main challenge in making the technology successful is in finding a coating composition that reflects a sufficient amount of UV light, yet does not interfere with the windows' visual functioning. The importance of this issue is reflected by the fact that the LEED green building rating system recently introduced a pilot credit for bird-collision deterrence as part of the new biodiversity category (Foster 2011).

An iconic example of a more *indirect* bio-inspired functional mechanism is the self-cleaning capability of Lotus leaves. After studying the physical principles, this advanced functionality has successfully been transferred in the form of coatings for building cladding and fenestration systems (Pacheco-Torgal and Jalali 2011; Solga et al. 2007). The deposition of a special hydrophobic layer prevents dirt from building up because it gets washed away by the almost spherical droplets of water (Parkin and Palgrave 2005).

Fig. 5.4 Algae facade, BIQ house at the International Building Exhibition 2013 in Hamburg (*Source* http://www.flickr.com/photos/_bundjugend/ under CC BY 2.0 license)

5.3.2.2 Function: Component

Aero Dimm (Fig. 5.5) is a good example of a biological principle that is *directly* transformed into a building component (Gruber 2011a). The façade component is inspired by coloration of cephalopod molluscs, and works via a pneumatic chamber that changes its volume between two elastic membranes. By controlling air pressure inside the cavities, a colour change and corresponding impact on radiant heat exchange is achieved. Doing the adaptation in a smart way, it is possible to realise a significant positive contribution in the building's energy balance.

An *indirect* example at the component level is the artificial green façade conceived by Šuklje et al. (2013). The façade system consists of bionic leaves, which are made of photovoltaic cells attached to evaporative matrices. These components serve the dual purpose of increasing photovoltaic system efficiency and cooling of the air in the microclimate that surrounds the building. The technical manifestation of this envelope system is only distantly related to evapotranspiration of actual vegetation in green facades. Results from the experiments nevertheless show that the evaporative cooling effect of the bionic façade approaches its natural counterpart, while introducing a 6.6 % increase in energy production compared to PV leaves without evaporative cooling.

Bio Skin, the façade construction of the Sony Research and Development Center in Tokyo, Japan, is inspired by transpiration to reject heat in summer, but is, in contrast to the sweating polymers from Sect. 5.3.2.1. packaged in the form of a more technological approach. The system uses a network of porous ceramic tubes, mounted on the exterior surface of the building. The tubes are fed with rainwater that is collected from the roof and nearby surroundings. By means of evaporation, the system provides a cooling effect due to latent heat transfer. Bio Skin's goal not only is to reduce cooling load of the building, but also to act as a mitigation strategy for the urban heat island effect by cooling the surrounding air (Yamanashi and Hatori 2011).

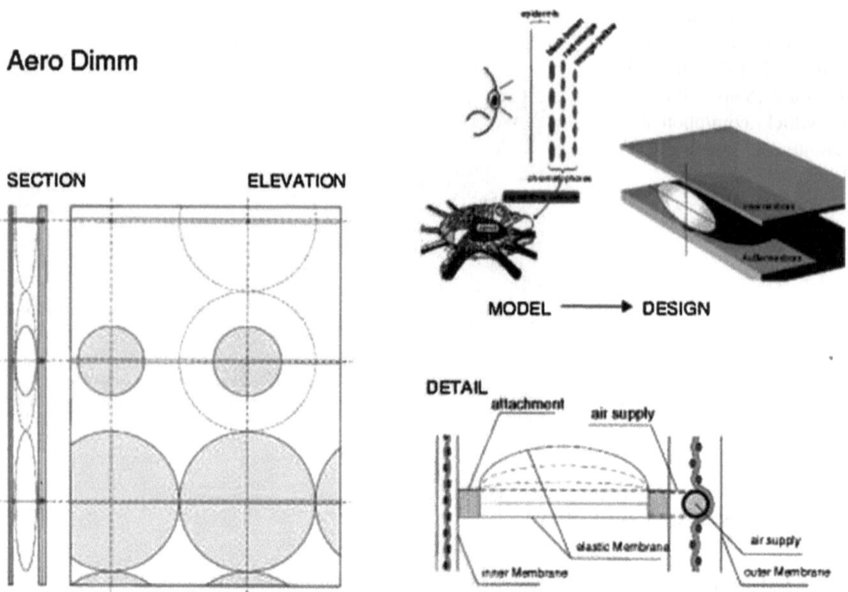

Fig. 5.5 Aero Dimm facade principle. (*Source* Gruber (2011a), reprinted with permission)

The internal vascular networks, found in most homeothermic living organisms, form the basis for an innovative adaptive window system, developed at the Wyss Institute for Biologically Inspired Engineering at Harvard University (Hatton et al. 2013). The window is supplemented with an array of very thin transparent water-carrying channels to act as microfluidic heat exchange layer for thermal control of the surface (Fig. 5.6). By adjusting the flow rate through the system, a dynamic thermoregulation effect is achieved. The researchers report a cooling effect of 7–9 °C at a relatively low flow rate, and expect a significant energy-saving potential with full-scale facade integration of the new technology (Hatton et al. 2013). A further iteration of this product aims at enhanced heat capacity, by enriching the fluid with phase change materials (PCM) in the form of nanoparticles (Alston 2014).

Despite the growing number of documented examples in the built environment, flexible, kinetic structures are still more frequently found in nature than in buildings. Especially the movement observed in plants offers a rich source of inspiration for technology transfer to buildings. Schleicher et al. (2014) propose a methodology for integrating these principles into new components for adaptive building skins, and demonstrate the potential with three case studies. One of these is Flectofin®, a new type of lamella system that provides adaptive solar shading through elastic deformation of flexible slats (Lienhard et al. 2011). In a slightly abstracted form, the full-scale integration of this type of bending technology has already been demonstrated: the "One Ocean" Pavilion at EXPO 2012 in Yeosu, South Korea (Knippers and Speck 2012).

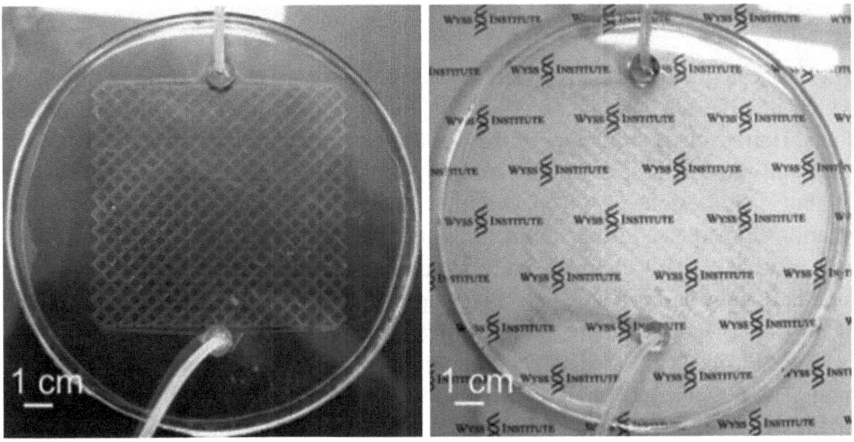

Fig. 5.6 Artificial vascular window system having a 10 × 10 cm^2 array of 1 mm wide channels, dry (*left*) and filled with water (*right*) (*Source* Hatton et al. (2013), reprinted with permission)

5.3.2.3 Function: Façade

Of the different scales and levels of abstraction discussed in this chapter, functional bio-inspiration for adaptation at the façade level is the most popular one. Probably this is because of the direct analogy that can be made with the skins of living organisms. Numerous biological principles have been investigated, with mixed success, for application in buildings, such as: animal fur (Webb et al. 2011, 2013), reptile skin (Zare and Falahat 2013), tree bark (Yowell 2011), bird feathers (Taghizade and Taraz 2013), plant species (Rezaei and Zare 2011), seeds (Fernandèz et al. 2013) or the hollow structure of polar bear hair (Stegmaier et al. 2009).

An exemplary case in this category is based on the optics of animal eyes and their responsiveness to light (Park and Dave 2013). The optical composition of reflecting superposition eyes is converted into a dynamic façade structure with square tube structures, embedded with optical systems. When integrated into the roof of large buildings, the technology is able to capture, concentrate and distribute natural daylight, to reach optimal lighting conditions and reduced energy consumption.

In terms of solar shading control for buildings, the Homeostatic façade, developed by DeckerYeadon, forms an interesting alternative compared to regular screens or blinds systems (Dahl 2013). The system regulates a building's indoor climate on the façade level by automatically and autonomously responding to environmental conditions. Through deformations of a shape-changing dielectric elastomer, the shading system expands under sunny conditions, and contracts when to allow for passive solar gains. The mechanism is modelled after the motion in muscles, and has an advantage over other systems because of its low power consumption and localized control (Minner 2011).

5.3.2.4 Function: Whole Building

The last category in this overview concerns examples of bio-inspired envelopes which involve not just the façade layer, but the whole building in three dimensions. Heliotrop, as a member of the class of rotating buildings, is a widely known and well-cited example in this group (Randl 2008). The building, located in Freiburg and designed by Rolf Disch architects mimics the directional growth or rotation in response to the sun in the same way as some plants species do (Spiegelhalter 2012). By choosing its orientation in a clever way, the building can modulate view and privacy, take advantage of solar gains when needed, or protect against overheating if there is an overheating risk.

The "Earth, Wind & Fire" building concept (Fig. 5.7) seeks analogy with the natural renewable potential of geothermal, wind and solar energy (Bronsema 2013a). In this climate-repsonsive, whole-building concept, the building skin plays a key role in creating a comfortable indoor environment. The gist of the design concept is in the thorough integration of building physics and HVAC functions, with the aim of obviating the use fossil fuel for fans and cooling machines. "Earth, Wind & Fire" consists of three main components for conditioning of ventilation air and creating sufficient pressure differences: (i) a venturi-shaped roof for enhancing the supply air flow potential for natural ventilation (Van Hooff et al. 2011), (ii) a gravity-actuated heat exchanging façade with mist spraying system for evaporative cooling, named Climate Cascade and (iii) a solar chimney for harvesting solar energy and inducing a buoyancy-driven ventilation flow. When integrated with appropriate thermal energy storage systems, a simulation study shows that the ambition of net positive energy building operation is within reach (Bronsema 2013a).

5.4 Design Support Tools

Biomimicry as a discipline is growing in the field of construction and architecture. As evidenced by the wide range of selected application examples presented in this chapter, there is already a considerable number of bio-inspired adaptable facade concepts making its way into practice. It is argued, however, that in many cases, the label 'bio-inspiration' was attached to the building envelope design in a rather ad-hoc and sometimes superficial way. Moreover, the bio-inspired premises on which these design concepts are based are not always fully understood, or do not always accurately reflect the biological principles. This mismatch is exemplified by the case of the well-known Eastgate centre in Harare, Zimbabwe. The building is said to be modeled after the climate-control features of termite mounds, but in fact is based on a flawed interpretation of how the biological system actually works (Turner and Soar 2008). Although not necessarily a bad thing in itself, this finding can be thought of as exemplary for the gap in domain knowledge that exists between building designers and biologists.

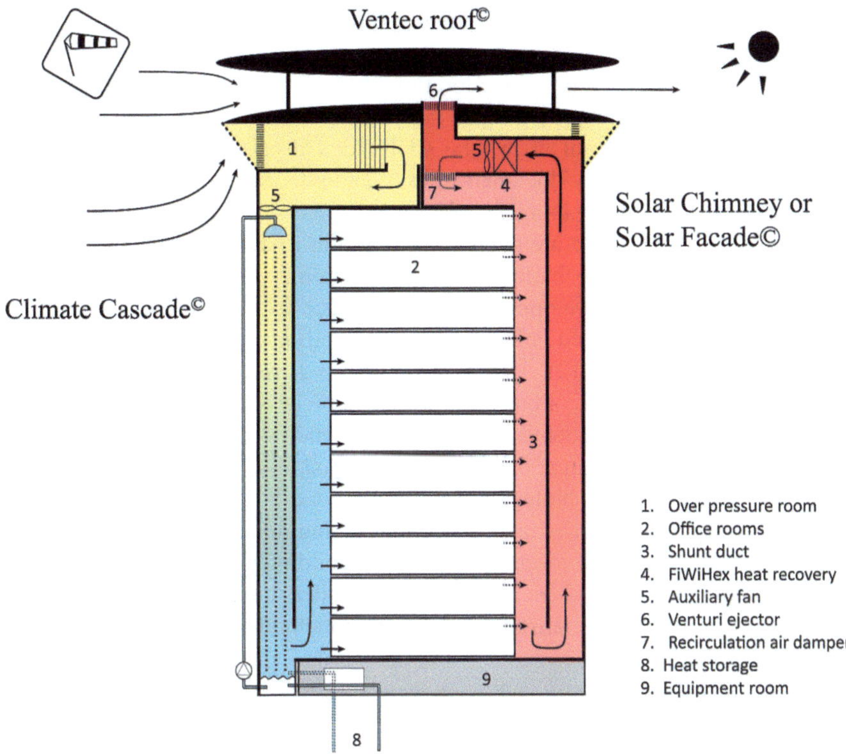

Fig. 5.7 Natural air-conditioning principles of Earth, Wind and Fire. (*Source* Bronsema (2013b), reprinted with permission)

In order to come to a more profound integration of biomimicry and architectural design, with potentially a higher impact on sustainable living, there is a need to make the translation process from nature to building envelope technologies more rigorous, systematic, and rational (Badarnah 2014). In literature (Badarnah 2012; Gamage and Hyde 2012; Pedersen Zari 2010; Knippers and Speck 2012; Mazzoleni 2010; Gruber 2011b), several barriers that currently hinder the transfer of relevant biological solutions have been identified:

- Difficulties with access to information that helps in narrowing down the enormous space of solution strategies found in nature;
- The sometimes large analogical distance between biology and building design and the lack of cross-domain knowledge;
- Conflicts between different requirements for aesthetics and functionality;
- Scaling issues to bridge the gap from small-scale observations to design principles on the human or whole-building level.

With these principles in mind, recently, a number of design methodologies and support tools have been proposed to stimulate the development of biomimetic building design.

On the one hand, these activities concentrate on categorization and organization. Noteworthy in this respect is the outcome of the Austrian BioSkin project (Gosztonyi 2011). After a broad exploration phase, 240 organisms with potential for functional translation into facade systems were identified. This led to the extraction and classification of in total 43 biological principles for the building skin, which are extensively documented in a database that is available for public perusal online (http://www.bionicfacades.net).

On the other hand, there are approaches that aim to develop design methodologies, capable of guiding building designers through the entire process from exploration to concept development. With reference to adaptable building envelopes, the recently completed PhD research by Lidia Badarnah at TU Delft is worth mentioning (Badarnah 2012). Her work develops a selective, *living envelope,* methodology for generation of bio-inspired adaptable facade concepts that creates a proper representation of biophysical information, and makes this more accessible for architects. The main feature of the exploration and investigation platform is that it offers a bridge between biological information and building envelope requirements, through clustering, abstracting and identifying dominant solution paths.

In bringing the field of bio-inspired adaptive building skin design forward in a way that transcends beyond the level of poorly understood, or simple metaphors, the application of such specialized design support methodologies will undoubtedly be of great value.

References

Aldersey-Williams H (2004) Towards biomimetic architecture. Nat Mater 3(5):277–279. doi:10.1038/nmat1119

Alston ME (2014) Energy adaptive glass matter. Architectural Eng Technol 3:115. doi:10.4172/2168-9717.1000115

Aries MBC, Aarts MPJ, van Hoof T (2014) Daylight and health: a review of the evidence and consequences for the built environment. Lighting Res Technol (in press) doi:10.1177/1477153513509258

Badarnah L, Kadri U (2014) A methodology for the generation of biomimetic design concepts. Arch Sci Rev (in press) doi:10.1080/00038628.2014.922458

Badarnah L (2012) Towards the LIVING envelope: biomimetics for building envelope adaptation. PhD Thesis, Delft University of Technology

Bakker LG, Hoes-Van Oeffelen ECM, Loonen RCGM, Hensen JLM (2014) User satisfaction and interaction with automated dynamic facades: a pilot study. Building Environ 78:44–52

Baetens R, Jelle BP, Gustavsen A (2010) Properties, requirements and possibilities of smart windows for dynamic daylight and solar energy control in buildings: a state-of-the-art review. Solar Energ Mater Solar Cells 94(2):87–105. doi:10.1016/j.solmat.2009.08.021

Bastiaansen CWM, Schenning A, Debije MD, Broer DJ (2013) Nano-textured polymers for future architectural needs. J Facade Des Eng 1(1–2):97–104. doi:10.3233/FDE-130002

Beatley T (2011) Biophilic cities—integrating nature into urban design and planning. Island Press, Washington DC

Braun DH (2008) Bionisch Inspirierte Gebäudehüllen. PhD Thesis, Stuttgart University

Bronsema B (2013a) Earth, wind & fire—natural air conditioning. PhD Thesis, Delft University of Technology
Bronsema B (2013b) Earth, wind & fire—natural airconditioning [1] research objectives and methods. In: Proceedings of CLIMA 2013—the 11th REHVA World congress. Prague, Czech Republic
Brownell B (2010) Transmaterial 3: a catalog of materials that redefine our physical environment. Princeton Architectural Press, New York
Chalupnik MJ, Wynn DC, Clarkson PJ (2013) Comparison of Ilities for protection against uncertainty in system design. J Eng Des 24(12):814–829. doi:10.1080/09544828.2013.851783
Chen PY, McKitrrick J, Meyers MA (2012) Biological materials: functional adaptations and bioinspired designs. Prog Mater Sci 57(8):1492–1704
Dahl R (2013) Cooling concepts: alternatives to air conditioning for a warm World. Environ Health Perspect 121(1):18–25
Dawson C, Vincent JFV, Rocca AM (1997) How pine cones open. Nature 390(1997):668
De Wilde P, Tian W (2010) Predicting the performance of an office under climate change: a study of metrics, sensitivity and zonal resolution. Energ Build 42(10):1674–1684. doi:10.1016/j.enbuild.2010.04.011
Dosier GK (2011) Methods for making construction material using enzyme producing bacteria. US Patent: US 20110262640 A1
Drake S (2007) The third skin: architecture, technology & environment. UNSW Press, Sydney
Ednie-Brown P (2013) bioMASON and the speculative engagements of biotechnical architecture. Architectural Des 83(1):84–91. doi:10.1002/ad.1529
Evans Ogden LJ (2014) Does green building come up short in considering biodiversity?: focus on a growing concern. Biosci 64(2):83–89
Ferguson S, Siddiqi A, Lewis K, De Weck O (2007) Flexible and reconfigurable systems: nomenclature and review. In: Proceedings of ASME 2007—international design engineering technical conferences and computers and information in engineering conference, Las Vegas
Fernandez ML, Rubio R, Gonzalez SM (2013) Architectural envelopes that interact with their environment. In: Proceedings of new concepts in smart cities: fostering public and private alliances (SmartMILE), 2013
Fisk WJ, Rosenfeld AH (1997) Estimates of improved productivity and health from better indoor environments. Indoor Air 7:158–172
Foruzanmehr A, Vellinga M (2011) Vernacular architecture: questions of comfort and practicability. Build Res Inf 39(3):274–285. doi:10.1080/09613218.2011.562368
Foster JM (2011) A reward for bird-friendly buildings. NY times blog: green—energy, the environment and the bottom Line. Accessed 2 Nov 2011
Gamage A, Hyde R (2012) A model based on biomimicry to enhance ecologically sustainable design. Architectural Sci Rev 55(3):224–235. doi:10.1080/00038628.2012.709406
Geiger J (2010) The living: surface tensions. Architectural Des 80(3):60–65
Gosztonyi S (2011) BioSkin—Bionische Fassaden: Potenziale Aus Der Bionik Für Adaptive Energieeffiziente Fassaden Der Zukunft. In: Proceedings of the 17th international Holzbau-forum, pp 1–14
Gruber P (2011a) Biomimetics in architecture [Architekturbionik]. In: Gruber P, Bruckner D, Hellmich C, Schmiedmayer HB, Stachelberger H, Gebeshuber IC (eds) Biomimetics—materials, structures and processes, Berlin, Heidelberg, pp 127–148. doi:10.1007/978-3-642-11934-7
Gruber P (2011b) Biomimetics in architecture—architecture of life and buildings. Springer, Vienna
Hatton BD, Wheeldon I, Hancock MJ, Kolle M, Aizenberg J, Ingber DB (2013) An artificial vasculature for adaptive thermal control of windows. Solar Energ Mater Solar Cells 117(October):429–436. doi:10.1016/j.solmat.2013.06.027
Hoes P, Trcka M, Hensen JLM, Hoekstra Bonnema B (2011) Investigating the potential of a novel low-energy house concept with hybrid adaptable thermal storage. Energ Conver Manage 52(6):2442–2447

Hofman M, Dujardin B (2008) European patent EP1644591: balcony that can be folded
Holmes M, Hacker J (2007) Climate change, thermal comfort and energy: meeting the design challenges of the 21st century. Energ Build 39(7):802–814. doi:10.1016/j.enbuild.2007.02.009
International Energy Agency (2012) Energy technology perspectives 2012—pathways to a clean energy system. International Energy Agency, Paris
Ip K, Lam M, Miller M (2010) Shading performance of a vertical deciduous climbing plant canopy. Build Environ 45(1):81–88. doi:10.1016/j.buildenv.2009.05.003
Jin Q, Overend M (2014) A prototype whole-life value optimization tool for façade design. J Build Perform Simul 7(3):217–232. doi:10.1080/19401493.2013.812145
Kasinalis C, Loonen RCGM, Cóstola D. Hensen JLM (2014) Framework for assessing the performance potential of seasonally adaptable facades using multi-objective optimization. Energ Build 79:106–113
Kellert SR, Heerwagen J, Mador M (2011) Biophilic design: the theory, science and practice of bringing buildings to life. John Wiley and Sons, New Jersey
Klem D (2009) Preventing bird—window collisions. Wilson J Ornithol 121(2):314–321
Klooster T, Boeing N, Davis S, Seeger A (2009) Smart surfaces: and their application in architecture and design. Birkhäuser, Basel
Knaack U, Klein T, Bilow M, Auer T (2007) Façade—principles of construction. Birkhäuser Architecture, Berlin
Knippers J, Speck T (2012) Design and construction principles in nature and architecture. Bioinspiration Biomimetics 7(1):015002. doi:10.1088/1748-3182/7/1/015002
Koch K, Bhusan B, Barthlott W (2009) Multifunctional surface structures of plants: an inspiration for biomimetics. Prog Mater Sci 54(2):137–178. doi:10.1016/j.pmatsci.2008.07.003
Leydecker S, Kölbel M, Peters S (2008) Nano materials in architecture, interior architecture, and design. Birkhäuser, Basel
Lienhard J, Schleicher S, Poppinga S, Masselter T, Milwich M, Speck T, Knippers J (2011) Flectofin: a hingeless flapping mechanism inspired by nature. Bioinspiration Biomimetics 6(4):045001. doi:10.1088/1748-3182/6/4/045001
Linn C (2014) Kinetic architecture: design for active envelopes. Images Publishing, Australia
Loonen RCGM, Singaravel S, Trcka M, Cóstola D, Hensen JLM (2014) Simulation-based support for product development of innovative building envelope components. automat Constr 45:86–95
Loonen RCGM (2014) Climate adaptive building shells. http://www.pinterest.com/CABSoverview/
Loonen RCGM, Trčka M, Hensen JLM (2011) Exploring the potential of climate adaptive building shells. In: Proceedings of building simulation 2011, pp 2148–2155
Loonen RCGM, Trčka M, Cóstola D, Hensen JLM (2013) Climate adaptive building shells: state-of-the-art and future challenges. Renew Sustain Energy Rev 25(September):483–493. doi:10.1016/j.rser.2013.04.016
Magnone G, van der Linden K (2014) Forest microclimates: investigating the performance potential of vegetation at the building space scale. Build Environ 73:12–23. doi:10.1016/j.buildenv.2013.11.012
Mayoral E (2011) Growing architecture through Mycelium and agricultural waste. Int J Construc Environ 1(4):87–132
Mazzoleni I (2010) Biomimetic envelopes. Disegnarecon 3(5):99–112
McLeod RS, Hopfe CJ, Kwan A (2013) An Investigation into future performance and overheating risks in Passivhaus Dwellings. Build Environ 70:189–209. doi:10.1016/j.buildenv.2013.08.024
Meijer F, Itard L, Sunnika-Blank M (2009) Comparing European residential building stocks: performance, renovation and policy opportunities. Build Res Inf 37(5–6):533–551. doi:10.1080/09613210903189376
Menges A, Reicher S (2012) Material capacity: embedded responsiveness. Architectural Des 82(2012):52–59
Minner K (2011) Moving homeostatic facade preventing solar heat gain. Archdaily. http://www.archdaily.com/?p=101578

Mlecnik E, Schütze T, Jansen SJT, de Vries G, Visscher HJ, van Hal A (2012) End-user experiences in nearly zero-energy houses. Energ Build 49:471–478. doi:10.1016/j.enbuild.2012.02.045
Montazeri H, Azizian R (2008) Experimental study on natural ventilation performance of one-sided wind catcher. Build Environ 43(12):2193–2202
Montazeri H, Blocken B, Janssen WD, van Hooff T (2013) CFD evaluation of new second-skin facade concept for wind comfort on building balconies: case study for the Park Tower in Antwerp. Build Environ 68:172–179. doi:10.1016/j.buildenv.2013.07.004
Moonen P, Defraeye T, Dorer V, Blocken B, Carmeliet J (2012) Urban physics: effect of the micro-climate on comfort, health and energy demand. Front Architectural Res 1(3):197–228. doi:10.1016/j.foar.2012.05.002
Ochoa CE, Aries MBC, van Loenen EJ, Hensen JLM (2012) Considerations on design optimization criteria for windows providing low energy consumption and high visual comfort. Appl Energ 95:238–245. doi:10.1016/j.apenergy.2012.02.042
Ottelé M (2014) Green facades and roofs. In: Pacheco-Torgal F, Labrincha JA, Diamanti MV, Yu CP (eds) Biotechnologies and biomimetics for civil engineering, Springer, Berlin
Pacheco-Torgal F (2014) Eco-efficient construction and building materials research under the EU framework programme Horizon 2020. Constr Build Mater 51(January):151–162. doi:10.1016/j.conbuildmat.2013.10.058
Pacheco-Torgal F, Jalali S (2011) Nanotechnology: advantages and drawbacks in the field of construction and building materials. Constr Build Mater 25(2):582–590. doi:10.1016/j.conbuildmat.2010.07.009
Park JJ, Dave B (2013) Bio-Inspired responsive façades. In: Proceedings of the 2nd central european symposium on building physics. Vienna, Austria
Parkin IP, Palgrave RG (2005) Self-cleaning coatings. J Mater Chem 15:1689–1695
Pedersen Zari M (2010) Biomimetic design for climate change adaptation and mitigation. Architectural Sci Rev 53(2):172–183. doi:10.3763/asre.2008.0065
Quesada G, Rousse D, Dutil Y, Badache M, Hallé S (2012) A comprehensive review of solar facades. Opaque solar facades. Renew Sustain Energy Rev 16(5):2820–2832. doi:10.1016/j.rser.2012.01.078
Ramponi R, Angelotti A, Blocken B (2014) Energy saving potential of night ventilation: sensitivity to pressure coefficients for different European climates. Appl Energ 123:185–195. doi:10.1016/j.apenergy.2014.02.041
Randl C (2008) Revolving architecture: a history of buildings that rotate, Swivel, and Pivot. Princeton Architectural Press, New york
Reichert S, Menges A, Correa D (2014) Meteorosensitive architecture: biomimetic building skins based on materially embedded and hygroscopically enabled responsiveness. Comput-Aided Des (in press) doi:10.1016/j.cad.2014.02.010
Rezaei D, Zare M (2011) Plants in arid climate as a pattern for bionic architecture in behavioral viewpoint. Can J Environ Constr Civil Eng 2(6):141–146
Rotzetter ACC, Schumacher CM, Bubenhofer SB, Grass RN, Gerber LC, Zeltner M, Stark WJ (2012) Thermoresponsive polymer induced sweating surfaces as an efficient way to passively cool buildings. Adv Mater 24(39):5352–6. doi:10.1002/adma.201202574
Schleicher S, Lienhard J, Poppinga S, Speck T, Knippers J (2014) A methodology for transferring principles of plant movements to elastic systems in architecture. Comput-Aided Des. doi:10.1016/j.cad.2014.01.005
Silver MR, de Weck O (2007) Time-expanded decision networks: a framework for designing evolvable complex systems. Syst Eng 10(2):167–186
Singh A, Syal M, Grady SC, Korkmaz S (2010) Effects of green buildings on employee health and productivity. Am J Public Health 100(9):1665–8. doi:10.2105/AJPH.2009.180687
Solga A, Cerman Z, Striffler BF, Spaeth M, Barthlott W (2007) The dream of staying clean: lotus and biomimetic surfaces. Bioinspiration Biomimetics 2(4):S126–34. doi:10.1088/1748-3182/2/4/S02

Spiegelhalter T (2012) Plus-energy building—designing with exergy-entropy processes. In: Proceedings of PLEA2012 conference, opportunities, limits and needs towards an environmentally responsible architecture Lima, Perú

Stegmaier T, Linke M, Planck H (2009) Bionics in textiles: flexible and translucent thermal insulations for solar thermal applications. Philosophical Transactions. Ser A Math Phys Eng Sci 367(1894):1749–58. doi:10.1098/rsta.2009.0019

Šuklje T, Medved S, Arkar C (2013) An experimental study on a microclimatic layer of a bionic façade inspired by vertical greenery. J Bionic Eng 10(2):177–185. doi:10.1016/S1672-6529(13)60213-9

Taghizade K, Taraz M (2013) Designing a mobile facade using bionic approach. Am J Mater Eng Technol 1(2):22–29

Taleghani M, Tenpierik M, van den Dobbelsteen A (2014) Energy performance and thermal comfort of courtyard/atrium dwellings in the Netherlands in the light of climate change. Renew Energ 63:486–497

Trame U (2001) Santiago Calatrava: Quadracci Pavilion, Milwaukee Art Museum. Editrice Compositori, Bologna

Turner JS, Soar RC (2008) Beyond biomimicry: what termites can tell us about realizing the living building. In: Proceedings of the international conference on industrialized, intelligent construction (I3CON)

Ürge-Vorsatz D, Novikova A (2008) Potentials and costs of carbon dioxide mitigation in the world's buildings. Energ Policy 36(2):642–661. doi:10.1016/j.enpol.2007.10.009

Van Dronkelaar C, Cóstola D, Mangkuto RA, Hensen JLM (2014) Heating and cooling energy demand in underground buildings: potential for saving in various climates and functions. Energ Build 71:129–136. doi:10.1016/j.enbuild.2013.12.004

Van Hooff T, Blocken B, Aanen L, Bronsema B (2011) A venturi-shaped roof for wind-induced natural ventilation of buildings: wind tunnel and CFD evaluation of different design configurations. Build Environ 46(9):1797–1807

Van Renterghem T, Hornikx M, Forssen J, Botteldooren D (2013) The potential of building envelope greening to achieve quietness. Build Environ 61:34–44. doi:10.1016/j.buildenv.2012.12.001

Weaver J, Wood KL, Jensen D (2008) Transformation facilitators: a quantitative analysis of reconfigurable products and their characteristics. In: Proceedings of ASME 2008—international design engineering technical conferences & computers and information in engineering conference

Webb M, Hertzsch E, Green R (2011) Modelling and optimization of a biomimetic facade based on animal fur. In: Proceedings of building simulation 2011

Webb M, Aye L, Green R (2013) Investigating potential comfort benefits of biologically-inspired building skins. In: Proceedings of building simulation 2013, pp 2634–2641

Wigginton M, Harris J (2002) Intelligent skins. Butterworth-Heinemann, Oxford

Wurm J (2013) Developing bio-responsive façades: BIQ House—the first pilot project. Arup J 2013(2):90–95

Xie X, Liu CH, Leung DYC (2007) Impact of building facades and ground heating on wind flow and pollutant transport in street canyons. Atmos Environ 41(39):9030–9049. doi:10.1016/j.atmosenv.2007.08.027

Yamanashi T, Hatori T (2011) Bio skin urban cooling facade. Architectural Design 81(6):100–107

Yowell J (2011) Biomimetic building skin: a phenomological approach using tree bark as a model. MSc Thesis, University of Oklahoma

Zare M, Falahat M (2013) Characteristics of reptiles as a model for bionic architecture. Adv Civil Environ Eng 01(3):124–135

Zhai Z, Previtali JM (2010) Ancient vernacular architecture: characteristics categorization and energy performance evaluation. Energ Build 42(3):357–365. doi:10.1016/j.enbuild.2009.10.002

Chapter 6
A Green Building Envelope: A Crucial Contribution to Biophilic Cities

Marc Ottelé

Abstract Throughout history, greening of outside walls and roofs of buildings has taken place. Reasons for doing so were the increase of insulation (keep cool in summer and keep cold out in winter), improved esthetics, improved indoor and outdoor climate, adsorption of particulate matter (PM_x), as well as increasing ecological values by creating habitats for birds and insects. Green façades and living walls systems can improve the (local) environment in cities. They offer more surfaces with vegetation and, at the same time, contribute to the improvement of the thermal performance of buildings. Although in the past, relatively little attention has been paid to these valuable opportunities of vegetation and its interaction with buildings. More and more attention is shifted to these so-called beneficial relations in especially dense urban areas, which can be considered as deserts in biological terms. This movement from a biophilic perspective point of view includes combining nature and natural elements in the built environment to ameliorate the negative impact of climate change as for example loss of biodiversity, mitigation of urban heat, or air pollution reduction.

6.1 Introduction

Cities and urban environments contain a variety of ecological and green assets, efforts are being made to further enhance the green elements and features of these living and work environments at the building scale. Integrating the positive aspects of greenery inside urban environments is called: Biophilic urbanism (Beatley and Newman 2013). A biophilic designed city is more than a biodiverse city; it is a place that learns from nature and emulates natural systems, incorporates natural

M. Ottelé (✉)
Faculty of Civil Engineering and Geosciences, Department of Materials & Environment, Delft University of Technology, Delft, The Netherlands
e-mail: M.Ottele@tudelft.nl

F. Pacheco Torgal et al. (eds.), *Biotechnologies and Biomimetics for Civil Engineering*,
DOI 10.1007/978-3-319-09287-4_6

forms and images into its buildings and cityscapes, and designs and plans in conjunction with nature; it transform cities from gray and lifeless to green and biodiverse (Beatley 2008).

Biophilic design is an innovative way of designing urban areas where we live, work, and learn, creating healthy and productive habitats for city dwellers. It is based on the theory of "biophilia" which contends that human health and well-being has a biologically-based need to affiliate with nature.

While parks have often been a part of cities, architects and designers today are incorporating nature into their designs through a variety of innovations such as green roofs and vertical gardens, a renewed focus on local and natural materials, and reclamation or restoration of spaces (Derr and Lance 2012).

This chapter will show within an ecological engineering context the impact of green roofs and green façades in (dense) urban areas. An overview and comparison of different types of horizontal and vertical green for housing, industrial, and other commercial buildings will be given. Some concrete examples will be elaborated to show possibilities of their multifunctionality.

There is a growing body of evidence of the positive physical and mental health benefits associated with greenery and green elements in the built environment (Beatley and Newman 2013). Realization of vegetated roofs and façades finds more and more frequent application in the building sector; although for large scale application, there is in general, still hesitation among policy makers and designers. That is a great pity as financial details show that applications of green roofs are not (any) more expensive than for example traditional flat roofs (Bohemen et al. 2009; Köhler 2012).

Greening of outside walls or facades of buildings gains also more interest in recent years. Although these concepts are not new (in the eighties of the twentieth century, different reports and books have been published, research into the use of green inside cities increased substantially). In particular, the amount of publications, articles, and research focused on the use of green roofs and green façades has increased in recent years (Köhler 2008). Despite the interest (under city dwellers, architects, city planners, policy makers, and scientists) in a green building envelope with corresponding positive claims, hard data about the effect of urban green is sometimes missing or not well studied yet. However, nowadays the environmental impact of buildings on the inner and outer climate becomes more and more apparent.

Green buildings are designed to reduce the overall impact of the built environment on human health and the natural environment. Buildings in which we spend a great part of our life to protect us from nature's extremes, yet they also affect our health and environment in countless ways (EPA 2010). Green building strategies not only stand for sustainable materials in their construction (e.g., reused, recycled-content, or made from renewable resources), but also by using of natural processes (e.g., shading effect of trees, insulation capacities of green roofs and green façades, mitigation of urban heat due to evapotranspiration). The green building strategy in the presented chapter focuses on one key aspect of the "greening process" namely the use of plants on and around urban buildings.

6.2 Green Building Envelope Strategy

The resilience of cityscapes against climate change is predominantly determined by the properties of their surfaces and the spatial arrangement of the buildings. These factors induce the occurrence of urban heat islands or flooding (Scharf et al. 2013). When global radiation reaches a surface it may be reflected (Albedo) or transformed to sensible or latent heat flux. While plants are able to transform the sun energy into biomass, oxygen, and air humidity, regular building surfaces (e.g., plaster) emit sensible heat flux. Plants regulate the urban microclimate, while conventional surfaces lead to microclimatic extremes and reduce the thermal comfort within cities (Scharf et al. 2013).

To deal with problems in dense urban areas often one-sided solutions are chosen. With the increased focus on ecological impacts of human activities on our environment the attention is shifted more and more to integrated solutions. Ecological engineering principles and biophilic design, can contribute to integrated solutions as it is applied and multidisciplinary science; it is integrating human activities with the natural environment, so that both can have advantage of designing and refurbishing of constructions. Conservation and the development of biodiversity by utilization of biological processes are central in the designing process. Dense and paved cities need an appropriate development, which incorporates an ecological approach to building and landscape design with respect to link functions such as water management, air pollution reduction, energy conservation, the recycling of waste (water), and nature conservation (biodiversity).

One promising option for dense urban cities is the greening of buildings (Johnston and Newton 2004; Ottelé 2011; Perini 2012). By strategically adding a "green skin", it is possible to create a new network of vegetation as roofs, walls, courtyards, streets, and open spaces. These networks, also called stepping stones, are particularly important in the city centers where vegetation may cover only about one third of the land surface, compared with 75–95 % in the outer suburbs (Johnston and Newton 2004). In these areas, there is less biodiversity and a lack of breeding and nesting possibilities for animals, besides paved surfaces collect a lot of heat, which negatively contributes to urban heat.

Application of plants rooted in the soil at the base of façades or on roofs by many architects and landscape architects is indicative of the value placed upon their presence in the urban landscape (Laurie 1977). Structures covered with green are a symbol of building in harmony with architecture and nature (Lambertini 2007). The garden-city movement at the end of the nineteenth century may be seen as one of the first ecological reactions to industrialization in urban areas (Kaltenbach 2008).

The many systems available on the market allow combining nature and built space to improve the environmental quality in urban areas (Fig. 6.1), and to retrofit the wide building heritage (which is often unsuitable and cause of relevant energy waste and discomfort conditions) with respect to architectural, functional, and performance aspects (Novi 1999; Nuzzo and Tomasinsig 2008; Dunnett and Kingsbury 2004). It is an important field to investigate since data show that

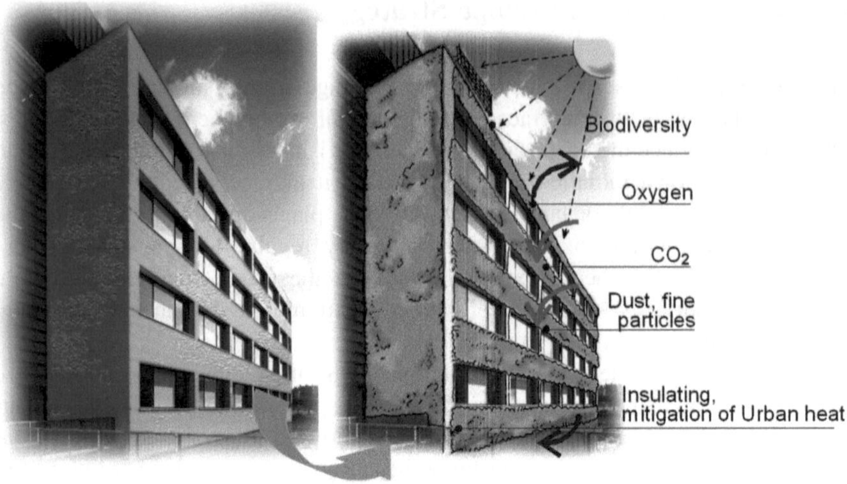

Fig. 6.1 A green building envelope strategy incorporates multidisciplinary environmental advantages for both city dwellers as nature

architecture plays an important role in the field of sustainability. In fact, the building sector has one of the greatest impacts on the environment; buildings consume a significant amount of energy over their life cycle and generate 40–50 % of the total output of greenhouse gases (Thormark 2002; Ardente et al. 2008; Prasad and Hill 2004).

A green building strategy offers the potential to learn from traditional architecture; the earliest form of vertical gardens dates from 2000 years ago in the Mediterranean region and ornamental roof gardens have been developed initially by the civilization of the Tigri and Euphrates River valleys (the most famous examples of which were the Hanging Gardens of Babylon in the seventh and eighth centuries B.C. (Köhler 2008; Dunnett and Kingsbury 2004). Several examples of green envelopes, back to eighteenth to nineteenth century, can be found in Northern European regions, such as climbing plants to shade vertical surfaces in Mediterranean regions, due to the cooling potential of vegetation and the insulation properties (thermal capacity). Nowadays, this kind of building envelope strategy also incorporates advanced materials and other technologies to promote sustainable building functions (Köhler 2008).

Greening the exterior of buildings (façades and roofs) provides numerous ecological and economic benefits, including storm water management, energy conservation, mitigation of the urban heat island effect, reducing air pollutants, increased longevity of building materials, as well as providing a more esthetically pleasing environment in which to work and live (Johnston et al. 2004; Dunnet and Kingsbury 2004; Getter and Rowe 2006; Minke and Witter1982; Krusche et al. 1982; Bohemen et al. 2009; Ottelé 2011; Perini 2012).

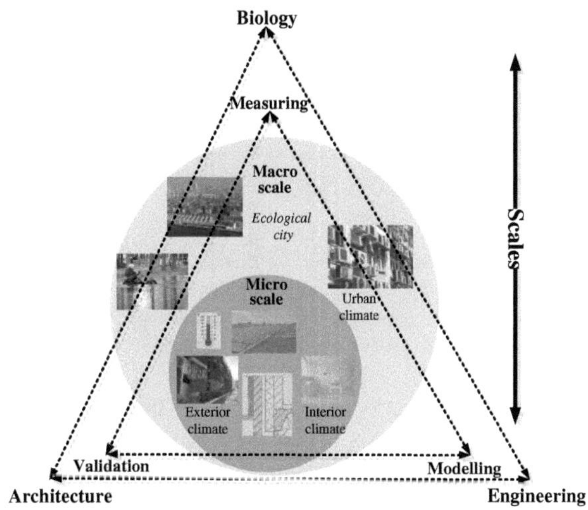

Fig. 6.2 Urban green and the application of it requires an holistic approach to encounter the negative impact of the build environment on (local) climate

The application of vegetation, especially in relation to the involved biological processes in urban areas, requires an holistic approach at different scales (Fig. 6.2), since the benefits operate at a range of different scales from the individual building, the domestic, or garden scale, to district and finally to the city scale (Beatly 2008). Different disciplines such as biology, architecture, and engineering comes together in this concept, as it should be applied, as an integral approach to optimize the efficiency of green structures and its surrounding.

Another important distinction can be made to private and public advantages. Private advantages should be found in the direction of energy savings, esthetic improvement, or for example extension of the life span of the waterproofing layer, while public advantages are associated with storm-water management, biodiversity, urban heat, and for example air pollution (Johnston and Newton 2004).

As mentioned earlier, green roofs and green facades are common techniques to implement urban greenery at the building level, and within these techniques there is a variety of concepts possible (Table 6.1).

Literature review done by Peck et al. (1999) and Köhler (2008) shows the benefit of green roofs on the inner temperature in buildings. Under a green roof

Table 6.1 Green building techniques for the urban area

Common urban green (techniques) at the building level			
Vertical green		Green roofs	
Extensive	Intensive	Living wall concepts	Green façade (traditional)
Thin substrate layer (5–15 cm). Vegetation mainly sedum, mosses, and herbs	Thicker soil layer (>25 cm). Vegetation mainly herbs, herbs, shrubs and trees	Modern technique based on grow able panels (mainly prefabricated), hydroponic system and nutrients needed	Consists out of self-adhesive creepers (tendrils)

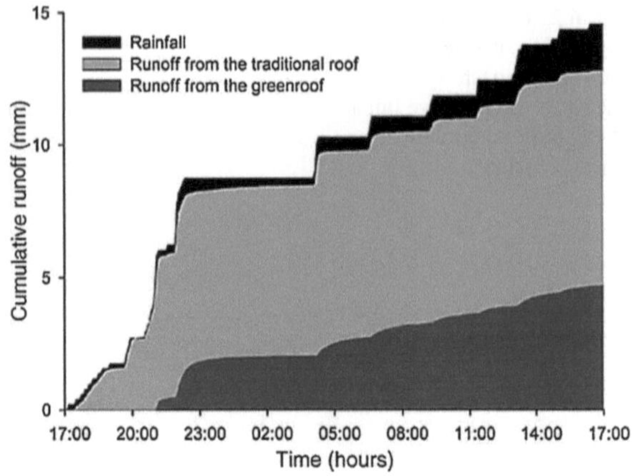

Fig. 6.3 Cumulative runoff as measured by Mentens et al. (2006) on an extensive green roof and a bare roof based on a 14.6 mm rain shower

without cooling, indoor temperatures were found at least 3–4 °C lower than the outdoor temperature, which was between 25 and 30 °C.

Green roofs can be used for the retention of water, and as a consequence, the sewer system can be tailored to lower the peak concentrations and to improve the water quality in the period of heavy rain. This subsequently can lead to savings on investments in sewer and water purification installations (Heidt and Neef 2008). The retention capacity can be between 60 and 100 % depending on the construction details of the green roof. According to Mentens et al. (2006) the reduction of the runoff consist in: delayment of the initial time of runoff due to adsorption of rainwater in the green roof system, reducement of the total runoff by retaining part of the rainfall, and by distributing the runoff over a longer period of time through a relative slow release of the excess water that is temporarily stored in the pores of the substrate. The effect of a green roof on the runoff compared to a traditional flat (bare) roof can be seen in Fig. 6.3. Green offers furthermore a variety of plants and animals; as a result of this many species can establish or maintain themselves in an urban environment. The biodiversity in cities is generally higher than in agricultural areas, but lower than in the rural area (Natuur balans 1999). The urban area offers a unique lodging to some specific types by the substrate (mostly brick, limestone, and masonry (Darlington 1981)) and the urban microclimate, such as wall vegetation and mosses. One of the characteristics that set a city apart from its rural surroundings is the altered climate that prevails over urban environments. Comparing rural areas with the urban areas differences can be found in solar input, rainfall patterns, and temperature.

The integration of buildings with vegetation, i.e., green roofs and vertical greening systems, is a constantly evolving research field and especially in the last decade a lot of technical developments are done. However, green envelopes (especially the most innovative vertical greening systems) are not yet fully accepted as an environmental quality restoration and energy saving method for the built environment, mainly due to the lack of data needed to quantify their effects, and to evaluate the real sustainability (environmental and economic) of these (Perini et al. 2012). The many greening systems available on the market allows combining nature and the built environment to improve the environmental quality in urban areas; for example, green façades and living wall systems offer more surfaces with vegetation and, at the same time, contributes to the improvement of the thermal performance of buildings.

As earlier stated, the largest uncovered surfaces in cities are rooftops, as these surfaces offer a great potential for urban agriculture. New York is an example where already many projects are established. The Brooklyn Grange, a rooftop farm business in Long Island (United States) is one of the largest of these projects. Also in the Netherlands a few projects exist. De Dakakker in Rotterdam and Zuidpark in Amsterdam are the most successful examples. Urban agriculture is a nice example to educate city dwellers and to connect them again with the whole food production chain (Dakakkers 2013). Urban agricultural projects are also a tool to stimulate social cohesion between citizens (Farming the city 2013).

6.3 Air Quality Improvement with Vegetation

All plants will help to ameliorate the effects of air pollution. This can be done at the microclimatic scale, but in the case of many green structures also at larger scale. Leaves of plants provide a large surface area (Fig. 6.4a, b), which is capable of filtering out particulate matter (PM_x) and other pollutants such as NO_x (conversion to nitrate (NO_3) and nitrite (NO_2)) and CO_2 in daytime. A green façade will block the movement of particulate matter particles along the side of a building and filter them (Minke and Witter 1982). Vegetation has a large collecting surface area and promotes also vertical transport by enhancing turbulence (Fowler et al. 2001;

Fig. 6.4 **a** Dust on European Holly (*Ilex aquifolium*) leaves near an unpaved road. **b** Micrograph (ESEM) of fine dust on common ivy (*Hedera helix*) leaf

Beckett et al. 2004). When concrete, brick, stone, glass, and asphalt surfaces are heated during the summer period, vertical thermal air movements (upward) are created and dust particles found on the ground are carried and spread into the air (Minke and Witter 1982). Particulate matter is adsorbed by the leaves, trunks, and twigs (Fig. 6.3a, b), and is an efficient sink for particulate matter (Fowler et al. 1989).

According to Hosker and Lindberg (1982) fine dust ($PM_{2.5}$ and PM_{10}) concentrations are reduced when particles are adhered to the leaves and stems of plants. Literature claims that by rainfall the adsorbed particulate matter is washed off into the soil or substrate below. However, results from a conducted simulated rainfall experiment (Ottelé 2011), shows that especially the fine and ultrafine particles are fixed on the leaf surface. Also falling of leaves in autumn contributes to particle binding. Research shows for example that plant barriers immediately along a roadside (daily traffic level 20.000–50.000 vehicles) are more beneficial in capturing lead (Pb) and cadmium (Cd) particles than plants investigated in the rural area (Bussotti et al. 1995).

Also Thönnessen (2002) found heavy metal concentrations and fine particles on leaves of a green façade (*Parthenocissus tricuspidata*) in the inner city of Düsseldorf (daily traffic level 12.500 vehicles). Sternberg (2010) found the same results by comparing *ivy* leaves from different sites (by counting particles on ivy leaves), the leaves from the sites exposed to a high daily traffic level, had collected a significant number of particles compared to the sites that are less exposed.

Research at the university of Bonn shows that mosses for example are excellent fine dust absorbers (Frahm and Sabovljevic 2007). One square meter of moss can "consume" 20 g of fine dust each year. Since moss is one of the easiest plants to use, for example, on a green roof, is an enormous advantage and cheap method against (local) air pollution.

Besides particle binding plants are also known to absorb gaseous pollutants through the stomata (CO_2 and NO_x). Via photosynthesis CO_2 is sequestered in the leaves (Minke and Witter 1982). The negative health effects of particulate matter pollution for human's stands for decreasing lung functions, increased respiratory problems, and other health care visits for respiratory and cardiovascular diseases (Pope et al. 2009).

Besides these effects also durability problems are involved and include accelerated corrosion of metals, as well as damage to paints, sculptures, and soil-exposed surfaces on man-made structures (United Nations 2007). The improved air quality by a green envelope has direct benefits for people who suffer a long disease. A decrease of smog formation will occur, and also durability or corrosion problems are reduced of urban infrastructure that is susceptible to damage from air pollution (United Nations 2007).

A study carried out by the University of Dresden (Schröder 2009) with regard to the organic balance of a greened façade with 1,000 m^2 *Hedera helix* pointed out that in one year: 1,019 kg of water and 2,351 kg of CO_2 is consumed and bound, respectively. In this reaction, 5,854 kg of organic mass (water content 4,409 kg and dry mass 1,415 kg) and 1,712 kg of O_2 is produced. With the assumption of an leaf area index (ratio between leaf surface in m^2 and covered wall surface in m^2)

for *Hedera helix* of 2.6 up to 7.7 m^2 leaf/m^2 wall (Bartfelder and Kohler 1987a, b), a leaf surface area of 2,600 up to 7,700 can be calculated.

Measurements carried out by Rath and Kießl (1989) on the effect of green façades on the SO_2 concentration show that the concentration of SO_2 was clearly lower between the foliage than in front of a non-greened façade.

Field measurements conducted by a national research program in the Netherlands (IPL 2006) to investigate the effect of a vegetation corridor on the reduction of PM_{10} levels near a highway (A50), show a minor contribution of the vegetation corridor on the concentration levels measured in the ambient air. They estimated the effect of vegetation smaller than 10–31 % on the traffic contribution of particulate matter, due to the high uncertainty of the used measuring equipment.

6.4 Temperature Regulation and Insulating Properties Due to a Vegetation Layer

Buildings consume roughly 36 % of total energy use and 65 % of the total electricity consumption. Kula (2005) suggests that a wide scale green roof implementation could significantly impact energy savings. According to Dunnet and Kingsbury (2004) every decrease of the internal building temperature with 0.5 °C may reduce the electricity use with 8 % for air conditioning in summer periods. Akabari et al. (2001) concluded that since 1940 the temperatures in urban areas have been increased by 0.5–3 °C. Akabari et al. (2001) also estimated that 5–10 % of the current electricity demand of cities is used to cool buildings just to compensate the 0.5–3 °C increased temperature.

Green roofs, living walls (LWS), and green façades create their own specific microclimate, quite different from surrounding conditions. Due to this specific microclimate, both around the building and at grade are affected. Depending on height, orientation, and the location of surrounding buildings, the façade is subjected to extreme temperature fluctuations (hot during the day and cool at night), with constant exposure to sunlight and wind. The climate on a roof or at a façade is comparable with an arid or alpine climate, and only suitable to specific types of plants. Most of the Sun's radiation that is adsorbed by concrete, bituminous materials, or masonry is reradiated as sensible heat. Asphalt, concrete, and masonry will reflect 15–50 % of the received radiation (Laurie 1977), greening paved surfaces with vegetation to intercept the radiation before it can hit hard surfaces can reduce the warming up of hard surfaces, especially in dense urban areas.

A study in Toronto carried out by Liu and Baskaran (2003) shows clearly the temperature difference between a bare and green roof. Temperatures up to 70 °C were measured for the bare roof, whereas the temperature on the green roof remains around 25 °C. The consequence of this is that the roof membrane (bituminous material) of the bare roof ages must faster due to UV light compared to a green roof, which is protected by the substrate and plant layer. Since UV light deteriorates material and mechanical properties of coatings, paints, plastics, etc., plants will also have an effect on durability aspects. This is a beneficial side effect, which will have a cost-effective effect on maintenance costs of buildings. The denser and thicker the plant layer on the green façade, the more beneficial these effects are. As for example, life expectancy of bare roofs is in general 15 years, whereas for green roofs this is up to 40 years.

Finally, due to the adsorption of heat during the day, the bare roof will reradiate the adsorbed heat during night contributing to the urban heat island effect(Eumorfopoulou and Aravantinos 1998). In an urban heat island effect situation, even night air temperatures are warmer because of built surfaces adsorb heat and radiate it back during the evening hours (Getter and Rowe 2006). Covering a building with vegetation prevents solar radiation from reaching the building skin (shading effect of leaves), and in the winter, the internal heat is prevented from escaping. By constructing green façades and green roofs, great quantities of solar radiation will be adsorbed for the growth of plants and their biological functions. Between façade and a dense vertical green layer (for both rooted in the subsoil as rooted in artificial soil-based systems) a stagnant air layer exist. Stagnant air has an insulating effect; green façades can therefore serve as an "extra insulation" of the building façade (Minke and Witter 1982; Krusche et al. 1982; Ottelé 2011). Also direct sunlight on the façade is blocked by the vegetation. This blocking of the sunlight ensures that the temperature will be less high inside a house. In winter, the system works the other way round, and heat radiation of the exterior walls is insulated by evergreen vegetation. In addition a dense foliage will reduce the wind speed along the façade, and thus also helps to prevent that the walls will cool.

The insulation value of vertical greened surfaces can be increased basically by different mechanisms (Peck et al. 1999; Rath and Kießl 1989; Pérez et al. 2011):

- by covering the building with vegetation, the summer heat is prevented from reaching the building skin (shadow), and in the winter, the internal heat is prevented from escaping, reflected, or absorbed.
- thermal insulation provided by vegetation, substrates, and configuration (mostly related to living wall concepts).
- by trapping an air layer within the plant foliage, since wind decreases the energy efficiency of a building by 50 %, a plant layer will act as a buffer that keeps wind from moving along a building surface.
- cooling of air due to evapotranspiration of plants and substrates (if used).

Green façades and roofs will cool local air temperatures in two different ways. As explained, first of all, walls behind greened surfaces absorb less heat energy from the sun (traditional façade and roof surfaces will heat up the air around them). This effect is clearly visible in Fig. 6.5a, b where uncovered parts of the façade are heated up (color red) and the parts covered with leaves considerable lower (color blue and green). Secondly, green façades and roofs will cool the heated air through evaporation of water (Wong et al. 2009) (for evaporation of 1 kg water, 2.5 MJ of energy is necessary); this process is also known as evapo-transpiration. Besides, hard surfaces encourage the runoff of rainwater into the sewage system. In urban

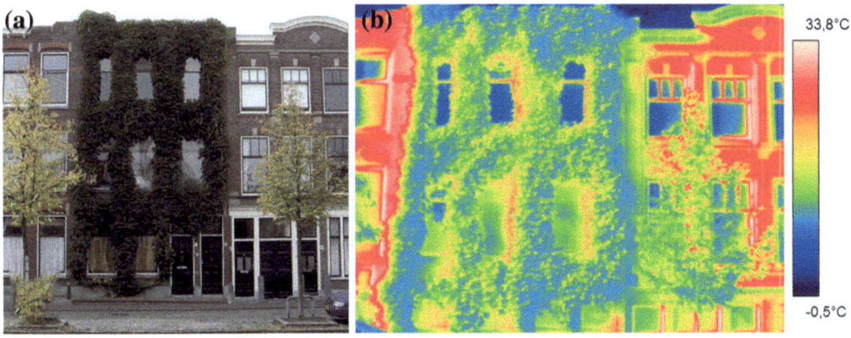

Fig. 6.5 **a** Boston ivy (*Parthenocissus*) rooted in the soil and applied directly against the façade in Delft summer 2009. **b** Photograph taken of the same façade with an infrared camera (FLIR) with ambient air temperature 21 °C

areas, the impact of evapotranspiration and shading of plants can significantly reduce the amount of heat that would be reradiated by façades and other hard surfaces. Plants buffer water on their leaf surfaces longer than building materials, and the processes of transpiration and evaporation, can add more water into the air. The result of this is a more pleasant (micro)climate in the urban area.

Field measurements performed by Bartfelder and Köhler (1987a, b) show a temperature reduction at the green façade in a range of 2–6 °C compared with a bare wall. Holm (1989) shows with field measurements and his DEROB computer model the thermal improvement potential of leaf covered walls. Also Eumorfopoulou and Kontoleon (2009) reported the temperature cooling potential of plant covered roofs and walls in a Mediterranean climate; the effect was up to 10.8 °C. Another study by Wong et al. (2009) on free standing walls in Hortpark (Singapore) with vertical greening types shows a maximum reduction of 11.6 °C. Also Ottelé (2011) shows that especially with living wall concepts high temperature reduction can be achieved, resulting in better insulation values.

Perini et al. (2011) show the influence of a green layer on the reduction of the wind velocity along the surface of a building. An extra stagnant air layer in optimal situations can be created inside the foliage, so that when the wind speed outside is the same as inside R_{exterior} can be equalized to R_{interior}. In this way, the building's thermal resistance can be increased by 0.09 m^2 K W^{-1}.

These results refer to the wind speed measured at a façade covered by a well grown direct greening system and a living wall system based on planter boxes; in the case of living wall systems the insulation properties change according to the materials used. The thermal resistance of a living wall system based on planter boxes is also influenced by the wind reduction, besides the thermal resistance of the system itself contributes to the thermal resistance and is estimated up to $R = 0.52$ m^2 K W^{-1}.

For both green façades and living wall systems these results imply potential energy savings for building envelopes in warmer and colder climates (Perini et al. 2011; Ottelé 2011). This "technical/thermal green" strategy of increasing exterior insulation properties of vertical surfaces stimulates upgrading or retrofitting of existing (under-insulated) façades without the added cost of interior or traditional exterior insulation systems.

An experimental research conducted by Ottelé (2011) was set up to in order to classify the thermal benefits of green façades or plant covered cladding systems under boundary conditions in a so-called hotbox testing facility. For this reason, an insulated (mineral wool) cavity wall with different (attached) vertical greening systems was built and tested in order to distinguish the thermal effect of the green systems. In total, there were two measurements performed with *Hedera helix* (direct and indirect to the wall) and four measurements were carried out with living wall systems (based on felt layers, planter boxes, mineral wool, and foam substrate.

In the study, it was found that, both for the direct and indirect greening principle lower surface temperatures of the exterior masonry were measured during summer conditions compared to the bare wall situation. The difference of temperature for the systems is reaching 1.7 and 1.9 °C, respectively, after 8 h of heating. The insulation material inside the bare wall moderates the prevailing temperature difference between the outside and inside climate chamber, resulting in no temperature difference for the inside climate chamber. The winter measurement after 72 h shows that the wall surface covered directly with *Hedera helix* is warmer compared to the bare wall, with a temperature difference of 1.7 °C. The air temperature of the inside climate chamber is lowered with 0.7 °C in the case of the bare wall, which means that the vegetation layer slows down the rate of heat flow through the façade, resulting in an improved *R*-value of the system. In the case of the indirect facade greening system the same trend was found; a temperature difference of 1.9 °C, compared with the bare wall was found and the interior air temperature is lowered with 1 °C in the case of the bare wall.

According to this measurement some conclusions can be drawn, namely, that the insulation material is superior compared with the green layer, and thus minimizes the effect indoor. However, since the green layer protects the heat accumulation in the outer layer of the masonry, less heat will be reradiated during the evening and night, which has a positive effect on the urban heat phenomena (lowering the urban temperature).

A stronger relation between temperature reduction and greenery was found for the living wall systems tested, a surface temperature reduction that can be achieved with the investigated living wall systems was between 7.2 and 10.3 °C during summer conditions. It can be noticed that the effect on the interior temperature is also higher as well as the relation between mitigation of the urban heat island effect.

For the winter measurements it was found that compared with the bare wall all the greening systems contributes to a better thermal resistance of the facade. Especially, in the case of the living wall systems higher interior room temperatures were measured up to 4 °C compared with the bare facade. Which means that the thermal resistance of the greened facades increased due to the extra material properties, air cavity, and plant tissue. Field measurements conducted by Mazzali et al. (2012) in a Mediterranean climate show comparable findings with laboratory tests and calculations conducted by Ottelé (2011). The facade covered with a living wall consisting an insulation layer (external side of the wall) shows a significant (66 %) reduction in cooling energy than a system where the insulation material is on the internal side (more heat accumulation in the massive facade); furthermore, they concluded that the most effective orientation of the (green) cladding, regardless the type of wall and the latitude, was the south side.

6.5 Utilization of Green Buildings

6.5.1 Green Roofs

Realization of green roofs is becoming a good construction practice in a lot of countries in Europe, especially in Germany, as well as in the USA (Osmundson 1999; Köhler et al. 2012); however, large scale implementation takes much more time and effort. In a report published by the municipality of Rotterdam (Anonymous 2007) a survey is given about the different types of green roofs with full financial details. Comparison of different types was needed to stimulate large scale application including suggestions for a system of subsidies (Anonymous 2007).

The advantages for the built environment as stated earlier by using vegetation on roofs are clear:

- increase of water buffering capacity (water management) instead off peak runoff to sewage system due to delayed runoff, transpiration, and evaporation.
- improvement of air quality (deposition of particulate matter on leaves for example).
- reduction of the heat island effect in urban areas. Energy savings (increase of insulation capacity—keep building cool in summer and keep cold out in winter).
- noise level reduction up to 10 dB(A).
- increase of lifetime of roofing material.
- increase of esthetic values.
- increase of ecological value and biodiversity.

In Scandinavia in the past, roofs were covered with a soil layer (sod) that was stripped from surrounding grass meadows (Donnelly 1992). Underneath the sod structurally heavy timber beams were interspaced with birch bark to act as a waterproofing layer.

Fig. 6.6 Modern version of a (Sedum planted) green roof (Texel, The Netherlands)

A range of different types of designs are now available on the market and realized: from very extensive (ecological roof and Sedum roof) to intensive roofs (gardens and parks) (Figs. 6.6, 6.7 and 6.8).

6.5.2 Intensive and Extensive Green Roofs

Vegetated roofs can, among other things, be categorized by the type of drainage system and their nominal thickness. These two properties determine the structural load, the maximum possible slope, the type of vegetation, and the water retention capacity (Köhler et al. 2012). Basically, we distinguish green roofs with a thin subtrate layer (extensive) and with a thicker substrate layer (intensive) see Fig. 6.9.

The surface layer thickness of an extensive green roof is typically 5 cm up to 15 cm (Table 6.2). The exact growing depth of plants differs per species, but in

Fig. 6.7 Extensive flat roof planted with sedum. After 5 years the first establishment of native Orchids (*Orchis praetermissa*) occurred (*photo* H. van Bohemen, The Netherlands)

Fig. 6.8 Extensive sloped green roof (Berghem, Belgium)

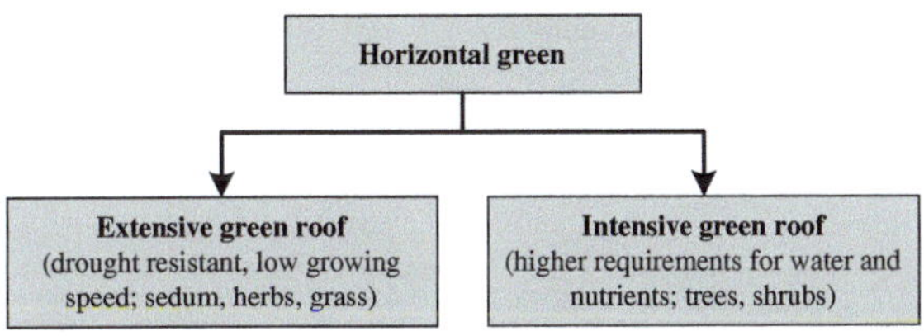

Fig. 6.9 Basic principles for a green roof strategy

general moss and sedum plants need the thinnest substrate thickness, and grasses and herbs need the upper limit of 15 cm. The growing medium should provide the vegetation of sufficient water, nutrients, and oxygen. Maintenance of extensive green roofs is low, since normally only mosses, sedum, or herbs are growing. Extensive roofs are limited accessible for people (only for maintenance purposes) and are less heavy (up to 170 kg/m^2) due to the low substrate thickness.

Intensive green roofs can be compared with parks or gardens in terms of plant diversity and application. They require deeper soil than extensive roofs and regular maintenance (Kadas 2006). This type of green roof, often accessible for people, requires a high carrying capacity of the structure of the building. The substrate layer varies from 25 cm or thicker, and must provide sufficient water, nutrients, and oxygen. The weight of intensive green roofs is usually more than 300 kg/m^2 (Table 6.2).

The boundary between intensive and extensive green roofs is vague and depends on perception. But a common rule of thumb is: non-accessible roofs with low vegetation are extensive roofs and accessible roof gardens with high vegetation are intensive green roofs.

Table 6.2 Overview of differences between extensive and intensive green roofs

Criteria	Extensive	Intensive
Field of application	Flat or sloped roof up to 45° (1–7 %)	Flat roof
Substrate height (cm)	5–15	>25
Layers	Multi-layered	Multi-layered
Weight of substrate (kg/m^2)	50–170	>300
Vegetation	Drought resistant, low growing speed, sedum, herbs, grass	Species with higher requirements for water and nutrients, grasses, trees, shrubs
Use	Habitat for animals	Additional living space for people
	Ecological compensation	Recreation area
	Rain water management	Meeting area
	Protection of roof material	Local food production
	Insulating capacity	Insulating capacity
Water retaining capacity	ca. 30–70 % of the annual precipitation	ca. 30–99 % of the annual precipitation
Maintenance	Low	High
Indication of costs (€/m^2)	20–30	>60

A green roof consists in essence of five different layers. The first layer on top of the regular roof construction is the waterproofing layer or membrane (1). This layer protects the roof against water leakages. Then there is a protection and storage layer (2), which prevents plant roots from growing through the roof package. This layer also keeps the whole green roof construction in place. The drainage and capillarity layer (3) buffers rainwater and drains surplus water. The root permeable filter layer (4) filters small particles out of the rainwater, to prevent them from ending up in the water drainage system where they might lead to blockages in the system. The final layer of the green roof is the growing media or substrate layer (5), in which plants can root. The thickness of this layer partly depends on the type of plants on the roof (Köhler et al. 2012). Table 6.2 gives an overview of the general differences between green roofs and is adapted from Köhler et al. (2012).

Brenneisen (2003) and Kadas (2006) conclude from their researches that green roofs contributes to preserving the local habitat. Green roofs are mainly inhabited by insects like beetles, ants, bees, and spiders. However, on some roofs uncommon and even rare species of spiders and beetles have been discovered (Brenneisen 2005). Furthermore, Brenneisen (2007) found also differences between extensive and intensive green roofs, as research done in Basel (Switzerland), shows that mainly because of the thin substrate layer (extensive green roof) less species can develop. A thin substrate layer is beneficial from a cost perspective, but for biodiversity it is harder to establish on an extensive roof than on an intensive roof (Brenneisen 2006). A variety of substrate thicknesses leads to different microclimates, and provides a wider potential for different species to establish. But in general can be stated that creating a green roof to foster biodiversity is a difficult task. Construction method, selection, and storage of local soil to create suitable substrate is crucial (Brenneisen 2006). Koster (2013) also emphasizes the importance of the substrate layer for the establishment of bees and in particular wild bees on green roofs. The composition of the substrate, the amount of nutrients in the soils, and the humidity of the soil determine which plants can grow there. For wild bees is the soil also directly important as nesting space, they especially like to nest in sandy soils (Koster 2013).

6.5.3 Greening of Outside Walls of Buildings

The same advantages of vegetation on roofs can be described for greening systems on walls. In recent years, different systems (Fig. 6.1) have been developed, like greening direct on the wall, greening systems before the wall, and greening possibilities incorporated within the construction of the wall (Hendriks 2000). Despite the range of possibilities there is still great hesitation in the building sector (from the originator, designer, architect, to the builder and the user) to increase the amount of outdoor wall greening. Probably mainly due to the possible disadvantages: the need for extra maintenance, falling of leaves, chance of damaging the wall structure, increase of the amount of insect and spiders in the house, and the expected extra costs involved.

By allowing and encouraging plants to grow on walls the natural environment is being extended into urban areas; the natural habitats of cliff and rock slopes are simulated by brick and concrete. There is a widespread belief that plants are harmful to building structures, ripping out mortar and prising apart joints with their roots (Johnston et al. 2004). The evidence suggests that these problems have been greatly exaggerated, except where decay has already set in and plants can accelerate the process of deterioration by the growing process. Certainly, there is little evidence that plants damage walls. In most cases the exact of opposite is true, with plant cover protecting the wall from the elements. Ancient walls still stand, despite centuries of plant growth (Johnston et al. 2004).

The leaves of climbing plants on walls provide a large surface area, which is capable of filtering out a lot of dust particles (particulate matter PM_x) and other pollutants such as NO_x and taking up CO_2 in daytime. Hard surfaces of concrete and glass encourage runoff of rainwater into the sewage system. Many plants hold water on their leaf surfaces longer than materials and processes of transpiration, and evaporation can add more water into the air. The result of this is a more pleasant climate in the urban area. Vegetation provides also nesting places for birds such as, blackbirds, song thrushes, and house sparrows.

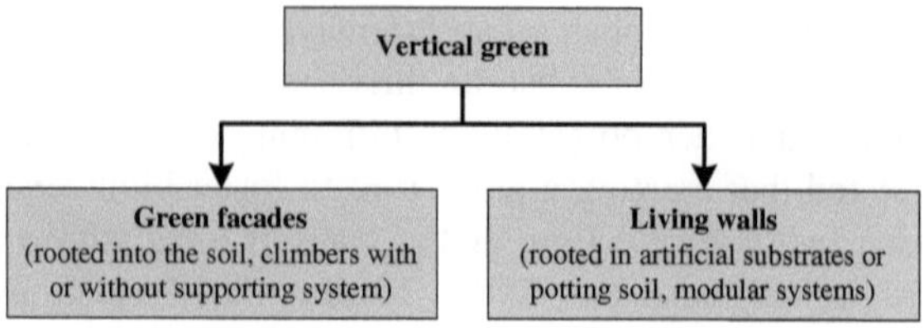

Fig. 6.10 Basic distinctions between greening principles

Table 6.3 Overview of differences between green facades and living wall concepts

Criteria	Green façade	Living wall concepts
Field of application	Sound barriers, facades (of typically older structures)	Every façade or vertical surface to create an "extra value"
Rooting medium/ substrate	Subsoil	Planter boxes, mineral wool, foam or felt layers
Layers	–	Multi-layered
Weight of substrate	–	>50 kg/m^2
Vegetation	Self-climbers (common ivy, boston ivy, etc.)	Climbers, shrubs, grasses, herbs, etc.
Use	Aesthetical reasons habitat for animals	Aesthetical reasons habitat for animals local food production insulating capacity ecological measure
Maintenance	Regular/low	High
Costs (€/m^2)	30–45	250–1,000

6.5.4 Overview of Vertical Green: Green Facades and Living Walls

Green façades, green walls, living walls, vertical green, and vertical gardens are descriptive terms, which are used to refer to all forms of vegetated wall surfaces. From the ground rooted traditional green façades and modern techniques to create green walls ensure that fundamental differences arise in vegetation types. Basically, one can understand systems rooted into the ground and based on hydroponic systems (not rooted into the ground). Green wall technologies may be divided therefore into two major categories (Fig. 6.10), namely: rooted into the ground and rooted in artificial substrates or potting soil.

Both basic principles can be classified according to their application form in practice (Table 6.3). Within the categories a distinction is made between whether if the greening system uses the façade as guide to grow upward (direct greening) or if the greening system and the façade are separated with an air cavity (indirect greening). The air space (cavity) between façade and greening system can be

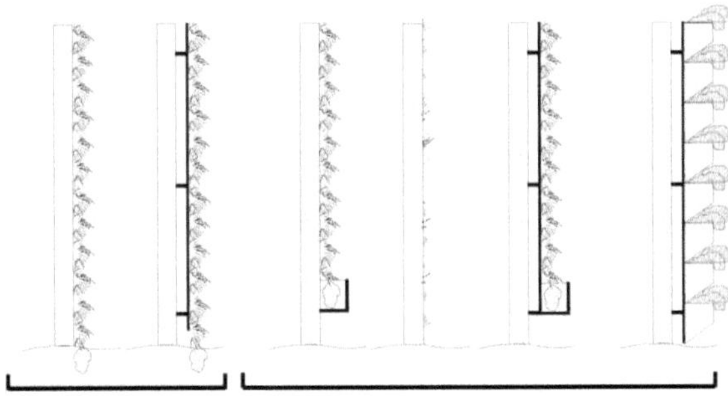

Fig. 6.11 Basic distinction between vertical green concepts, either rooted into the ground or rooted in artificial substrates (like mineral wool, foam, etc.) or potting soil

created by supporting systems, spacers, planter boxes, or by modular substrate systems. Figure 6.11 shows differences between direct and indirect façade greening and possible forms of their application.

The systems that are based on "artificial substrates and potting soil" principles are dependent on irrigation systems and adding nutrients to the substrate. These systems are known as living wall (LWS) concepts. Characteristic for this greening principle is the use of planter boxes filled with artificial substrate/potting soil or modular prefabricated panels equipped with artificial substrate. Additionally, an irrigation system is needed to keep the green system in the right condition. The used substrates and composition of living wall concepts can vary by manufacturer of the product. In general, one can distinguish systems based on (Fig. 6.12). The plants used for LWS are different type of evergreen small shrubs, offering much more creative, and aesthetical potential (Figs. 6.13, 6.14, 6.15, 6.16 and 6.17).

From a functional point of view, most of the living walls systems demand a more complex design compared to green façades; as a greater number of variables

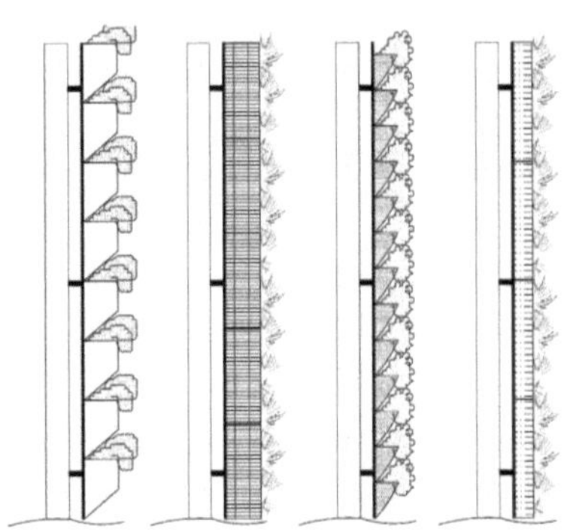

Fig. 6.12 Typical configurations of LWS concepts (based on planter boxes, foams, laminar layers of felt and mineral wool as substrate)

Fig. 6.13 Examples of green facades (rooted into the ground) *left photo* Dordrecht (The Netherlands), *right photo* Frederikshaven (Denmark)

Fig. 6.14 *Left* indirect greening system based on climbing plants attached to the façade in planter boxes. *Right* planterboxes system filled with small shrubs

must be considered, several layers are involved, and there are more supporting materials, and control of water and nutrients must be carried out. Living wall systems in fact are often very expensive, energy consuming, and difficult to maintain (Ottelé et al. 2011a, b). Furthermore, it is also important to take also the durability aspects of the systems into account. The durability of living wall

Fig. 6.15 Cross section of a living wall system based on felt layers. Due to the minimum rooting thickness very vulnerable for dehydration. Irrigation and nutrient system necessary

Fig. 6.16 Cross section of a living wall system based on mineral wool. Thicker rooting medium resulting in more redundancy against dehydration. Irrigation and nutrient system necessary

systems varies according to the type of system available. Living wall systems with panels based on felt layers have an average life expectancy of 10 years, and living wall systems based on planter boxes last more than 50 years. A thorough design (details of window ledges, doors, etc.) is always necessary to avoid damages, as corrosion or rot, caused by leakage of water and nutrients (Ottelé et al. 2011a, b). The green layer also results in a shading effect, which reduces the amount of UV light that will fall on building materials; since UV light deteriorates the material and mechanical properties of coatings, paints, plastics, etc., plants will also have an effect on durability aspects (Wong et al. 2009). Greening the building envelope with living wall systems is a suitable construction practice for new building and retrofitting (Ottelé 2011; Perini 2012). In both situations, it is possible to have a higher integration within the building envelope by combining functionalities. For example, in the case of the conventional bare walls constructed by several layers it is possible to avoid building the outer façade element since the protection against the environmental parameters is ensured by a living wall system. For retrofitting projects an external insulation material can be easily covered with LWS panels (Ottelé et al. 2011a, b).

Fig. 6.17 *Left photo* example of a living wall (felt layers) attached to a façade in the inner city of Antwerp (Belgium). *Right photo* living wall system in the inner city of Madrid (Spain)

6.6 Conclusions and Reflection

In the previous article, the author attempted to demonstrate the value of reintroducing vegetation to the surfaces of urban buildings and their related spaces. The ecological and environmental benefits of a green building envelope include the improvement of air quality and the reduction of pollution. The advantages are mainly related to the reduction of fine dust levels (Ottelé et al. 2011a), increased biodiversity (Köhler 1993), and reduction of the heat island effect in urban areas (Taha 1997; Onishi et al. 2010). This occurs thanks to the lower amount of heat reradiated by greened surfaces and humidity affected by the evapotranspiration caused by plants (Scudo and Ochoa De La Torre 2003). This process also allows indirectly to save considerable energy supplied to the building (Dunnett and Kingsbury 2004; Köhler 1993) as the plants and the growing medium provide insulation and shade, which can reduce, especially in the Mediterranean area, energy needed for cooling (Wong et al. 2009). Besides these benefits, social and economical advantages of greening systems are also involved, including: the real estate market, greater durability of buildings, and the better psychological state of citizens (Köhler 1993; Dunnett and Kingsbury 2004; Johnston and Newton 2004).

We have suggested that, far from being a radical or fashionable solution, this is simply the reinterpretation of an approach with a long and distinguished history.

Before the full benefits of green buildings could be scientifically proven, the principle was already accepted and practiced in cities all over the world. Green building is most effective as part of an integrated green approach to cities. Such an approach demands a much closer cooperation between architects, ecologists, developers, and green planners than has so far taken place. The integration of vegetation in the built space can be an opportunity to improve the environmental conditions of dense urban areas and to reduce the energy demand of buildings, especially in Mediterranean area due to its cooling capacity. This is an important field to investigate considering the growing interest on these systems, which is not only connected to a more sustainable approach to construction, but also to esthetic intentions (Perini 2012).

To guarantee sustainable practices, benefits, and performances obtainable thanks to greening systems, have to be considered along with the environmental burden produced during the life span of greening systems, and with the possible problems connected to maintenance demand. Also costs have to be considered for a wider diffusion of these systems. Measurements carried out in the field by many researchers show not only the potential of vertical green on the thermal performance, but also under laboratory conditions (Ottelé 2011). The positive effect on the thermal resistance (i.e., summer and winter) is mainly caused by the materials used, extra cavity, water availability, and metabolism of the plant tissue. Through (significant) less heat accumulation by the masonry in combination with evapotranspiration caused by the plant material, a positive effect to lower/mitigate the urban heat island effect can be taken into account.

As discussed throughout the article, many aspects have to be considered to avoid that green only plays an esthetic role with respect to sustainability. Characteristics, components, and materials of vertical and horizontal greening systems can have an influence on the environmental burden and environmental benefits, etc. Some systems, as the living wall ones described, offer much more creative and aesthetical potential, but due to the material used and durability in some cases cannot be considered as sustainable. Material choice and durability aspects are important (environmental impact) when the energy demand of a building can be reduced or when the multifunctionality of the construction due to the integration of vegetation can be increased. These aspects have been considered also through a life cycle analysis (Ottelé et al. 2011a, b). As suggested by Henry and Frascaria-Lacoste (2012) the adoption of LCA analysis for the labeling of green products could increase their use since it has the potential to boost the confidence of consumers. Therefore, this could lead to particular focus being placed on specific green elements, which could potentially further homogenize natural features within cities, with possible negative impact on other benefits of green, such as biodiversity (Henry and Frascaria-Lacoste 2012). However, a LCA could lead to deeper consideration by manufacturers of the environmental burden produced by their systems to improve the balance between benefits and burden for a more sustainable built environment.

To stimulate biophilic design in architecture modern greening concepts as vertical gardens and green roofs, should be considered as a "building material" with multifunctional properties (ecological, social, mitigation of urban heat, etc.) compared to our traditional cladding and roofing materials (masonry, concrete, marble, glass, bitumen, etc.). Material choices are often underestimated by designers, manufacturers of greening systems, and architects. An optimal balance have to be found between durability aspects, materials really needed (also by mass, i.e., in this case less is more!) service life, and lifespan. In the case of a new design, try to integrate a greening concept into the building envelope instead to add an "extra" green layer to a conventional solution. Besides, it is mandatory to be aware of the cooling and insulation potential of green structures related to energy savings, it contributes to a lower energy demand at the building level, and must not be underestimated. This "total" awareness, which is related to a wider research in this field, will lead to a more eco-friendly and sustainable design of cities.

References

Akbari H, Pomerantz M, Taha H (2001) Cool surfaces and shade trees to reduce energy use and improve air quality in urban areas. Sol Energy 70(3):295–310

Anonymous (2007) Toepassingen van groene daken in Rotterdam. Uitgave Gemeente Rotterdam. Verkrijgbaar via gemeente Rotterdam

Ardente F, Beccali M, Cellura M, Mistretta M (2008) Building energy performance: a LCA case study of kenaf-fibres insulation board. Energy Build 40(2002):1–10

Bartfelder F, Köhler M (1987a) Experimentelle untersuchungen zur function von fassadenbegrünungen, Berlin

Bartfelder F, Köhler M (1987b) Experimentelle untersuchungen zur function von fassadenbegrünungen, Abbildungen, tabellen und literaturverzeichnis, Berlin

Beatley T (2008) Toward biophilic cities: strategies for integrating nature into urban design. In: Kellert SR, Heerwagen J, Mador M (eds) Biophilic design: theory, science and practice. Wiley, Hoboken, NJ, pp 277–295

Beatley T, Newman P (2013) Biophilic cities are sustainable, resilient cities. Sustainability 5:3328–3345. http://www.mdpi.com/2071-1050/5/8/3328

Beckett KP, Freer-Smith PH, Taylor G (2004) Deposition velocities to Sorbus aria, Acer campetre, Populus deltoides x trichocarpa 'Beaupre, Pinus nigra and x Cupressocyparis leylandii for coarse, fine and ultra-fine particles in the urban environment. Environ Pollut 133(1):157–167

Bohemen HD, van Ottelé M, Fraaij ALA (2009) Ecological engineering; green roofs and the greening of vertical walls of buildings. Landscape Archit 1:042–049 (Proceedings of Ecocity 2008, Mexico)

Brenneisen S (2003) The benefits of biodiversity from green roofs: key design consequences in Greening Rooftops for Sustainable Communities. A. Loder and J. Sprout. Chicago, Green Roofs for Healthy Cities

Brenneisen S (2005) research project report on the use of extensive green roofs by wild bees. The Natural Roof (NADA) University of Wadenswil (translated by Waldbaum H, edited by Gedge D)

Brenneisen S (2006) Space for Urban wildlife: designing green roofs as habitats in Switzerland. Urban Habitats 4(1):27–36 ISSN 1541–7115

Bussotti F, Grossoni P, Batistoni P, Ferretti M, Cenni E (1995) Preliminary studies on the ability of plant barriers to capture lead and cadmium of vehicular origin. Aerobiologia 11(1995): 11–18

Dakakkers (2013) Stadslandbouw, Ruimte zat in steden voor het telen van voedsel. http://www.dakakkers.nl/index.php?subonderwerp_ID=79. Accessed on Dec 2013

Darlington A (1981) Ecology of wall. Heinemann Education books, London

Derr V, Lance K (2012). Biophilic boulder: children's environments that foster connections to nature. Children Youth Environ 22(2):112–143. http://www.colorado.edu/journals/cye

Donnelly MC (1992) Architecture in the Scandinavian countries. Massachusetts Institute of Technology, Cambridge

Dunnet N, Kingsbury N (2004) Planting green roofs and living walls. Timber Press, Oregon

EPA (2010) United States environmental protection agency. http://www.epa.gov/greenbuilding/pdf/2010_fed_gb_report.pdf

Eumorfopoulou EA, Aravantinos D (1998) The contribution of a planted roof to the thermal protection of buildings in Greece. Energy Build 27(1):29–36

Eumorfopoulou EA, Kontoleon KJ (2009) Experimental approach to the contribution of plant covered walls to the thermal behaviour of building envelopes. Build Environ 44(2009): 1024–1038

Farming the City (2013) Farming the city, http://www.trancity.nl/publicaties/farming-the-city.html, Accessed on Dec 2013

Fowler D, Cape JN, Unsworth MH (1989) Deposition of atmospheric pollutants on forests. Philos Trans R Soc Lond 324:247–265

Fowler D, Coyle M, AsSimon HM, Ashmore MR, Bareham SA, Battarbee RW, Derwent RG, Erisman JW, Goodwin J, Grennfelt P, Hornung M, Irwin J, Jenkins A, Metcalfe SE, Ormerod SJ, Reynolds B, Woodin S, Hall J, Tipping E, Sutton M, Dragosits U, Evans C, Foot J, Harriman R, Monteith D, Broadmeadow M, Langan S, Helliwell R, Whyatt D, Lee DS, Curtis C (2001) National expert group on transboundary air pollution. Transboundary air pollution: acidification, eutrophication and ground-level ozone in the UK, NEGTAP 2001

Frahm JP, Sabovljevic M (2007) Feinstaubreduzierung durch Moose. Immissionsschutz, (4), 2007

Getter KL, Rowe BD (2006) The Role of extensive green roofs in sustainable development. HortScience 41(5):1276–1285

Heidt V, Neef M (2008) Benefits of urban Green space for improving urban climate. Ecol Plan Manag Urban For Part I: 84–96. doi:10.1007/978-0-387-71425-7_6

Hendriks CF, Bijen JMJM, Felix F, Fraaij ALA, Janse H, de Munck ED, Reintjes RC, Schutte-Postma ET, Stroeven P, Vogtlander JG, van der Wegen GJL (2000) Durable and sustainable construction materials. ISBN 90-75-365-30-6

Henry A, Frascaria-Lacoste N (2012) Comparing green structures using life cycle assessment: a potential risk for urban biodiversity homogenization? Int J Life Cycle Assess 17:949–950. doi:10.1007/s11367-012-0462-3

Holm D (1989) Thermal improvement by means of leaf cover on external walls—a simulation model. Energy Build 14(1989):19–30

Hosker RP, Lindberg SE (1982) Review article: atmospheric deposition and plant assimilation of airborne gases and particles. Atmos Environ 16:889–910

Innovatie project luchtkwaliteit (IPL) (2006) Kennisdocument Vegetatie-luchtkwaliteit ten behoeve van het uitvoeren van een pilotproject langs rijkswegen rapportnummer DWW 2006-094 /IPL 06.00019

Johnston J, Newton J (2004) Building green: a guide to using plants on roofs, walls and pavements. Greater London Authority City Hall. ISBN 1 85261 637 7

Kadas G (2006) Rare invertebrates colonizing green roofs in London. Urban Habitats 4(1): 66–86. ISSN 1541-7115

Kaltenbach F (2008) Living walls, Vertical gardens—from the flowerpot to the planted system façade. Detail nr 12(2008):1454–1463

Köhler M (1993) Fassaden- und Dachbegrünung. Ulmer Fachbuch Landschafts- und Grunplanung, Stuttgart. ISBN 3-8001-5064-6
Köhler M (2008) Green façades—a view back and some visions. Urban Ecosyst 11:423–436. doi:10.1007/s11252-008-0063-x
Köhler M, Ottelé M, Ansel W, Appl R, Betzler F, Mann G, Wunschmann S (2012) Handbuch Bauwerksbegrunung, planung-konstruction-ausfurung. Published by Rudolf Muller. ISBN 978-3-481-02968-5
Koster A (2013) Adviesgroep Vegetatiebeheer. www.bijenhelpdesk.nl. Accessed on June–July 2013
Krusche P, Krusche M, Althaus D, Gabriel I (1982) Ökologisches bauen. Herausgegeben vom umweltbundesamt, Bauverlag
Kula R (2005) Green roofs and the LEED green building rate system. In: Proceedings of 3rd North American green roof conference: greening rooftops for sustainable communities, Washington D.C., pp 141–153
Lambertini A (2007) Vertical gardens: bringing the city to life. Thames and Hudson, London. ISBN 978-0-500-51369-9
Laurie IC (1977) Nature in cities, the natural environment in the design and development of urban green space. Wiley, Chichester. ISBN 0 471 99605 X
Liu K, Baskaran B (2003) Thermal performance of green roofs through field evaluation. Greening rooftops for sustainable communities. In: Proceedings of the first North American green roofs conference, Chicago
Mazzali U, Peron F, Scarpa M (2012) Thermo-physical performances of living walls via field measurements and numerical analysis. Eco-architecture IV. Harmonisation between architecture and nature. WIT Trans Ecol Environ 165:239–250. doi:10.2495/ARC120011. ISBN: 978-1-84564-614-1
Mentens J, Raes D, Hermy M (2006) Green roofs as a tool for solving the rainwater runoff problem in the urbanized 21st century? Landscape Urban Plann 77(3):217–226
Minke G, Witter G (1982) Häuser mit grünen pelz. Ein handbuch zur hausbegrünung
Natuur balans (1999) Rijksinstituut voor volksgezondheid en milieu. RIVM Rapport 408663001. ISBN 90 140 6228 1
Novi F (1999) La riqualificazione sostenibile. Alinea Editrice, Firenze
Nuzzo E, Tomasinsig E (2008) Recupero ecoefficiente del costruito. Edicom Edizioni, Monfalcone (Gorizia)
Onishi A, Cao X, Ito T, Shi F, Imura H (2010) Evaluating the potential for urban heat-island mitigation by greening parking lots. Urban For Urban Green 9(2010):323–332
Osmundson T (1999) Roof gardens; history, design and construction. W.W. Norton and Company, New York, London
Ottelé M (2011) The green building envelope, vertical greening. ISBN 978-90-90-26-217-8
Ottelé M, van Bohemen H, Fraaij ALA (2011a) Quantifying the deposition of particulate matter on climber vegetation on living walls. Ecol Eng 36(2010):154–162
Ottelé M, Perini K, Fraaij ALA, Haas EM, Raiteri R (2011b) Comparative life cycle analysis for green facades and living wall systems. Energy Build. doi:10.1016/j.enbuild.2011.09.010
Pérez G, Rincón L, Vila A, González JM, Cabeza LF (2011) Behaviour of green facades in Mediterranean Continental climate. Energy Convers Manag 52(4):1861–1867
Peck SW et al (1999) Greenbacks from green roofs: forging a new industry in Canada, Status report on benefits, barriers and opportunities for green roof and vertical garden technology diffusion, environmental adaptation research group, Canada
Perini K (2012) L'integrazione di vegetazione in architettura. Metodi e strumenti innovativi – The integration of vegetation in architecture. Innovative methods and tools. Tesi di Dottorato, Università degli Studi di Genova. ISBN 9788890692505
Perini K, Ottelé M, Haas EM, Fraaij ALA, Raiteri R (2011) Vertical green systems and the effect on air flow and temperature near the façade. Building and Environment (2011), doi: 10.1016/j.buildenv.2011.05.009

Pope AC, Ezzati M, Dockery DW (2009) Fine-particulate air pollution and life expectancy in the United States. The new England journal of medicine. N Engl J Med 360:376–386

Prasad D, Hill M (2004) The construction challenge: sustainability in developing countries, London Royal Institution of Chartered Surveyors (RICS) series, Leading Edge Series

Rath J, Kießl K (1989) Auswirkungen von Fassadenbegrünungen auf den Wärme- und Feuchtehaushalt von Aussenwänden ubd schadensrisiko. Fraunhofer-Institut für Bauphysik, IBP-Bericht Ftb-4/1989

Scharf B, Pitha U, Trimmel H (2013) Resilience of cityscapes In: IFLA, Davies R, Menzies D (eds), Proceedings for the 50th IFLA World Congress. ISBN 978-0-473-24360-9

Schröder FG (2009) Automatisierte, biologische, senkrechte, städtische Fassadenbegrünung mit dekorativen funktionellen Parametern; Abschlussbericht zum Kooperationsprojekt im Rahmen von PRO INNO II; Hochschule für Technik und Wirtschaft Dresden, 2009

Scudo G, Ochoa De La Torre JM (2003) Spaziverdiurbani, Se, Napoli, Italy

Sternberg (2010) Dust particulate absorption by ivy (Hedera helix L) on historic walls in urban environments. Sci Total Environ 409(1):162–168 (Troy Sternberg, Heather Viles, Alan Cathersides, Mona Edwards)

Taha H (1997) Urban climates and heat islands: albedo, evapotranspiration, and anthropogenic heat. Energy Build 25:99–103

Thönnessen M (2002) Elementdynamik in fassaden begrünendem wilden Wein, Kölner Geographischer Arbeiten, Heft 78, Köln

Thormark C (2002) A low energy building in a life cycle. Its embodied energy, energy need for operation and recycling potential. Build Environ 37(4):429–435

United Nations (2007) Economic and social council. Recent results on corrosion trends and the protection of cultural heritage from air pollution. ECE/EB.AIR/WG.1/2007/8

Wong NH et al (2009) Thermal evaluation of vertical greenery systems for building walls, building and environment. doi:10.1016/j.buildenv.2009.08.005

Chapter 7
Architectural Bio-Photo Reactors: Harvesting Microalgae on the Surface of Architecture

Rosa Cervera Sardá and Javier Gómez Pioz

Abstract This chapter presents innovative construction systems developed by the authors, integrated into the surfaces of cities and architecture—facades, roofs and pavements—giving them a supplement to traditional construction, aesthetic and enclosure, for the production of bio-energy. These systems allow the movement of fluids with microalgae, becoming Architectural Photo-Bioreactors that, by absorbing sunlight and converting it through photosynthesis of microalgae, produce biomass, biogas, bio-fertilizers and other value-added products to the pair that captures CO_2. The grand innovation of the project is the combination of two fields, constructive and biological, highly disruptive and uniquely for a sustainable city in a holistic view of energy, for growing algae in architecture.

7.1 Introduction

7.1.1 Urban Consumption and Ecological Consciousness

The awareness that humanity is throwing away at a vertiginous speed the heritage for which the Earth needed millions of years to accumulate is something really new. During the recent times a barbarous attack against the environment has been constant. However, all these consequences were welcome as the advanced society, in exchange, reached degrees of comfort and welfare unknown until then. It is now in the first decades of the twenty-first century when ecological attention extends to all scales, making us conscious that the city is one of the big focuses of energy consumption, of land use and one of the biggest sources of polluting emissions.

R. Cervera Sardá (✉)
Co-founder Cervera and Pioz Architects, Alcala University, Madrid, Spain
e-mail: cyp@cerveraandpioz.com

J. Gómez Pioz
Co-founder Cervera and Pioz Architects, Polytechnic University of Madrid, Madrid, Spain

F. Pacheco Torgal et al. (eds.), *Biotechnologies and Biomimetics for Civil Engineering*,
DOI 10.1007/978-3-319-09287-4_7

The surprising growth of urban settlements runs in parallel with the surprising waste of our resources and of our energy that the above demand, generating thus an imbalance between the availability of production in Nature and the capability of plunder of the environment by humans. With more than half the world's population living in cities, it is crucial to reduce greenhouse gas emissions and a new model of city is required.

7.1.2 City as an Energy Bio-Factory: Cultivating Algae on the Surface of Architecture

Reliance on energy is at the core of twenty-first century society. There is a steady increase in demand and a foreseeable shortage of supply set to worsen as existing resources continue to be drained. Consequently, a change in existing habitat models is vital. According to this statement Cervera and Pioz architects have researched on a new proposal with an aim to transform the envelope or surface of our buildings and cities into "Energy Factories". The purpose is to combine the regeneration, reutilisation and rehabilitation of existing urban fabric with the creation of freshly conceived green building typologies able to generate energy on-site and to establish a self-sustaining 24 h multifunctional city area. As per the notion of "redefining what is possible in the places we live and work and play", Energy Factories suggest a timely holistic approach that defies the detrimental belief in continuously consuming the new and discarding the old, which rates so high in environmental costs.

Research on algae is a field of growing interest, development and experimentation in the world and is presented as a valuable alternative to meet the needs of mankind in the twenty-first century. The using of living matter as part of the construction and use of buildings as a base or support for the cultivation of living matter is a promising path that now begins its journey. The works presented here come from the project entitled "VIDA-BIOCAS, A Bio Self-Sufficient City from Algae" developed by a consortium of companies specialized in energy, algae cultivation and, in our case, innovative sustainable architecture. This long-term and large scope project (13 companies, 4 years and 18 million euros)[1] is co-financed by the Spanish Government as a result of a competitive process and opens highly groundbreaking proposals that overcome the current technology of microalgae production in architecture, still at a very initial stage (Fig. 7.1).

The basis of the proposal is the combination of the living with the inert, the biological with the architectonical, giving rise to a new concept that could be baptized as "Living Architecture". The project is based on the integration of:

---

[1] www.cenitvida.es.

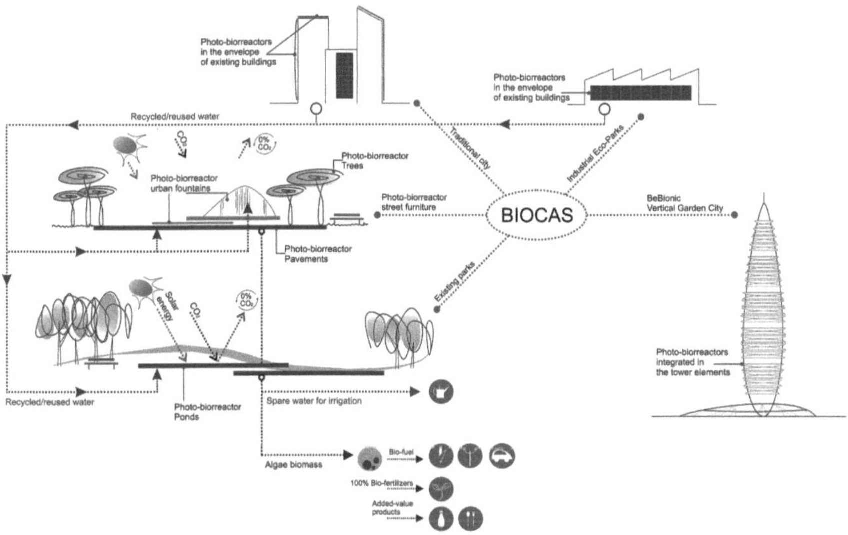

Fig. 7.1 A bio self-sufficient city from Algae (R. Cervera and J. Pioz)

- On one hand, algae, as autotrophic organisms, collectors of carbon, hydrogen, oxygen, nitrogen and phosphate and producers of sugars, fats, proteins, nucleic acids, enzymes, vitamins and oxygen, and therefore, the raw material for food manufacturing, both human and animal; medicines; bio-energy; biomass; bio-fertilizers, etc., and secondly, as heterotrophic organisms, acting as purifying agents of contaminants.
- On the other hand, current technologies for building surfaces and architectural envelopes (facades, roofs, floors, etc.) and the available materials, combined in solutions capable of conducting liquids with microalgae, CO_2 and nutrients and able to allow the passage of light for photosynthesis and conversion of solar energy into other energy.

The proposal presented here has therefore considerable impact on the way we must design our buildings and cities. There is an increasing need to gain self-sufficiency in architecture and cities in order to move on from “cities that consume” to “cities that produce”. The search for innovative alternatives is on, and includes the emerging technology of microalgae cultivated via photo-bioreactors. Architecture has become susceptible of taking on the role of a Bio-Factory for energy and food and other products.

Enclosed surfaces of buildings are increasingly versatile for balancing their enclosure with other function such as to capture energy or CO_2. This vision of multi-functionality of architecture has a dual contribution as it benefits the society to use the architecture for a function of energy and environmental value in addition to their usual constructive missions and providing greater benefits like building insulation. The great potential of this project is the transformation of cities into sustainable, holistic energy environments. Thus a technology becomes tangible global awareness by inversion of I + D + i.

7.2 An Overview on Microalgae Photo-Bioreactors

7.2.1 What Is a Microalgae Photo-Bioreactor?

The photo-bioreactors piping systems consist of water with algae, nutrients and CO_2, whether these are in closed tubes or well sandwiched or in an open pond, provided with materials that let light to penetrate for the process of photosynthesis (Figs. 7.2 and 7.3).

An alga effectively uses the sunlight by capturing its energy by means of the photosynthesis process that converts inorganic matter into simple sugars. There are two principal methodologies used for its cultivation: ponds or open systems and closed photo-bioreactors. Cultivation in ponds is the cheapest method, although is less efficient and not easily controllable. Whereas for closed photo-bioreactors, the plants are grown in a controlled environment. For maximum efficiency photo-bioreactors need to be cleaned from time to time, which means it is to be thoroughly distributed dividing into sectors or modules that could be easily maintained. As mentioned above, in both cases, what exactly the algae require are sunlight, water with nutrients and CO_2. Algae could be harvested when fully grown by separating the slurry parts from the water, then pressing to be followed

Fig. 7.2 Industrial microalgae photo-bioreactors, Alga Energy company installations (Provided by Alga Energy)

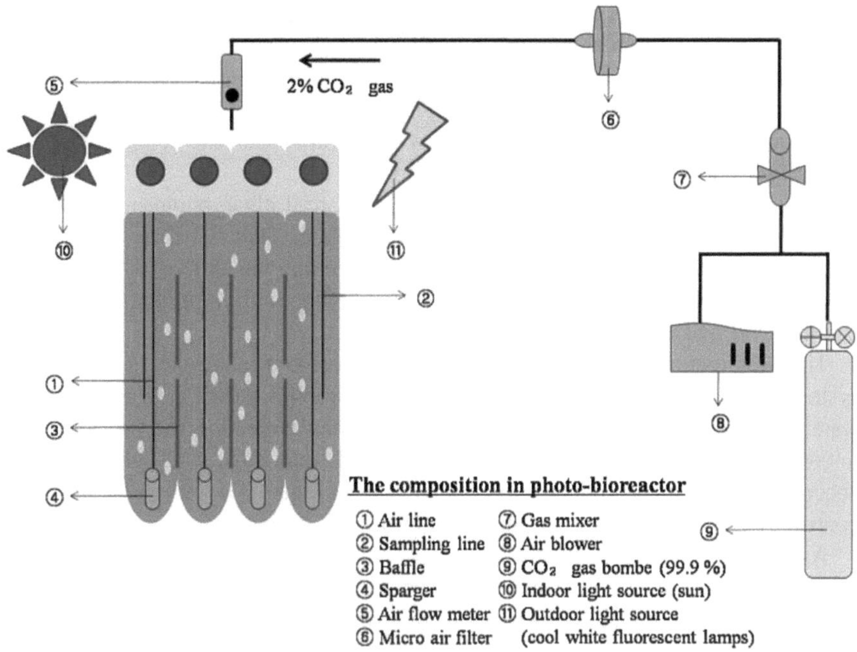

Fig. 7.3 Schematic diagram of overall microalgal cultivation system with a developed vertical tubular photo-biorreactor (Yoo et. al. 2013)

by centrifugation. As a result of this process, we are able to get products like algae oil and biomass.

One of the major advantages that algae inculcate is a rapid growth rate that generates a full crop cycle every 24 h. The biomass it produces can truly become a valuable alternative to that of cereals or other products used for several purposes by humans. Although things are gradually improving, the current production cost for oil and bio fuel is comparatively expensive, so economically speaking it can hardly stand as a competitive fuel option. The potential area where microalgae cultivation can be widely accepted and developed is that of highly valued pharmaceutical products, foods and cosmetics, all of which offer excellent value for money. And now new fields of application are coming for bio-energy, bio-fertilizers and to capture CO_2.

7.2.2 What Is an Architectural Photo-Bioreactor (A-PhBR)?

The A-PhBR, Architectural Photo-Bioreactors, are photo-bioreactors similar to the industrial ones, and in the same way, fulfil the functioning of cultivating microalgae, leading the same energy needs and benefits of producing added-value products, clean energy, environmental improvement, etc.

The big difference is given by its "Architecture" specificity. The innovation is its integration into architecture. This makes the industrial element transform into architectural and vice versa. The synergy generated by summation "profits" turn the architecture into iconic, environmentally didactic, active energetically, surface saving and environmentally friendly, all with minimal increase of their cost, that is, with very low investment and high returns.

In short, this is a new reinterpretation of industrial algae photo-bioreactors to become "skin" or constructive architectural surface. That is to say, to turn the "back-stage" production to the "front-row", converting the up to now industrial in a new type of outer layer of the architectural envelope with important additions to energy level and environmental values.

The architectural photo-bioreactors are integrated, in addition, in a complete system that involves collecting the algae, which implies a duct with impulsion of water with microalgae, nutrients and CO_2 and then collecting the production of microalgae, with the help of storage tank and centrifuge and other deposits according to the final product to be obtained from algae. All with the appropriate control system, automation and monitoring of the cultivation process (Fig. 7.4).

As a particular feature for the integration of photo-bioreactors with the buildings, the overall described system can be combined with other specific constructive systems to obtain added benefits as described below:

- The water supply system may be combined with the system of sewage collection, since they provide precisely the chemicals and nutrients needed for algal growth. Thus it produces a closed water cycle so that the grey water can be commonly drained into the sewage system which could be now recycled, to minimize the impact on the sewerage.
- The system to supply the CO_2, which is necessary for the growth of algae, can come from all the machinery in a construction producing CO_2, so that instead of emitting CO_2 in the free environment, these emissions could be let in a closed circuit which goes to the cultivation of algae.
- This system is also capable of combining with CO_2 from power plants or industries in a symbiotic process.
- The algae photo-bioreactor also can be converted into an excellent insulator for architecture which can replace the usual insulators. It can modify the building systems to replace the traditional "fill" insulation of external walls by the outer layer of the construction formed in this case by the photo-bioreactor itself and water containing organic material, acting this thermoregulatory and as insulation.

The diversity of types of construction requires specific technology proposed for each type of building and constructive circumstance or environment. And in turn the various materials offer opportunities for a range of solutions that are addressed in this project.

The new approach provides a number of tangible and intangible benefits that make the product unbeatable and highly competitive in various ways.

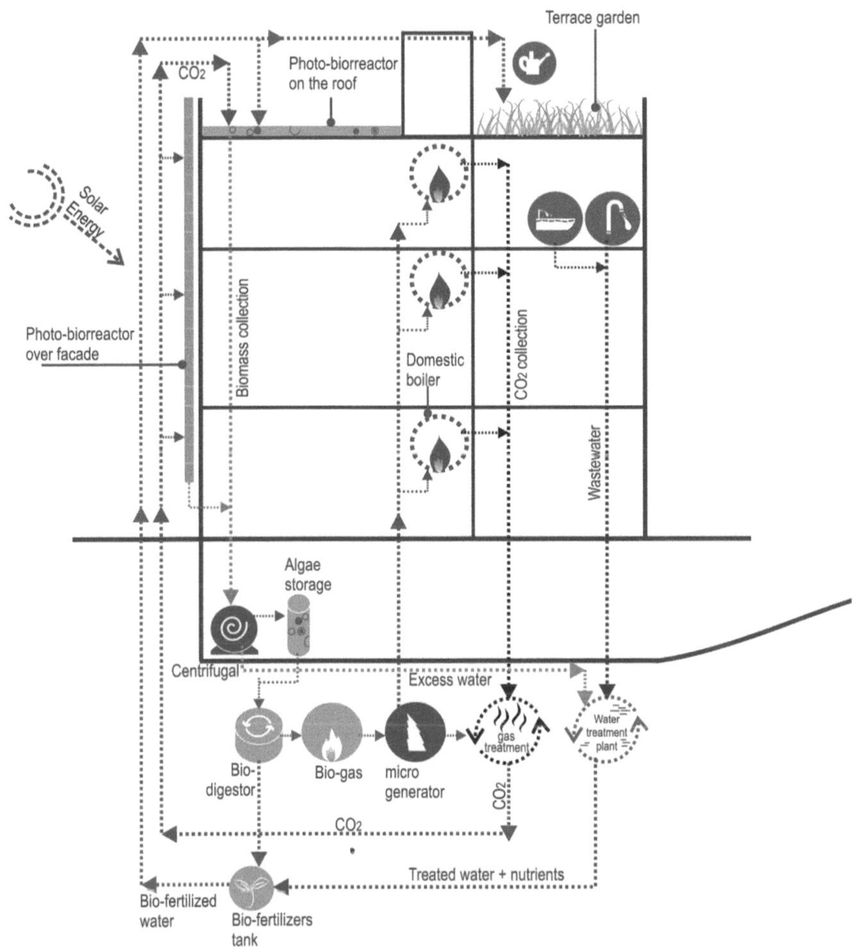

Fig. 7.4 Building installation for A-PhBR (R. Cervera and J. Pioz)

- The A-PhBR, to be integrated into the architecture and urban real estate, do not consume space or territory, reducing the footprint of farming and therefore having an ethical attitude in biomass production.
- Installation costs are much lower than the industrial PhBR as for the A-PhBR, as part of the enclosures or architectural envelopes, are amortized over their own constructive role.
- The A-PhBR consumes CO_2 in the city itself, helping to improve the air quality wherever needed.
- It has aesthetic and extraordinary iconic value, which is assessed by the combination of the two functions, architecture and culture, causing the actual value (i.e., values that are willing to pay companies and institutions for it) which is equivalent to the enclosures with very high end (e.g. curtain walls).

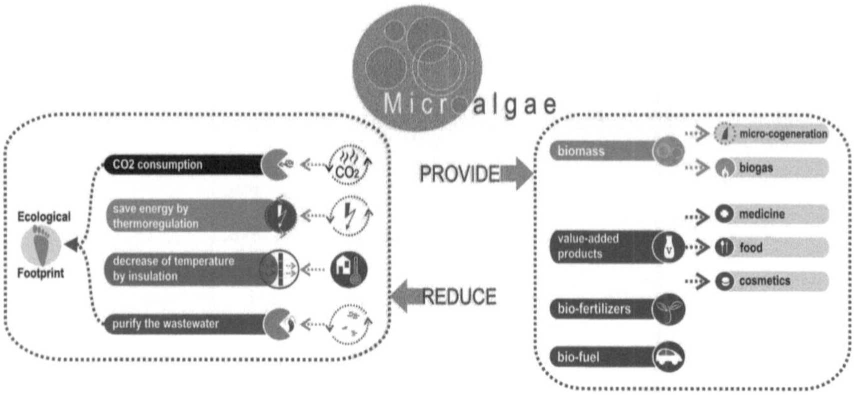

Fig. 7.5 Contributions and reductions in the use of microalgae (R. Cervera, J. Pioz)

This makes associated facilities also depreciated. Considered in this way, the cost of A-PhBR would be amortized from the first minute construction.

- The educational value of the A-PhBR has a remarkable aspect, showing society the life cycle of the algae and promoting environmental awareness.
- Contribute to the image and positioning of ecological spirit, which is an intangible that many organizations appreciate.

Cultivation of microalgae in architecture on one hand makes or produces and on the other reduces or minimizes. What biomass provides is biomass that can be used to obtain different products depending on the conditions of cultivation and further processing. The microalgae biomass can be the basis for production of biogas and electricity through co-generation; which can be transformed into value-added products, such as pharmaceuticals, cosmetics and alimentation for humans and animals; can be used as bio-fertilizer and bio-stimulant in the agriculture and, ultimately, through a process of further development and still uncompetitive, in bio-fuel. Subtractive or fundamental action -minimizing algae and is unchecked technology to other renewable—is its ability to capture CO_2 and sink effect. Other qualities to be considered are those relating to the ability of algae to purify water, thus contributing to the cycle of the same, and for your help to save energy by thermal regulation as a living organism. All this makes the A-PhBR today really economical, ecological alternative, and valuable image, which is competitive for medium to large spaces, especially recommended two market segments such as public institutions and large companies (Fig. 7.5).

7.3 Building Scenarios of Application for A-PhBR

The city presents various opportunity scenarios for integrating bio-energy production and various other products from algae. The greatest potential for insertion of photo-bioreactors is in buildings; however, the urban scenario also offers interesting opportunities as described below:

Fig. 7.6 A-PhBR integrated on building surfaces and louvers (Design by R. Cervera and J. Pioz)

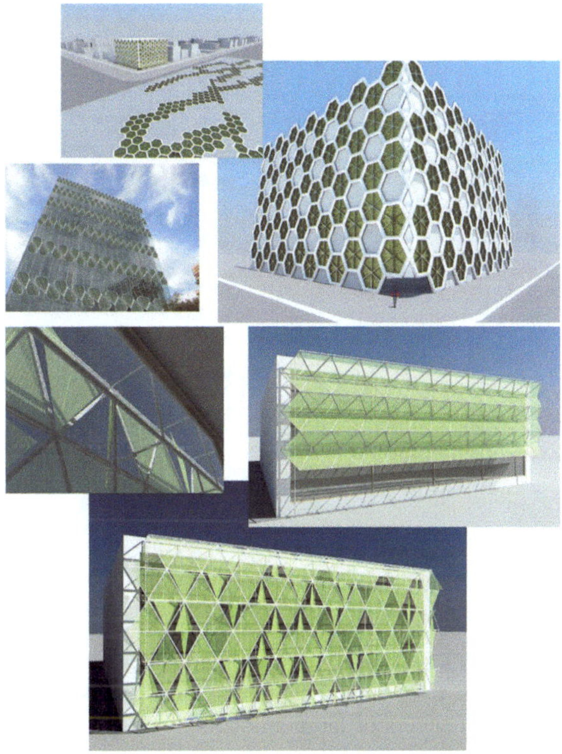

1. In the case of architectural surfaces (Fig. 7.6):
    a. A-PhBR over building facades: They are developing from the model of closed circuits photo-bioreactors and always from materials that allow the passage of light: glass, metacrylate and various plastics, ETFE, etc.; over the assumptions of standard construction.
    b. A-PhBR forming the facade by themselves. In this case we refer to curtain wall facades and glass bricks facades.
    c. A-PhBR that overlap existing facades. Can be conceived as circuits attached to opaque walls, as if it were an outer layer of the facade, or as circuits and louvers on glazed exterior walls.
    d. A-PhBR on building roofs: In this case the photo-bioreactor system in an open pool.

2. In the case of urban scenario (Figs. 7.7 and 7.8):
    a. As urban pavements, semi-closed circuit type, allowing people walking and at the same time cultivating microalgae.
    b. As ponds and fountains.

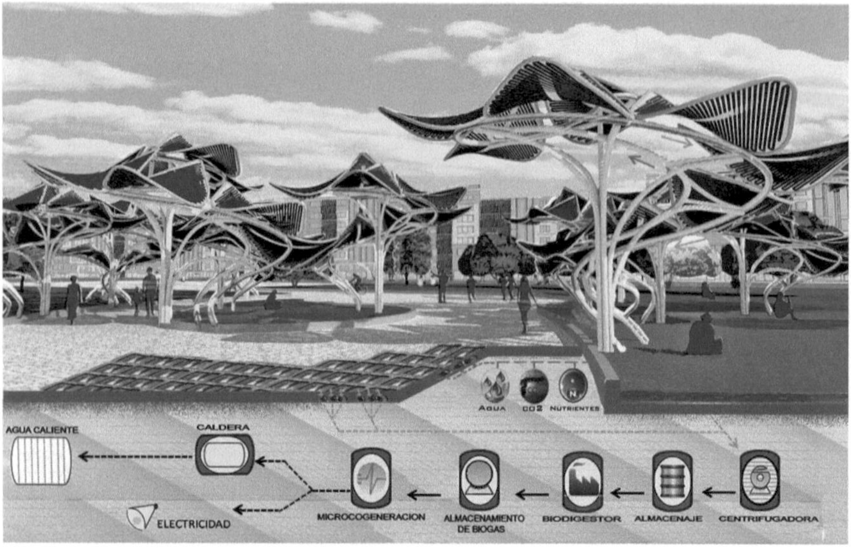

Fig. 7.7 Photo-biorreactor urban trees (Design by R. Cervera and J. Pioz)

c. As street furniture, i.e., marquees and canopies.
d. As CO_2 urban capturers, near industries, polluting activities or places producers of large amounts of CO_2.

A considerable aspect of these solutions is their ability to function both in new buildings and already built ones, especially those that are not up to current standards of thermal and acoustic insulation. For the latter, the A-PhBR not only involves an "active" energy advantage (energy production) but also a "passive" (insulation and energy saving), which makes it especially competitive its incorporation.

The diagrams shown have been designed with this dual functionality: can be fully adapted to all constructive technologies available today and at the same time work as a complement of any type of construction of any period, since they can be added as layers.

This means that the use of algae in facade and roof is an effective tool for real reform at a time when we need more the re-building than the new-building. In short, this is a particularly novel strategy of "recycling the city", one among others that our team has been studying and publishing for years.

We must also highlight the fact that they can be applied to both the scale of the building or cluster of buildings or even neighborhood. The higher the scale is the larger the yield energy production and profitability of the whole.

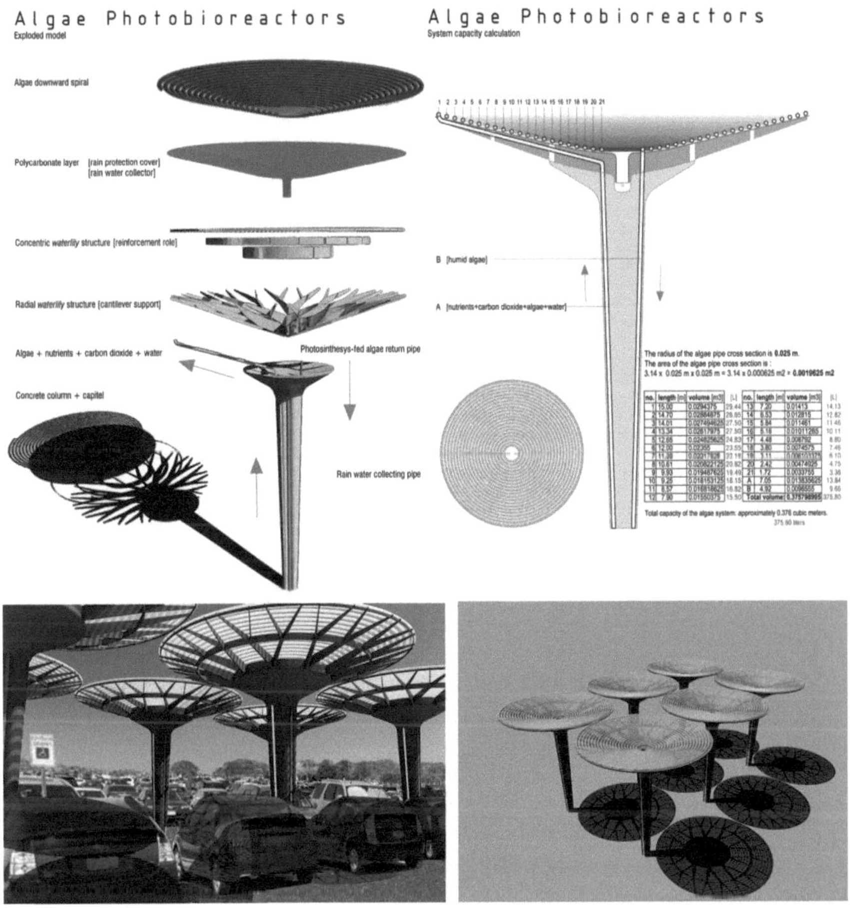

Fig. 7.8 Photo-biorreactor integrated in the street furniture, canopies (Design by R. Cervera and J. Pioz)

7.4 Molded Translucent Glass Blocks: Photo-Bioreactor in Facades

As a particular case, we would like to show a single application of Ph-BRA that has produced a patent and is able to form a facade. It is a modular system of hollow glass with double sided translucency, moulded with high mechanical strength, forming a duct system with closed loop, which can form facades of buildings and that allows the circulation of fluids inside a mode of photo-bioreactor for cultivating microalgae. Depending on its size, the modules form a single duct

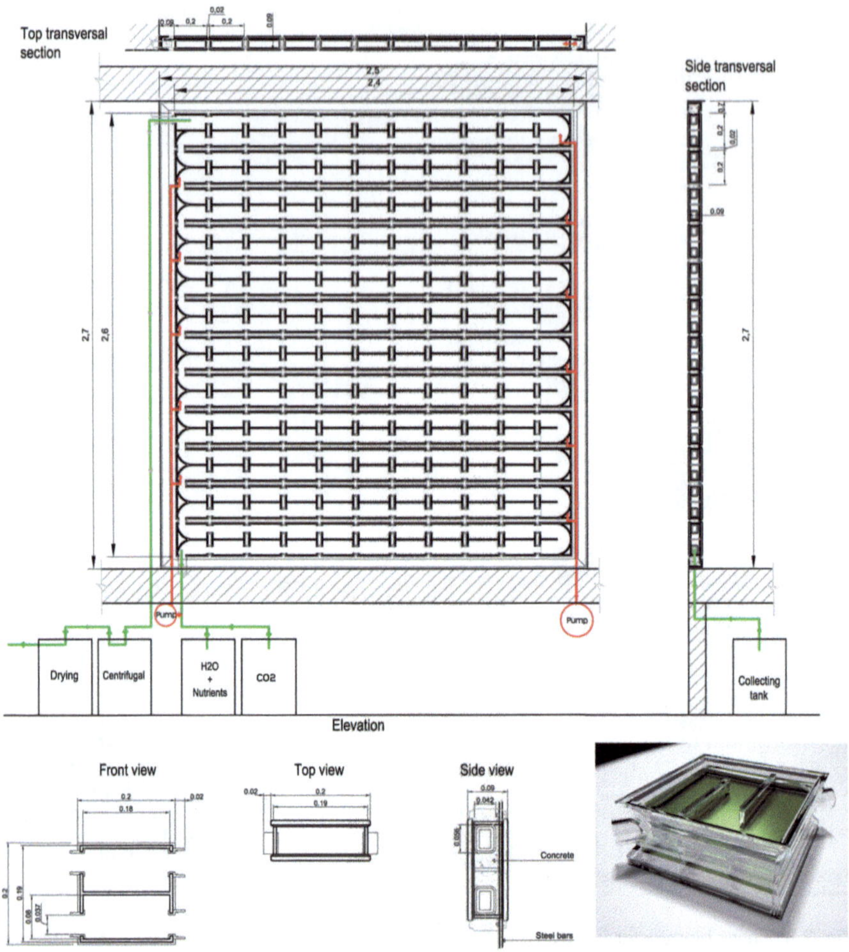

Fig. 7.9 Glass block photo-biorreactor. Technical details (Patent by Rosa Cervera, Javier Pioz and Antonio Ruiz de Elvira)

or through inner plates of glass or plastic that separates its interior into two or more chambers, multiple conduits and are designed so as to allow the modular coupling with each other by fasteners bayonet, and passage through those conduits, with pure liquid, a solution of algae or detergent for cleaning, and the entry of essential nutrients and carbon dioxide for photosynthesis, allowing the output of liquid with algae and oxygen in photo-bioreactors is a waste product (Figs. 7.9 and 7.10).

The system is complemented by the technical elements of water supply, microalgae, nutrients and CO_2 and harvested, including nutrient reservoir, preparation tank, collection tank, gas treatment unit, wastewater treatment unit and a control and electrical system.

Fig. 7.10 Glass block photo-biorreactor. Lab tests

The advantage of this system is that the façade itself, with no added external building or constructive elements, combined with biological products can function as a photo-bioreactor. In short, biomass is grown within the architectural construction itself. This significantly reduces the price of A-PhBR because with certain modifications, it is the facade itself that acts as an industrial A-PhBR.

Thus, the facades of molded glass or glass block, thanks to its transformation into A-PhBR, acquiring an iconic value converting into a facade of high value comparable to very high end range, while maintaining a very competitive price in the market.

The system is complemented by the technical elements of water supply, microalgae, nutrients and CO_2 and harvested, including nutrient reservoir, preparation tank, collection tank, gas treatment unit, wastewater treatment unit, as a control and electrical system.

Technical equipments vary by product obtained from cultivation of microalgae. Biomass obtained from the architectural photo-bioreactors can get heat energy from the biogas from the anaerobic digestion of biomass, which can obtain electrical energy from cogeneration of biogas and producing different products. The products obtained depend on the degree of processing and control. The least amount of intervention required is fertilizers, and then with an intermediate degree of processing are agricultural bio-stimulants. At the other extreme, i.e. with high quality processing control, are value-added products such as medicinal and those for human consumption. Depending on the treatment capacity of the biomass, it will choose one product over another (Fig. 7.11).

In general, we can say that with a system under control and processing is feasible to produce heat and electricity for self consumption. The system is viable from 500 m^2 of facade photo-bioreactor and advisable for areas of 1,000 m^2 on. From this measure the economic impact of the transformation of the facade photo bioreactor has very low impact (Fig. 7.12).

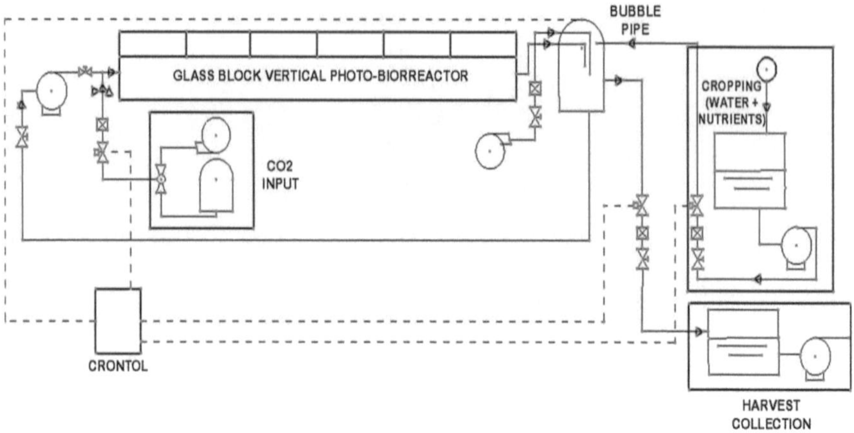

Fig. 7.11 Glass block photo-biorreactor. Diagram (R. Cervera and J. Pioz)

Today, the most profitable product obtained from the microalgae is bio-fertilizer and agricultural bio-stimulants with almost immediate return on investment. We must take into the consideration that the level of energy that balances chemical fertilizers, especially nitrogen, is responsible for a large part of energy consumption in agriculture. In particular, we can say that in developing countries fertilizers account for up to 70 % of total energy consumed in agriculture and in developed countries up to 40 %. Specifically in Spain the impact of bio-fertilizers in agriculture is 45 %.

7.4.1 Horizontal Photo-Bioreactor for Roof and Urban Fountains: Covered

This system is easily applicable to sources or urban landscaped ponds and transforming these elements in microalgae photo bioreactors. It is an innovative combination of cultivation of microalgae in building systems with indoor architectural function for energy and other products from biomass and microalgae and the fixation of CO_2 and improvement of the water cycle (Fig. 7.13).

The system supports the cover with simple and inexpensive materials such as plastic film. In this case it becomes semi-closed photo-bioreactor with a slight increase in productivity.

It is a shallow pool built by waterproofing sheet and piping system for water supply, nutrients and collecting the harvest. The contribution of CO_2 from the atmosphere may come or be brought directly.

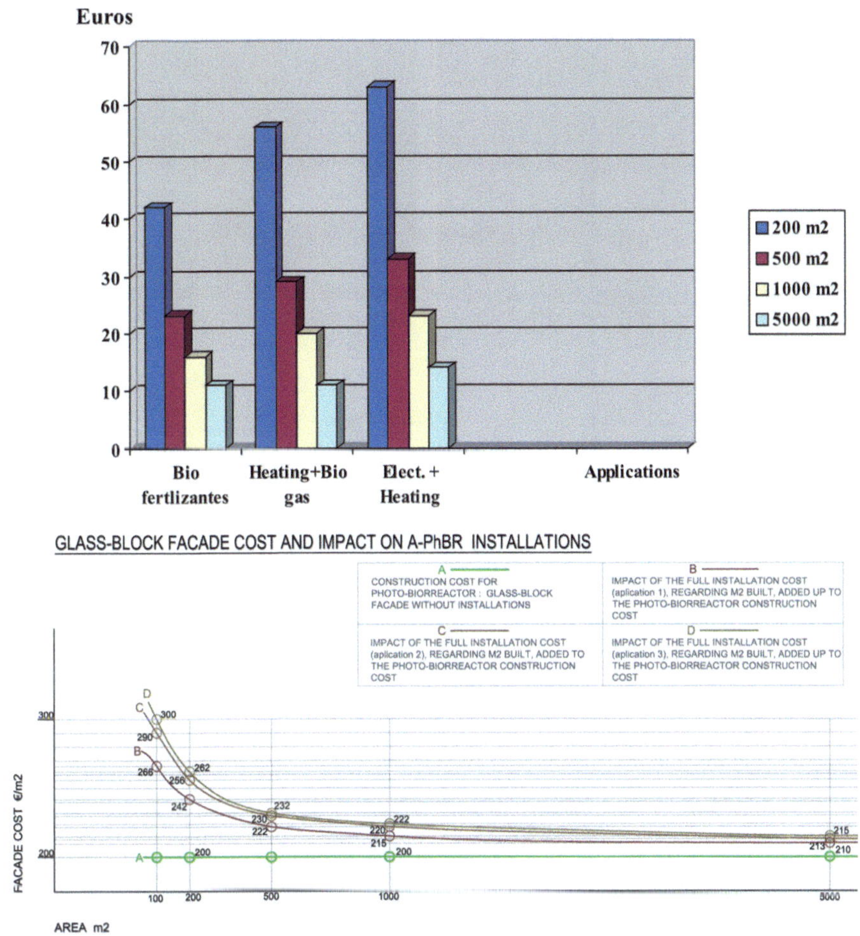

Fig. 7.12 Comparison between costs, application and surfaces (R. Cervera and J. Pioz)

We present a real project designed to be built on the outskirts of Madrid, is an urban fountain to be implanted in a roundabout. It is a type Open Pond photo-bioreactor formed by circular ponds with stirring blades in the centre to ensure the movement of the crop. The source also has decorative elements that in turn work with the operation of the photo-bioreactor as water curtain arches with LED lights (Fig. 7.14).

The source moves a volume of 23 m^3 of water with algae and achieves bio-fixation every year to about 337 kg/year, which represents emissions of a gasoline car that travels at about 1,700 km (Fig. 7.15).

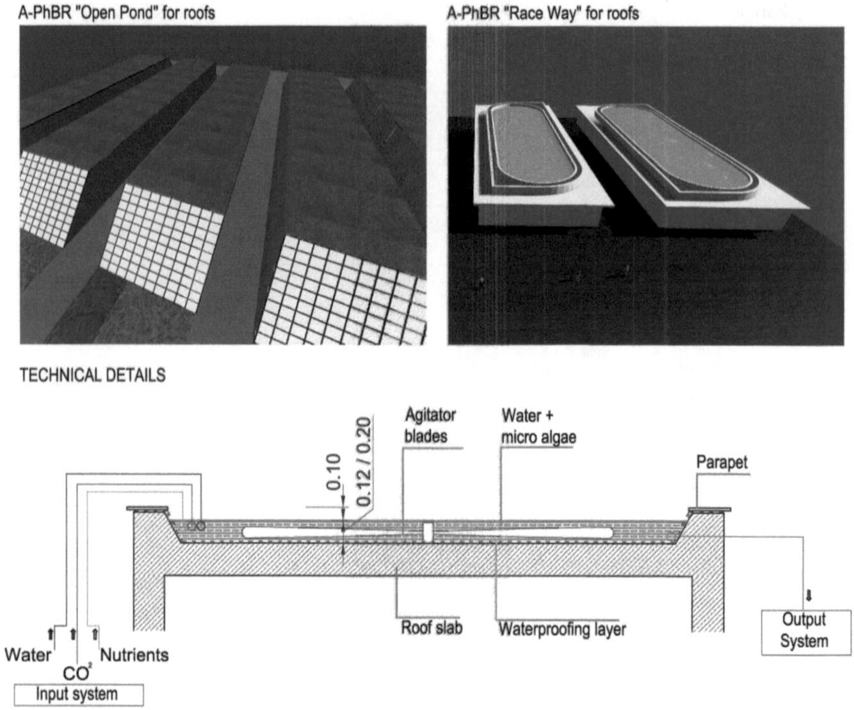

Fig. 7.13 Horizontal photo-bioreactor for roofs (Design by R. Cervera and J. Pioz)

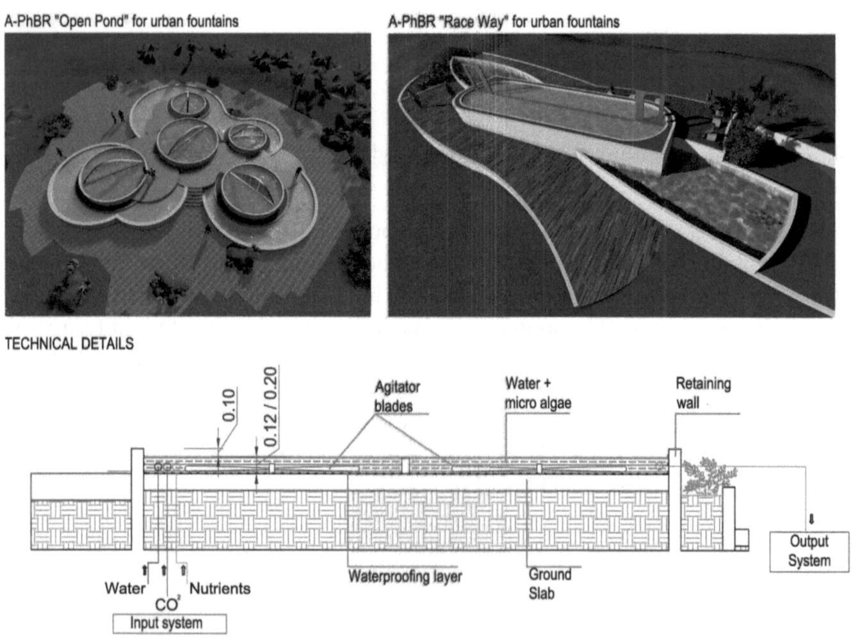

Fig. 7.14 Horizontal photo-bioreactor for fountains (Design by R. Cervera and J. Pioz)

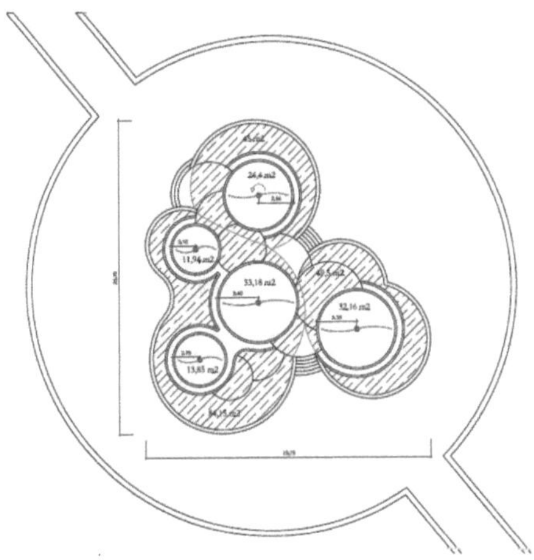

TECHNICAL SPECIFICATIONS

CULTIVATED VOLUME (M^3)	23,0
WATER CONSUMPTION FLOW (M^3/DAY)	2,3
ESTIMATED IRRIGATION FLOW (M^3/DAY)	2,3
BIOMASS PRODUCTION (KG/YEAR)	170,0
CO_2 BIOFIXATION (KG/YEAR)	337,0

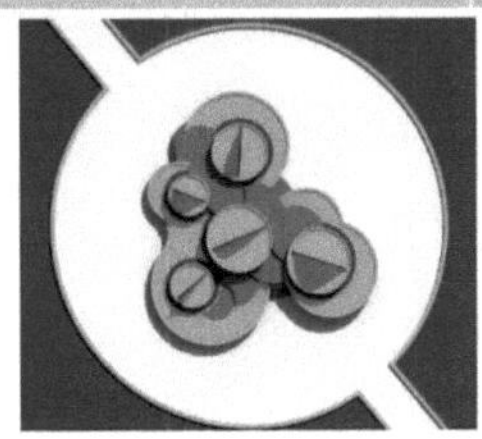

Fig. 7.15 Horizontal photo-bioreactor for fountains. Calculated technical data (Given by Alga Energy Company)

7.5 Conclusions

The proposal presented by the authors develops innovative prototypes of photo-bioreactors, integrated into the surfaces of cities and architecture—facades, roofs and pavements. Through this we are producing energy up to 10–15 % of a total energy consumption of a building. Parallely we are capturing CO_2 emissions in the place where those emissions are produced, being algae the only source of renewable energy able to fix carbon dioxide. The additional benefits are saving of land for algae cultivation, reducing urban temperature by self-thermoregulation and increasing architectural insulations—up to 10 % because of the effect of water. In this fashion, the architecture itself becomes an energy producing envelope, thus becoming a recycling form of architecture and introducing the nature into urban landscape.

Acknowledgments We acknowledge the financial support by a competitive pre-industrial research program given by CDTI (Centro para el Desarrollo Tecnológico Industrial) and the Government of Spain, Economic and Competitiveness Ministry. It has also been funded by Cervera and Pioz Architects S.L.P. We also acknowledge the cooperation from the University of Alcala (UAH).

References

Yoo JJ et al (2013) Development of thin-film photo-biorreactor and its application to out door culture of microalgae. doi:10.1007/s00449-0130898-2

Chapter 8
Reducing Indoor Air Pollutants Through Biotechnology

Fraser R. Torpy, Peter J. Irga and Margaret D. Burchett

Abstract Indoor environmental quality is a growing concern, as populations become more urbanised and people spend a greater proportion of their lives indoors. Volatile organic compounds outgassing from synthetic materials and carbon dioxide from human respiration have been major indoor air quality concerns. The growing use of energy-efficient recirculating ventilation solutions has led to greater accumulation of these pollutants indoors. A range of physiochemical methods have been developed to remove contaminants from indoor air, but all methods have high maintenance costs and none reduce CO_2, which some biological systems can achieve effectively with the additional benefit of the self-sustaining capacity of biological material. Bacteria are the major organisms involved in bioremediation of VOCs, although green plants may help sustain the bacterial community and add the capacity for CO_2 reduction to a system. The main problems faced by indoor air bioremediation systems is the extremely low concentrations of VOCs present indoors and the possibility of microbial release. Simple, passive biofiltration with potted green plants may be the simplest and most effective system for indoor air cleaning, but further research into substrate types, ventilation, and the microbiology of biodegradation processes is required to reveal their ultimate potential. Purely microbial systems have potential for the bioamelioration of high concentrations of toxic gases, but not without significant maintenance costs. Despite many years of study and substantial market demand, a proven formula for indoor air bioremediation for all applications is yet to be developed.

F.R. Torpy (✉) · P.J. Irga · M.D. Burchett
Plants and Environmental Quality Research Group School of the Environment,
University of Technology, Sydney, Australia
e-mail: Fraser.Torpy@uts.edu.au

F. Pacheco Torgal et al. (eds.), *Biotechnologies and Biomimetics for Civil Engineering*,
DOI 10.1007/978-3-319-09287-4_8

8.1 Growing Concerns About Indoor Air Pollution

As a result of the ongoing urbanisation of the Western world, around 80 % of people now live in urban areas, where around 90 % of time is spent indoors (Newton 2001; Environment Australia 2003). The quest for sustainable urban communities must therefore include the achievement and maintenance of a healthy 'building ecology'. Thus urban air pollution (UAP) has become a worldwide health concern. Ninety per cent of UAP comes from fossil fuel emissions, comprising a mixture of carbon dioxide (CO_2) and monoxide (CO), nitrogen oxides (NO_x), sulfur oxides (SO_x), volatile organic compounds (VOCs), ozone and fine to very fine particulate matter (PM10 and PM2.5).

Global healthcare and associated costs due to outdoor air pollution have been estimated at US$828 trillion for developed countries, while indoor air pollution costs for these countries is likely to be just under US$90 trillion (Hutton 2013). Health costs associated with indoor air pollution may be as high as US$9.4 billion in the USA alone, leading to as many as 150,000 mortalities per year (Guieysse et al. 2008).

The air quality of indoor spaces has now become a dominant heath concern (Brown 1997; WHO 2000; Environment Australia 2003). Indoor air quality (IAQ) has been designated as a significant health concern in both the USA and Europe (Morey et al. 2001; Bernstein et al. 2008a). Although not generally recognised, indoor air pollution is almost always higher than outdoors. This is because as outdoor-sourced contaminated air enters through natural or mechanical ventilation systems, it mixes with indoor-sourced pollutants, in particular VOCs outgassing from synthetic furnishings, finishes, paints, solvents and other materials derived from petrochemicals (Environment Australia 2003; Sakai et al. 2004; Barro et al. 2009; Chan et al. 2009), and higher CO_2 concentrations produced by human respiration and gas appliances (Norbäck and Nordström 2008). The tendency of buildings constructed since the worldwide oil crisis in the 1970s to be sealed against the outside environment and rely wholly on mechanical ventilation for air replenishment has exacerbated problems associated with pollutant concentration in indoor air. This issue is discussed further in Sect. 8.2. Unless or until such time as a return to primarily natural ventilation becomes the norm, there is a critical need to develop means of maintaining a habitable indoor environment, whilst simultaneously improving building energy use and lowering greenhouse gas emissions in an increasingly energy aware world. Hence it would appear that biological, botanical and biotechnological approaches to solving this growing problem are the most promising avenues to follow.

8.2 Current Practice

Replacing contaminated indoor air with outdoor air by increasing building ventilation rates is the simplest and probably most effective means of improving IAQ. A key reason that IAQ is increasingly becoming a major concern is the increasing reliance in modern buildings on heating, ventilation and air conditioning systems (HVACs) for the maintenance of thermal comfort, and a growing dependence on reducing ventilation rates to save energy. Since the 1970s, building designers have increasingly sealed buildings to reduce energy expenditure from having to adjust the temperature of outdoor air used for ventilation (Darlington et al. 2000). To maintain IAQ, building designers have come to rely on HVAC systems to recirculate indoor air (Fadzli Haniff et al. 2013) with a greatly reduced outdoor air dilution level, leading to the accumulation of VOCs and CO_2. Noise, security and declining outdoor air quality are other reasons that increased outdoor ventilation is declining as a means of regulating IAQ (Guieyesse et al. 2008). If these low external ventilation rates are to be maintained, air cleaning technology will become essential for maintaining adequate IAQ whilst reducing energy expenditure.

8.3 The History of Bioremediation of Indoor Air

The early impetus for developing biological systems to mitigate the deterioration of IAQ came from space research, in particular the move to develop long-term habitations outside of the earth's atmosphere. 'Biological life support systems' (BLSS) have been proposed over a number of decades as a means for supporting sustainable human habitation in the vacuum of space (e.g. Myers 1954).

Early studies on potted plants and air-pollution reduction were carried out by Wolverton and colleagues for NASA and, subsequently, for the US Interior Plant Growers Association (Wolverton et al. 1984, 1989). These studies investigated potential uses of potted plants in space stations. It was found that these plants could absorb substantial concentrations of the VOCs formaldehyde, trichloroethylene and benzene from chamber air (Wolverton et al. 1984, 1989). Wolverton (1997) subsequently conducted screening chamber studies of the VOC removal ability of 50 indoor plant species, and found they all showed some capacity to reduce VOC concentrations. This early work was developed by Wood et al. (2002) and Orwell et al. (2004, 2006), who established that it was the microflora associated with the growth substrate, rather than the potted plant itself that was the active component of the system for removing VOCs. These studies demonstrated VOC (benzene) removal by potted plants at very high rates for initial VOC concentrations of up to 163,000 $\mu g.m^{-3}$, and also at more indoor air realistic levels of $\sim$800 $\mu g.m^{-3}$ for *n*-hexane, toluene and xylene.

Numerous further laboratory test-chamber studies demonstrated the potential for significant improvement in IAQ through the passive use of potted plants, and, to date, approximately 200 species have been tested for VOC removal capacity, all with positive results (e.g. Liu et al. 2007; Yang et al. 2009; Kim et al. 2008). However a number of criticisms have appeared in the literature, in particular challenging the validity of extrapolating static chamber data to real-world environments, most notably regarding the capacity of passive potted plants to quantitatively influence the extremely low normal airborne concentrations of VOCs in most indoor environments (e.g. Llewellyn and Dixon 2011; Soreanu et al. 2013). Simple potted plant based-biolfiltration systems, nonetheless, clearly have the potential to be effective in the built environment. In a large study of university offices, Wood et al. (2006) demonstrated that the presence of an indoor plant was associated with significant reductions in ambient total volatile organic compound (TVOC) levels. Unfortunately, there has been little subsequent research performed under field conditions to test the efficacy of passive potted plant systems, and possibly as a result of this, active, and other biofilter type systems have dominated the research in this field in more recent years.

8.4 Indoor Air Pollutants

8.4.1 Volatile Organic Compounds

Of the different types of indoor air pollutants, VOCs have received the great majority of research attention. Even at imperceptible levels (<200 ppb), mixtures of VOCs are thought to cause symptoms of ‘sick-building-syndrome’ or ‘building-related-illness’ (Jaakola et al. 2007; Liu et al. 2007; Epstein 2008).

VOC is a general term encompassing various classes of carbon-containing, low boiling-point compounds that are gaseous at room temperature. The combination of VOC sources within buildings result in the occupant typically being exposed to 50–300 different VOCs, at levels usually in the $\mu g/m^{-3}$ (or ppbv) range (Bernstein et al. 2008b). VOCs have been linked to an array of adverse health responses. Even at levels well below human perception, VOCs can contribute to symptoms that resemble those of SBS, and at high levels have the potential to be hematotoxic, neurotoxic, leukemogenic and, in the case of some compounds, such as benzene, also carcinogenic (Vaughan et al. 1986; Wallace 2001; Wolkoff and Nielsen 2001). The review of Guieysse et al. (2008) provides a detailed review of the health effects of indoor VOCs. Whilst there is some debate as to the magnitude of the health effects of $\mu g/m^{-3}$ concentrations of VOCs on building occupants (Wolkoff 2013), and how those effects vary with different VOC mixtures and relative concentrations, it is generally accepted that exposure to VOCs indoors has significant negative health outcomes.

Apart from the occasional report documenting trivial VOC removal capacity by plant tissues (Kim et al. 2008, Treesubsuntorn and Thiravetyan 2012), or bacteria on the leaf surface or phyllosphere (Sandhu et al. 2007), the quantitative bioremediation of VOCs is now generally accepted to primarily occur through root zone microbial activity (Wood et al. 2002; Orwell et al. 2004; Kim et al. 2008), where the VOCs are used directly as carbon sources by microorganisms. One of the primary challenges for the biological removal of VOCs is that alone, their concentrations in indoor air are probably insufficient to sustain microbial growth (Guieyesse et al. 2008). Thus major efforts in the development of biological processes to ameliorate

VOCs in indoor air have targeted means to expose active, contained microbial communities or populations to large volumes of air. With the high airflow rates required with such systems, the release of microorganisms back into the air has been recognised as a potential concern for microbial air remediation processes (see Sect. 8.12).

8.4.2 Carbon Dioxide

Indoors, excess CO_2 is produced mainly by human respiration. It is not normally considered to be a toxic air contaminant, but with increasing levels above the current outdoor ambient concentration (currently ~397 ppmv; Tans 2014) it can act as a simple narcotic, and has been associated with adverse symptoms related to the mucous membranes (dry eyes, sore throat, nose congestion, sneezing) and to the lower respiratory tract (tight chest, short breath, cough and wheezing) (Erdmann and Apte 2004). At concentrations between 2,500 and 5,000 ppmv, CO_2 can cause headache, with serious health consequences arising at concentrations greater than 50,000 ppmv (Milton et al. 2000; Harris and Moore 2009; Gurjar et al. 2010). It is well known that elevated CO_2 concentrations in office buildings are associated with increased illness symptoms among occupants (Milton et al. 2000; Erdmann and Apte 2004; Seppänen and Fisk 2004), and student academic performance and workplace productivity have both been shown to decline with increased CO_2 levels (Bakó-Biró et al. 2004; Seppänen et al. 2006; Shaughnessy et al. 2006). In most cases, inadequate building ventilation system is found to be the cause of CO_2 accumulation indoors (Redlich et al. 1997). The American Society of Heating, Refrigeration and Air-Conditioning Engineers (ASHRAE) recommends that the maximum concentration of CO_2 should not exceed 1,000 ppm (ASHRAE 2011), but even at this level some reductions in workplace performance can be expected.

The bioremediation of indoor CO_2 has received limited research attention, although several studies have tested the potential of indoor plants for mitigating excess CO_2. Oh et al. (2011) demonstrated chamber CO_2 reduction capacity for three indoor plant species. Pennisi and van Iersel (2012) examined the CO_2 draw-down capacity of common indoor plants and concluded that the limited photosynthetic rate of indoor plants would lead to the necessity for an impractical volume of indoor plants to make a substantial difference to indoor CO_2 levels. Contrasting with these results, Tarran et al. (2007) performed a study using city offices and found that three or more potted plants were associated with a 10 % reduction in CO_2 concentrations in an air-conditioned building, and a 25 % reduction in a non-air-conditioned building. The greatest potential use of phytotechnology for CO_2 mitigation thus possibly lies in replacing a proportion of a building's ventilation system, and thus saving energy. It has been estimated that the use of appropriate green plant design could reduce HVAC energy loads by 10 % (Afrin 2009), whilst Rodgers et al. (2012) estimated an 86 % energy saving from an active 'Biowall' system.

Torpy et al. (2014) performed further chamber experiments, revealing that by manipulating the light level and plant species the net efficiency of the botanical CO_2 mitigation system could be significantly improved, although it was also noted that due to the respiration of substrate, microorganisms can reduce the potential effectiveness of any plant-based CO_2 removal system. Irga et al. (2013) detected improved CO_2 removal efficiency by replacing the potting mix substrate with hydroculture medium, which also lead to a slight reduction in the VOC removal efficiency of the system. Future research to determine the practicality of biological CO_2 mitigation may involve green wall technology, as the higher density of photosynthetic material, along with the ease of increasing light intensity and thus photosynthetic rate should improve the effectiveness of the system.

8.4.3 Other Pollutants

There are a range of other indoor air pollutants that can lead to health problems, notably NO_x, SO_x, ozone, radon and particulate matter including bioparticles such as mould spores, suspended bacteria and viruses and microbial cell components (Harris and Moore 2009). Whilst the biotechnological mitigation of pollutants other than CO_2 and VOCs have received little research attention, indoor plants have been shown to have removal capacity for NO_2 (Coward et al. 1996; Yoneyama et al. 2002), ammonia (Yoneyama et al. 2002), SO_2 (Lee and Sim 1999), ozone (Elkiey and Ormrod 1981) and particulate matter (Lohr and Pearson-Mims 1996). Whilst industrial processes for the reduction of many other pollutants are well known, the potential for the abatement of these materials indoors is currently undeveloped, and will undoubtedly receive growing attention in the future. The generation of microbial air pollution is often claimed as a major drawback to the use of biological air quality mitigation systems. This aspect will be discussed in Sect. 8.12.

8.5 Physiochemical Versus Biological Methods

Much of the conceptual understanding on the advantages and disadvantages of the competing physiochemical and biological systems has once again come from space research. The key driver for such applications is based on Equivalent System Mass (ESM) estimates, whereby projections are made on the duration of space habitation required for a biological system to become more efficient on a system weight basis than the mechanical, physiochemical alternatives (Bamsey et al. 2009). It is generally thought that sustained human presence outside of the earth's atmosphere will require BLSS, owing to their subsidiary value in food and water production, along with atmospheric management (Bamsey et al. 2009). Direct technology transfer from this research has led to the development of biofilters which have been installed in several public buildings in Canada (Bamsey et al. 2009).

Sorption filtration is the most commonly used air cleaning process (Chen et al. 2005). It involves the removal of pollutants (typically VOCs) by adsorption on solid media. Common adsorbent substrates include activated carbon, zeolite and activated alumina. Adsorbents generally suffer from VOC saturation after long-term use, which can be as little as 3–6 months (Chen et al. 2005; Wang and Zhang 2011). Furthermore, re-emission of the adsorbed VOCs is possible from some systems (Chen et al. 2005), leading to an increased necessity for maintenance.

Ultraviolet-photocatalytic oxidation (UV-PCO) removes gaseous contaminants via chemical reactions on a semiconductor catalyst surface under UV irradiation. Whilst the system has been shown to be effective in reducing a range of VOCs (Zhao and Yang 2003; Chen et al. 2005; Mo et al. 2009), this system has not yet been widely commercialised, although it shows considerable potential (Guieyesse et al., 2008; Mo et al. 2009). Photocatalysts may also be added to interior paint (Auvinen and Wirtanen 2008), although such systems are yet to be developed in an effective form.

Ozone-based air cleaners work on the principle of VOC oxidation. However, ozone is itself hazardous and the oxidation process may in some cases lead to the generation of harmful by-products (Chen et al. 2005; Kwong et al. 2008).

Air ionizers also work on the principle of the oxidation of VOCs, however, through the action of radicals rather than ozone. This process suffers from the same disadvantage as ozone air cleaners, and may, in fact, generate ozone (Chen et al. 2005).

All physiochemical air cleaners may employ a high efficiency particulate air (HEPA) filter and various types of pre-filters within the systems for reduction of particulate matter.

Chen et al. (2005) compared 15 different air cleaning technologies for their capacity to reduce airborne concentrations of 16 VOCs. The technologies evaluated included sorption filtration, UV-PCO, ozone oxidation and air ionization (plasma decomposition), which were compared to an undescribed proprietary botanical process. Twelve of the air cleaners were portable and three were residential scale in-duct systems. The study found that sorption based air cleaners were the most effective for both VOCs and ketones (formaldehyde and acetaldehyde), although some UV-PCO methods were also quite effective. The botanical filter had little effect on chamber concentrations of VOCs other than *n*-hexanal and ketones, although these were removed at a substantially lower rate than the sorption processes except those using only activated carbon, which did not reduce ketone concentrations. Ozone generating methods were ineffective, and are not recommended by the authors. Saturation effects and ozone generation were observed for several systems, reducing their practical application for air cleaning.

Whilst it is not known how effective the proprietary botanical process in Chen et al.'s (2005) study was relative to other biological systems, it would thus appear that for a simple, short-term comparison, physiochemical methods are more effective for general removal of VOCs, and certainly particulates, from indoor air. However when the long-term effectiveness of the systems are taken into account, along with the simultaneous capacity to remove CO_2, reduced maintenance and subsidiary benefits such as workplace efficiency improvement (see Sect. 8.10), horticultural systems become competitive. Biological air cleaners may also be more effective at removing the most volatile VOCs, such as formaldehyde than physiochemical processes that do not use activated carbon adsorption (Chen et al. 2005). More recent work (e.g. Torpy et al. 2013a, b) has shown that there is a large and mostly unexplored potential for horticultural biotechnology to improve the efficacy of biological systems.

Thus physiochemical methods of IAQ amelioration, whilst effective in the short term, have a number of crucial disadvantages. Most are incapable of removing all of the key gaseous pollutants simultaneously, some are unsafe due to the emission of ozone, and all require expensive regular maintenance and are costly to install (Soreanu et al. 2013). Research into biotechnological alternatives thus may offer a realistic alternative to engineering solutions for the maintenance of acceptable IAQ.

8.6 Hybrid Physiochemical–Biological Systems

As noted previously, physiochemical air quality improvement systems require regular replacement of the filtration medium to remain effective, and to prevent the re-emission of absorbed gases. To address this problem, Wang and Zhang (2011) used activated carbon as a support for a microbial biofilter, postulating that in such a system the microbial community may constantly degrade accumulating VOCs, alleviating the need for constant replacement. Their system, installed in a new 96.8 m^2 office, was highly effective: 5 % outdoor air plus biofiltration using plants lead to similar indoor formaldehyde and toluene concentration level as 25 % outdoor air without biofiltration.

Aydogan and Montoya (2011) compared the performance of three hydroponic supports: growstone, expanded clay and activated carbon, with and without plants, to remove formaldehyde. Activated carbon was the most effective, removing $\sim$98 % of the formaldehyde in 10 h, a rate somewhat more efficient than that achieved by four hydroponic indoor plant species, although the authors unfortunately did not describe the hydroponic support used for this assessment.

Such hybrid physiochemical–biological systems may have major potential for development as the most effective means of non-ventilation-mediated removal of VOCs from indoor air.

8.7 Phytoremediation and Horticultural Biotechnology

Phytoremediation, the process whereby plants are used to ameliorate a pollution source, is a well-developed and economical technology for remediation of soil and water contaminated with heavy metals, fertilisers or hydrocarbons (Pilon-Smits

2005). The development of phytoremediation systems for mitigating contaminated air is somewhat less developed. Whilst effective IAQ improvement has been recorded (Darlington et al. 2000; Wood et al. 2006; Wang and Zhang 2011), there are relatively few fully tested, commercial systems for integration into common building schemes, although the systems that have been developed have considerable potential, and will be discussed in a later section.

The most widely researched system for botanical air cleaning is by passive filtration using potted plants. Research over the last three decades has demonstrated that passive biofiltration with indoor plants can significantly reduce concentrations of most types of urban air pollutants (Wolverton et al. 1989; Coward et al. 1996; Lee and Sim 1999; Yoneyama et al. 2002; Orwell et al. 2004; Wood et al. 2006; Yoo et al. 2006; Kim et al. 2008; Irga et al. 2013). The primary advantages of passive phytoremediation over active biofilters are cost and flexibility of installation, as in many cases no specific infrastructure changes are required to install potted plants in most indoor spaces (Soreanu et al. 2013).

The biologically active component of botanical air filtration systems for hydrocarbon VOC biodegradation is the microorganisms associated with the roots of the plants or the growing substrate of the plants, whether it be potting mix or a hydroponic or hydroculture material. Whilst older studies hypothesised on the VOC bio-degradative capacity of plant-associated bacteria (e.g. Wood et al. 2002), Zhang et al. (2013) provided empirical evidence for this effect by demonstrating toluene removal by bacteria isolated from the rhizosphere of potted indoor plants. Although some VOC removal has been recorded across plant leaves with the rhizosphere having been excluded (Ugrekhelidze et al. 1997; De Kempener et al. 2004), the quantities removed were trivial relative to the amounts removed by whole potted plants, or even potting mix without plants present (e.g. Wood et al. 2002; Orwell et al. 2004). Reported formaldehyde removal rates by plant tissues alone were found to be so low as to be insignificant for IAQ improvement purposes (Schmitz et al. 2000), whilst substantial formaldehyde removal has been detected for individual bacterial species associated with an active biofilter (Wang and Zhang 2011), and also for indoor plants when the substrate was also present (Xu et al. 2011). The removal of several other contaminants, however, such as CO_2 (Oh et al. 2011, Torpy et al. 2014), SO_x, NO_x (Elkiey and Ormrod 1981; Esguerra et al. 1983) and ozone (Elkiey and Ormrod 1981) appear to be mostly or wholly plant mediated, and are taken up directly through the stomates (gas exchange pores) of the green shoots, which in most species are open only during daylight hours, and not in the dark. Particulate matter (Lohr and Pearson-Mims 1996) is effectively reduced by living plants by deposition through the extensive boundary layer area present around leafy tissue.

It is generally thought that, with respect to VOC removal, the value of the potted plant in passive systems is mainly to provide a supply of nutrients to the microbial community as root exudates (e.g. Orwell et al. 2004, 2006), although this has not been directly tested. Further, the biochemical pathways by which VOCs are degraded and what bacterial or fungal taxa are directly involved have not been determined. However, although the biology of the process by which

VOCs are degraded in plant–rhizosphere microbial community systems is generally poorly described, it is probable that it resembles the processes whereby liquid or solid phase hydrocarbon water and soil contaminants are biodegraded, although in many cases the details of the biochemical pathways involved are also still unknown (e.g. Haritash and Kaushik 2009; McGenity 2014).

Research suggests that all potted plant /substrate combinations can remove VOCs at similar rates for a given volume of growth substrate (Orwell et al. 2004), with slightly reduced rates for plants grown in hydroculture, probably due to the lower microbial density in the substrate (Irga et al. 2013). Two recent studies have reported significant variation in VOC removal rates between plant species (Yang et al. 2009; Liu et al. 2007), however, in each study rates were computed on the basis of removal per unit leaf area, rather than on pot volume, which reveals removal rates of the rhizosphere. A large number of potted plant species and varieties have been tested to date (approx. 120: Soreanu et al. 2013), and all have shown VOC removal capacity. However, there is still a clear need for further research into how plants influence the biodegradative capacity of their rhizosphere microflora.

The identification of bacteria as the primary agents in the biodegradation of VOCs leads to the potential for improvement of the system's capacity. Zhang et al. (2013) identified the rhizospheric bacteria directly associated with toluene removal, and postulated that the specific bioaugmentation with these species may increase the rate of VOC removal. Torpy et al. (2013a) demonstrated the microbial community-level activity specific to biodegradation, and that the benzene removal efficiency of the bacterial community can be manipulated with the addition of specific bacterial nutrients, leading to a simple means of increasing the capacity of the system. Once the concept of using bacterial communities for the bioremediation of indoor air pollutants has become widely accepted, it is clear that the potential for developing very large capacities for VOC removal will be the next target in the development of these systems for air cleaning.

The great majority of the botanical VOC removal literature describes manipulative laboratory experiments, where potted plants are placed in small, sealed chambers, into which one or more VOCs are introduced and the rate of decay determined by analysis of the chamber atmosphere (Fig. 8.1). Most studies correct for background VOC losses with no plant or substrate in the chambers, which are substantial in most studies (e.g. Yoo et al. 2006). It is obvious that chamber studies cannot realistically be extrapolated directly to real environments in buildings (e.g. Llewellyn and Dixon 2011; Irga et al. 2013; Soreanu et al. 2013), primarily because the plant density per unit volume of chamber atmosphere is far higher than would be possible in buildings. Future research thus needs to test objectively the relationship between the chamber performance of potted plants and their performance in buildings.

Recent research has attempted to estimate the efficacy of biological air cleaning systems by determining their outdoor air ventilation rate equivalence. An active botanical air cleaning system (Wang and Zhang 2011) was found to be equivalent to a ventilation rate with 20 % outdoor air, measured on the basis of VOC

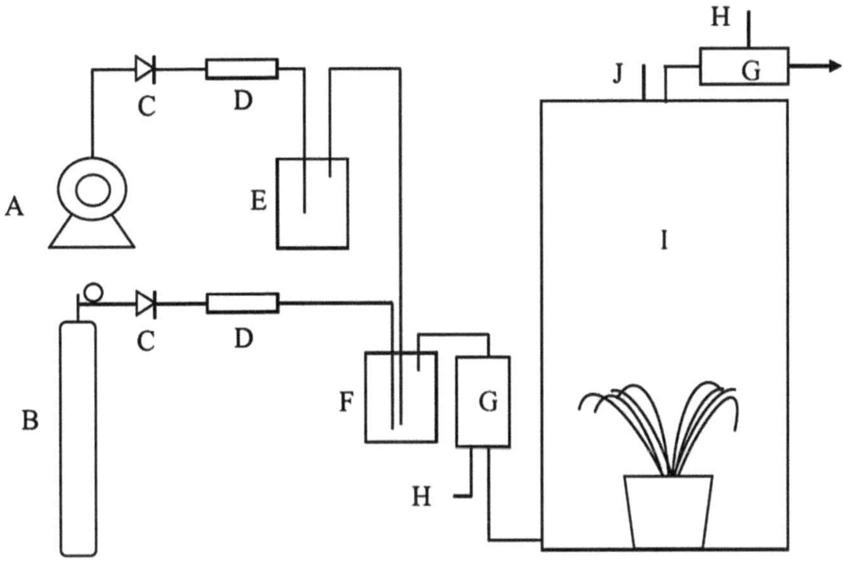

Fig. 8.1 Scheme of the experimental chamber apparatus used by Xu et al. (2011). (*A*) air pump, (*B*) N_2 gas cylinder, (*C*) needle vale, (*D*) flowmeter, (*E*) formaldehyde solution (37 %) vessel, (*F*) mixing vessel, (*G*) buffer vessel, (*H*) gas sampling port, (*I*) dynamic chamber, (*J*) water injecting port

concentrations. The authors, however, did not assess CO_2 concentrations, thus leading to ambiguity as regards the true effect of the system as a replacement for HVAC. Rodgers et al. (2012) installed an active 'Biowall' into a mock residential house, to investigate its potential energy saving capacity. This system was integrated into an air-conditioning system drawing outside air only, and yielded an 86 % reduction in total energy expenditure for ventilation, when compared to the air-conditioning system alone.

The only in situ study to show significant reduction in airborne TVOC levels using passive botanical air cleaning systems is by Wood et al. (2006), who found that three or six standard potted plants could produce a steady state in University offices where indoor VOC concentrations were maintained below 100 ppbv. Thus the plants only had a measurable effect when VOC concentrations were >100 ppbv, but were capable of reducing any greater concentrations than this to a minimum level. These findings are in contrast to both mass balance estimates (e.g. Guieyesse et al. 2008) and some experimental findings (e.g. Chen et al. 2005), leading to Guieyesse et al.'s (2008) suggestion that the current mathematical evaluation models are inadequate, and there are sources of variability in the biological studies which the mathematical models do not capture.

Potted plants therefore clearly have some capacity to ameliorate indoor VOC levels, although there are significant anomalies between empirical evidence and theoretical modelling, and research is urgently needed in this area to determine how VOC phytoremediation systems can be designed as feasible installations for general IAQ improvement in real buildings. It is thus hoped that future research funding for IAQ improvement is directed towards disentangling this problem.

It has been proposed that the capacity-limiting step for biological air cleaning is the rate at which gaseous air contaminants can infiltrate the substrate upon which the biological component, whether plants or microorganisms, are growing (Guieyesse et al. 2008). Active or dynamic air cleaning systems operate on the principle of using either pressure to increase the volume of air to which a biofiltration system is exposed, or a mobile aqueous phase into which contaminants are accumulated and thus transferred to a bioreactor. Both methods thus potentially increase the rate at which contaminants are removed. Passive systems rely on the diffusion of contaminants to a static bioremediation system, such as the rhizosphere of a potted plant or green wall, which could be expected to be slow for low concentrations of pollutants in spaces with poor air circulation. Some green wall systems, with constant trickling nutrient solutions but no pressure assistance to increase airflow (e.g. Darlington et al. 2000), might thus be considered as hybrid passive–active systems.

Chen et al. (2005) found that whilst a plant-based air cleaning system was effective in removing VOCs, its applicability was restricted by its inability to process high rates of airflow, a disadvantage not suffered by most sorption filtration systems. Darlington et al. (2001) assessed the effect of airflow rate on the capacity of his active botanical biofiltration system to remove airborne toluene, ethylbenzene and *o*-xylene. They found that, whilst the single pass efficiency of the biofilter was greater when the airflow was very low (0.025 ms^{-1}), the maximum removal capacity of the system occurred at the highest airflow tested (0.2 ms^{-1}). The authors proposed that it was the diffusion of VOCs into the aqueous phase of the system that was rate-limiting to biodegradation, suggesting that systems that increase the airflow rate even further may have further increased efficiency. However converse to these observations, the compost-based biofilter tested by Delhoménie and Heitz (2003) had a gross toluene removal efficiency that was proportional not to the airflow rate, but the residence time within the biofilter column. This system was tested at relatively high VOC concentrations only.

To maximise the capacity of biofiltration systems, there is thus a need to process as large a volume of contaminated air as possible, whilst exposing the air to the biological material for the critical time period over which *sufficient* pollutant removal will occur to allow air contaminants to be reduced to habitable levels. These two variables: airflow versus residence time are clearly the key attributes to maximise the efficiency of any system. It would thus be useful if, whilst developing and testing new biological air cleaning systems, these data could be included in published results, as they allow a simple and direct comparison of new systems for their efficacy.

Darlington et al. (2000) investigated a system installed in a 160 m^2 room with a very low outdoor air change rate (0.2 air changes per hour). Air from the test room was circulated through an 'ecologically complex' biofilter consisting of a 'bioscrubber', 30 m^2 of hydroponic plants and a 3,500 L aquarium containing aquatic and semi-aquatic plants. It was hypothesised that the very large biomass to building space ratio utilised would lead to reduced VOCs due to bioremediation from both the bioscrubber and plants, but may have led to the release of new VOCs and biological particles, such as mould spores, associated with the biological material. The bioscrubbers comprised multiple 2.4 m^2 fibreglass supports faced with lava rock, and colonised with mosses. Indoor air was actively exposed to the bioscrubbers with variable speed fans, maintaining a low airflow rate (0.01 ms^{-1}) across the modules. Air was passed through the canopy of a small hydroponic garden before reaching the bioscrubbers. A substantial non-integrated (i.e. passive) green wall and hydroponic plantings were also resident in the room. Plants were supplied with supplementary lighting. The aquarium was integrated in the irrigation system for both the hydroponics and the green wall. The biofilter appeared to substantially reduce VOCs including formaldehyde. Unfortunately, the authors (Darlington et al. 2000) did not record CO_2 levels in their study, as hydroponic plants with supplementary lighting could realistically be expected to remove large amounts of this gas, based on the findings of Irga et al. (2013), who found strong CO_2 removal for small hydroponic plants in sealed chambers.

Wang and Zhang (2011) developed an active biofiltration system combining both physiochemical and biological bioremediative capabilities, by using activated carbon as a hydroponic substrate for indoor plants. They tested the system both in a chamber and fitted to the HVAC system of a 97 m^2 newly built office, thus allowing a reasonable comparison between closed-system and *in situ* conditions. They found that their system was equivalent to a clean air delivery rate of 476 m^3/h outdoor air with respect to VOC (formaldehyde and toluene) removal. After demonstrating the efficacy of their system over a 300 d period, the authors estimated a 10–15 % reduction in ventilation energy costs in cold climate conditions. The function of biofilters to reduce building energy consumption will become a major focus of future research (e.g. Rodgers et al. 2013).

An active biofiltration system using a column containing inert substrates and compost, and supporting the growth of spider plant (*Chlorophytum comosum*), a species known to be capable of formaldehyde biodegradation (Giese et al. 1994), was tested by Xu et al. (2010), and found to be very effective at removing high concentrations of formaldehyde. Active systems thus may have potential to substantially increase the effectiveness of botanical biofilters, although at the cost of some energy. The potential for microorganism release when pressurised air is used may also be a concern (see Sect. 8.12). The biotrickling filter system used by Darlington et al. (2000), utilised plants as an adjunct to the biologically active microorganisms is his system, and is thus an alternative active phytoremediation system.

Active biofiltration is now a widely used air pollution control technology for industrial waste gases and odours (e.g. Elmrini et al. 2001; Vergara-Fernández et al. 2007; De Visscher and Li 2008; Mudliar et al. 2010; Lebrero et al. 2012). Systems are now becoming highly developed (e.g. Estrada et al. 2013b), and due to low running costs, high removal efficiency for a range of organic and inorganic gaseous pollutants and lack of secondary pollutant production are now competitive with physiochemical processes (Lee et al. 2013). With this strong background, biofilters are beginning to emerge as an accepted alternative to ventilation for the maintenance of IAQ.

Whilst it is usually assumed that active biofilters will remove higher quantities of air pollutants than passive systems due to the increased rate of airflow over the biodegradative surfaces (e.g. Soreanu et al. 2013), there appears to be no literature comparing an active system to an otherwise identical passive arrangement, and thus there is no empirical evidence to show that actively increasing the airflow to a system increases biodegradation over simple diffusion. Also, the potential generation of CO_2 from substrate microorganism respiration, and the emission of microbial particles are other issues that have been inadequately addressed in the literature. Clearly, there is a need for greater research on the correlation between the rate of airflow and all types of air quality for active IAQ bioremediation systems.

8.8 Health Benefits of Indoor Plants Unrelated to Air Quality

Indoor plants have also been shown to yield directly measurable benefits to the health and wellbeing of building occupants. Fjeld et al. (1998, 2000) found that staff sick leave was reduced by over 60 % when indoor plants were installed. They also found less sick leave among school children with plants in their classroom, and that staff with plants in offices showed significantly fewer health and discomfort problems, including 37 % less coughing, 30 % less fatigue and 23 % in symptoms such as headaches, sore eyes, nose or throat, 'heavy-headedness' or lowered concentration. Studies by Lohr et al. (1996) also showed productivity gains and reductions in perceptions of pain and discomfort when plants were present. A Texan survey with over 400 respondents (Dravigne et al. 2008) found that job satisfaction rose significantly on all 10 criteria tested among staff with indoor plants and that they preferred them to planted window views.

The effect of indoor plants on psychological health may be due to a surrogate effect akin to exposure to 'nature', with the clinical effect described as 'attention restoration theory' (e.g. see Bringslimark et al. 2009). The magnitude of the psychological benefits of botanical material may be proportional to the volume of plant material in an indoor space, presence in the line of sight of an occupant and duration of exposure. Whist the magnitude of the effects of plants on psychological health are subject to some controversy (Brigslimark et al. 2009), the bulk of the literature shows a quantifiable effect.

As some of the current biotechnological approaches to botanical air cleaning, notably green walls, are often associated with a substantial increase in the amount of plant material in a building (Fig. 8.2), the potential exists that these systems could provide corollary benefits to psychological health and worker productivity

Fig. 8.2 A green wall installation at the University of Technology, Sydney (image by the authors)

along with their air quality effects. Alternately, biofiltration systems that are remote from the indoor space their service would presumably provide less psychological benefits over traditional potted plants. Whilst the primary goal of the biotechnological approaches discussed in this paper are clearly air quality improvement, it would be advantageous to quantify the potential side effects on other aspects of building occupant workplace quality when assessing the overall efficacy of any system.

8.9 Microbial Systems

Microbial systems that operate without the synergy of higher plants have received more research than the latter systems, and there is thus a clearer understanding of their performance potential. These systems rely on the well-recognized capacity of microorganisms to use VOCs as carbon sources, and sometimes with the addition of physical filtration to remove particulate matter to complete their air cleaning process. Bacteria are the group of microorganisms generally utilised in these systems. Whilst fungi have been shown to have very high removal capacity for some VOCs due to the large surface area created by aerial growth (Kennes and Veiga 2004; Vergara-Fernández et al. 2008), the resistance to air flow created by fungal mycelia restricts their utility in microbial biofiltration systems, although biofilter packing material technology (e.g. Gutiérrez-Acosta et al. 2012) may effectively address this problem in future developments. The ability of individual fungal species to degrade specific VOCs also appears to be variable (see García-Peña et al. 2008). Resistance to low pH, water and nutrient availability are key advantages of fungal systems over bacteria-based biofilters, and these may emerge as key constraints in future bacterial biofiltration systems (Estrada et al. 2013a).

Microbial air cleaning systems have been classified into three general types (Guiyesse et al. 2008): *bioscrubbers*, where pollutants are collected in a mobile aqueous phase and exposed to microorganisms in a bioreactor (Fig. 8.3);

Biotrickling filters where microorganisms are grown on an inert support such as activated carbon or geological materials such as vermiculite, supplied with nutrients from a gravity fed solution and exposed to the air either passively or actively (Fig. 8.4);

Biofilters where air is passed through a porous support on which microorganisms are growing, and which may or may not receive an exogenous nutrient supply (Fig. 8.5). However, systems that utilise natural microbial communities such as those found in potted plant substrates (Fig. 8.6), obviate the need for a continually provided nutrient supply and are self-supporting over a period of months to years (e.g. Wood et al. 2002), thus having substantial maintenance advantages.

It seems that virtually all organic pollutants are biodegradable by bacteria, although the rates at which they are degraded vary widely based on a number of factors, notably their polarity (which determines their diffusion coefficient in aqueous environments, and thus their accessibility to microorganisms) and toxicity effects, which could be expected to be minimal at the low concentrations found in

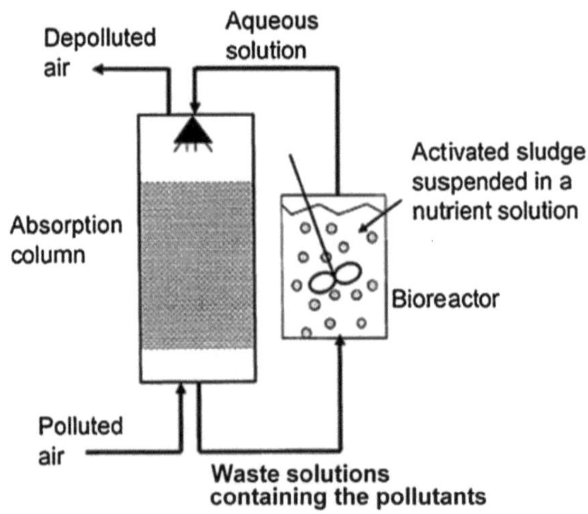

Fig. 8.3 Simplified representation of a bioscrubber system (Mudliar et al. 2010)

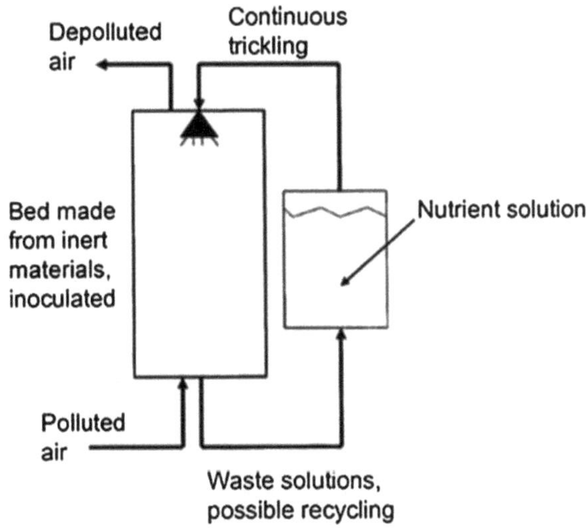

Fig. 8.4 Simplified representation of a biotrickling filter system (Mudliar et al. 2010)

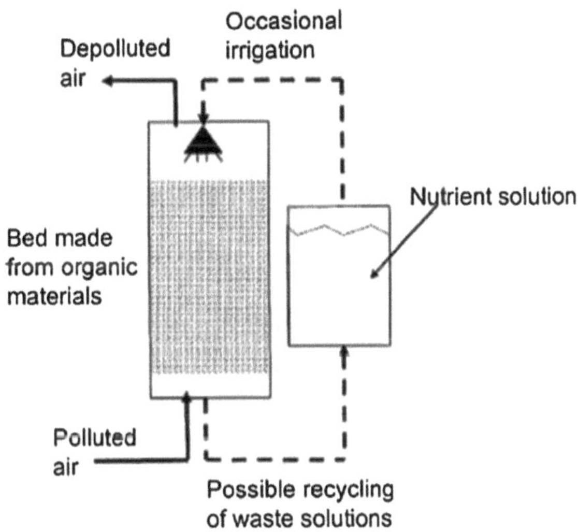

Fig. 8.5 Simplified representation of a biofilter unit that does not utilise plants (Mudliar et al. 2010)

indoor air (Guieyesse et al. 2008). Due to both the variety of molecular structures of VOCs, and the general tendency of microorganisms to degrade xenobiotic materials by consortial activity (e.g. Torpy et al. 2013a), microbial biofilters must support a diverse community of microorganisms and expose a large, moist surface area to maximise the solubilisation of airborne molecules (Beveridge et al. 1997). Apart from natural media, such as plant growth substrates, fixed biofilm reactors are considered to be probably the most efficacious means of air contaminant biodegradation (Singh et al. 2006).

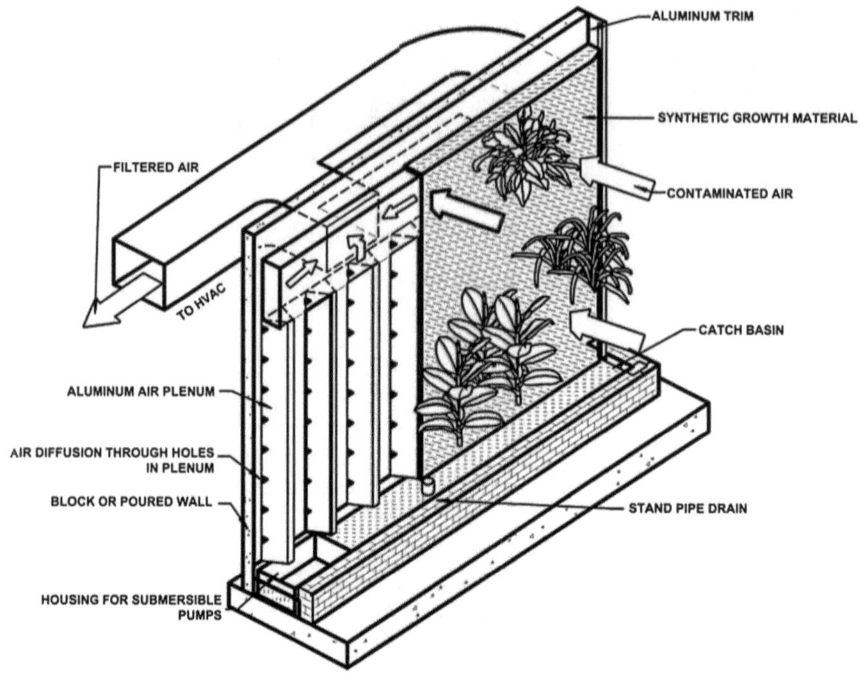

Fig. 8.6 Schematic of an indoor air biofilter that utilizes plants (From Soreanu et al. 2013). The biofilter consists of a vertical green wall on a porous support, which is irrigated by circulation from the catch basin. Indoor air is circulated through the green wall, where all air cleaning processes occur

As stated previously, VOCs are present in indoor air at extremely low concentrations: far lower than would appear to be necessary to sustain microorganisms if they were the sole carbon sources (Guieyesse et al. 2008). Effective microbial air cleaning systems may thus rely on microorganisms co-metabolising VOCs amongst other sources of energy, i.e. the microbial communities need to be sustained independently of the supply of VOCs if they are not to become carbon source limited. Fortunately, the bacteria that degrade most indoors VOCs of concern appear to be ubiquitous, at least in soils (Torpy et al. 2013a), and so do not need to be specifically isolated and cultured, as has been proposed as a limiting factor for the use of microbial air cleaners (Guieyesse et al. 2008). Certainly, most research suggests that there is little evidence that natural substrata, such as soils, lack microorganisms capable of degrading common VOCs. A hydroponic plant growth system, with axenic plant stock and no specific microbial inoculation other than that gained from the air and tap water at the time of transfer to the hydroculture pots, was found to be capable of bioremediating gaseous benzene, albeit at a slightly reduced rate than that achieved by potting mix (Irga et al. 2013), providing evidence that ubiquitous air- and waterborne bacterial species have VOC biodegradation capacity. An earlier study by Chen et al. (2005) did not detect toluene removal from an undescribed, 'botanical' air cleaning system, although the constituents and construction details of the system were not supplied, so it cannot be speculated on what, if any, microbial community was present in the apparatus used. Surprisingly, Ondarts et al. (2012) did not detect trichloroethylene removal by a compost-based biofilter, unless activated carbon was added to the substrate.

An alternative use for microbial biofilters that bypasses problems associated with the low background concentrations of indoor airborne VOCs is to use the systems as an initial clean-up process to remove high VOC levels in new buildings (Lu et al. 2010). This can be achieved by using an initial bake-out process, whereby a newly constructed room is heated to accelerate VOC emissions which can then be effectively removed in the biofilter (Lu et al. 2012a). However, such technology would appear to have little application in the ongoing maintenance of building air quality.

Another putative concern raised with respect to microbial air cleaners is that they would lose their capacity to remove trace VOCs over time if not constantly exposed to them (Guieyesse et al. 2008). Whilst empirical research has shown that the rate of VOC biodegradation increases markedly on repeated exposure, and that the rate of removal is proportional to the concentration of VOC present, there is also evidence that the basic capacity to biodegrade some VOCs is innate and not lost in the absence of the gases (Orwell et al. 2004, 2006).

The capacity of microbial biofilters to remove VOCs in high concentrations is inarguable. Delhoménie et al. (2002) developed a novel vegetable and sewage sludge compost-based biofiltration system and tested its capacity to remove high concentrations of toluene. Contaminated air was exposed to the biofilter at 0.4–1 m^3h^{-1}. The compost column was fed with a nutrient solution to sustain the microbial population. Whilst the system removed toluene with high efficiency, the system also produced significant quantities of CO_2, and the release of microorganisms from the system may also have been possible (see Sect. 8.12) but was not tested. Lu et al. (2010) detected very high rates of biodegradation of formaldehyde, benzene, toluene and xylene from a biotrickling filter dominated by four common bacterial genera. Ondarts et al. (2012) used a compost-based system, which had very high removal efficiency for many VOCs, including the highly hydrophobic undecane and limonene, leading to the hypothesis that an abiotic sorption process was important in the overall biodegradation efficiency of the system, possibly due to the hydrophobic components of the organic matter.

Logically, the rate of removal of any air pollutant will be limited by the rate at which it is exposed to any air filtration apparatus, whether biological or physical. This has been seen as a major constraint on the development of biological air cleaning systems (e.g. Guieyesse et al. 2008). The large magnitude of the biofilter: room volume ratio for the effective air cleaning system developed by Darlington et al. (2001) reinforces the difficulties in gaining effective air cleaning performance with a reasonably compact system. Future research into increasing the effective air exposure of biofilters, whether by examining airflow rates over the filters or by increasing their effective surface area, would be highly valuable.

Apart from its direct role in improving IAQ, biotechnology has also been used to indicate poor indoor environmental conditions. As noted previously, the high degree of environmental sealing of modern buildings has led to the proliferation of conditions that favour poor IAQ. As well as increasing levels of some air pollutants, the reduction in airflow and increase in humidity from poorly designed building ventilation systems may lead to increased fungal growth, and a community shift in the types of fungi present. Both increased fungal spore loads indoors, and a shift from 'normal' fungal communities to ones strongly dominated by single species such as *Aspergillus* and *Stachybotrys* spp. High concentrations of spores of these species are known to lead to symptoms of 'sick building syndrome' (Cooley et al. 1998; Cabral 2010), thus indoor assessment of these organisms can act as an indicator of poor IAQ in general. Fungi may also produce their own range of VOCs (Douwes 2009). There is a strong correlation between high indoor humidity, fungal growth and human respiratory pathology (Cabral 2010). Thus an assessment of the fungal community composition in a building can be a useful indicator of whether moisture accumulation is at a level where poor health outcomes may result. Fungal sampling may be performed on precipitated dust or air samples, and the community may be described by culture or molecular methods (Pitkäranta et al. 2008).

8.10 Do Biological Air Filtration Methods Lead to Microbial Biopollution?

As both botanical and exclusively microbial biofiltration systems, whether active or passive, are primarily dependent on microorganisms for their function, as well as presenting excellent conditions for general microbial growth, valid concerns arise over whether the air cleaning benefits of these systems may be negated by the emission of viable or non-viable bioparticles into the indoor environment (e.g. Lu et al. 2012b). Of the microorganisms likely to lead to any significant health problems, fungi are of primary concern (Torpy et al. 2013b). A primary cause of fungal amplification indoors is relative humidity, with levels over 80 % leading to potential problems (Adan and Samson 2011). All current biological air filtration systems thus provide possible point sources of the ideal conditions for fungal proliferation.

It has been proposed that indoor plants could indeed act as a significant source of pathogenic fungal inocula (Staib et al. 1978; Summerbell et al. 1989; Hedayati et al. 2004; Engelhart et al. 2009). These studies have generally been conjectural or correlative, where the pathogenic fungal species implicated in a particular disease case has been isolated from or associated with a potted plant, with no direct testing of aetiology performed. A number of more recent studies, however, have more directly investigated the potential for potted plants to support and release significant fungal material.

Darlington et al. (2000) assessed the effect of his biofiltration system on the indoor airborne microbiota. The two-stage Anderson sampler was used to assess both bacteria and viable fungi from the room in which his system was operating, and samples were taken one year apart to test the long-term effect of the system. Total fungal, yeast, mould and bacterial numbers declined in the second year's samples, suggesting that this active biofiltration system went through a stabilisation period over which microbial release declined, probably due to a reduction in the quantity of compostable organic matter. Whilst the biofilter increased fungal spore numbers relative to adjoining rooms and buildings, the spore density detected (mean 115 CFU/m^{-3}) was considered within the range expected from normal buildings by Darlington et al. (2000), and is certainly surprisingly low given the large quantity of biological material used in the system. The community present in the biofiltered room was dominated by *Penicillium* spp., suggesting the indoor sources predominated over outdoor sourced genera, which are normally dominated by *Cladosporium* spp. (Cabral 2010). In contrast, Wang and Zhang (2011) performed a pilot trial for their previously described biofiltration systems. Whilst they could not culture any microbial growth in the outlet air of their system, the authors used a single culture medium only, and recognised the need for further research into whether long-term use may lead to the release of microorganisms from their system. The presence of non-viable bioparticles was also not assessed by either study. Microbial by-products, such as ergosterol and lipopolysaccharide, as well as non-living cellular material are well known to lead to disease states such as allergy, headache, coughing and dermatitis (Mendell et al. 2011).

A study into the effect of passive botanical air cleaning systems on the indoor airborne fungal community was performed by Torpy et al. (2013b). The air of 54 offices both with and without potted plants was tested with an impaction air sampler and fungal-specific growth media for viable fungal spores. The authors found that whilst potted plants were associated with an increase in total spore numbers, the maximal level reached with two large indoor plants was approximately an order of magnitude lower than matched outdoor air samples. The absence of any major fungal pathogens (e.g. *Aspergillus fumigatus*) in the air samples provided further evidence that the passive system tested was generally safe for most building occupants with regard to viable fungi.

As potted plants have become more popular as installations within buildings, some concerns have been raised as to whether the plants and their growth substrates could be a reservoir for *Legionella* spp, the bacteria responsible for Legionnaires disease. The conditions for *Legionella* proliferation within man-made environments include stagnant water, temperatures around 38 °C, and some form of carbohydrate nutritional source (Grimes 1991). Once bacterial growth has taken place, the water then needs to be aerosolised or come into immediate human contact for infection to occur (Verissimo et al. 1990). As most buildings maintain temperatures well below optimal for *Legionella* growth, and current indoor plant husbandry practices utilise little or no excess exposed water, the potential for the proliferation and transmission of *Legionella* is limited (Burchett et al. 2007). However, with active green wall and biofiltration technologies, there is potential for increased temperatures due to both mechanical water circulation pumps or integrated lighting systems, coupled with increased quantities of biological material and the potential requirement for larger watering and draining basins. These factors may thus lead to significant concern for the rapid growth of aquatic bacterial pathogens. If such systems also have the potential for these bacterial suspensions to be aerosolized, such systems may be of concern and should be further researched. Legionnaires disease caused by *L. longbeachae* has on rare occasions been implicated in potting-mix handling scenarios (Patten et al. 2010); however, poor hand washing procedures and long-term smoking were found to be better determinants of the disease than exposure to the potting-mix itself (O'Conner et al. 2007), so major concerns over this disease may not be warranted.

Few studies to date on indoor biofiltration have included any assessment of bacterial emissions from the tested systems. Lu et al. (2012b) addressed concerns over bacterial emission from an active biofiltration system by using a calcium alginate gel to immobilise the active bacterial strain in their system (*Pseudomonas putida*). The study detected no bacterial release, whilst effectively biodegrading formaldehyde. Whilst the system required a constant supply of liquid nutrients, the amount required was relatively low and no loss of biodegradation efficiency was detected over 10 days. The pressure drop across the filtration matrix was higher than optimal, leading to a relatively high energy usage for this system. If the economics of energy use and nutrient supply can be effectively addressed, this technology may have some potential for preventing bacterial release from active biofilters, which may find use in buildings that have demanding microbial standards, such as hospitals. The system, however, would be unlikely to control fungal spore release due to the aerial growth and sporulation of mould fungi.

Thus there appears to be some potential that biological air cleaning systems can lead to increased microbial density in the treated air. Whilst no reliable research has indicated the potential for danger to health, there is nonetheless a need for both the assessment of existing systems and a consideration of microbial safety when developing new systems. It is probable that the greatest potential for microbial material release from any biofiltration system will occur when the medium becomes excessively dry (Pasanen et al. 1991). With increased airflow rates across media with uncontrolled microbiological communities, and the use of supports that have the potential for rapid desiccation if irrigation fails, some of the active systems thus have the potential for major microbial release. It is thus suggested that future research assess these systems in both their functional state and at reduced moisture levels to assess the potential for health (and thus legal) problems that may arise in failure situations. The integration of HEPA into both active microbial and botanical IAQ cleaning systems would solve the problems of bio-pollutant emission, although at the cost of further increasing pressure drop.

8.11 Commercial Systems

A small number of biological indoor air cleaning systems are currently available on the market (Table 8.1); however, it is clear that this industry is expanding. Of the available systems, all use plants, and most have an interior design component to increase their acceptability to purchasers.

Table 8.1 A selection of commercially available biological air cleaning products applicable for indoor applications

Company name	Country of origin	Mode of action	Website
Phytofilter Technologies, Inc	USA	Plant substrate consisting of activated carbon and inert materials colonised with microorganisms. The system is incorporated into a HVAC system, where air is pumped under pressure through the substrate	http://phytofilter.com
Naturaire/ Nedlaw Living Walls	Canada	Indoor air is actively drawn through the plant wall by either the HVAC system or on-board fans and then returned to the occupied space	http://naturaire.com
ANDREA	France	A personal and portable air purifier, consisting of a plant with fan fed air to the root zone	http://www.andreaair.com/
CASE	USA	Modular hydroponic plant system that has in built fans that increase air flow across the plants' root surface	www.case.rpi.edu/
Junglefy	Australia	Modular plant system that has in built fans that increases air flow across the plants' root surface	http://junglefy.com.au/

A prototype device (Mittelmark et al. 2009) has been developed by Phytofilter Technologies, Inc., which uses a plant substrate consisting of activated carbon and inert materials colonised with microorganisms. The system is active: the system is incorporated into an HVAC system, where air is pumped under pressure through the substrate, which is colonised with the indoor plant Pothos (*Epipremnum aureum*). The system has been demonstrated to have high removal efficiency for formaldehyde and toluene, and this removal efficiency was maintained over 300 + days of continuous operation. Furthermore, the system was found to have good ventilation energy saving potential based on VOC removal, although the system does not appear to have been tested for CO_2 emission, which may reduce its ventilation replacement effectiveness. Phytofilter Technologies is currently looking to develop a portable system incorporating this technology (http://phytofilter.com/announcements/?p=64).

The leading developer of commercial phytoremediation technologies is probably Naturaire (www.naturaire.com/function), led by Alan Darlington. The green wall systems this company and its antecedents have installed in the University of Guelph in Canada (Naturaire 2014) and the Canada Life Assurance headquarters building (Toronto, ON, Canada; Darlington et al. 2001) are archetypes of integrated green wall systems specifically installed for indoor phytoremediation.

ANDREA (http://www.andreaair.com/), also known as Bel Air, is the result of a collaboration between French designer Mathieu Lehanneur and Harvard academic David Edwards, and is a desktop "mobile greenhouse". Essentially a plant with fan feeding air to the root zone, this personal portable air purifier is a new innovation in this market. Although its air cleaning potential has not been demonstrated in the scientific literature, the company claims that the rate of gas removal by ANDREA is over 10 times faster than for passive indoor plants. The system is affordable, retailing at £125 in the UK in 2014.

The Centre for Architecture Science and Ecology (www.case.rpi.edu/) has developed an active, hydroponic green wall system called the 'Active Modular Phytoremediation Wall System' (AMPS). The system uses a building's ventilation system for increasing the exposure of air to the hydroponic medium.

The Australian company, Junglefy (www.junglefy.com) has developed an active botanical air cleaning system for use in both new buildings and as a retrofit. Like the AMPS system, Junglefy uses a modular design that allows flexible installation in different areas of buildings.

8.12 Conclusions

The importance of IAQ to the industrialised world is increasing dramatically. The U.S. market for IAQ was valued at nearly $7 billion in 2011 and is projected to reach nearly $9.2 billion in 2017, a 5-year compound annual growth rate of 4.6 %. The equipment component of the industry is expected to account for $4.7 billion in 2017 (BCC Research 2014). Thus the acceleration of research into biological air cleaning systems is timely. Whilst there is no doubt these systems have functional potential, issues such as microbial safety, long-term efficiency, CO_2 emissions, maintenance costs and energy use for active systems are yet to be sufficiently addressed to be able to recommend systems to meet all applications. The rapid development of industrial waste gas biofilters has demonstrated that with a concerted research effort, rapid progress can be made. However, the primary current problem facing research to improve on simple potted plants as bioremediators of IAQ is that the process is, in a way, answering a problem no one has yet asked. The general public has yet to fully embrace the concept of botanically based air purification systems in their simplest form: claims of further developments largely fall on deaf ears.

References

Afrin S (2009) Green skyscraper: integration of plants into skyscrapers. Kungliga Tekniska Hogskolan, Stockholm

Adan OCG, Samson RA (2011) Fundamentals of mold growth in indoor environments and strategies for healthy living. Wageningen Academic Publishers, Wageningen

ASHRAE (2011) GreenGuide: the design, construction, and operation of sustainable buildings, 3rd edn. http://www.ashraeregion12.org/?p=105. Accessed 15 Dec 2013

Auvinen J, Wirtanen L (2008) The influence of photocatalytic interior paints on indoor air quality. Atmos Environ 42:4101–4112

Aydogan A, Montoya LD (2011) Formaldehyde removal by common indoor plant species and various growing media. Atmos Environ 45:2675–2682

Bakó-Biró Z, Wargocki P, Weschler CJ, Fanger PO (2004) Effects of pollution from personal computers on perceived air quality, SBS symptoms and productivity in offices. Indoor Air 14:178–187

Bamsey M, Graham T, Stasiak M, Berinstain A, Scott A, Vuk TR, Dixon M (2009) Canadian advanced life support capacities and future directions. Adv Space Res 44:151–161

Barro R, Regueiro J, Llompart M, Garcia - Jares C (2009) Analysis of industrial contaminants in indoor air: part 1. Volatile organic compounds, carbonyl compounds, polycyclic aromatic hydrocarbons and polychlorinated biphenyls. J Chromatogr A 1216:540–566

Bernstein JA, Alexis N, Bacchus H, Bernstein IL, Fritz P, Horner E et al (2008a) The health effects of nonindustrial indoor air pollution. J Allergy Clin Immunol 121(3):585–591

BCC Research (2014) US Market report: indoor air quality. http://www.bccresearch.com/market-research/environment/indoor-air-quality-us-market-env003d.html. Accessed 14 Feb 2014

Bernstein JA, Alexis N, Bacchus H, Bernstein IL, Fritz P, Horner E, Li N, Mason S, Nel A, Oullette J, Reijula K, Reponen T, Seltzer J, Smith A, Tarlo SM (2008b) The health effects of nonindustrial indoor air pollution. J Allergy Clin Immunol 121:585–591

Beveridge TJ, Makin SA, Kadurugamuwa JL, Li Z (1997) Interaction between biofilms and the environment. FEMS Microbiol Rev 20:291–303

Bringslimark T, Hartig T, Patil GC (2009) The psychological benefits of indoor plants: a critical review of the experimental literature. J Environ Psychol 29:422–433

Brown SK (1997) Indoor air quality, Australia: state of the environment technical paper series (Atmosphere). Department of the Environment, Sport and Territories, Canberra

Burchett MD, Tarran J, Torpy FR (2007) Sources and avoidance of Legionnaires disease commissioned report to national indoor Plantscapers Association. University of Technology, Sydney, Australia

Cabral JPS (2010) Can we use indoor fungi as bio-indicators of indoor air quality? Historical perspectives and open questions. Sci Total Environ 408:4285–4295

Chen W, Zhang JS, Zhang Z (2005) Performance of air cleaners for removing multiple volatile organic compounds in indoor air. ASHRAE Trans OR-05-17-2:1101–1104

Chan W, Lee S-C, Chen Y, Mak B, Wong K, C-S C, Zheng C, Guo X (2009) Indoor air quality in new hotels' guest rooms of the major world factory region. Int J Hosp Manag 28:26–32

Cooley JD, Wong WC, Jumper CA, Straus DC (1998) Correlation between the prevalence of certain fungi and sick building syndrome. Occup Environ Med 55:579–584

Coward M, Ross D, Coward S, Cayless S, Raw G (1996) Pilot study to assess the impact of green plants on NO_2 levels in homes, building research establishment note N154/96. Watford, UK

Darlington A, Chan M, Malloch D, Pilger C, Dixon MA (2000) The biofiltration of indoor air: implications for air quality. Indoor Air 10:39–46

Darlington A, Dat JF, Dixon MA (2001) The biofiltration of indoor air: air flux and temperature influences the removal of toluene, ethylbenzene, and xylene. Environ Sci Technol 35:240–246

Delhoménie M-C, Heitz M (2003) Elimination of chlorobenzene vapors from air in a compost-based biofilter. J Chem Technol Biotechnol 78:588–595

De Kempeneer L, Sercu B, Vanbrabant W, Van Langenhove H, Verstraete W (2004) Bioaugmentation of the phyllosphere for the removal of toluene from indoor air. Appl Microbiol Biotechnol 64:284–288

De Visscher A, Li GQ (2008) Toluene removal biofilter modeling: optimization and case study. Process Saf Environ Prot 86:277–282

Dravigne A, Waliczek TM, Lineberger RD, Zajicek JM (2008) The effect of live plants and window views of green spaces on employee perceptions of job satisfaction. HortScience 43:183–187

Douwes J (2009) Building dampness and its effect on indoor exposure to biological and nonbiological pollutants. WHO guidelines for indoor air quality: dampness and mould. WHO Europe, Copenhagen, pp 7–29

Elkiey T, Ormrod DP (1981) Absorption of ozone, sulphur dioxide, and nitrogen dioxide by Petunia plants. Environ Exp Bot 21:63–70

Elmrini H, Kerdouss F, Jorio H, Heitz M (2001) Biofiltration of air contaminated with toluene. Environ Technol 22:927–940

Engelhart S, Rietschel E, Exner M, Lange L (2009) Childhood hypersensitivity pneumonitis associated with fungal contamination of indoor hydroponics. Int J Hyg Environ Health 212:18–20

Environment Australia (2003) BTEX personal exposure monitoring in four Australian Cities. Technical paper No. 6 (Australia E ed.), Canberra, ACT, Australia

Epstein Y (2008) Sick Building Syndrome. Harefuah 147:607–608

Erdmann CA, Apte MG (2004) Mucous membrane and lower respiratory building related symptoms in relation to indoor carbon dioxide concentrations in the 100-building BASE dataset. Indoor Air 14:127–134

Esguerra C, Santiago E, Aquino N, Ramos M (1983) The uptake of SO_2 and NO_2 by plants. Sci Diliman 2:45–56

Estrada JM, Hernández S, Muñoz R, Revah S (2013a) A comparative study of fungal and bacterial biofiltration treating a VOC mixture. J Hazard Mater 250–251:190–197

Estrada JM, Quijano G, Lebrero R, Muñoz R (2013b) Step-feed biofiltration: a low cost alternative configuration for off-gas treatment. Water Res 47:4312–4321

Fadzli Haniff M, Selamat H, Yusof R, Buyamin S, Sham Ismail F (2013) Review of HVAC scheduling techniques for buildings towards energy-efficient and cost-effective operations. Renew Sustain Energy Rev 27:94–103

Fjeld T, Veiersted B, Sandvik L, Riise G, Levy F (1998) The effect of indoor foliage plants on health and discomfort symptoms among office workers. Indoor Built Environ 7:204–206

García-Peña I, Ortiz I, Hernández S, Revah S (2008) Biofiltration of BTEX by the fungus *Paecilomyces variotii*. Int Biodeterior Biodegradation 62:442–447

Giese M, Bauer-Doranth U, Langebartels C, Sandermann H (1994) Detoxification of formaldehyde by the spider plant (*Chlorophytum comosum* L.) and by soybean (*Glycine max* L.) suspension cultures. Plant Physiol 104:1301–1309

Grimes D (1991) Ecology of estuarine bacteria capable of causing human disease. Estuaries 14:334–360

Gurjar B, Molina L, Ojha C (2010) Air pollution: health and environmental impacts. CRC Press, Florida

Gutiérrez-Acosta OB, Arriaga S, Escobar-Barriosa VA, Casas-Flores S, Almendarez-Camarilloc A (2012) Performance of innovative PU-foam and natural fiber-based composites for the biofiltration of a mixture of volatile organic compounds by a fungal biofilm. J Hazard Mater 201–202:202–208

Guieysse B, Hort C, Platel V, Munoz R, Ondarts M, Revah S (2008) Biological treatment of indoor air for VOC removal: Potential and challenges. Biotechnol Adv 26:398–410

Haritash AK, Kaushik CP (2009) Biodegradation aspects of polycyclic aromatic hydrocarbons (PAHs): a review. J Hazard Mater 169:1–15

Harris R, Moore D (2009) Indoor work and living environments: health and safety and performance. In: Hedge A (ed) Indoor air quality and productivity. Science Publisher Inc, Ithaca, New York

Hedayati MT, Mohseni-Bandpi A, Moradi S (2004) A survey on the pathogenic fungi in soil samples of potted plants from Sari hospitals, Iran. J Hosp Infect 58:59–62

Hutton G (2013) Air pollution: global damage costs. In: Lomborg B (ed) How much have global problems cost the world?. Cambridge University Press, New York, USA

Irga PJ, Torpy FR, Burchett MD (2013) Can hydroculture be used to enhance the performance of indoor plants for the removal of air pollutants? Atmos Environ 77:267–271

Jaakola M, Yang L, Ieromnimon A, Jaakola J (2007) Office work exposures and respiratory and sick building syndrome symptoms. Occup Environ Med 64:178–184

Kennes C, Veiga MC (2004) Fungal biocatalysts in the bio-filtration of VOC-polluted air. J Biotechnol 113:305–319

Kim KJ, Kil MJ, Song JS, Yoo EH, Son KC, Kays SJ (2008) Efficiency of volatile formaldehyde removal by indoor plants: contribution of aerial plant parts versus the root zone. J Am Soc Hortic Sci 133:1–6

Kwong CW, Chao CYH, Hui KS, Wan MP (2008) Removal of VOCs from indoor environment by ozonation over different porous materials. Atmos Environ 42:2300–2311

Lebrero R, Rodríguez E, Estrada JM, García-Encina PA, Muñoz R (2012) Odor abatement in biotrickling filters: Effect of the EBRT on methyl mercaptan and hydrophobic VOCs removal. Bioresour Technol 109:38–45

Lee J-H, Sim W-K (1999) Biological absorption of SO_2 by Korean native indoor plants. In: Burchett MD, Tarran J, Wood R (eds) Towards a new millennium in people-plant relationships—Contributions from the international people-plant symposium, Sydney, July, 1998. UTS Printing Services, Sydney, NSW, Australia, pp 101–108

Lee S-H, Li C, Heber AJ, Ni J, Huang H (2013) Biofiltration of a mixture of ethylene, ammonia, *n*-butanol, and acetone gases. Bioresour Technol 127:366–377

Llewellyn D, Dixon M (2011) Can plants really improve indoor air quality? In: Murray M-Y (ed) Comprehensive biotechnology, 2nd edn. Academic Press, Burlington, pp 331–338

Lohr VI, Pearson-Mims CH (1996) Particulate matter accumulation on horizontal surfaces in interiors: influence of foliage plants. Atmos Environ 30(14):2565–2568

Lohr VL, Pearson-Mims CH, Goodwin GK (1996) Interior plants may improve worker productivity and reduce stress in a windowless environment. J Environ Hortic 14:97–100

Liu Y-J, Mu Y-J, Zhu Y-G, Ding H, Arens NC (2007) Which ornamental plants species effectively remove benzene from indoor air? Atmos Environ 41:650–654

Lu Y, Liu J, Lu B, Jiang A, Wan C (2010) Study on the removal of indoor VOCs using biotechnology. J Hazard Mater 182:204–209

Lu Y, Liu J, Yoshino H, Lu B, Jiang A, Li F (2012a) Use of biotechnology coupled with bake-out exhaust to remove indoor VOCs. Indoor Built Environ 21:741–748

Lu N, Pei J, Zhao Y, Qi R, Liu J (2012b) Performance of a biological degradation method for indoor formaldehyde removal. Build Environ 57:253–258

McGenity TJ (2014) Hydrocarbon biodegradation in intertidal wetland sediments. Curr Opin Biotechnol 27:46–54

Mendell MJ, Mirer AG, Cheung K, Tong M, Douwes J (2011) Respiratory and Allergic Health Effects of Dampness, Mold, and Dampness. Related Agents: A Review of the Epidemiologic Evidence. Environ Health Perspect 119:748–756

Milton DK, Glencross PM, Walters MD (2000) Risk of sick leave associated with outdoor air supply rate, humidification, and occupant complaints. Indoor Air 10:212–221

Mittelmark M, Zhang J, Wang Z (2009) Air cleaning technologies for indoor air quality (ACT-IAQ). growing fresh and clean air. final report: centre of excellence-technology application and development (COE-TAD). Syracuse Centre of Excellence in Environmental and Energy Systems, Syracuse, USA

Mo J, Zhang Y, Xu Q, Lamson JJ, Zhao R (2009) Photocatalytic purification of volatile organic compounds in indoor air: a literature review. Atmos Environ 43:2229–2246

Morey PR, Crawford GN, Rottersman RB (2001) Indoor air quality in nonindustrial occupational environments. In: Patty's Industrial Hygiene. Wiley, New York

Mudliar S, Giri B, Padoley K, Satpute D, Dixit R, Bhatt P, Pandey R, Juwarkar A, Vaidya A (2010) Bioreactors for treatment of VOCs and odours—A review. J Environ Manag 91:1039–1054

Myers J (1954) Basic remarks on the use of plants as biological gas exchangers in a closed system. J Aviat Med 25:407–411

Naturaire (2014) www.naturaire.com. Accessed 14 Feb 2014

Newton P (2001) Human settlements. Australian State of the environment report 2001. CSIRO Publishing on behalf of the Department of the Environment and Heritage, Canberra, Australia

Norbäck D, Nordström K (2008) Sick building syndrome in relation to air exchange rate, CO_2, room temperature and relative air humidity in university computer classrooms: an experimental study. Int Arch Occup Environ Health 82:21–30

O'Connor B, Carman J, Eckert K, Tucker G, Givney R, Cameron S (2007) Does using potting mix make you sick? Results from a *Legionella longbeachae* case-control study in South Australia. Epidemiol Infect 135:34–39

Oh G, Jung G, Seo M, Im Y (2011) Experimental study on variations of CO_2 concentration in the presence of indoor plants and respiration of experimental animals. Hortic Environ Biotechnol 52:321–329

Ondarts M, Hort C, Sochard S, Platel V, Moynault L, Seby F (2012) Evaluation of compost and a mixture of compost and activated carbon as biofilter media for the treatment of indoor air pollution. Environ Technol 33(3):273–284

Orwell RL, Wood RA, Tarran J, Torpy FR, Burchett M (2004) Removal of benzene by the indoor plant/ substrate microcosm and implications for air quality. Water Air Soil Pollut 157:193–207

Orwell RL, Wood RA, Burchett MD, Tarran J, Torpy F (2006) The potted-plant microcosm substantially reduces indoor air VOC pollution: II. Laboratory study. Water Soil Air Pollut 177:59–80

Patten S, Sur E, Sundaram R, Weinhardt B (2010) Dangers in the garden. The Lancet 376:844

Pasanen AL, Pasanen P, Jantunen MJ, Kalliokoski P (1991) Significance of air humidity and air velocity for fungal spore release into the air. Atmos Environ Part A. Gen Topics 25:459–462

Pennisi SV, van Iersel MW (2012) Qunatification of carbon assimilation of plants in simulated and in situ interiorscapes. HortScience 47:468–476

Pilon-Smits E (2005) Phytoremediation. Annu Rev Plant Biol 56:15–39

Pitkäranta M, Mekin T, Hyvärinen A, Paulin L, Auvinen P, Nevalainen A, Rintala H (2008) Analysis of fungal flora in indoor dust by ribosomal DNA sequence analysis, quantitative PCR, and culture. Appl Environ Microbiol 74(1):233–244

Redlich C, Sparer J, Cullen M (1997) Sick-building syndrome. Lancet 349:1013–1016

Rodgers K, Hutzel W, Dana M, Handy R (2012) Can plants save money: a look at the Biowall. In: International high performance buildings conference, West Lafayette, Indiana

Rodgers K, Handy R, Hutzel W (2013) Indoor air quality (IAQ) improvements using bio-filtration in a highly efficient residential home. J Green Build 8(1):22–27

Sakai K, Norbäck D, Mi Y, Shibata E, Kamijima M, Yamada T, Takeuchi Y (2004) A comparison of indoor air pollutants in Japan and Sweden: formaldehyde, nitrogen dioxide, and chlorinated volatile organic compounds. Environ Res 94:75–85

Sandhu A, Halverson LJ, Beattie GA (2007) Bacterial degradation of airborne phenol in the phyllosphere. Environ Microbiol 9:383–392

Schmitz H, Hilgers U, Weidner M (2000) Assimilation and metabolism of formaldehyde by leaves appear unlikely be of value for indoor air purification. New Phytol 147(2):307–315

Seppänen OA, Fisk WJ (2004) Summary of human responses to ventilation. Indoor Air 14:102–118

Seppänen O, Fisk WJ, Lei QH (2006) Ventilation and performance in office work. Indoor Air 16:28–36

Shaughnessy RJ, Haverinen-Shaughnessy U, Nevalainen A, Moschandreas D (2006) A preliminary study on the association between ventilation rates in classrooms and student performance. Indoor Air 16:465–468

Singh R, Debarati PD, Jain RK (2006) Biofilms: implications in bioremediation. Trends Microbiol 14:389–397

Soreanu G, Dixon M, Darlington A (2013) Botanical biofiltration of indoor gaseous pollutants—A mini-review. Chem Eng J 229:585–594

Staib F, Tompak B, Thiel D, Blisse A (1978) *Aspergillus fumigatus* and *Aspergillus niger* in two potted ornamental plants, cactus(*Epiphyllum truncatum*) and clivia (*Clivia miniata*). Biol Epidemiol Aspects Mycopathol 66:27–30

Summerbell RC, Krajden S, Kane J (1989) Potted plants in hospitals as reservoirs of pathogenic fungi. Mycopathologia 106:13–22

Tans P (2014) NOAA/ESRL. www.esrl.noaa.gov/gmd/ccgg/trends/. Accessed 3 Jan 2014

Tarran J, Torpy F, Burchett MD (2007) Use of living pot-plants to cleanse indoor air. Research Review. In: 6th International conference on indoor air quality, ventilation and energy conservation, sustainable built environment, Sendai, Japan, pp 249–256

Treesubsuntorn C, Thiravetyan P (2012) Removal of benzene from indoor air by *Dracaena sanderiana*: Effect of wax and stomata. Atmos Environ 57:317–321

Torpy FR, Irga PJ, Moldovan D, Tarran J, Burchett MD (2013a) Characterization and biostimulation of benzene biodegradation in the potting-mix of indoor plants. J Appl Hortic 15(1):10–15

Torpy FR, Irga PJ, Brennan J, Burchett MD (2013b) Do indoor plants contribute to the aeromycota in city buildings? Aerobiologia 29:321–331

Torpy FR, Irga PJ, Burchett MD (2014) Profiling indoor plants for the amelioration of high CO_2 concentrations. Urban Forestry and Urban Greening Article in press, as corrected proof

Ugrekhelidze D, Korte F, Kvesitadze G (1997) Uptake and transformation of benzene and toluene by plant leaves. Ecotoxicol Environ Saf 37:24–29

Vaughan TL, Strader C, Davis S, Daling JR (1986) Formaldehyde and cancers of the pharynx, sinus and nasal cavity: II Residential exposures. Int J Cancer 38:685–688

Vergara-Fernández A, Hernández S, Revah S (2008) Phenomenological model of fungal biofilters for the abatement of hydrophobic VOCs. Biotechnol Bioeng 101:1182–1192

Vergara-Fernández A, Molina LL, Pulido NA, Aroca G (2007) Effects of gas flow rate, inlet concentration and temperature on the biofiltration on toluene vapors. J Environ Manag 84:115–122

Veríssimo A, Vesey G, Rocha GM, Marrão G, Colbourne J, Dennis PJ, da Costa MS (1990) A hot water supply as the source of *Legionella pneumophila* in incubators of a neonatology unit. J Hosp Infect 15:255–263

Wallace L (2001) Human exposure to volatile organic pollutants: implications for indoor air studies. Annu Rev Energy Env 26:269–301

Wang Z, Zhang JS (2011) Characterization and performance evaluation of a full-scale activated carbon-based dynamic botanical air filtration system for improving indoor air quality. Build Environ 46:758–768

WHO (2000) the right to healthy indoor air. In: Report on a WHO meeting. European Health Targets, Bilthoven

Wolkoff P, Nielsen GD (2001) Organic compounds in indoor air—their relevance for perceived indoor air quality? Atmos Environ 35(26):4407–4417

Wolkoff P (2013) Indoor air pollutants in office environments: assessment of comfort, health, and performance. Int J Hyg Environ Health 216:371–394

Wolverton BC (1997) How to grow fresh air: 50 houseplants that purify your home or office. Penguin Books, New York

Wolverton BC, McDonald RC, Watkins E Jr (1984) Foliage Plants for Removing Indoor Air Pollutants from Energy-Efficient Homes. Econ Bot 38:224–228

Wolverton BC, Johnson A, Bounds K (1989) Interior landscape plants for indoor air pollution abatement. National Aeronautics and Space Administration, David-sonville

Wood RA, Orwell RL, Tarran J, Torpy FR, Burchett MD (2002) Potted-plant/growth media interactions and capacities for removal of volatiles from indoor air. J Hortic Sci Biotechnol 77(1):120–129

Wood RA, Burchett MD, Alquezar R, Orwell RL, Tarran J, Torpy F (2006) The potted-plant microcosm substantially reduces indoor air VOC pollution: I Office field-study. Water Air Soil Pollut 175(1–4):163–180

Xu Z, Quin N, Wang J, Tong H (2010) Formaldehyde biofiltration as affected by spider plant. Bioresour Technol 101:6930–6934

Xu Z, Wang L, Hou H (2011) Formaldehyde removal by potted plant–soil systems. J Hazard Mater 192:314–318

Yang DS, Pennisi SV, Son K-C, Kays SJ (2009) Screening indoor plants for volatile organic pollutant removal efficiency. HortScience 44(5):1377–1381

Yoneyama T, Kim HY, Morikawa H, Srivatava HS (2002) Metabolism and detoxification of nitrogen dioxide and ammonia in plants. In: Omasa K, Saji H, Youssefian N, Kondo N (eds) Air pollution and plant biotechnology—Prospects for Phytomonitoring and Phytoremediation. Springer, Tokyo, pp 221–234

Yoo MH, Kwon YJ, Son K-C, Kays SJ (2006) Efficacy of indoor plants for the removal of single and mixed volatile organic pollutants and the physiological effects of the volatiles on the plants. J Am Soc Hortic Sci 131:452–458

Zhang H, Pennisi SV, Kays SJ, Habteselassie MY (2013) Isolation and identification of toluene-metabolizing bacteria from rhizospheres of two indoor plants. Water Air Soil Pollut 224:1648–1662

Zhao J, Yang X (2003) Photocatalytic oxidation for indoor air purification: a literature review. Build Environ 38:645–654

Chapter 9
Bioinspired Self-cleaning Materials

Maria Vittoria Diamanti and MariaPia Pedeferri

Abstract Among nature-inspired phenomena, the interactions of nanostructured surfaces with water are probably the most studied ones, as well as the most mimicked by science: geckos and spiders that can stick on smooth surfaces, beetles that collect fog in the desert, gerridae that walk on water—which is the reason why they are also called water striders, or pond skaters; all of these creatures owe their characterizing properties to the influence of surface nanostructuring on their affinity to water. Still, the most popular example of "nature-created" nanotechnology is the self-cleaning one, given by the onset of either superhydrophilicity, superhydrophobicity, or superoleophobicity. This is allowed by particular conditions of surface (photo)chemistry and structuring: the former is typical of TiO_2-containing surfaces, while the latter is based on the formation of air layers between water and the surface nanometric protrusions, preventing the liquid from wetting it. This chapter is dedicated to the mechanisms underlying bioinspired self-cleaning and to the fields of application of these effects.

9.1 Water and Surfaces

9.1.1 An Introduction to Self-cleaning

Self-cleaning is clearly a fascinating concept, which reflects the capability of a material to maintain its aspect unaltered in time, avoiding the deposition of dirt. Such property first attracted producers and consumers working in the construction

M.V. Diamanti (✉) · M. Pedeferri
Department of Chemistry, Materials and Chemical Engineering
"Giulio Natta", Politecnico di Milano, Via Mancinelli 7, 20131 Milan, Italy
e-mail: mariavittoria.diamanti@polimi.it

M. Pedeferri
e-mail: mariapia.pedeferri@polimi.it

F. Pacheco Torgal et al. (eds.), *Biotechnologies and Biomimetics for Civil Engineering*,
DOI 10.1007/978-3-319-09287-4_9

world, owing to the important savings that self-cleaning could bring—and has actually brought—in the maintenance of buildings, especially in the presence of wide glass façades: the use of self-cleaning glass and other cladding materials surely has an impact on the overall building life cycle cost. One example is provided on Phys.Org as a result of a market analysis performed by the Singapore Institute of Manufacturing Technology (SIMTech), a research institute under A*STAR (Agency for Science, Technology and Research) (Phys.Org 2008). In this analysis, the typical cleaning cycles of commercial buildings in Singapore are estimated as frequent as 1–4 per year, with unit costs for energy and detergents ranging from 7 to 40 kUSD, which can be reduced even by 75 % in the presence of self-cleaning materials. The possibility of making surfaces more resistant to soiling has important spin-offs not only in the built environment, but also in micro-circuitry, where the presence of contaminants can reduce the device lifetime, and—for the same reason—in the most recent application field of solar panels, where even a slight reduction in transmittance can cause severe efficiency losses (Zhu et al. 2010).

Still, the term self-cleaning may be misleading at times. In fact, it generally refers to two situations (Fig. 9.1):

- surfaces may show a superhydrophobic character and impede dust and soot adhesion and adsorption, due to peculiar morphological features and/or chemical affinity; or
- surfaces with photoactivated properties can decrease the adhesion strength of dust and particles—as well as degrade bacterial contaminants—while byproducts are removed by water, owing to a superhydrophilic surface state.

Moreover, some mixed mechanisms—pertaining to hydrophilic surfaces with a relevant oleophobic, or superoleophobic, character—have been recently considered in the development of self-cleaning surfaces.

Considering first the superhydrophilicity condition, the triggering element of such mechanism is photoactivity, which can be intrinsic of the whole material or conferred to the sole surface by applying a suitable coating. This is the case of titanium dioxide (TiO_2) surfaces, whose hydrophilicity is greatly enhanced by irradiation of suitable energy (generally in the UV range) until a superhydrophilic state is reached; similar results can be provided by similar wide band semiconductors, such as zinc oxide (ZnO) (Fujishima et al. 2008; Khranovskyy et al. 2012; Zhang et al. 2012). Although no natural structure exhibits such self-cleaning

Fig. 9.1 Example of superhydrophobic (**a**) and superhydrophilic (**b**) surfaces. Reprinted with permission from Men et al. (2010)

character, these materials are included in biomimicry owing to the mechanism of photoactivation, whose origin traces back to plants photosynthesis. It goes without saying that in absence of light, these surfaces do not show any self-cleaning property, which probably makes "easy-cleaning" a more precise definition of their functionality.

On the other hand, proper self-cleaning should not need any external stimulus: this can be achieved by suitable intrinsic surface chemistry, morphology, or both. In general, these conditions always induce the opposite behavior—that is, superhydrophobicity, sometimes combined with oleophobicity to give superomniphobicity: for a surface to be self-cleaning, the requisite is to strongly repel water, reducing at maximum the water drops contact area, and therefore the chances of contamination. This surface behavior is typical of several living organisms, from both the Plant and the Animal kingdoms, as further described in the next paragraphs (Qu et al. 2010; Genzer et al. 2008). Here, what most matters is the formation of hierarchical structures at the micro or nanoscale, which force water to interact with air entrapped between adjacent surface features more than with the surface itself (Li and Amirfazli 2008). In this sense, the development of nanotechnology has surely boosted laboratory and industrial research, allowing first the observation of such morphological features—and therefore the understanding of the phenomenon—and subsequently their replication. The most common example of this behavior is the lotus effect, named after the Indian lotus (*Nelumbo nucifera*) leaf, on which rain drops have such a small contact area that they result in bouncing and rolling, collecting dust and pathogens and hence cleaning the leaf surface.

Such efficient self-cleaning is ensured by a combination of a waxy, highly hydrophobic surface and its nanostructured patterning, and is observed on several plant leaves, as represented in Fig. 9.2 (Barthlott and Neinhuis 1997; Solga et al. 2007). Another case is a special fish scale covered by riblets that helps reducing friction in water, which in turn decreases the adhesion of other microorganisms: this is the case of shark skin (Schumacher et al. 2007).

Latest advancements in nature-mimicking nanotechnology take inspiration from the binary periodic structure of some insects eyes, whose hierarchical structuring induces a superhydrophobic surface state (Yang et al. 2013). A wider description of natural superhydrophobic surfaces for self-cleaning is reported in the following paragraphs.

9.1.2 Hydrophilicity and Hydrophobicity

Understanding the interaction of surfaces with water molecules is essential to achieve the control of the material response in most applications, especially considering the exposure to an external environment and its humidity. Having a suitably hydrophilic or hydrophobic character plays a crucial role in nature and decides your fate in extreme environments, as in the case of the desert Stenocara beetle, whose hydrophilic/superhydrophobic patterned elytra with its hierarchical

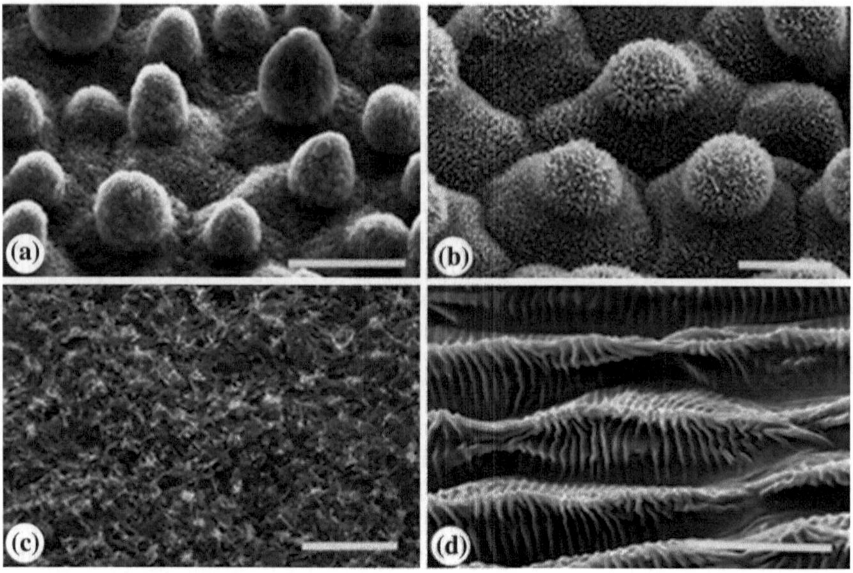

Fig. 9.2 Scanning electron micrographs of the adaxial leaf surface of rough, water-repellent leaf surfaces. *Nelumbo nucifera* (**a**) and *Colocasia esculenta* (**b**) are characterized by papillose epidermal cells and an additional layer of epicuticular waxes. *Brassica oleracea* leaves (**c**) are densely covered by wax crystalloids without being papillose, and the petal surfaces of *Mutisia decurrens* (**d**) are characterized by cuticular folds. *Bars* 20 μm. Adapted with permission from Barthlott and Neinhuis (1997)

structure allows water to be collected from morning mist, creating a unique water reservoir—the key to their surviving (Parker and Lawrence 2001).

To meet this aim, it is useful to remind some common concepts about surface wettability.

The parameter that identifies the wetting properties of a surface is contact angle θ (Fig. 9.3), defined as the angle between the tangent to the curved water surface at the point of contact with the solid surface and the plane of the surface on which the drop is resting, measured through the water (Marmur 2010).

This parameter is the result of the equilibrium between three components of surface tension, represented by three forces acting at the liquid–solid (γ_{sl}), solid–gas (γ_{sg}), and liquid–gas interfaces (γ_{lg}). Water droplets form high contact angles (>90°) on hydrophobic materials on account of the energy increase upon contact between surface and liquid, which favors the minimization of contact area between the two. This is explained by the extremely low surface energies of such materials ($\sim$10–50 mN/m). On the contrary, hydrophilic materials have a strong affinity with water, which thus seeks to maximize its contact area, forming low contact angles on their highly energetic surfaces, ranging from 500 to 5,000 mN/m (de Gennes et al. 2002).

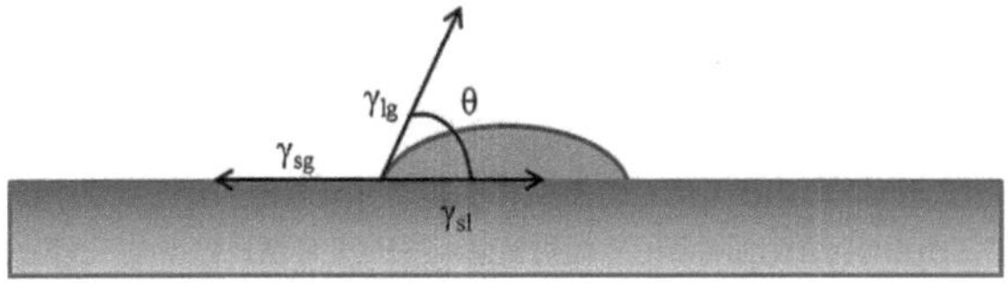

Fig. 9.3 The contact angle θ formed by a liquid drop on a surface is a consequence of equilibria among solid–gas (γ_{sg}), liquid–gas (γ_{lg}) and solid–liquid (γ_{sl}) surface tension

In both cases, the static contact angle θ_0 that a liquid forms on a surface can be derived by applying Young's equation (Eq. 9.1):

$$\cos\theta_0 = \frac{\gamma_{sg} - \gamma_{sl}}{\gamma_{lg}} \tag{9.1}$$

and is valid for any triple liquid–solid–gas interphase, with the gas phase generally being air. Yet, from the point of view of self-cleaning, another fundamental aspect of surface wettability must be considered, that is, hysteresis.

In fact, the static contact angle can only be observed on ideally flat surfaces, while most commonly a range of values will be observed, from a minimum value defined as receding contact angle θ_{rec}, to a maximum advancing contact angle θ_{adv}: the difference between the two represents contact angle hysteresis. The smaller this parameter, the easier the movement of a drop on the surface; this concept is particularly important in the case of superhydrophobicity-driven self-cleaning, where the cleaning effect is assured by water rolling on the surface, collecting dust deposited on it and carrying it away (see Sect. 9.2.2). DCA (dynamic contact angle) is generally used to quantify it and different methods can be applied. Although the correct observation method of hysteresis would be the one schematized in Fig. 9.4a—i.e., a drop deposited on a tilted surface, generally, first a drop is dispensed to read the advancing angle, while the receding one is measured by drop retraction (Fig. 9.4b). Additionally, the material can be immersed directly

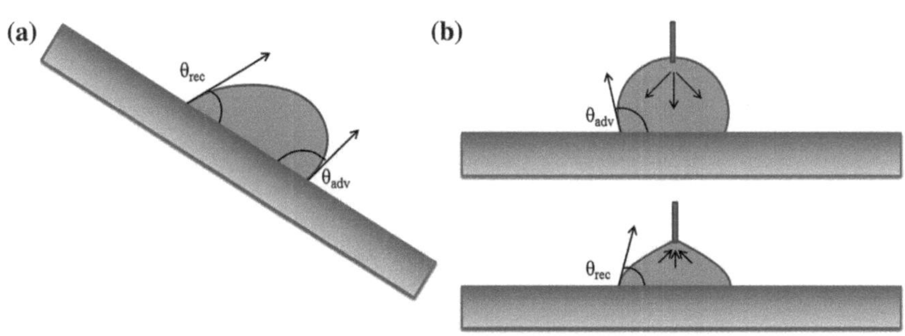

Fig. 9.4 Methods for determining advancing and receding contact angles: **a** tilted plane and **b** dynamic contact angle

in water (or other liquid), which produces the advancing contact angle, and subsequently extracted, generating the receding contact angle.

While the characterization of a superhydrophilic surface ($\theta \sim 0°$) is rather simple—being related to the spreading of water on the surface to cover the largest area possible, superhydrophobic surfaces ($\theta > 150°$) are more complex to describe. In fact, based on surface tension equilibria and Young's equation, no known chemistry allows water to configure with a contact angle larger than 120° on a smooth surface (Hunter 2010): therefore, surface morphology must necessarily be involved to reach a superhydrophobic state.

When surface roughness comes into play, two models are generally adopted to describe surface wettability, that is, Wenzel state and Cassie-Baxter state (Fig. 9.5) (Wenzel 1949; Cassie et al. 1944). In 1936 Wenzel stated that adding surface roughness enhances surface wettability, as shown in Fig. 9.5a, b, where the contact angle of a water drop on the same hydrophobic surface is apparently higher when roughness is introduced, and vice versa on hydrophilic surfaces (Fig. 9.5c), since it is measured against an average line that represents the apparent solid surface. This variation can be calculated by applying Eq. 9.2:

$$\cos\theta_W = r_\varphi \cos\theta_0 \qquad (9.2)$$

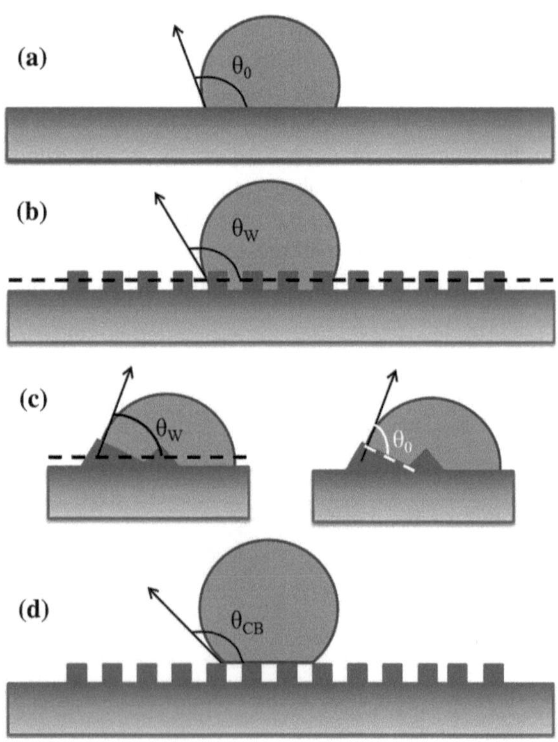

Fig. 9.5 From *top* to *bottom*: **a** water drop on a smooth hydrophobic surface; **b** Wenzel state; **c** detail of apparent decrease in contact angle predicted by Wenzel model (θ_W) compared to the one calculated by Young's equation (θ_0) on hydrophilic surfaces; **d** Cassie-Baxter state

where θ_W is the contact angle observed on a rough surface following Wenzel's model, θ_0 is the contact angle on an ideally smooth surface of same chemistry, and r_ϕ is the average roughness ratio, i.e., the ratio between actual surface in contact with the liquid and projected area of the wetted region (Wenzel 1936). Wenzel's theory has been extended during years to take into account the different apparent contact angles that may be observed on surfaces with nonuniform roughness (Johnson and Dettre 1964; Wolansky et al. 1999; Gao et al. 2007; Liu et al. 2012). Wenzel's model refers to a complete wetting of the surface, and the hysteresis in contact angle measurements can reach 100°.

The second model, proposed by Cassie and Baxter (1944), considers the formation of air pockets in roughness valleys under the drop, which leads to higher contact angle values and—theoretically—almost null hysteresis, as shown in Fig. 9.5d (Johnson and Dettre 1964; Callies et al. 2005; Verplanck et al. 2007). Similarly to Wenzel's model, a change in contact angle is computed owing to the presence of air-liquid interfaces at the droplet-roughness interface. Again, a parameter is identified to quantify the excess interfacial surface free energy per unit area, substituting the term $\gamma_{sg} - \gamma_{sl}$ of Young's equation with terms that consider ϕ_s—the fraction of solid surface area available for direct contact—and r_ϕ—as in the previous model, the average roughness ratio. The Cassie-Baxter equation then results to be (Cassie et al. 1944; Marmur 2003; Choi et al. 2009):

$$\cos\theta_{CB} = r_\varphi \varphi_s \cos\theta_1 + (1 - \varphi_s)\cos\theta_2 \tag{3}$$

where θ_1 and θ_2 represent the equilibrium contact angle of the liquid on solid (θ_0) and air (180°), respectively. $r_\phi \phi_s$ and $(1-\phi_s)$ come to represent the fractions of drop surface in contact with the solid and with air, respectively.

Indeed, the Cassie-Baxter model also required corrections and extensions in order to justify the hysteresis that is observed on rough, nonwetting surfaces (Choi et al. 2009, and references therein).

Another important consideration to understand surface wetting phenomena is the metastability of most Cassie-Baxter states, with consequent transitions to the Wenzel state: water tends to penetrate the air pockets to minimize surface energy, gradually filling all roughness cavities starting from the center of the drop until it reaches its outer limits. This phenomenon is subject of several research studies aimed at achieving a controlled and reversible Cassie-Baxter to Wenzel state transition (Peters et al. 2009; Giacomello et al. 2012; Bormashenko et al. 2013).

In spite of all theories, validations and confutations of Cassie-Baxter and Wenzel models that may be found on the topic(Extrand 2002; Gao and McCarthy 2007, 2009; Panchagnula et al. 2007; Nosonovsky 2007; McHale 2007; Whyman et al. 2008), what really causes self-cleaning on superhydrophobic surfaces—in practical terms—is the achievement of the spontaneous bouncing and rolling of water drops on a horizontal surface, as will be described in next paragraph (Morra et al. 1989; Solga et al. 2007; Balu et al. 2008; Myint et al. 2014).

9.2 The Different Mechanisms of Self-cleaning

As previously mentioned, self-cleaning is a wettability-related property that can be achieved on both superhydrophilic and superhydrophobic surfaces, being the obtaining of an extreme wettability state the ruling condition. Here the different bioinspired self-cleaning mechanisms will be described in more details, starting from the one that is most commonly applied in the field of building materials (Diamanti and Pedeferri 2013), based on one of the most studied functional materials of the last decades: titanium dioxide (Fig. 9.6).

9.2.1 Self-cleaning on Photoactive Surfaces: TiO_2

The biomimetic nature of TiO_2 self-cleaning mechanism lies in the mimicking of green plants photosynthesis processes, where sunlight is used to produce highly energetic chemical species that modify the oxide surface properties and reactivity, as first described by Fujishima and coworkers in 1970s and further investigated in the following decades.

In fact, transition metal oxides can be photoactivated by light absorption of suitable wavelength. This generates the promotion of an electron from the valence band to the conduction band, where a hole (positive charge) is left: such reactive species can recombine or react with the surrounding environment, giving rise to a series of important photoactivated functionalities that contribute to the material self-cleaning ability (Fujishima et al. 1972; Carp et al. 2004).

One of the most intriguing—yet not easily governable—consequences of photoactivation is photoacatalysis, i.e., the reaction of photogenerated holes and electrons with water molecules present in the atmosphere to produce radicals and

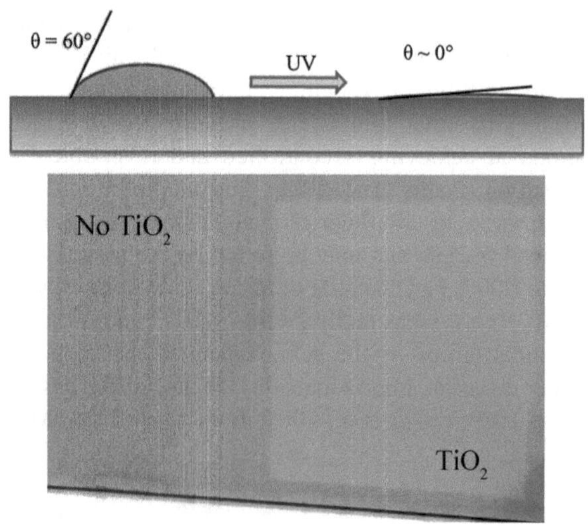

Fig. 9.6 *Top* photoinduced formation of a superhydrophilic TiO_2 surface (reversible upon dark storage). *Bottom* ceramic panel, partially treated with TiO_2 and exposed to urban atmosphere for 1 year

ions with great redox power, which in turn react with organic and inorganic chemical species adsorbed on the oxide surface and induce their decomposition (Fujishima et al. 2008). However, the majority of industrial applications involving photoactivity consists of the production of self-cleaning surfaces (Ganesh et al. 2011; Zhang et al. 2012), with applications ranging from self-cleaning windows, to antifogging rear view mirrors, to tiles, cements, and other building materials, to textiles (Drelich et al. 2011; Diamanti and Pedeferri 2013).

The effect relies on the coexistence of two mechanisms: photocatalysis degrades the organic functional groups by which pollutants adhere to a surface, while superhydrophilicity forces water to spread completely on the surface, carrying away particulate matter and degraded contaminants (Fig. 9.6) (Wang et al. 1997; Ganesh et al. 2011).

Superhydrophilic TiO_2 surfaces were first described in 1997 by Wang et al. who observed a complex modification of surface chemistry, with Ti^{4+} ions reduction and oxygen vacancies formation, leading to the formation of a highly hydroxylated surface layer. The increase of van der Waals forces and hydrogen bonding interactions between water and hydroxyl groups accounts for the extreme wetting properties under UV irradiation (Wang et al. 1999; Drelich et al. 2011). As an effect, drops formation on superhydrophilic surfaces is avoided, which precludes stains due to slow water evaporation from the surface; for the same reason, an antifogging mechanism is enabled (Lai et al. 2012).

The onset of superhydrophilicity has also been ascribed to the photodegradation of hydrocarbon contaminants present on TiO_2 surfaces: recent XPS investigations do not reveal substantial formation of OH groups on its surface upon irradiation (Wang et al. 2003), while several theoretical and experimental studies address the effects of water adsorption onto TiO_2 (Jribi et al. 2009; Diamanti et al. 2013).

The debate in the scientific community on the actual mechanisms of TiO_2 photoinduced superhydrophilicity is still open, although both water dissociation and adsorption and photocatalytic decontamination surely bring a positive contribution to surface hydrophilicity.

9.2.2 Learning from Carnivorous Plants

A fascinating example in nature of a potential self-cleaning surface with hydrophilic character is represented by *Nepenthes* pitcher plants (Fig. 9.7). The pray-trapping technique of such carnivorous plants is based on a cup-folded leaf filled with nectar juices and digesting enzymes: insects, attracted by the sweet nectar, lean on the inner rim of the leaf and find an extremely slippery surface, which forces them to fall inside the cavity, with no hope for climbing the leaf back.

The reason for this unexpected trapping lies in the interplay between surface hydrophilicity and its hierarchical structure (Gaume et al. 2002; Bohn et al. 2004).

The peristome surface has a very regular microstructure consisting of first- and second-order radial ridges formed by straight rows of epidermal cells, each

Fig. 9.7 *Nepenthes* pitcher plant: on the *left*, folded leaves that create pitcher traps (Copyright © 2013 Alessandro Pelizzi). On the *right*, *Nepenthes* pitcher and peristome morphology highlighting the porous cross-sectional structure and the two levels of ridges (Adapted with permission from Bohn and Federle 2004; Copyright © 2004 National Academy of Sciences, U.S.A.)

overlapping the adjacent cell, so that the surface contains a series of steps toward the pitcher inside and is anisotropic. In some species, this surface is covered with wax; still, the actual trapping mechanism is based on the formation of a layer of water on top, owing to the presence of hygroscopic nectar—in fact, their trapping ability is boosted in rainy days. Surface wettability is anisotropic, following ridges direction (Bohn and Federle 2004; Bauer et al. 2009).

This peculiar structuring has recently become focus of several research studies dedicated to the production of slippery surfaces, with applications in the field of self-cleaning, anti-icing, antifouling, and fluid transport (Lafuma and Quéré 2011; Kim et al. 2012b; Epstein et al. 2012). The first biomimicking of pitcher plants was proposed by Wong et al. (2011), who labeled their newly developed system as SLIPS—slippery liquid infused porous surface. SLIPS are based on micro-nano-structured porous surfaces that are infused with lubricating liquids, which induce a superhydrophobic character until the lubricant stably wets the surface: currently, the major issue of such self-cleaning and anti-icing mechanism is indeed the effect durability (Rykaczewski et al. 2013).

9.2.3 Hydrophilic and Superoleophobic Plants and Animals

Pitcher plants are just the first example of natural systems used as reference in biomimetic self-cleaning materials. Several systems take inspiration from hydrophilic surfaces, with different mechanisms. Recent investigations on some microstructured surfaces, such as snail shells (Fig. 9.8a), brought to light a peculiar superoleophobic behavior that induces extraordinary self-cleaning performances: also in this case, the presence of water is crucial for the mechanism to work.

Fig. 9.8 **a** snails show outstanding self-cleaning behavior in wet environments; **b** scale pattern of a Galapagos shark (Adapted with permission from Reif 1985)

In fact, snail shells are naturally oleophilic—they show a contact angle of oil in air of approximately 10°—but they can be turned into superoleophobic, with contact angles of oils higher than 150° when immersed in water, or when covered by a water layer. Once more, surface chemistry and microstructure are responsible for this behavior: the protein-coated shell has three ranges of millimeter and micrometer wide grooves that create a hierarchical pattern, where water can be trapped, hindering the direct contact with oil (Nishimoto and Bhushan 2013).

Analogous results can be achieved by mimicking fish scales, and specifically shark skin. Initially studied for the low drag properties that account for sharks speed (Bechert et al. 2000; Krieger 2004; Dean and Bhushan 2010), the scales microstructure (Fig. 9.8b) was recently connected to the possibility to create self-cleaning surfaces. In fact, each scale presents a precise texture of riblets parallel to swimming direction, which deviate water flow and vortices while swimming at high speed, reducing the contact area with turbulent flow and therefore minimizing friction and shear stresses. In parallel to an improvement of drag properties, this microstructure decreases microorganisms adhesion and therefore biofouling, and increases surface washing (Bixler and Bhushan 2012; Nishimoto and Bhushan 2013). In 2007, Schumaker and coworkers proposed several substrate topographies inspired to shark skin, and evidenced the dependence of biofouling reduction—in terms of zoospores adhesion—on the dimensions of geometrical protrusions.

9.2.4 The "lotus effect"

In spite of the growing scientific, technological, and industrial interest in the natural systems previously described, the most representative case of biomimicry remains the "lotus effect," named after the lotus—*Nelumbo nucifera*—leaf self-cleaning capability, which now counts hundreds of scientific publications and several dedicated websites.

Although lotus has always been considered an icon of purity in ancient oriental societies, thanks to its perpetual cleanliness even in muddy waters, Barthlott and coworkers were pioneers in investigating lotus surfaces and identifying the correlated self-cleaning mechanism (Barthlott and Neinhuis 1997). Their study unveiled a first level of micrometric asperities, 20–40 μm apart, and a more complex roughness at the nanoscale induced by the presence of epicuticular wax crystals (Fig. 9.2). This mixture of surface chemistry and hierarchical micro-nanostructuring, which has then been identified on several other plants and leaves, accounts for surface superhydrophobicity (Fig. 9.9), producing a water contact angle of 160° and sliding angle lower than 4°, and consequently allowing water rolling and bouncing on the surface. (Cheng et al. 2006; Su and Chen 2008; Koch et al. 2009). Since the presence of micro- and nano-asperities drives surface wettability, Quéré in 2002 depicted the behavior or water on these surfaces by introducing the concept of "fakir droplets" (Quéré 2002).

Other plant leaves show analogous, or similar, hierarchical arrangement of micro and nano texture. Taro, or *Colocasia esculenta*, surfaces also exhibit the same hierarchical structure, the only difference being the shape of wax nano-structures (Koch et al. 2008; Xiu and Wong 2010). Rice leaves of all rice categories show the same binary structure of lotus leaves, with 5–8 μm papillae parallel to the leaf edge and nanoscale pins distributed on the whole surface sublayer: this explains the peculiar superhydrophobicity of rice leaves, which is anisotropic—with sliding angles of 4° in the papillae direction and 12° perpendicularly. Other examples of self-cleaning superhydrophobic surfaces coming from the Plant kingdom include purple setcreasea petals, indian canna leaves, perfoliate knotweed, wild pansy flowers (Fig. 9.9c) (Guo and Liu 2007; Schulte et al. 2011). Another interesting plant showing superhydrophobicity is the tropical aquatic fern *Salvinia molesta* (Barthlott et al. 2010; Bhushan 2012), where the effect is related to multicellular surface structures consisting of crown-like hairs of hundreds of micrometers—up to

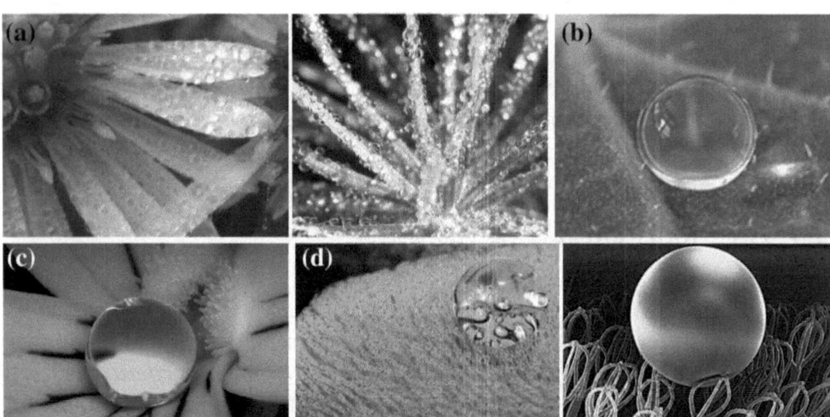

Fig. 9.9 **a** Examples of superhydrophobic daisies and grass (Copyright © 2013 Giorgio Re); **b** An almost *ballshaped* water droplet on a nonwettable plant leaf (Adapted with permission from Blossey 2003); **c** Macro photo of a water droplet on a flower of the wild pansy (*Viola tricolor*) (Adapted with permission from Schulte et al. 2011); **d** *Salvinia Molesta* SEM micrographs of hair structure (Reprinted with permission from Barthlott et al. 2010)

millimeters of length: this macroscopic hierarchical structure retains air, which induces a Cassie-Baxter wetting state (Fig. 9.9d).

Finally, some superhydrophobic rose petals show unexpected adhesion of water drops, which remain adherent to the surface even if tilted to 180°. This effect is driven by microstructures spacing: if microsized pillars are far enough, it is possible for water to fill the air gaps, causing a Cassie-Baxter to Wenzel transition and increasing surface adhesion (Nosonovsky and Bhushan 2012).

Animals are not outdone in terms of variety of interactions with water and hierarchical microstructuring. For instance, water striders (Gao and Jiang 2004), like water spiders (Yang et al. 2008) and several other insects, present thin hydrophobic wax-coated hair that allow them to trap air, analogously to a Cassie-Baxter surface, and float on water (Fig. 9.10a, b). Mosquito's eyes have antifogging properties, generated by the hexagonal nonclose-packed nanonipples arrays that cover hundreds of microhemispheres, which in turn are arranged in hexagonal close-packing, as shown in Fig. 9.10c (Gao 2012; Xiu and Wong 2010). Similarly, cicada wings consist of wax-coated hexagonally packed pillars whose height is approximately 250 nm and spatial distribution is slightly higher than 100 nm (Fig. 9.10d) (Nishimoto and Bhushan 2013).

Butterfly wings are covered with hierarchical scales, which induce once again the desired self-cleaning effect, with an anisotropic structure that drives water along the radial outward direction, while high resistance to water droplets movement is opposed in the inward direction (Bixler and Bhushan 2012; Wong et al. 2013).

9.2.5 Gecko

Geckos are renowned in the field of biomimicry for the outstanding adhesive properties of their toes, based on dry adhesion, which gives them the ability to cling to any surface and detach at will. The means that allows such powerful

Fig. 9.10 **a** SEM image of a water strider leg showing oriented microsetae (Reprinted with permission from Gao and Jiang 2004); **b** water spiders have thin hydrophobic (wax-coated) hairs with nanoscale roughness which trap air and enhance hydrophobicity (Reprinted with permission from Yang et al. 2008); **c** SEM image of a mosquito eye (Reprinted with permission from Xiu and Wong 2010); **d** SEM image of a cicada wing (Reprinted with permission from Xiu and Wong 2010)

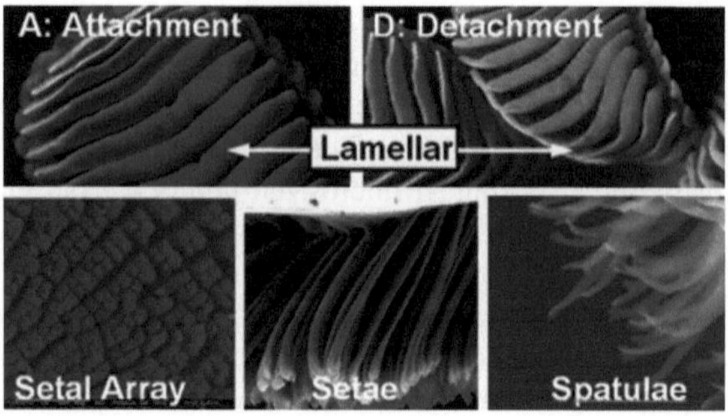

Fig. 9.11 Hierarchical structure of a gecko and its initial contact to a surface: lamellae (attachment and detachment), and setal arrays to the spatulae are highlighted (Adapted with permission from Tian et al. 2013)

adhesion is again the presence of hierarchical structures: each toe is covered with rows of overlapping pads (lamellae) each consisting of millions of densely packed microscale hair, called setae, further branching off in nanoscale spatulae (Fig. 9.11). This creates a huge contact area—ultimately, gecko adhesion is primarily pushed by van der Waals forces (Ruibal and Ernst 1965).

Still, a crucial aspect that was not considered in early studies on gecko spatulae is their ability of remaining clean from contaminants (Hansen and Autumn 2005): this has recently become the focus of a new branch of biomimetic self-cleaning surfaces.

The mechanism diverges from those previously described, which all involve the action of water. In this case, a disequilibrium in adhesive forces between dirt, substrate and gecko spatulae has been indicated as the source of such phenomenon. In spite of the huge adhesive strength of full setae, dirt particles—due to their small size—can only adhere to few spatulae: as a consequence, the substrate-particle adhesion strength remains higher than the setae-particle one, and can even help restoring clean spatulae in case of contamination by simply moving the setae on the surface, which generally occurs within few steps (Hansen and Autumn 2005). In fact, gecko feet do accumulate dirt, but the essence of their self-cleaning is the ability to clean toe pads while in motion.

Self-cleaning of gecko—and other insects, such as *Hymenopterans, Coccinellids* and *Phasmatodeas*—was further studied thoroughly by several research groups (Clemente et al. 2009; Hu et al. 2012; Orchard et al. 2012), in order to get a deeper insight of the process and replicate it, by monitoring the variations of adhesive strength of their feet—which lose stickiness when contaminated, and recover it when clean. Hu et al. proposed in 2012 a dynamic digital hyperextension mechanism: setae can recover up to 80 % of their adhesive strength in only four steps in hyperextension mode, that is, with sudden releases, while the recovery limits to 40–50 % in case of absence of digital hyperextension motion.

9.3 Production Techniques and Applications

It should now be clear how technologically important the understanding and development of self-cleaning surfaces is. Plants developed such capability to protect themselves from hazardous microorganisms such as bacteria, and to maintain their energy source unaltered during their lifetime: in fact, covering leaves with mud, dust, soot, or any other contaminant would hinder photosynthesis, which requires solar energy to reach the leaf surface and be absorbed and converted by chlorophyll-containing proteins. On the other hand, artificial self-cleaning surfaces applied to the built environment could bring major energy savings, starting from indirect costs—i.e., the reduction of maintenance and cleaning interventions—and affecting also the overall material performance, with a prolonged maintaining of its reflectance properties and therefore of the energy gain through solar irradiation. As a consequence, a more reflective surface—especially in the infrared spectrum—would decrease the amount of absorbed energy, and therefore the need for air conditioning in hot season, attenuating the urban heat island phenomenon (Levinson et al. 2005; Sleiman et al. 2011). Moreover, as previously cited, several research works are now dedicated to the implementation of self-cleaning functionalities on solar cells, to avoid efficiency losses due to atmospheric soiling.

Efforts in biomimicking the abovementioned self-cleaning mechanisms are huge, and include all mechanisms described, from SLIPS to photoinduced superhydrophilicity, to superhydrophobic surfaces (Genzer and Marmur 2008; Liu and Jiang 2012). Both the production techniques that can be adopted and the materials used as starting building blocks are numerous. For instance, a comprehensive description summarizing the realization of superhydrophobic surfaces with lotus effect can be found in Bhushan et al. (2010); details on the use of self-cleaning superhydrophilic coatings are given in Drelich et al. (2011), and Nakata and Fujishima (2012). In the latter case, the methods most commonly applied to generate TiO_2, ZnO, WO_3 or SiO_2 superhydrophilic surfaces involve the deposition of a coating, generally by sol-gel spray-coating or dip-coating deposition, as well as by sputtering and physical (PVD) or chemical (CVD) vapor deposition (e.g., in the case of glass), or alternatively the massive addition of nanoparticles to the substrate itself (e.g., in the case of paints, or cementitious materials). On the other hand, superhydrophobic surfaces are not only based on surface chemistry, but also—and prevalently—on a hierarchical surface structure. As a consequence, production techniques are necessarily more elaborated, and follow two approaches: either a surface with suitable morphology is chemically modified to achieve low surface tension, or a hydrophobic material is used to build hierarchical structures (Guo et al. 2011). Examples of artificial hierarchical structures are reported in Fig. 9.12.

These approaches belong to lithography, etching, deposition and self-assembly, and include plasma and chemical etching, electropolymerization and chemical polymerization, nanoimprint, electrodeposition, self-assembly, as well as sol-gel and CVD. Coatings generally consist of low surface tension polymers—such as Teflon, PDMS or fluorinated polymers, silicon, silica, carbon nanotubes or

(a) (b) (c)

Fig. 9.12 **a** A designed rough surface (Reprinted with permission from Blossey 2003); **b** SEM images of nanopillar array manufactured on quartz by mask-free lithography with PMMA resist, *inset* contact angle (Adapted with permission from Kim et al. 2012a); **c** SEM picture of a micro-patterned polymer (KratonTM D-1102CS) film with gap size 5 μm and pillar height 10 μm (Reprinted with permission from Peters et al. 2009)

composite systems; superhydrophobicity was also achieved by controlled etching of metallic surfaces, such as steel, copper, and aluminum (Bhushan et al. 2010; Guo et al. 2011).

Here we list some outstanding examples of biomimetic self-cleaning surfaces, and of how they were created.

In a work by Kwon et al. (2009) silicon surfaces were uniformly modified by combining two etching techniques, generating a superhydrophobic nonclosely packed roughness with contact angles higher than 170°, low contact angle hysteresis and tilting angles lower than 2°. The authors observed multiple bouncing of water droplets on the surface, proving the onset of a Cassie nonwetting state.

A large-area substrate was converted by Lee et al. (2010) into a superhydrophobic surface by combining PMMA microspheres and a silicon grease, which was turned into a nanostructured ceramic through electron irradiation: the obtained microporous structure with ceramic nanobumps showed good mechanical stability to standard tests.

In a recent work by Zhang et al. (2013b) fluoro-silicone nanofilaments were combined with Krytox liquids to produce a SLIPS structure and achieve antiwetting coatings inspired to *Nepenthes* pitcher plants (Fig. 9.13). Contact angle increased with surface tension of liquids investigated, and a sliding angle of 4° was recorded with both polar (water) and nonpolar (n-hexadecane, n-decane) liquids.

Groten and Rühe (2013) investigated more in-depth the effect of multiscale hierarchical structures on surface wetting, by producing nanoscale, microscale or combined micro and nanoscale roughness on silicon substrates by etching, and coating them with a fluoropolymer. Nanorough surfaces showed contact angles close to 180° and almost no contact angle hysteresis, but at the same time poor mechanical stability was observed, while multiscale hierarchical structures with identical chemical composition allowed both to reach superhydrophobicity and to confer wear resistance to the surface.

Nevertheless, the latest advances in the field of self-cleaning surfaces were proposed by Bird et al. (2013) who designed a fluorosilane-coated, laser-ablated

Fig. 9.13 Preparation of the composite fluoro-silicone nanofilaments/Krytox anti-wetting coatings inspired to *Nepenthes* pitcher, and various liquids contact angle (Adapted with permission from Zhang et al. 2013b)

silicon wafer with superhydrophobic character and demonstrated the reduction of contact time of water on the surface by breaking the drop dynamics symmetry. This surface is now described as the most waterproof material ever. Investigations from the same research group led in 2012 to LiquiGlide™, a patented liquid-impregnated, nontoxic coating based on SLIPS concept that has wide applications in consumer packaged goods to decrease the adhesion of fluids—food, cosmetics, drugs—to the container, thus improving product evacuation and reducing wastes (Smith et al. 2013).

Still, in the building industry—where large surface areas must be treated—the most diffuse wetting modification treatments are those based on superhydrophobic carbon nanotubes (CNTs) or SiO_2 (Guo and Wang 2010; Bu and Oei 2010) and superhydrophilic TiO_2 (Chen and Poon 2009; Diamanti and Pedeferri 2013). As for the latter case, several works describe the obtaining of self-cleaning, superhydrophilic glass, tiles and construction materials in general, with hundreds of scientific publications and a growing number of self-cleaning products developed and commercialized (Diamanti 2012). Current evolutions in this research field are related to combining micro and nanotextures with the photoinduced superhydrophilicity of TiO_2, as in the study proposed by Nakata et al. (2011), where PET films with moth-eye-like surfaces were modified to achieve self-cleaning and antireflective coatings for solar cells.

On the other hand, superhydrophobic treatments mainly involve glass and textiles. Epoxy resin micropillars coated with 10 nm SiO_2 nanoparticles or multiwall CNTs showed excellent superhydrophobicity under 1 kPa water jet (contact angles higher than 160°, increasing hysteresis with exposure to water jet), combined with mechanical stability and wear resistance (Jung and Bhushan 2009; Ebert and Bhushan 2009).

Additionally, Joshi et al. (2012) proposed dip-coating and layer-by-layer self-assembly of nanosilica and nanoclay particles to develop superhydrophobic cotton through properly aligned nanoroughness. Once again, matching an adequate

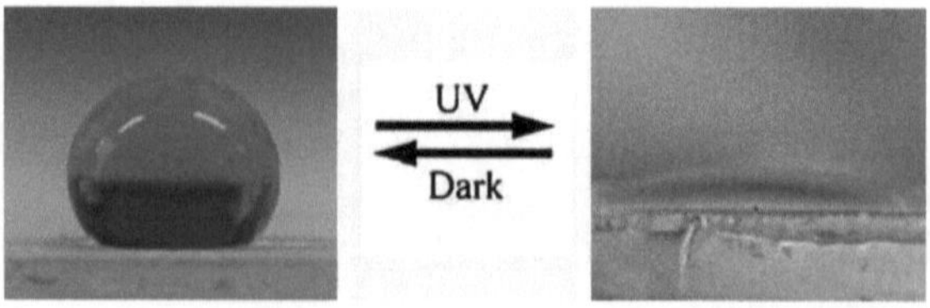

Fig. 9.14 Optical images of a water droplet on the surface of a film composed of ZnO hollow microspheres with 2D nanosheet stacks (Adapted with permission from Sun et al. 2013)

surface nanostructuring with low surface tension—conferred by coating the modified surface with a fluorocarbon-based emulsion—allowed to reach and exceed contact angles of 150° and low sliding angles.

CNT-modified glass surfaces were produced with 80 % transmittance in the visible region, overcoming the issue of transparency related to the use of CNTs coatings (Bu and Oei 2010; Meng and Park 2010).

The possibility of achieving switchable superhydrophilic/superhydrophobic surfaces is also of great interest, in order to produce surfaces with controllable interactions with water: this was achieved with CTNs (Men et al. 2010), as well as with ZnO (Sun et al. 2013) (Fig. 9.14) and TiO_2 (Lai et al. 2012), which allowed to maintain a transparency of 80 %.

SiO_2 nanoparticles were also employed by Gao et al. (2011) to produce the opposite effect—i.e., superhydrophilic glass surfaces with contact angles varying in the range 0° ÷ 20° depending on the formation conditions of SiO_2 layers. SiO_2/TiO_2 composite films also find wide application in this field, owing to the simultaneous obtaining of self-cleaning and antireflective conditions (Zhang et al. 2013a).

9.4 Concluding Remarks

Nature is a constant source of inspiration for technological innovations, as described in this review on bioinspired self-cleaning surfaces. Such materials have already proved their valuable help in reducing surface soiling, which is of particular interest in the built environment, where building envelope materials undergo a constant soiling operated by atmospheric contaminants. The development of self-cleaning materials—and of the related properties of antifogging, low adhesion, antifouling, and possibly antireflection—contribute to the field of green technologies through a reduction in maintenance costs and energetic consumptions.

The working mechanism of the different self-cleaning technologies discussed is summarized in Fig. 9.15 in order to provide a comparison among their potential efficiency.

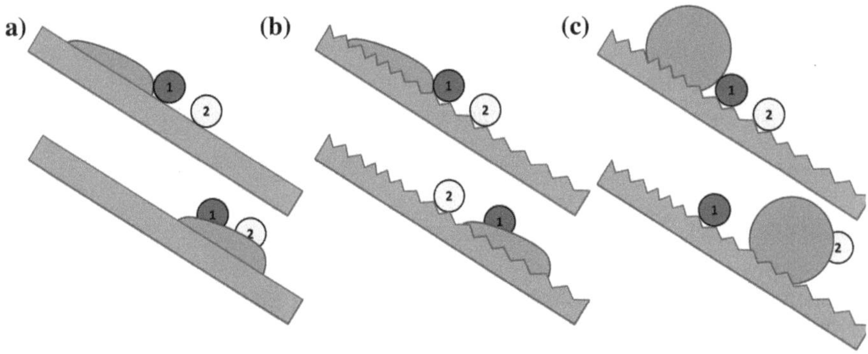

Fig. 9.15 Mechanisms of self-cleaning on **a** superhydrophilic surfaces, **b** hydrophilic, superoleophobic surfaces (SLIPS), and **c** superhydrophobic surfaces, and related ability to remove hydrophilic (*1*) or hydrophobic (*2*) contaminants

A first observation concerns the role of water, which is vital in all situations to carry away particles from the surfaces. Furthermore, the nature of contaminants also plays a key role: while water can easily carry away any type of contaminant, be it inorganic, organic, oil-based, or hydrophilic—owing to the possibility of partial contaminant degradation offered by the conjugated photocatalytic activity—superhydrophobic materials only offer water a bouncing surface, therefore hydrophobic contaminants are hardly removed. Opposite considerations apply to hydrophilic, superoleophobic surfaces, where hydrophilic sources of soiling are retained by the surface.

Superhydrophilicity should therefore appear to be the most promising technology to produce effective and reliable self-cleaning surfaces; nevertheless, its drawback lies in the need for photoactivation, which makes it purposeless in the absence of a suitable irradiation source—in most cases, UV light is required, limiting its use in outdoor applications. Current research is dedicated to the development of doped TiO_2 components with visible light activity, which may find application indoor as well.

One last note must be dedicated to durability. The use of a surface modification technique to alter the wettability properties of building materials cannot set aside the treatment lifetime, as such materials are generally designed for service lives of several decades. Moreover, atmospheric dust and particles may exert some erosive effect on the material surface. In this respect, the use of coatings with enhanced, extremely controlled micrometric and nanometric roughness could be detrimental, unless the coating is proved to exhibit excellent wear resistance. If a superhydrophilic coating is employed, this issue is mitigated, as only the complete removal of the coating—rather than its partial damage—would hinder its functioning. On the other hand, such photoactive coatings may suffer from deactivation due to irreversible adsorption of reaction products released during contaminants degradation.

It is therefore clear how the choice of the most suitable self-cleaning technology must necessarily take into account the surrounding environment, in terms of water availability (e.g., rain), level and type of pollution, and possibility of erosion (e.g., in sandy environments, in the presence of wind).

References

Balu B, Breedveld V, Hess DW (2008) Fabrication of “roll-off” and “sticky” superhydrophobic cellulose surfaces via plasma processing. Langmuir 24:4785–4790

Barthlott W, Neinhuis C (1997) Purity of the sacred lotus, or escape from contamination in biological surfaces. Planta 202:1–8

Barthlott W, Schimmel T, Wiersch S, Koch K, Brede M, Barczewski M, Walheim S, Weis A, Kaltenmaier A, Leder A, Bohn HF (2010) The Salvinia paradox: superhydrophobic surfaces with hydrophilic pins for air-retention under water. Adv Mater 22:1–4

Bauer U, Federle W (2009) The insect-trapping rim of Nepenthes pitchers: surface structure and function. Plant Signal Behav 4:1019–1023

Bechert DW, Bruse M, Hage W (2000) Experiments with three-dimensional riblets as an idealized model of shark skin. Experim Fluids 28:403–412

Bhushan B (2012) Fabrication and characterization of micropatterned structures inspired by *Salvinia molesta*. In: Bhushan B (ed) Biomimetics—bioinspired hierarchical-structured surfaces for green science and technology. Springer, Berlin, pp 179–186

Bhushan B, Jung YC, Nosonovsky M (2010) Lotus effect: surfaces with roughness-induced superhydrophobicity, self-cleaning, and low adhesion. In: Bhushan B (ed) Springer handbook of nanotechnology. Springer, Berlin, pp 1437–1524

Bird JC, Dhiman R, Kwon HM, Varanasi KK (2013) Reducing the contact time of a bouncing drop. Nature 503:385–388

Bixler GD, Bhushan B (2012) Bioinspired rice leaf and butterfly wing surface structures combining shark skin and lotus effects. Soft Matter 8:11271–11284

Blossey R (2003) Self-cleaning surfaces—virtual realities. Nat Mater 2:301–306

Bohn HF, Federle W (2004) Insect aquaplaning: nepenthes pitcher plants capture prey with the peristome, a fully wettable water-lubricated anisotropic surface. Proc Natl Acad Sci USA 28:14138–14143

Bormashenko E, Pogreb R, Sagi B, Aurbach D (2013) Electrically controlled membranes exploiting cassie-wenzel wetting transitions. Sci Rep 3:3028

Bu IYY, Oei SP (2010) Hydrophobic vertically aligned carbon nanotubes on Corning glass for self cleaning applications. Appl Surf Sci 256:6699–6704

Callies W, Quéré D (2005) On water repellency. Soft Matter 1:55–61

Carp O, Huisman CL, Reller A (2004) Photoinduced reactivity of titanium dioxide. Prog Solid State Rep 32:33–177

Cassie ABD, Baxter S (1944) Wettability of porous surfaces. Trans Faraday Soc 40:546–551

Chen J, Poon CS (2009) Photocatalytic construction and building materials: from fundamentals to applications. Build Environ 44:1899–1906

Cheng YT, Rodak DE, Wong CA, Hayden CA (2006) Effects of micro- and nano-structures on the self-cleaning behaviour of lotus leaves. Nanotechnology 17:1359–1362

Choi W, Tuteja A, Mabry JM, Cohen RE, McKinley GH (2009) A modified Cassie-Baxter relationship to explain contact angle hysteresis and anisotropy on non-wetting textured surfaces. J Colloid Interface Sci 339:208–216

Clemente CJ, Bullock JMR, Beale A, Federle W (2009) Evidence for self-cleaning in fluid-based smooth and hairy adhesive systems of insects. J Exp Biol 213:635–642

de Gennes P-G, Brochard-Wyart F, Quere D (2002) Gouttes, bulles, perles et ondes. Collection Echelles, Belin, Paris
Dean B, Bhushan B (2010) Shark-skin surfaces for fluid-drag reduction in turbulent flow: a review. Philos Trans R Soc A 368:4775–4806
Diamanti MV (2012) Advances (and limits) in photocatalytic building materials. In: Nanotechnology 2012: Advanced materials, CNTs, particles, films and composites, vol 1. CRC Press, pp 491–494
Diamanti MV, Gadelrab KR, Pedeferri MP, Stefancich M, Pehkonen SO, Chiesa M (2013) Nanoscale investigation of photoinduced hydrophilicity variations in anatase and rutile nanopowders. Langmuir 29:14512–14518
Diamanti MV, Pedeferri MP (2013) Concrete, mortar and plaster using titanium dioxide nanoparticles: applications in pollution control, self-cleaning and photo sterilization. In: Pachego Torgal F, Diamanti MV, Nazari A, Granqvist CG (eds) Nanotechnology in eco-efficient construction. Woodhead Publishing Ltd, Cambridge, pp 299–326
Drelich J, Chibowski E, Desheng Meng D, Terpilowski K (2011) Hydrophilic and superhydrophilic surfaces and materials. Soft Matter 7:9804–9828
Ebert D, Bhushan B (2009) Durable lotus-effect surfaces with hierarchical structure using micro- and nanosized hydrophobic silica particles. J Colloid Interface Sci 368:584–591
Epstein AK, Wong TS, Belisle RA, Boggs EM, Aizenberg J (2012) Liquid-infused structured surfaces with exceptional anti-biofouling performance. Proc Natl Acad Sci USA 109:13182–13187
Extrand CW (2002) Model for contact angles and hysteresis on rough and ultraphobic surfaces. Langmuir 18:7991–7999
Extrand CW (2003) Contact angles and hysteresis on surfaces with chemically heterogeneous islands. Langmuir 19:3793–3796
Fujishima A, Honda K (1972) Electrochemical photolysis of water at a semiconductor electrode. Nature 238:37–38
Fujishima A, Zhang XT, Tryk DA (2008) TiO_2 photocatalysis and related surface phenomena. Surf Sci Rep 63:515–582
Ganesh VA, Raut HK, Nair AS, Ramakrishna S (2011) A review on self-cleaning coatings. J Mater Chem 21:16304–16322
Gao L, McCarthy TJ (2007) How Wenzel and Cassie were wrong. Langmuir 23:3762–3765
Gao L, McCarthy TJ (2009) An attempt to correct the faulty intuition perpetuated by the Wenzel and Cassie "laws". Langmuir 23:3762–3765
Gao N, Yan Y, Chen X, Mee DJ (2011) Nanoparticle-induced morphology and hydrophilicity of structured surfaces. Langmuir 28:12256–12265
Gao X (2012) Antifogging properties in mosquito eyes. In: Bhushan B (ed) Encyclopedia of nanotechnology. Springer, Berlin. Springer Reference (www.springerreference.com). Accessed 5 Dec 2013
Gao X, Jiang L (2004) Biophysics: water-repellent legs of water striders. Nature 432:436
Gaume L, Gorb S, Rowe N (2002) Function of epidermal surfaces in the trapping efficiency of Nepenthes alata pitchers. New Phytol 156:479–489
Genzer J, Marmur A (2008) Biological and synthetic self-cleaning surfaces. MRS Bull 33:742–746
Giacomello A, Chinappi M, Meloni S, Casciola CM (2012) Metastable wetting on superhydrophobic surfaces: continuum and atomistic views of the cassie-baxter-wenzel transition. Phys Rev Lett 109:226102
Groten J, Rühe J (2013) Surfaces with combined microscale and nanoscale structures: a route to mechanically stable superhydrophobic surfaces? Langmuir 29:3765–3772
Guo Y, Wang Q (2010) Facile approach in fabricating superhydrophobic coatings from silica-based nanocomposite. Appl Surf Sci 257:33–36
Guo Z, Liu W (2007) Biomimic from the superhydrophobic plant leaves in nature: binary structure and unitary structure. Plant Sci 172:1103–1112

Guo Z, Liu W, Su B-L (2011) Superhydrophobic surfaces: from natural to biomimetic to functional. J Colloid Interface Sci 353:335–355
Hansen WR, Autumn K (2005) Evidence for self-cleaning in gecko setae. Proc Natl Acad Sci USA 102:385–389
Hu S, Lopez S, Niewiarowski PH, Xia Z (2012) Dynamic self-cleaning in gecko setae via digital hyperextension. J R Soc Interface 9:2781–2790
Hunter L. (2010). The C-F bond as a conformational tool in organic and biological chemistry. Beilstein J Org Chem 6:1–14
Johnson RE Jr, Dettre RH (1964) Contact angle hysteresis. In: Fowkes FM (ed) Contact angle, wettability, and adhesion. Advances in chemistry series, vol 43. American Chemical Society, Washington, pp 112–144
Joshi M, Bhattacharyya A, Agarwal N, Parmar S (2012) Nanostructured coatings for super hydrophobic textiles. Bull Mater Sci 35:933–938
Jribi R, Barthel E, Bluhm H, Grunze M, Koelsch P, Verreault D, Sondergard E (2009) Ultraviolet irradiation suppresses adhesion on TiO_2. J Phys Chem C 113:8273–8277
Jung YC, Bhushan B (2009) Mechanically durable carbon nanotube-composite hierarchical structures with superhydrophobicity, selfcleaning, and low-drag. ACS Nano 3:4155–4163
Khranovskyy V, Ekblad T, Yakimova R, Hultman L (2012) Surface morphology effects on the light-controlled wettability of ZnO nanostructures. Appl Surf Sci 258:8146–8152
Kim JS, Jeong HW, Lee W, Park BG, Kim BM, Lee KB (2012a) A simple and fast fabrication of a both self-cleanable and deep-UV antireflective quartz nanostructured surface. Nanoscale Res Lett 7:430
Kim P, Wong TS, Alvarenga J, Kreder MJ, Adorno-Martinez WE, Aizenberg J (2012b) Liquid-infused nanostructured surfaces with extreme anti-ice and anti-frost performance. ACS Nano 6:6569–6577
Koch K, Bhushan B, Barthlott W (2008) Diversity of structure, morphology, and wetting of plant surfaces. Soft Matter 4:1943–1963
Koch K, Bhushan B, Jung YC, Barthlott W (2009) Fabrication of artificial Lotus leaves and significance of hierarchical structure for superhydrophobicity and low adhesion. Soft Matter 5:1386–1393
Krieger K (2004) Do pool sharks really swim faster? Science 305:636–637
Kwon Y, Patankar N, Choi J, Lee J (2009) Design of surface hierarchy for extreme hydrophobicity. Langmuir 25:6129–6136
Lafuma A, Quéré D (2011) Slippery pre-suffused surfaces. Europhys Lett 96:56001
Lai Y, Tang Y, Gong J, Gong D, Chi L, Lin C, Chen Z (2012) Transparent superhydrophobic/superhydrophilic TiO_2-based coatings for self-cleaning and anti-fogging. J Mater Chem 22:7420–7426
Lee EJ, Kim JJ, Cho SO (2010) Fabrication of porous hierarchical polymer/ceramic composites by electron irradiation of organic/inorganic polymers: route to a highly durable, large-area superhydrophobic coating. Langmuir 26:3024–3030
Levinson R, Berdahl P, Berheb AA, Akbari H (2005) Effects of soiling and cleaning on the reflectance and solar heat gain of a light-colored roofing membrane. Atmos Environ 39:7807–7824
Li W, Amirfazli A (2008) Hierarchical structures for natural superhydrophobic surfaces. Soft Matter 4:462–466
LiquiGlide^{TM} (2012) Slippery surface solutions. http://www.liquiglide.com. Accessed 4 Dec 2013
Liu J, Jiang K (2012) Bio-inspired self-cleaning surfaces. Annu Rev Mater Res 42:231–263
Liu J, Xia R, Zhou X (2012) A new look on wetting models: continuum analysis. Sci China-Phys Mech Astron 55:2158–2166
Marmur A (2003) Wetting on hydrophobic rough surfaces: to be heterogeneous or not to be? Langmuir 19:8343–8348
McHale G (2007) Cassie and Wenzel: were they really so wrong? Langmuir 23:8200–8205

Men XH, Zhang ZZ, Yang J, Wang K, Jiang W (2010) Superhydrophobic/superhydrophilic surfaces from a carbon nanotube based composite coating. Appl Phys A 98:275–280

Meng L-Y, Park S-J (2010) Effect of fluorination of carbon nanotubes on superhydrophobic properties of fluoro-based films. J Colloid Interface Sci 342:559–563

Morra M, Occhiello E, Garbassi F (1989) Contact-angle hysteresis in oxygen plasma treated poly(tetrafluoroethylene). Langmuir 5:872–876

Myint MTZ, Hornyak GL, Dutta J (2014) One pot synthesis of opposing 'rose petal' and 'lotus leaf' superhydrophobic materials with zinc oxide nanorods. J Colloid Interface Sci 415:32–38

Nakata K, Fujishima A (2012) TiO_2 photocatalysis: design and applications. J Photochem Photobiol C 13:169–189

Nakata K, Sakai M, Ochiai T, Murakami T, Takagi K, Fujishima A (2011) Antireflection and self-cleaning properties of a moth-eye-like surface coated with TiO_2 particles. Langmuir 27:3275–3278

Nishimoto S, Bhushan B (2013) Bioinspired self-cleaning surfaces with superhydrophobicity, superoleophobicity, and superhydrophilicity. RSC Adv 3:671–690

Nosonovsky M (2007) On the range of applicability of the Wenzel and Cassie equations. Langmuir 23:9919–9920

Nosonovsky M, Bhushan B (2012) Lotus versus rose: biomimetic surface effects. In: Nosonovsky M, Bhushan B (eds) Green tribology. Springer, Berlin, pp 25–40

Orchard MJ, Kohonen M, Humphries S (2012) The influence of surface energy on the self-cleaning of insect adhesive devices. J Exp Biol 215:279–286

Panchagnula MV, Vedantam S (2007) Comment on how Wenzel and Cassie were wrong by Gao and McCarthy. Langmuir 23:13242

Parker AR, Lawrence CR (2001) Water capture by a desert beetle. Nature 414:33–34

Peters AM, Pirat C, Sbragaglia M, Borkent BM, Wessling M, Lohse D, Lammertink RGH (2009) Cassie-Baxter to Wenzel state wetting transition: scaling of the front velocity. Eur Phys J 29:391–397

Phys.org (2008) Photo-catalytic, self cleaning coating for building exteriors. http://phys.org/news146916753.html. Accessed 18 Nov 2013

Qu M, He J, Zhang J (2010) Superhydrophobicity, learn from the lotus leaf. In: Mukherjee A (ed) Biomimetics learning from nature. INTECH, pp 325–342. http://www.intechopen.com/articles/show/title/superhydrophobicity-learn-from-the-lotus-leaf. Accessed 20 Nov 2013

Quéré D (2002) Fakir droplets. Nat Mater 1:14–15

Reif W-E (1985) Functions of scales and photophores in mesopelagic luminescent sharks. Acta Zool 66:111–118

Ruibal R, Ernst V (1965) The structure of the digital setae of lizards. J Morphol 117:271–293

Rykaczewski K, Anand S, Subramanyam SB, Varanasi KK (2013) Mechanism of frost formation on lubricant-impregnated surfaces. Langmuir 29:5230–5238

Schulte AJ, Droste DM, Koch K, Barthlott W (2011) Hierarchically structured superhydrophobic flowers with low hysteresis of the wild pansy (Viola tricolor)—new design principles for biomimetic materials. Beilstein J Nanotechnol 2:228–236

Schumacher JF, Aldred N, Callow ME, Finlay JA, Callow JA, Clare AS, Brennan AB (2007) Species-specific engineered antifouling topographies: correlations between the settlement of algal zoospores and barnacle cyprids. Biofouling 23:307–317

Sleiman M, Ban-Weiss G, Gilbert HE, Francois D, Berdahl P, Kirchstetter TW, Destaillats H, Levinson R (2011) Soiling of building envelope surfaces and its effect on solar reflectance—Part I: Analysis of roofing product databases. Sol Energy Mater Solar Cells 95:3385–3399

Smith JD, Dhiman R, Anand S, Garduno ER, Cohen RE, McKinley GH, Varanasi KK (2013) Droplet mobility on lubricant-imprenated surfaces. Soft Matter 9:1772–1780

Solga A, Cerman Z, Striffler BF, Spaeth M, Barthlott W (2007) The dream of staying clean: lotus and biomimetic surfaces. Bioinsp Biomim 2:S126–S134

Su C, Chen Q (2008) Research progresses of the surface similar to lotus leaves. Chem Bull 71:24–31

Sun Z, Liao T, Liu K, Jiang L, Kim JH, Dou SX (2013) Robust superhydrophobicity of hierarchical ZnO hollow microspheres fabricated by two-step self-assembly. Nano Res 6:726–735

Tian Y, Wan J, Pesika N, Zhou M (2013) Bridging nanocontacts to macroscale gecko adhesion by sliding soft lamellar skin supported setal array. Sci Rep 3:1382

Verplanck N, Coffinier Y, Thomy V, Boukherroub R (2007) Wettability switching techniques on superhydrophobic surfaces. Nanoscale Res Lett 2:577–596

Wang CY, Groenzin H, Shultz MJ (2003) Molecular species on nanoparticulate anatase TiO_2 film detected by sum frequency generation: trace hydrocarbons and hydroxyl groups. Langmuir 19:7330–7334

Wang R, Hashimoto K, Fujishima A, Chikuni M, Kojima E, Kitamura A, Shimohigoshi M, Watanabe T (1997) Light-induced amphiphilic surfaces. Nature 388:431–432

Wang R, Sakai N, Fujishima A, Watanabe T, Hashimoto K (1999) Studies of surface wettability conversion on TiO_2 single-crystal surfaces. J Phys Chem B 103:2188–2194

Wenzel RN (1936) Resistance of solid surfaces to wetting by water. Ind Eng Chem 28:988–994

Wenzel RN (1949) Surface Roughness and Contact Angle. J Phys Chem 53:1466–1467

Whyman G, Bormashenko E, Stein T (2008) The rigorous derivation of Young, Cassie-Baxter and Wenzel equations and the analysis of the contact angle hysteresis phenomenon. Chem Phys Lett 450:355–359

Wolansky G, Marmur A (1999) Apparent contact angles on rough surfaces: the Wenzel equation revisited. Colloids Surf A 156:381–388

Wong TS, Kang SH, Tang SKY, Smythe EJ, Hatton BD, Grinthal A, Aizenberg J (2011) Bioinspired selfrepairing slippery surfaces with pressure-stable omniphobicity. Nature 477:443–447

Wong TS, Sun T, Feng L, Aizenberg J (2013) Interfacial materials with special wettability. MRS Bull 38:366–371

Xiu Y, Wong CP (2010) Biomimetic lotus effect surfaces for nanopackaging. In: Wong CP, Moon KS, Li Y (eds) Nano-bio- electronic, photonic and MEMS packaging. Springer, New York, pp 47–85

Yang C, Tartaglino U, Persson BNJ (2008) Nanodroplets on rough hydrophilic and hydrophobic surfaces. Eur Phys J E 25:139–152

Yang H, Dou X, Fang Y, Jiang P (2013) Self-assembled biomimetic superhydrophobic hierarchical arrays. J Colloid Interface Sci 405:51–57

Zhang H, Fan D, Yu T, Wang C (2013a) Characterization of anti-reflective and self-cleaning SiO_2–TiO_2 composite film. J Sol-Gel Sci Technol 66:274–279

Zhang J, Wang A, Seeger S (2013b) Nepenthes pitcher inspired anti-wetting silicone nanofilaments coatings: preparation, unique anti-wetting and self-cleaning behaviors. Adv Funct Mater (in press). doi:10.1002/adfm.201301481

Zhang L, Dillert R, Bahnemann D, Vormoor M (2012) Photo-induced hydrophilicity and self-cleaning: models and reality. Environ Sci 5:7491–7507

Zhu J, Hsu C-M, Yu Z, Fan S, Cui Y (2010) Nanodome solar cells with efficient light management and self-cleaning. Nano Lett 10:1979–1984

Chapter 10
Bio-inspired Bridge Design

Nan Hu and Peng Feng

Abstract This chapter reviews the development of the bio-inspired concept on bridge design in the past two decades from two major forms: stationary forms and movable forms. The objective is to show how the inspiration from the biological world has influenced recent bridge designs and discusses how the bio-inspired idea could transform into a new language for the future bridge design industry. Four major challenges of the marriage between biology and engineering were discussed and latest endeavor on each aspect are presented. Thus, a close multidisciplinary collaboration may help engineers build more sustainable and smart structural systems for bridges in the twenty-first century.

10.1 The Inspiration from Nature

A paradigm on biomimetics has emerged in the past two decades featuring the translation of biological inspiration into a powerful problem-solving tool (Bar-Cohen 2011, 2012; Bhushan 2009; Bonser 2006; Petra 2008; Vincent 2006; Vincent and Mann 2002; Yoseph 2006). Bio-inspired design is perhaps the oldest methodology throughout the thousands of years on man-made construction history. In ancient times, the designs of early architecture were more or less inspired from the natural forms even without being aware of such influence. Later in history, pioneer architects led by Antoni Gaudi and Frank Wright started to actively consider nature as design prototypes and incorporated that philosophy into much of their uniquely creative designs. The major drawback of biological inspiration in

N. Hu (✉)
Department of Civil and Environmental Engineering, Michigan State University,
East Lansing, USA
e-mail: aaronx26@gmail.com

P. Feng
Department of Civil Engineering, Tsinghua University, Beijing, China
e-mail: fengpeng@tsinghua.edu.cn

F. Pacheco Torgal et al. (eds.), *Biotechnologies and Biomimetics for Civil Engineering*,
DOI 10.1007/978-3-319-09287-4_10

architectural design is that only the geometrical aspect of biological creatures that commonly aimed to achieve visual expression were learned without much attention to efficiency and economy. Review articles about the development of bio-inspired architecture can be found in papers (Aldersey-Williams 2004; John et al. 2005; Knippers and Speck 2012; Petra 2008).

Architectural design and structural design are professions that have a historical misunderstanding, even though the core of both professions is shaping man-made structures. That was not the case during most of history because up until the end of the eighteenth century, the two professions were as one. After the Industrial Revolution, engineering developed as a discipline in its own right with a scientific basis and moved increasingly further away from architecture. The rationale behind the history of structural design did not change much no matter the scale of a structure, which featured a search on a cost-effective and performance-efficient design without losing its elegance. A structural design can be regarded as a work of art by the following three tenets: economy, efficiency, and elegance Billington (1983) Bridge design, in particular, is a unique regime where engineers evolved a dominant role over architects and are capable of designing a work of art in their discipline.

Biological inspiration on bridge design is not a recent topic. Ancestors started to build ancient bridges after recognizing how the natural world used structural forms to span physical obstacles. In fact, all basic types of modern bridge forms (beam, arch, and suspension) can find their ancient prototypes in nature (Tang 2007). The only difference is that ancient people built the bridges with natural materials (wood, vines, stone, and ropes) while modern technology uses steel, concrete, etc. Even new structural forms show connections with natural forms. For example, the Hacking Ferry Bridge's three-way deck (45 m on each side) in Fig. 10.1b not only provides sufficient structural stability but also necessary clearance for the channel. The Webb Bridge (Fig. 10.1d) at Melbourne has a curve layout instead of a straight one, but provides an interesting experience to the bicyclists and pedestrians. Architects or engineers may not directly learn from natural forms, but it does show that biological creatures can be regarded as a significant resource with a variety of interesting designs.

Limited papers are found on the discussion of bio-inspired bridges. Hu et al. (2013) summarizes five main aspects in nature (Table 10.1) that may inspire the development of future bridge systems: geometry, structure, kinetic mechanism, energy efficiency, and intelligence

It is shown that biological inspiration is a very promising route to inspire bridge design not only in traditional stationary forms but also in emerging movable form. Thus, the objective of this chapter is to present a review on recent bridge projects to show how the bio-inspired philosophy has impacted conceptual design in the past and will bring a new language in the future that potentially could encourage a marriage between biological knowledge and bridge design.

Fig. 10.1 The biological shape and structural shape in bridge **a** shape of starfish (*photo* Wikimedia Commons); **b** Hacking Ferry Bridge (*photo* Wilkinson Eyre Architects); **c** earthworm shape (*photo* Wikimedia Commons); **d** Webb bridge (*photo* bridgeworld.net)

Table 10.1 Toward the bio-inspired bridge

Prototype in nature	Bio-inspired design goals
Geometry	Eye-catching and signature bridge
Structures	Material-adapted supporting and deck system
Mechanism	Deployable and moving bridge
Energy	Sustainable and multifunctional bridge
Intelligence	Self-control and smart bridge

10.2 Bio-inspired Form-Finding

Biological inspiration on engineering design has developed in various aspects, but still faces new challenges. For example, the tree is a classic prototype in bio-inspired research (Mattheck 1998). The trunk and supporting branches of a tree is a brilliant example of structural optimization. From a morphological perspective, many trees are subjected to wind load and evolutionary transform into an

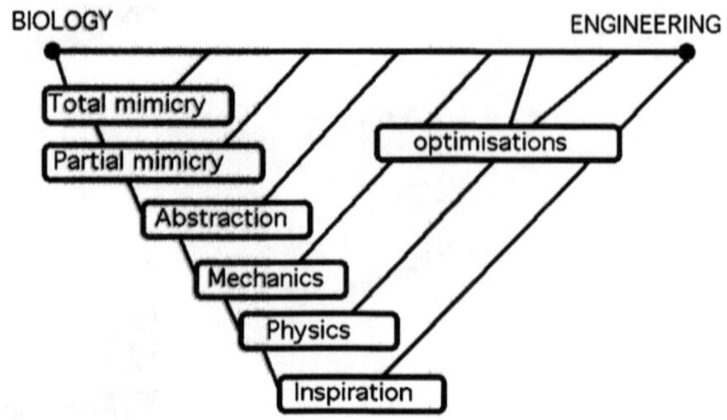

Fig. 10.2 Biomimetics "map" (Vincent 2001)

adaptable structural form. From a sustainability perspective, engineers were also trying to rediscover intelligent features of trees, such as the ability to self-adapt, be self-controlled, and be energy efficient, which leads to some interesting sustainable designs. Vincent (2001) developed a biomimetic "map" to clarify the transfer from biology to engineering, as shown in Fig. 10.2.

The more one can move further down from the natural origin (top left), the more powerful the concept will be. It illustrates that inspiration from biology at the top level can be more adaptable within the engineering discipline, even though the final product may vary far away from the prototype founded in nature. There are many versions of this diagram which could be drawn for any engineering design problems.

For structural design, some engineers may claim with buildings and bridges as large-scale man-made structures, it is difficult to reach the efficiency as biological creatures. However, many engineers have been sparked by the bio-inspired philosophy and are dedicated to the development of bridges in future generations. After simply borrowing geometrical feature from nature for centuries, recent efforts in bridge design and construction by both architects and engineers have shown inspiration from the mechanistic and the material level. Table 10.2 shows many recently built bridge projects with a varied level of inspiration from different aspects in the nature. Most designs are still within the low level of inspiration (total mimicry, partial mimicry, and abstraction), sometimes resulting in many awkward shapes rather than efficient structural forms. Fewer designs featured a high-level of inspiration, in which design problems are addressed by biomimetics. A logical explanation is that those engineers are aware of bio-inspired idea, but they are not capable of finding tangible resources that could revolutionize their designs.

10.2.1 Stationary Forms

Traditionally, bridge design is still a classical static problem searching for efficient forms across existing roads and rivers. The best bridge design in history has been regarded as structural art, because it shows very good balance on economy, efficiency, and elegance (Billington 1983).

Table 10.2 Recent bridge projects with varied level of bio-inspiration (Hu et al. 2013)

Bridge name	Built	Location	Main bio-inspired aspect	Inspiration level
Campo Volantin bridge	1997	Bilbao, Spain	Structure of beam	Mechanics
Foldable bridge	1997	Kiel, Germany	Kinematics of beam	Mechanics
Butterfly bridge	1998	Bedford, UK	Structure of arch	Partial mimicry
Nesenbach valley bridge	1999	Vaihingen, Germany	Structure of beam and pier	Mechanics
Millennium bridge	2001	Gateshead, UK	Kinematics of main arch	Abstraction
Ribble Way bridge	2002	Lancashire, UK	Shape of layout	Abstraction
Webb bridge	2003	Melbourne, Australia	Shape of layout	Abstraction
Rolling bridge	2004	London, UK	Kinematics of beam	Mechanics
Dragon bridge	2008	Recklinghausen, Germany	Shape of superstructure	Total mimicry
Chords bridge	2008	Jerusalem, Israel	Cable-stayed structure	Abstraction
Henderson bridge	2008	Singapore	Shape of beam	Mechanics
Double Helix bridge	2009	Singapore	Structure of beam	Partial mimicry
Chunhua footbridge	2011	Shenzhen, China	Shape of layout	Total mimicry
Peace bridge	2012	Calgary, Canada	Structure of beam	Abstraction
Dragon Eco bridge	2012	Chongqing, China	Structure of arch	Partial mimicry

Not coincidentally, structural forms in nature also integrate these three tenets, ranging from a plant stem to animal skeletons. The awareness of learning from natural shapes pushed architects and engineers to think out of the box in many recent bridge designs, practically in the design of small-scale structure such as pedestrian bridge. Some architects have realized the benefit of using bio-inspired philosophy to improve visual impact and esthetic value of a pedestrian bridges, but this leads to the tendency of defining the structural forms by architects which leads to extravagance in pedestrian bridge design (Gauvreau 2002). Two examples in Singapore are shown in Fig. 10.3a, b. The curvilinear steel "ribs" along the side of the Henderson Road Bridge may be inspired from natural waves, while the double helix bridge was inspired from the shape of the DNA. Both have become a new landmark for the community. However, a partial mimicry of the geometric shape from nature led to awkward designs with difficult construction and high cost. The efficiency and economy of these designs are questionable. The Chunhua footbridge (Fig. 10.3c) at Shenzhen, China, copied the shape of a flower, but it is not a cost-effective design compared to a regular design. Another recent controversial design

Fig. 10.3 Bio-inspired shape in recent bridges: **a** Henderson bridge (*photo* RSP Architects); **b** double helix bridge (*photo* Cox Architects; **c** Chunhua footbridge (*photo* China Foto Press); **d** Peace bridge (*photo* Nelson Hein)

in Calgary is the Peace Bridge by Santiago Calatrava (Fig. 10.3d), which was inspired by fishbone structures.

The public criticized that this design was too modern in the old downtown. Given that most engineers are not capable of finding inspiration, bionic research should try to establish a database with potential engineering solutions, such as an illustrated guide to flowering plant morphology for learning plant forms.

The efficient natural forms feature optimal material distribution along with elegant shapes. Historical studies on many bridge structures showed that there are two necessary conditions that contributed to the birth of a masterpiece design, i.e., the understanding of structural principles and the efficient use of construction materials. In the past 2 decades, some leading engineers have become increasingly aware of learning from nature so that a variety of bridges were built with inspirations on the mechanical principles of nature. For example, the twin inclined steel arches as the supporting system of the Butterfly Bridge (Fig. 10.4a) were inspired from butterfly wings. The use of arch hangers is similar to the veins in a butterfly wing. In contrast, the Dragon Bridge (Fig. 10.4b) is totally mimicry of a dragon-like skeleton form in which the actual structural system is the girder and column underneath the deck. Santiago Calatrava is a well-known architect with a strong bio-inspired motivation in his designs. Like his predecessor Spanish fellow

Fig. 10.4 Bridge design cases of bio-inspired structural system: **a** Butterfly bridge (*photo* Wilkinson Eyre architects); **b** Dragon bridge (*photo* Structure Website, Id 108893); **c** Campo Volantin footbridge (*photo* Structure Website, Id 141499); **d** Jerusalem Chords bridge (*photo* Wikimedia Commons)

architect Antonio Gaudi, many bridge designs by Calatrava were inspired from the structural forms of plants and animals. The main girder of the Campo Volantin Bridge was inspired from the spine structure as shown in Fig. 10.4c, while the tower of the Chords Bridge (Fig. 10.4d) was created with inspiration from the leg. Knippers and Speck (2012) used design principles in natural structure characterized by heterogeneity (geometric differentiation of their elements), anisotropy (fiber-reinforced composite materials), hierarchy (multilevel structure with independent functional properties), and multifunctionality.

It is worth mentioning that both structure and material in nature often appear in a composite way. Thus, the principle of more efficiency and less material in nature also indicates the trend of composite materials and structures for future bridge design.

The innovation of structural forms was reliant on the development of new materials. History clearly demonstrates that new materials are only successful when the substitution phase has been surmounted and when a transition to new material-tailored structural concepts and construction processes occurs (Dooley 2004). Great builders were then liberated from tradition and tried new materials in the old forms: cast iron in a wood arch, steel in a wood truss, and reinforced

Fig. 10.5 Bio-inspired design proposals: **a** Douglas river bridge (*photo* Exploration Architecture); **b** Spiral bridge

concrete in a stone arch (Billington 2003). Therefore, future structural form must be material-adapted to achieve high efficiency.

Figure 10.5 shows three bio-inspired proposals of new structural systems. A bridge crossing the River Douglas in Lancashire (Fig. 10.5a) was turning biomimetic research into pioneering lightweight structures with small steel elements and a pressurized air beam.

This design was inspired from the internal pressures seen in nature which create self-supporting structures. Another interesting design (Fig. 10.5b) has a series of intersecting spirals as the structural system. The designers incorporated a seedpod spiral shape from the tree Tipuanatipu and a rectilinear lattice seen in the Euplectella as bridge cover (Dollens 2005). The spirals also reflect the colors of the trees, light so that they will merge into the environment. It is noted that the optimized structural form in future bridge designs should utilize appropriate material-adapted form and certain manufacturing technology.

Energy costs on man-made structures are always a big concern. Future bridges may also play an important role relating to sustainability and the use of green energy. "Green" bridges are designed to reduce the overall impact of the built environment on the natural environment. The idea of "running on natural energy"

Fig. 10.6 The solar panel on Blackfriars bridge (Hu et al. 2013)

should not only be embodied in the period of construction, but also be extended to the whole life cycle. Photovoltaic solar panels for example could be installed on the bridge so that it could provide sustainable electricity for any use. The renovation project of the Blackfriars Bridge over the Thames River led this old bridge to a very notable landmark as shown in Fig. 10.6.

The power produced by the photovoltaic panels on the roof can support Blackfriars station's overall power requirement. The critical issue for developing future bridges is how to make best use of natural energy. Future proposals have taken the sustainable issue into account. Wilkinson Eyre Architects proposed a botanic bridge in 2002 at the gateway to the University of Newcastle, which provides an ecological space above the bridge, as shown in Fig. 10.7a. Bridge structures are built to minimize material use and recycled natural resources. Figure 10.7b shows an interesting proposal about the installation of wind turbines between the bridge piers for some heavy wind locations. It is noted that future bridges may also have multifunctional features on energy issues, not only for energy use but also for energy harvesting, energy absorption, and energy storage. The design of such smart structures will require a higher level inspiration that designs the structure in a more comprehensive way.

10.2.2 Moveable Forms

Compared to most stationary forms, another interesting bridge design problem, moveable form, can also be inspired from the kinetic mechanisms in nature. Movable bridges represent an economical way for traffic to cross an active waterway granting passage with little vertical clearance. Wallner and Pircher (2007) summarized three prevailing structural systems of moveable bridges such as bascule, swing, and lift bridges. Those examples of movable bridge in all types, in fact, represent not only a unique mechanism, but also a self-adapting way to the

Fig. 10.7 Environmental-friendly bridge proposals: **a** Newcastle Botanic bridge (*photo* Wilkinson Eyre Architects); **b** "Windy" design (*photo* Francesco Colarossi Architecture)

natural world. Figure 10.8 exhibits Paddington Rolling Bridge with a span of 12.9 m, which has served both pedestrians and the channel in an interesting way.

Although designers may not have completely agreed with the statement that structural form is like a rolling insect, its shape and structural principle coincides with the biological creature. If engineers can try to invite kinematic principles from nature during the conceptual design phase, mechanisms for guiding movements in movable bridges may be better defined. One example (Fig. 10.9) is the Gateshead Millennium Bridge in Newcastle, UK. The designers adopted an unconventional open mode to adapt to channel traffic in the Tyne River. Another example in Fig. 10.10 is the Keil-Horn Folding Bridge with a remarkable trailing design case where three segments are held by cables when it opens for channel traffic.

Some interesting topics on the development of movable bridges includes the use of deployable form (Alegria Mira et al. 2014; Friedman and Ibrahimbegovic 2013; Russell and Thrall 2013; Thrall et al. 2012) and tensegrity form (Korkmaz et al. 2012; Rhode-Barbarigos et al. 2010).

Fig. 10.8 Paddington Rolling bridge (Hu et al. 2013)

Fig. 10.9 Gateshead Millennium bridge (Hu et al. 2013)

Fig. 10.10 Folding bridge at Kiel (*photo* Schlaich Bergermann and Partner)

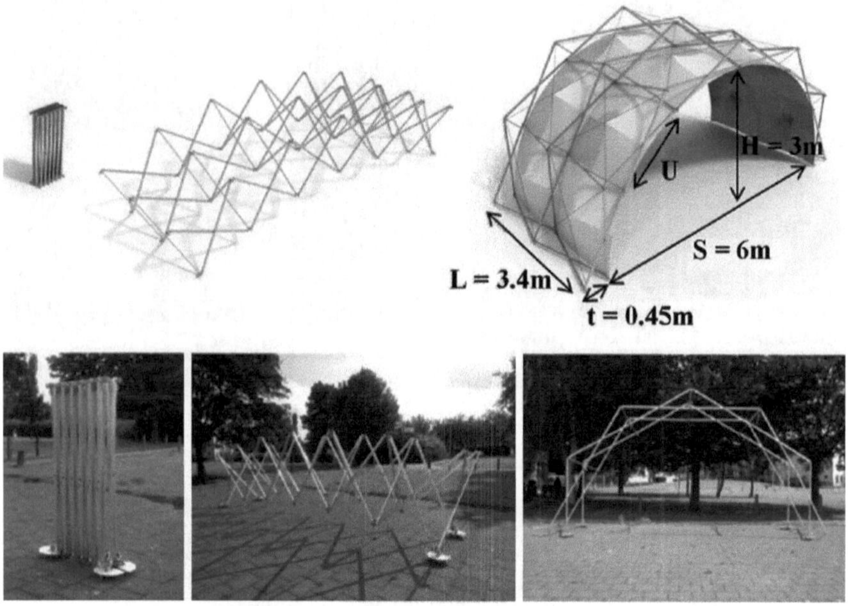

Fig. 10.11 The concept of deployable scissor arch prototype (Alegria Mira et al. 2014)

Today, deployable structure played a variety of roles in both military and relief operations from emergency to reconstruction. Prototype lightweight forms are studied such as deployable bridge (Ario et al. 2013) and deployable shelter (Alegria Mira et al. 2014), see Fig. 10.11.

Tensegrity structure is spatial structures composed of tensile and compression components in a self-equilibrated state of prestressing. Such concept recently has already been used in bridge design, see Fig. 10.12.

According to Vincent's biomimetic map, these cases discussed above represent the mechanical level of inspiration. In particular, those footbridges provide a more adaptable way than regular footbridge to span a river through combining technologies of machinery and automation.

The development of intelligent and smart bridges is another encouraging goal for the design of active bridges in the near future. The transfer from a "passive" to an "active" bridge through learning from the biological world is a great breakthrough for future smart and intelligent structures. Primary driving factors of the information technology age (such as nanotechnology, microelectronics, and biotechnology) will support the development of intelligent bridge systems and materials (Chong 2004).

The Innenhafen Bridge in Fig. 10.13 shows a remarkable design case where the traditional suspension form is embedded with an intelligent system. It is an adaptive form subjected to different environmental and traffic service conditions. Schlaich (2004) defined the level of intelligence in a structural system by

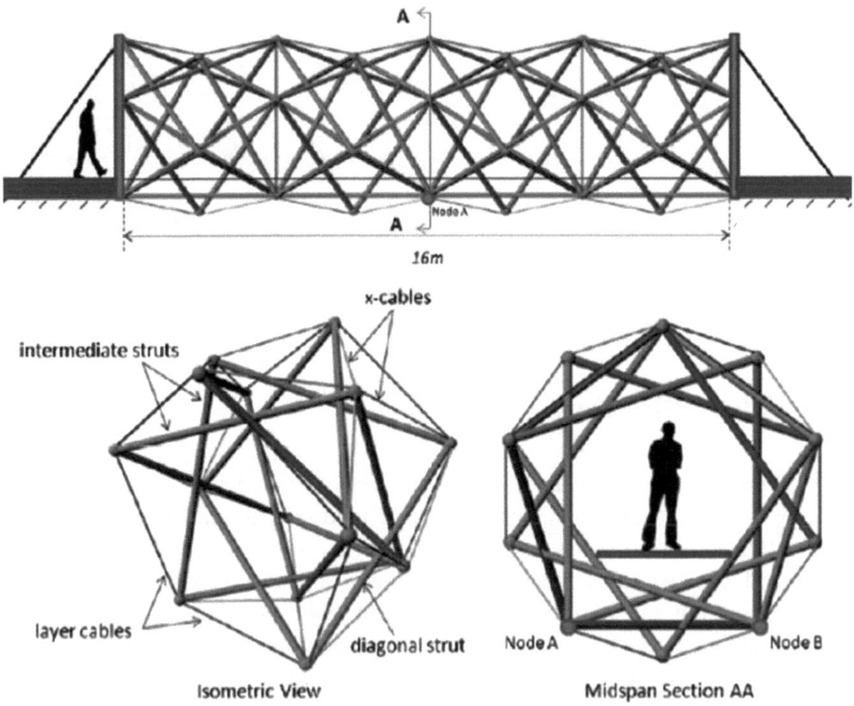

Fig. 10.12 The concept of tensegrity bridge (Korkmaz et al. 2012)

Fig. 10.13 Innenhafen Bridge (Hu 2010)

comparing the degree of those electronic devices. A proposal is shown using computer-controlled jacks to actively control the function of stay-cables (Fig. 10.14). This initially solved the problem of uneven deflection when trains ran through the railway bridge. Future intelligent bridges could soon have the capacity of self-control, self-adjust, and even self-repair. Recent endeavors on smart infrastructure can be found (Hoult et al. 2009; Michaël and Dominique 2013; Spencer and Cho 2011; Stajano et al. 2010).

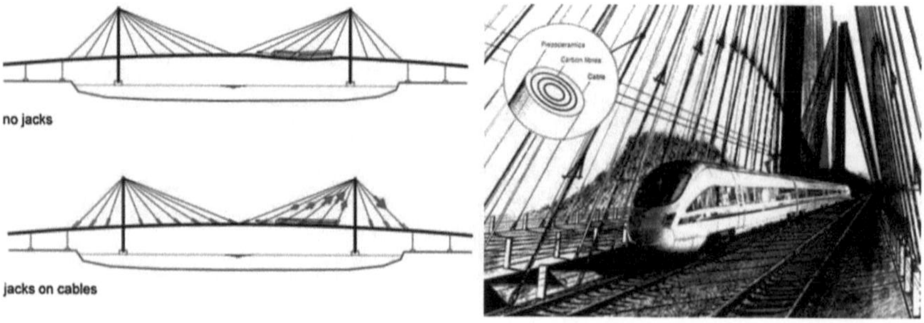

Fig. 10.14 Self-adjusted cable-stayed bridge (Hu 2010)

10.3 Future Directions

A brief project review in the previous section illustrated that bio-inspired philosophy is a promising route for the design industry. Instead of mimicry, both architects and engineers have adopted the mechanistic and material level inspiration and utilized those ideas in the design and construction of man-made structures during the past decade. The latest published book on biologically inspired design by Goel et al. (2014) pointed out four major challenges of the marriage between biology and engineering: (1) solve real problems and document successful applications; (2) develop bio-inspired design theory; (3) develop computational methods and tools; (4) educate new generation students.

For a cross-disciplinary environment in the near future, the bio-inspired design requires the cooperation between engineers and biologists. Future engineering students should not be trapped in professional boundaries and must expand their interdisciplinary knowledge. Figure 10.15 describes a typical process for bio-inspired design, which refers to a regular scientific research procedure that includes experimental investigation, numerical analysis and testing, etc. The most important step in this process is the selection of the appropriate biological prototype. Bionic researchers need to provide resources of prototypes that allow engineers to reference for design inspiration. For example, Biomimicry 3.8, a nonprofit corporation dedicated to biomimicry education launched the world's first digital library of nature's solutions "AskNature." To develop a feasible study on a prototype, it is necessary to work with biologists so that more interdisciplinary knowledge can be transferred. For the past fifty years, many bio-based technologies have been illustrated, such as tough composites based on fiber orientations in wood, tough ceramics based on mother-of-pearl, deployable structures based on flowers and leaves, underwater glues based on mussel adhesive, drag reduction based on dermal riblet on shark skin, flight mechanisms based on insect flight, etc. Recent efforts on bio-inspired design on multiscale smart structures can be found in Barbarino et al. (2011), Espinosa et al. (2009), Giurgiutiu (2007), Hurlebaus and Gaul (2006), Kovač (2013), Liu and Jiang (2011), Mudupu et al. (2008),

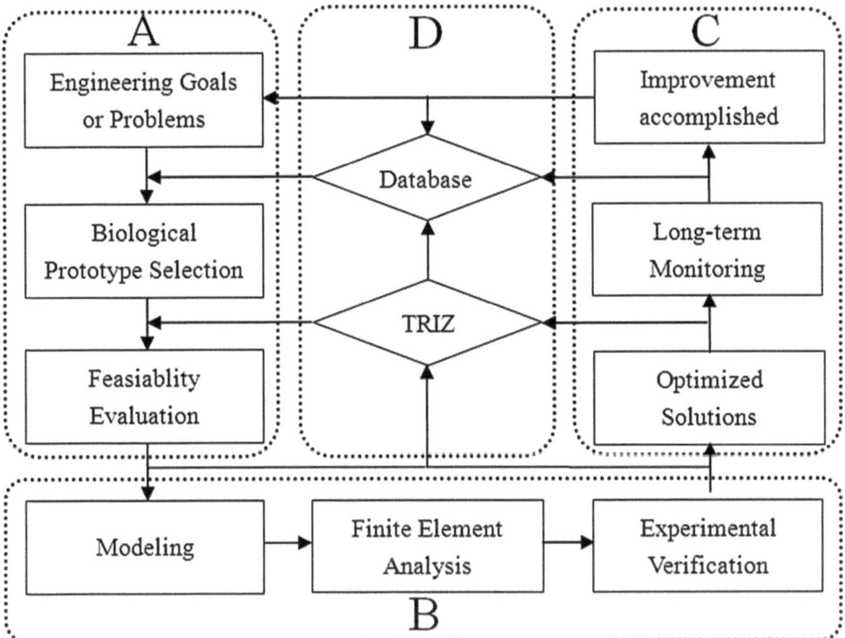

Fig. 10.15 The procedure of the bio-inspired strategy (Hu et al. 2013)

Singh et al. (2012), Stephen and Christopher (2012), Thomas et al. (2012), Yeesock et al. (2010).

Beyond the success of learning a particular biological structure or mechanism, the search for biological analogies in the conceptual design process also requires design theory to aid in systematically generating new and novel ideas. The Theory of Inventive Problems Solving (TRIZ) has become a significant method that effectively analogized the interaction between engineering and biology. Any technical contradictions can be quantified by transforming the application into a measurable and mathematical algorithm. The latest works for helping engineers to connect and identify biological phenomena can be found in Mak and Shu (2008), Nagel and Stone (2011), Nagel et al. (2010), Sartori et al. (2010), Shu et al. (2011), Srinivasan et al. (2011), Tomiyama et al. (2009), Trotta (2011), Wilson et al. (2010), Xing and Chen (2011).

Traditionally, bio-inspired design products are an outcome of individual effort. Today, advanced computational methods and tools can also assist engineers in modeling bio-inspired forms. Meanwhile, the advances in fabrication tools can be a catalyst to search optimal and elegant structural form. For example, topology optimization has evolved into a computer-aided technique for modern engineering design and practice. The use of this technique in recent studies (Adriaenssens et al. 2012; Asadpoure et al. 2011; Guest 2009) has helped engineers seek the best distribution of a given amount of material or the optimal geometry of a given domain of structure via an iterative numerical process under specific constraints. An up-to-date summary on recent topology optimization techniques and applications in a variety of disciplines can be found in Deaton and Grandhi (2013).

Nonuniform rational B-Splines (NURBS)-based isogeometric analysis is another geometric numerical modeling technique that optimize the form and the material distribution either in separate or in simultaneous fashions (Bletzinger et al. 2010; Firl and Bletzinger 2012; Nagy et al. 2013; Wall et al. 2008). Other recent case studies on the form-finding techniques support a variety of design innovation on structural forms, which could achieve efficiency, economy, and elegance (Bel Hadj Ali et al. 2010; Bletzinger and Ramm 2001; Burry et al. 2005; Li et al. 2010; Richardson et al. 2013; Xu and Luo 2010).

Another reason for slow pace to meaningful bio-inspired application on the civil infrastructure is the lack of such courses in the engineering curriculum. The good news is that some universities do offer a bio-inspired course to teach students to work in a multidisciplinary environment and learn the principle from nature. For example, a course titled "Exploring Bio-Inspired Systems in Architecture" at the University of Minnesota co-taught by an architecture professor (Marc Swackhamer) and a biology professor (Neil Olszewski).

They followed the traditional architectural teaching approach in terms of design, modeling, and fabrication, but architectural students will receive interesting input from a biology professor that may help them expand their thinking on conceptual design. MIT offered a course in biologically inspired digital design and fabrication course aimed at introducing the emerging interface of engineering and biology and to explore the logic and principles of natural systems through multiscale modeling and experiment. A "Bio-Inspired Design" course at TU Delft gives an overview of nonconventional mechanical approaches in nature and shows how this knowledge can lead to more creativity in mechanical design.

The University of Bath provides a graduate-level course "Biomimetics" to encourage student to extract design principles by introducing a variety of natural materials, structures, and mechanisms. A similar course can also be found at Georgia Institute of Technology. The latest endeavors on teaching and education aspects of promoting bio-inspired studies can be found in many studies (Bruck et al. 2006; Glier et al. 2011; Helms et al. 2009; Jenkins 2011; Santulli and Langella 2011; Stone et al. 2014; Wiltgen et al. 2011).

10.4 Summary

The multidisciplinary environment in the design field is pushing engineers to come up with new innovations in structural design. Compared to bio-inspired philosophy in architectural design and mechanical design, the development of bio-inspired design on bridge structures is far behind. This chapter reviewed some recent projects to highlight those endeavors made in this field. It is shown that the bio-inspired method can help engineers address both traditional issues (efficiency, economy, and elegance) and emerging issues (sustainability, energy use, etc.). However, it is noted that most of completed projects only reach a low or middle inspiration level according to the Vincent's biomimetic map. The difficulty of using the bio-inspired method in bridge design is due to the inefficiency of searching biological analogies and the lack of support from a biomimetic researcher during the design process. In modern bridge design, we need more interesting forms for both stationary structures and moveable structures. A close multidisciplinary collaboration may help engineers build more sustainable and smart structural systems for bridges in the twenty-first century.

Acknowledgments The authors would like to acknowledge Prof. Pacheco-Torgal for his support and feedback on this chapter and Dr. William C. Taylor, a civil engineering emeritus professor from Michigan State University for checking the language. The authors also appreciate Elsevier and Journal of Bionic Engineering for their reprint permission and Structure for the use of photos.

References

Adriaenssens S, Ney L, Bodarwe E, Williams C (2012) Finding the form of an irregular meshed steel and glass shell based on construction constraints. J Architect Eng 18(3):206–213. doi:10.1061/(ASCE)AE.1943-5568.0000074

Aldersey-Williams H (2004) Towards biomimetic architecture. Nat Mater 3(5):277–279

Alegria Mira L, Thrall AP, De Temmerman N (2014) Deployable scissor arch for transitional shelters. Autom Constr 43:123–131

Ario I, Nakazawa M, Tanaka Y, Tanikura I, Ono S (2013) Development of a prototype deployable bridge based on origami skill. Autom Constr 32:104–111

Asadpoure A, Tootkaboni M, Guest JK (2011) Robust topology optimization of structures with uncertainties in stiffness—application to truss structures. Comput Struct 89(11–12):1131–1141. doi:http://dx.doi.org/10.1016/j.compstruc.2010.11.004

Bar Cohen Y (2011) Biomimetics: nature-based innovation. CRC press, Boca Raton

Bar-Cohen Y (2012) Nature as a model for mimicking and inspiration of new technologies. Int J Aeronaut Space Sci 13(1):1–13

Barbarino S, Bilgen O, Ajaj RM, Friswell MI, Inman DJ (2011) A review of morphing aircraft. J Intell Mater Syst Struct 22(9):823–877. doi:10.1177/1045389x11414084

Bel Hadj Ali N, Rhode-Barbarigos L, Pascual Albi AA, Smith IFC (2010) Design optimization and dynamic analysis of a tensegrity-based footbridge. Eng Struct 32(11):3650–3659. doi:http://dx.doi.org/10.1016/j.engstruct.2010.08.009

Bhushan B (2009) Biomimetics: lessons from nature–an overview. Philos Trans Roy Soc A Math Phys Eng Sci 367(1893):1445–1486. doi:10.1098/rsta.2009.0011

Billington DP (1983) The tower and the bridge: the new art of structural engineering. Basic Books, New York

Billington DP (2003) The Art of Structural Design. A Swiss Legacy. Yale University Press, New Haven

Bletzinger K-U, Firl M, Linhard J, Wüchner R (2010) Optimal shapes of mechanically motivated surfaces. Comput. Methods Appl Mech Eng 199(5–8):324–333. doi:http://dx.doi.org/10.1016/j.cma.2008.09.009

Bletzinger K-U, Ramm E (2001) Structural optimization and form finding of light weight structures. Comput Struct 79(22–25):2053–2062. doi:http://dx.doi.org/10.1016/S0045-7949(01)00052-9

Bonser RHC (2006) Patented biologically-inspired technological innovations: a twenty year view. J Bionic Eng 3(1):39–41. doi:http://dx.doi.org/10.1016/S1672-6529(06)60005-X
Bruck H, Gershon A, Golden I, Gupta S, Gyger Jr LS, Magrab E, Spranklin B (2006) New educational tools and curriculum enhancements for motivating engineering students to design and realize bio-inspired products. In: Design and Nature III: Comparing Design in Nature with Science and Engineering, vol 87, p 325
Burry J, Felicetti P, Tang J, Burry M, Xie M (2005) Dynamical structural modeling a collaborative design exploration. Int J Architect Comput 3(1):27–42. doi:10.1260/1478077053739595
Chong KP (2004) Nanotechnology and information technology in civil engineering. In: Paper presented at the towards a vision for information technology in civil engineering
Deaton J, Grandhi R (2013) A survey of structural and multidisciplinary continuum topology optimization: post 2000. Struct Multi Optim, 1–38. doi:10.1007/s00158-013-0956-z
Dollens DL (2005) A system of digital-botanic architecture. Leonardo 38(1):15–21. doi:10.1162/leon.2005.38.1.15
Dooley S (2004). The development of material-adapted structural form. (Ph.D. Dissertation), École polytechnique fédérale de Lausanne
Espinosa HD, Rim JE, Barthelat F, Buehler MJ (2009) Merger of structure and material in nacre and bone—perspectives on de novo biomimetic materials. Prog Mater Sci 54(8):1059–1100. doi:http://dx.doi.org/10.1016/j.pmatsci.2009.05.001
Firl M, Bletzinger K-U (2012). Shape optimization of thin walled structures governed by geometrically nonlinear mechanics. Comput Methods Appl Mech Eng 237–240(0):107–117. doi:http://dx.doi.org/10.1016/j.cma.2012.05.016
Friedman N, Ibrahimbegovic A (2013) Overview of highly flexible, deployable lattice structures used in architecture and civil engineering undergoing large displacements. YBL J Built Environ 1(1):85–103
Gauvreau P (2002) The three myths of bridge aesthetics. In: Developments in short and medium span bridge engineering–2002
Giurgiutiu V (2007) Concepts of adaptronic structures adaptronics and smart structures, pp 9–28. Springer, Berlin
Glier MW, McAdams DA, Linsey JS (2011) Concepts in biomimetic design: methods and tools to incorporate into a biomimetic design course. In: Paper presented at the ASME 2011 international design engineering technical conferences and computers and information in engineering conference
Goel AK, McAdams DA, Stone RB (2014) biologically inspired design: Computational Methods and Tools. Springer
Guest JK (2009) Topology optimization with multiple phase projection. Comput Meth Appl Mech Eng 199(1–4):123–135. doi:http://dx.doi.org/10.1016/j.cma.2009.09.023
Helms M, Vattam SS, Goel AK (2009). Biologically inspired design: process and products. Des Stud 30(5):606–622. doi:http://dx.doi.org/10.1016/j.destud.2009.04.003
Hoult N, Bennett PJ, Stoianov I, Fidler P, Maksimović Č, Middleton C, Soga K (2009) Wireless sensor networks: creating 'smart infrastructure'. In: Paper presented at the proceedings of the ICE-civil engineering
Hu N (2010) The bio-inspired method in design and optimization of bridge structures. (M.S.), Central South University, China
Hu N, Feng P, Dai G (2013) The gift from nature: bio-inspired strategy for developing innovative bridges. J Bionic Eng 10(4):405–414. doi:http://dx.doi.org/10.1016/S1672-6529(13)60246-2
Hurlebaus S, Gaul L (2006) Smart structure dynamics. Mech Syst Signal Process 20(2):255–281. doi:http://dx.doi.org/10.1016/j.ymssp.2005.08.025
Jenkins CH (2011) Doing BiE: Lessons Learned From Teaching Bio-Inspired Engineering. In: Paper presented at the ASME 2011 international mechanical engineering congress and exposition

John G, Clements-Croome D, Jeronimidis G (2005) Sustainable building solutions: a review of lessons from the natural world. Build Environ 40(3):319–328. doi:http://dx.doi.org/10.1016/j.buildenv.2004.05.011

Knippers J, Speck T (2012) Design and construction principles in nature and architecture. Bioinspiration Biomimetics 7(1):015002

Korkmaz S, Ali NBH, Smith IF (2012) Configuration of control system for damage tolerance of a tensegrity bridge. Adv Eng Inform 26(1):145–155

Kovač M (2013) The bioinspiration design paradigm: a perspective for soft robotics. Soft Robot 1(P):28–37

Li Y, Feng X-Q, Cao Y-P, Gao H (2010) A Monte Carlo form-finding method for large scale regular and irregular tensegrity structures. Int J Solids Struct 47(14–15):1888–1898. doi:http://dx.doi.org/10.1016/j.ijsolstr.2010.03.026

Liu K, Jiang L (2011) Bio-inspired design of multiscale structures for function integration. Nano Today 6(2):155–175. doi:http://dx.doi.org/10.1016/j.nantod.2011.02.002

Mak TW, Shu LH (2008) Using descriptions of biological phenomena for idea generation. Res Eng Des 19(1):21–28. doi:10.1007/s00163-007-0041-y

Mattheck C (1998) Design in nature: learning from trees. Springer, Heidelberg

Michaël P, Dominique S (2013) Piezoelectric energy harvesting from traffic-induced bridge vibrations. Smart Mater Struct 22(9):095019

Mudupu V, Trabia MB, Yim W, Weinacht P (2008) Design and validation of a fuzzy logic controller for a smart projectile fin with a piezoelectric macro-fiber composite bimorph actuator. Smart Mater Struct 17(3):035034

Nagel JK, Stone RB (2011) A systematic approach to biologically-inspired engineering design. In: Paper presented at the ASME 2011 international design engineering technical conferences and computers and information in engineering conference

Nagel JK, Stone RB, McAdams DA (2010) An engineering-to-biology thesaurus for engineering design. In: Paper presented at the ASME 2010 international design engineering technical conferences and computers and information in engineering conference

Nagy AP, Ijsselmuiden ST, Abdalla MM (2013) Isogeometric design of anisotropic shells: Optimal form and material distribution. Comput Methods Appl Mech Eng 264(0):145–162. doi:http://dx.doi.org/10.1016/j.cma.2013.05.019

Petra G (2008) The signs of life in architecture. Bioinspiration Biomimetics 3(2):023001

Rhode-Barbarigos L, Hadj Ali NB, Motro R, Smith IF (2010) Designing tensegrity modules for pedestrian bridges. Eng Struct 32(4):1158–1167

Richardson JN, Adriaenssens S, Filomeno Coelho R, Bouillard P (2013) Coupled form-finding and grid optimization approach for single layer grid shells. Eng Struct 52(0):230–239. doi:http://dx.doi.org/10.1016/j.engstruct.2013.02.017

Russell B, Thrall A (2013) Portable and rapidly deployable bridges: historical perspective and recent technology developments. J Bridge Eng 18(10):1074–1085. doi:10.1061/(ASCE)BE.1943-5592.0000454

Santulli C, Langella C (2011) Introducing students to bio-inspiration and biomimetic design: a workshop experience. Int J Technol Des Educ 21(4):471–485. doi:10.1007/s10798-010-9132-6

Sartori J, Pal U, Chakrabarti A (2010) A methodology for supporting "transfer" in biomimetic design. AI EDAM 24(4):483–506

Schlaich M (2004) Future engineering structures for the urban habitat and infrastructure. In: Paper presented at the IABSE symposium report

Shu LH, Ueda K, Chiu I, Cheong H (2011) Biologically inspired design. CIRP Ann Manuf Technol 60(2):673–693. doi:http://dx.doi.org/10.1016/j.cirp.2011.06.001

Singh AV, Rahman A, Sudhir Kumar NVG, Aditi AS, Galluzzi M, Bovio S, Parazzoli D (2012) Bio-inspired approaches to design smart fabrics. Mater Des 36(0):829–839. doi:http://dx.doi.org/10.1016/j.matdes.2011.01.061

Spencer B, Cho S (2011) Wireless smart sensor technology for monitoring civil infrastructure: technological developments and full-scale applications. In: Paper presented at the proceeding of the world congress on advances in structural engineering and mechanics (ASEM'11 Plus)

Srinivasan V, Chakrabarti A, Pal U, Ranjan BSC, Ojha SP, Ranganath R (2011) Supporting process and product knowledge in biomimetic design. Int J Des Eng 4(2):132–158. doi:10.1504/IJDE.2011.045132

Stajano F, Hoult N, Wassell I, Bennett P, Middleton C, Soga K (2010) Smart bridges, smart tunnels: transforming wireless sensor networks from research prototypes into robust engineering infrastructure. Ad Hoc Netw 8(8):872–888

Stephen JE, Christopher AS (2012) The cochlea as a smart structure. Smart Mater Struct 21(6):064001

Stone R, Goel A, McAdams D (2014) Charting a course for computer-aided bio-inspired design. In: Goel AK, McAdams DA, Stone RB (eds) Biologically inspired design, pp 1–16. Springer, London

Tang M-C (2007) Evolution of bridge technology. Paper presented at the IABSE symposium report

Thomas S, Massimilano V, Gunnar T (2012) Design and development of a deployable self-inflating adaptive membrane. In: 53rd AIAA/ASME/ASCE/AHS/ASC structures, structural dynamics and materials conference. American Institute of Aeronautics and Astronautics

Thrall A, Adriaenssens S, Paya-Zaforteza I, Zoli T (2012) Linkage-based movable bridges: design methodology and three novel forms. Eng Struct 37:214–223

Tomiyama T, Gu P, Jin Y, Lutters D, Kind C, Kimura F (2009) Design methodologies: industrial and educational applications. CIRP Ann Manuf Technol 58(2):543–565. doi:http://dx.doi.org/10.1016/j.cirp.2009.09.003

Trotta MG (2011) Bio-inspired design methodology. Int J Inform Sci 1(1):1–11

Vincent JFV (2001) Stealing ideas from nature. In: Pellegrino S (ed) Deployable structures. Springer, Vienna

Vincent JFV (2006) Applications—influence of biology on engineering. J Bionic Eng 3(3):161–177. doi:http://dx.doi.org/10.1016/S1672-6529(06)60020-6

Vincent JFV, Mann DL (2002) Systematic technology transfer from biology to engineering. Philos Trans Roy Soc Lond Ser A Math Phys Eng Sci 360(1791):159–173. doi:10.1098/rsta.2001.0923

Wall WA, Frenzel MA, Cyron C (2008) Isogeometric structural shape optimization. Comput Methods Appl Mech Eng 197(33–40):2976–2988. doi:http://dx.doi.org/10.1016/j.cma.2008.01.025

Wallner M, Pircher M (2007) Kinematics of movable bridges. J Bridge Eng 12(2):147–153. doi:10.1061/(ASCE)1084-0702(2007)12:2(147)

Wilson JO, Rosen D, Nelson BA, Yen J (2010) The effects of biological examples in idea generation. Des Stud 31(2):169–186. doi:http://dx.doi.org/10.1016/j.destud.2009.10.003

Wiltgen B, Vattam S, Helms M, Goel AK, Yen J (2011) Learning functional models of biological systems for biologically inspired design. In: Paper presented at 11th IEEE international conference on advanced learning technologies (ICALT), 2011

Xing D, Chen W (2011) Systematic method of applying structural characteristics of natural organisms to mechanical structures. Trans Tianjin Univ 17:293–297

Xu X, Luo Y (2010) Form-finding of nonregular tensegrities using a genetic algorithm. Mech Res Commun 37(1):85–91. doi:http://dx.doi.org/10.1016/j.mechrescom.2009.09.003

Yeesock K, Changwon K, Reza L (2010) Novel bio-inspired smart control for hazard mitigation of civil structures. Smart Mater Struct 19(11):115009

Yoseph B-C (2006) Biomimetics—using nature to inspire human innovation. Bioinspiration Biomimetics 1(1):P1

Chapter 11
Bio-inspired Sensors for Structural Health Monitoring

Kenneth J. Loh, Donghyeon Ryu and Bo Mi Lee

Abstract Structural systems are susceptible to damage throughout their operational lifetime. Thus, structural health monitoring technologies and, in particular, sensors that could monitor structural performance and detect damage are needed. While there exist a variety of different sensing platforms, this continues to be an active area of research due to the many challenges associated with identifying and quantifying structural damage, which is inherently very complex. This chapter discusses an emerging area of sensors research in which sensor design or functionality is inspired by biological systems. By borrowing concepts from and learning how nature's creations sense and interact with its environment, the goal is to create novel sensors with unparalleled performance as compared to the current state-of-art. This chapter is not meant to be an exhaustive literature review on this topic. Rather, only a small selection of published work is sampled and presented to showcase different ideas and the breadth of research. Topics ranging from bio-inspired algorithms, creature-like robots, and skin-like sensors are presented.

11.1 Introduction

Infrastructure systems, such as bridges, buildings, cranes, dams, and pipelines, among others, could incur damage throughout their service lifetime. Natural disasters, excessive loads, accidents, and/or environmental degradation could cause various types of damage. Damage (e.g., corrosion or fatigue cracks) could escalate over different time- and length-scales and while the structure remains in service. If left undetected, accumulated damage could diminish structural performance, serviceability, and safety.

K.J. Loh (✉) · D. Ryu · B.M. Lee
Department of Civil and Environmental Engineering, University of California, Davis, CA 95616, USA
e-mail: kjloh@ucdavis.edu

F. Pacheco Torgal et al. (eds.), *Biotechnologies and Biomimetics for Civil Engineering*,
DOI 10.1007/978-3-319-09287-4_11

For example, a fatal rupture and explosion of a natural gas pipeline in Carlsbad, NM in 2000 was a result of deterioration caused by corrosion and other factors (NTSB 2003, Peekema 2013). Another example is cracking in steel girders of the Interstate I-794 Daniel Webster Hoan Bridge (Milwaukee, WI) in 2000 that caused one of its spans to sag by more than 1.2 m (4 ft); as a result, a total of US$7.8 million was invested for repair and retrofit (Fisher et al. 2001). The catastrophic collapse of the Sgt. Aubrey Cosens V.C. Memorial Bridge (Latchford, ON, Canada) in 2003 was due to fatigue rupture of hanger connections within the arch bridge's enclosed steel boxes (Biezma and Schanack 2007). As evident from these historic events, the timely detection of damage is crucial for facilitating the necessary structural repairs, ensuring optimal performance of structures, and enhancing public safety.

Routine visual inspection by trained technicians continues to be the predominant methodology adopted for evaluating the performance and safety of existing structural systems. For instance, the National Bridge Inspection Standards, set forth by the U.S. Federal Highway Administration, mandates that all highway bridges be inspected every 24 months (FWHA 2004). Due to their inherent time-, labor-, and cost-intensiveness (Hartle et al. 1990; Moore et al. 2001), tethered sensing systems are sometimes used to supplement visual inspection (Zhang et al. 2007; Sumitro et al. 2005; Celebi 2006). While simple sensors such as accelerometers and strain gages offer quantitative measures of structural response, their high costs to install and maintain the great lengths of coaxial cables have limited the number and density of sensors installed per structure (Celebi 2006). The end result is a sparsely distributed monitoring system (relative to the size of the structure) that is, at times, poorly scaled with structural damage, which is inherently highly localized.

On the other hand, structural health monitoring (SHM) integrates structural response data collected from distributed sensors with feature extraction algorithms for identifying the presence of damage relative to its *healthy* or pristine state. Statistical analysis is then required for relating accumulated damage to structural functionality, resistance to various loading scenarios, and overall safety (Farrar and Worden 2007). According to Farrar and Worden (2007) and Rytter (1993), damage detection involves a five-step process of increasing complexity, namely: (1) existence; (2) location; (3) type; (4) extent; and (5) prognosis. One of the major considerations in which any or all of these five steps could be achieved relies on the careful selection and deployment of sensors. These sensors need to be strategically instrumented and need to provide high signal-to-noise data that are indicative of damage, where damage and detection threshold is defined based on the end-user's needs.

In fact, a plethora of emerging technologies have been proposed over the past few decades for autonomous SHM and damage detection. Examples include fiber optics and fiber Bragg gratings for distributed strain and temperature measurements (Tsuda et al. 1999; Kersey 1996), wireless sensors for densely distributed system identification (Lynch and Loh 2006; Spencer Jr. et al. 2004), and piezoelectric sensor and actuator arrays for active sensing and acoustic emissions

(Tadigadapa and Mateti 2009; Yu et al. 2008; Zhao et al. 2007), among others (Boller 2000). Despite these advancements, these SHM systems possess certain limitations that sometimes make it challenging for them to diagnose a structure's *health*. While limitations vary between sensor platforms, in most cases, sensors are discrete transducers that only measure data at instrumented locations. Spatial structural response is usually obtained using a dense network of sensors and interpolation. Most sensors only measure structural response, and physics-based models or statistical methods are needed for inferring information about damage.

Unlike the aforementioned manmade sensing technologies, nature and its diverse creations have perfected biological assemblies and functionalities to enable the five senses of touch, visual, auditory, olfactory, and taste. In fact, humans have historically relied on biological inspirations to create artificial systems that mimicked the functionalities of many creatures. For example, camouflage tactics used in the military (i.e., planes, ships, and tanks) have been motivated by an octopus' ability to match its colors to its surroundings (Bar-Cohen 2006). With the advent of modern computing technology, electronics, and nanotechnology, these bio-inspired systems have become increasingly more complex and higher performance. For example, McGary et al. (2006) and Liu (2007) have fabricated artificial hair cell arrays for magnetostrictive acoustic sensing and fluid flow drag force sensing, respectively. Other examples include bio-inspired auto-adaptive and autonomous engineering systems (Tomizuka et al. 2007), plant-inspired actuators (Philen et al. 2007), phospholipids, and microfluidic membranes (Horsley et al. 2008), among many others (Loh et al. 2009; Li et al. 2004; Kao et al. 2007; Lin and Sodano 2009).

Instead of presenting an exhaustive review of bio-inspired sensing systems, this book's chapter uses select examples of published works to highlight the breadth of cutting-edge research. In particular, this chapter focuses on three different types of bio-inspired sensors for SHM. First, Sect. 11.2 presents bio-inspired algorithms for SHM. These algorithms could be used with current sensor networks for detecting damage or sensor faults. Second, Sect. 11.3 discusses robotic sensors that mimic the functionality of various creatures. Like geckos, these sensors could crawl on various structural surfaces and collect information from locations inaccessible to inspectors. Then, Sect. 11.4 highlights recent advances in designing conformable sensors based on nanotechnology. These thin films could be coated onto structural surfaces, and like skin, be able to detect damage over large spatial areas. This chapter concludes with a brief summary and discussion of future trends and needs.

11.2 Bio-Inspired Computational Tools

The development of various data management, signal processing, and damage detection algorithms has been inspired by biology. One particular prominent area is computational methods inspired by the biological immune system (BIS). In particular, BIS has provided a conceptual basis for deriving an artificial immune system (AIS) to improve the reliability and security of sensor networks for SHM (Drozda et al. 2011a; Twycross and Aickelin 2009; Dasgupta 2006).

One of the interesting features of the BIS is homeostasis, which helps maintain a normal operation level in environments where errors and changes could occur (Drozda et al. 2011a). One should know that BIS consists of two components (i.e., innate and adaptive immune systems) to protect the biological host from threatening pathogens. Here, the innate immune system immediately responds to the known pathogen, and the system does not change. On the other hand, the adaptive immune system involves learning and memory that take place over relatively longer periods of time (e.g., over several days). When unknown pathogens affect the host, the adaptive immune system recognizes and memorizes the new threat or change.

The detailed mechanism that enables the adaptive immune system response is that lymphocytes experience either positive or negative selection processes with a minimum self-reactivity or autoimmunity. When autoimmunity happens, the host cells undergo self-attack. It should be noted that first generation AIS considered only one type cell (i.e., B cell) for stabilization of the network (Hunt et al. 1999). When antigens were presented to B cells, and if affinities between B cells were strong, the B cells were cloned and contributed to stabilization of the network.

A second generation AIS was proposed by introducing different cell types (e.g., B and T cells) to the conventional AIS to enhance performance of error detection (Twycross and Aickelin 2009). Research focused on the communication between the different cell types. The innate immune system took part in error detection, which involves classifying the error and providing context on errors triggering damage. AIS' error detection structure was composed of adaptive and context classification modules. The two modules interacted in a feedback loop. By mimicking biological adaptive immunity, the structure offered functions of selection, memory, and learning for detecting new errors and changes in existing errors in the monitored sensor network.

The context classification module, inspired by innate immunity, evaluates the quality of service, and provides feedback to the adaptive module after error detection. Using the immune system-inspired error detection algorithm, it was shown that various objectives, such as energy efficiency, final false positives rate, adaptivity, and novel error detection, could be taken into account. By simulating the experiments conducted by Drozda et al. (2011b), improvements in the adaptive strategy-based error detection method were validated with minimum impact on energy efficiency and intact false positives control (Drozda et al. 2011a).

Bio-inspiration has also been adopted to improve the performance of complicated sensor networks for SHM. In addition, as these sensor networks evolve, they become even more complicated, with increasing number of sensors and dimensions. In particular, complex sensor networks with large number of sensors require the use of novel algorithms and strategies for optimum sensor placement to avoid coverage holes and unbalanced data flows.

For instance, genetic algorithms (GA) were investigated by numerous researchers for fault detection, as well as for monitoring spatial lattice structures and large space structures (Worden and Burrows 2001; Liu et al. 2008; Yao et al. 1993).

Javadi et al. (2005) developed a hybrid GA by integration of back-propagation neural network to improve convergence performance and the quality of the solution. Inspired by natural evolution, evolutionary algorithms (EA) have been used for optimizing mobile sensor deployment with optimum sensor coverage in a randomly distributed wireless sensor network (Abbasi et al. 2014). To simulate social behavior, particle swarm optimization (PSO) was implemented, which was suggested by Kennedy and Eberhart (1995). PSO was further enhanced by introducing chaos into an accelerated PSO by Gandomi et al. (2013). An overview of sensor node movement methodologies was provided by Abbasi et al. (2014) to improve network coverage as well as network lifetime, and the bio-inspired algorithms were shown to enhance data details, timeline, and reliability.

To shorten computational run-time and improve convergence performance in complicated sensor networks, Yi et al. (2012) developed an asynchronous-climb monkey algorithm (AMA) for optimizing sensor placement. The AMA mainly consists of monkeys' mountain-climbing process (i.e., climb, watch, jump, and somersault) (Fig. 11.1). In particular, AMA mimics monkeys' social behavior to find the optimum positions of sensors, in which monkeys exchange information with neighbor monkeys and were guided by the monkey king (i.e., the first step of AMA). This procedure was repeated until the best objective value (in this case the binary vector of sensor location) was obtained. The objective value was continuously updated with the position vector of monkeys by considering each monkey's own experience and the social experience from neighbor monkeys. The best objective value was designated as a "monkey king," and the Darwinian principle of natural selection was incorporated to accommodate replacing the monkey king with the next best one.

Fig. 11.1 The process by which monkeys climb, watch, jump, and somersault provides the conceptual basis for the asynchronous-climb monkey algorithm (Yi et al. 2012). (Image provided courtesy of IOP publishing)

Furthermore, AMA was improved by combining PSO, which was expected to be useful for civil infrastructures instrumented with large number of sensors. The proposed AMA was verified through a trial involving different sensor placements for characterizing the vibrational behavior of the Canton Tower (Shanghai, China). Modal assurance criterion (MAC) was chosen to yield two types of objective functions, and for comparison with AMS, a conventional monkey algorithm (MA) was used with a dual-structure coding method. Two simplified finite element (FE) models of the Canton Tower were established with and without the antenna mast. For all cases of objective functions and FE models, AMA exhibited the best performance in optimizing sensor placement.

Data processing and transferring techniques have also been developed by adopting biological concepts with the purposes of improving accuracy in data detection and realizing real-time data processing for large-scale civil infrastructures (Lin et al. 2010; Peckens and Lynch 2013). A new damage extraction and classification method for SHM based on the concept of a deoxyribonucleic acid (DNA) array was implemented (Lin et al. 2010). The purpose was to improve damage detection accuracy. Damage characteristics were extracted using a double-tier regression model to establish the autoregressive (AR)-autoregressive exogeneous (ARX) database, which was regarded as analogous to DNA of biological creatures. Just as how DNA arrays are classified based on DNA patterns using naïve Bayesian (NB) algorithm for detecting cancer cells, AR-ARX arrays could also be classified using an NB algorithm and further improved through optimization using a likelihood selection method. The novel DNA-inspired SHM system was verified through experiments involving a six-story building under ambient vibration. It was shown that the optimized SHM system achieved 90–95 % accuracy in terms of damage identification. To enable real-time data processing, the auditory signal compression and transferring concepts were used (Peckens and Lynch 2013). Inspired by the cochlea, high signal compression ratio was achieved with a reasonable reconstruction error as compared to two other conventional compressive techniques, namely wavelet transforms and compressed sensing (Fig. 11.2) (Peckens and Lynch 2013).

There have been other efforts in using bio-inspired concepts for improving damage detection and SHM (Loh and Azhari 2012; Salowitz et al. 2013; Kirikera et al. 2008). Fatigue monitoring has been considered challenging, particularly due to its long time-scales and high-frequency signals. Inspired by the concept of tree ring data tracking, an intelligent fatigue monitoring system was developed (Bai et al. 2014). Fatigue characteristics (i.e., the amplitude of strain, the number of cycles, and the stress state) were acquired with a high degree of accuracy, which could be transmitted in real-time via a wireless network. The wireless fatigue monitoring system was validated using cantilever bending fatigue tests. Maximum strain of the test specimen was tracked using either strain gage or polyvinylidene fluoride, and the obtained digital signals were processed to extract the number of loading cycles, among other fatigue characteristics. Feature extraction was performed using digital

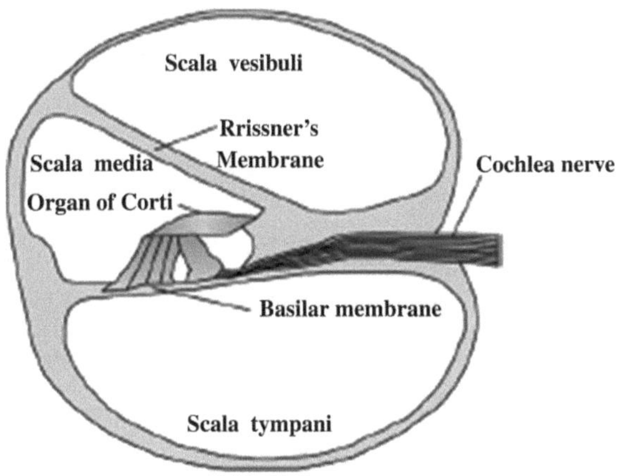

Fig. 11.2 A cross-section of the cochlea is shown. The cochlea is a part of the mammalian inner ear and plays an important role in processing and transmitting auditory signals (Peckens and Lynch 2013). (Image provided courtesy of IOP publishing)

signal processing technology integrated with a rain-flow counting method. It was shown that loading cycles were accurately measured with <5 % error, and fatigue life was calculated.

11.3 Creature-like Robotic Sensors

While bio-inspired algorithms offer the possibility of optimizing sensor instrumentations, in many cases, large structures still require densely distributed sensors for SHM. It has already been shown by Celebi (2002) that the cost to install sensors in tall buildings could exceed US$5,000 per channel. Although one of the major goals (and advantages) of wireless sensors was to lower costs by eliminating sensor dependence on a tethered connection (Lynch and Loh 2006), the need for a dense sensor instrumentation could make a wireless SHM system cost prohibitive for many end-users and applications.

Mobile or robotic sensors that are inspired by different creatures' ability to navigate around the natural world offer unique benefits for SHM. In particular, the fundamental benefit is that mobile sensors are no longer tied to their instrumented locations, as is the case for static sensor networks. Instead, a small number of mobile sensors could crawl around a bridge or building and record structural response data at many measurement points. The robotic sensor could be wirelessly commanded, collect data at a variety of measurement points over large spatial domains, and as frequently as desired (given the limitations of onboard power). The end result would be the possibility of obtaining densely distributed structural response data using only a handful of sensors. These robots could also navigate to

Fig. 11.3 A mobile sensor prototype (developed by researchers at the Georgia Institute of Technology, USA) is crawling on a steel pedestrian bridge. (Photo courtesy of Prof. Yang Wang, Georgia Institute of Technology, USA)

hard-to-reach areas that an inspector or engineer would otherwise not be able to access conveniently.

A prime example of a bio-inspired mobile sensor for SHM is work by Zhu et al. (2010). The mobile sensor was fabricated by connecting two two-wheeled vehicles with a flexible beam (Fig. 11.3). The wheels were made with magnets so that it could climb ferromagnetic structures. The design of the robot also included infrared sensors and Hall effect sensors for the purposes of detecting boundaries and for locomotion, respectively. For SHM, a Silicon Designs 2260-010 accelerometer was located on the center of the robot's center flexible beam, and the mobile sensor could control the attachment and detachment of the accelerometer onto the structure (i.e., by moving the two two-wheeled cars closer or farther away from each other, respectively). Two different experimental laboratory tests were conducted in the laboratory for validating their performance. The mobile sensors successfully crawled around a steel portal frame, and two robots collected structural vibration response (excited using impact hammer strikes) at 11 different measurement points. Damage was simulated by adding a mass block or by loosening bolts at connections. By analyzing the transmissibility function, damage location was successfully identified. Other research groups have also developed various types of crawling robots (Huston et al. 2005; Akiba et al. 2013; Oh et al. 2009).

While the aforementioned mobile sensor was more mechanical and resembled that of a car, the design of the Geckobot was inspired by the gecko's ability to climb almost any type of surface (Unver et al. 2006). In particular, the Geckobot featured synthetic dry adhesive feet, made with polydimethyl siloxane (PDMS) elastomer that mimicked the gait and climbing mechanism of real geckos. Like a real gecko, when the PDMS toe was controlled to detach from the surface, the PDMS was bent and peeled off in a way that minimized the detaching force. Another unique design was its active tail that enhanced the robot's mobility during climbing. With the existence of a tail, it could provide additional force that secured

Fig. 11.4 Polyurethane micro-hair arrays mimic the hairs (or setae) on a gecko's feet that enable vertical climbing (Wu et al. 2013). (Images provided courtesy of Springer)

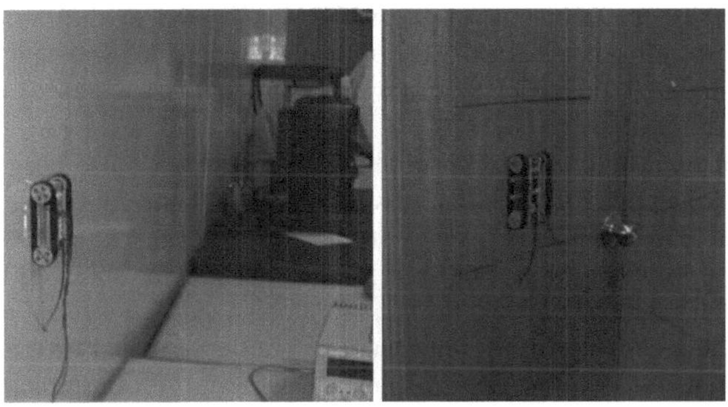

Fig. 11.5 A gecko-inspired robot is shown climbing vertical walls of different materials. Vertical climbing is enabled with the robot's polyurethane micro-hair array on its wheels (Wu et al. 2013). (Images provided courtesy of Springer)

the Geckobot to the surface. Experimental studies demonstrated that the mobile sensor could walk at a speed of up to 5 and 1 cm/s on a flat and inclined surface, respectively. However, one limitation was that the robot could not climb a normal or 90° surface.

In contrast, Wu et al. (2013) proposed a climbing robot with two pedrails, in lieu of individual wheels or robotic feet. A unique feature of the pedrail was that it featured dry adhesive pedrails modeled after the gecko's feet. The surface of the pedrail was patterned with micro-fiber hair arrays, very much similar to the spatulae that populates setae found on Gecko's feet. The micro-fiber hair array was fabricated using polyurethane and is shown in Fig. 11.4. The prototype gecko-inspired robot successfully climbed surface materials and of different inclinations (Fig. 11.5).

Robots inspired by other creatures, including snakes, have also been explored. For example, Enner et al. (2012, 2013) proposed a snake-like robot for monitoring pipeline systems. The main function of the snake-inspired robot was to measure the pipe's diameter, since its diameter could change as a result of corrosion or damage. The robot could wrap itself inside the pipe or along the outer surface.

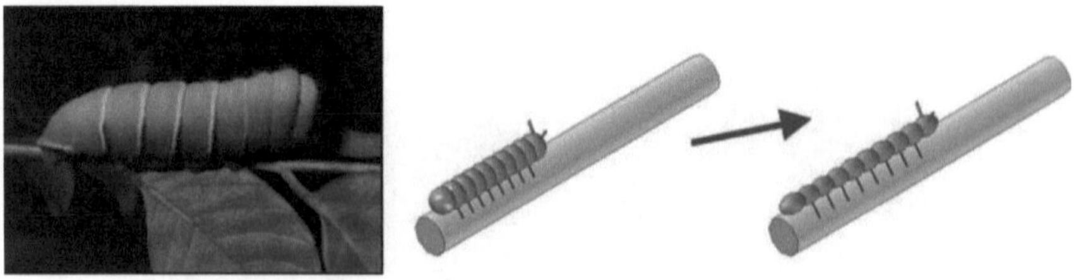

Fig. 11.6 The ability of a worm to shrink and extend its own body to propel itself forward has inspired the creation of artificial counterparts that mimic this behavior (Balaguer et al. 2005). (Images provided courtesy of Springer)

For both cases, the geometry of the robot was related to the pipe radius (or diameter), assuming that the centerline of the robot coincided with that of the pipe. Therefore, information about the pipe diameter could be obtained directly for SHM applications.

It should be mentioned that other climbing robots have also been developed. Examples include the ROMA I that was inspired by the locomotion of caterpillars that shrink and extend its body to move (Fig. 11.6) (Balaguer et al. 2005). Another example is the ROMA II that utilized vacuum suction for climbing on different surfaces (Balaguer et al. 2005). In fact, the mobile sensor consisted of two arms and 10 vacuum cups. By controlling suction on either arm and by moving and/or rotating the unattached arm, motion could be achieved. Bio-inspired mobile sensors continue to be an important area of research with commercial prototypes emerging in the marketplace.

11.4 Skin-Inspired Sensors

The human skin is another prime example of an impressive biological system that is capable of densely distributed sensing. The sensing and monitoring of changes in temperature, deformation, flow, pressure, and damage (i.e., injuries) is achieved by the 640,000 sensory receptors distributed throughout the entire system (Schmidt 1986). Unlike mobile sensors that need to navigate to different locations and collect data at different points in time, the skin is able to resolve changes throughout the entire system simultaneously.

In fact, the skin has inspired the development of various tactile sensors, as have been described in review articles by Lee and Nicholls (1999) and Yousef et al. (2011). These tactile sensors are particularly important considerations for the development of next-generation human-interactive robotic systems as an example Mukai et al. (2008). In the context of SHM, Loh and Azhari (2012) have presented a review of skin-inspired sensors.

Extensive research in flow sensors inspired by, for instance, human hairs and the lateral line system in fish, has been conducted. Velocity-, acceleration-, or pressure-sensitive neuromasts in the lateral line allow these creatures to measure flow and changes in motion (Coombs 2001). Hairs on the skin or on the surface of insects (like crickets) also enable biological systems to monitor changes in fluid flow (Pfatteicher and Tongue 2002) or acoustic signals (Wiegerink et al. 2007). Hairs or whiskers on mice are also used for object identification and avoidance.

Research by Dijkstra et al. (2005) and Wiegerink et al. (2007) have shown that biomimetic filiform hairs could be used for sensing flow. Viscous flow would tilt these micro-machined hairs and induce measureable capacitance changes. These artificial hairs have also been organized into a large array for enhanced sensitivity and performance. An array is also useful for enhancing sensing resolution, as well as differentiating flow direction. Besides measuring capacitance changes, others have used the measurement of changes in electrical current (Sarles et al. 2011), field-effect response (Kim et al. 2009), piezoelectricity (Yu et al. 2010), or magnetostriction (McGary et al. 2006). These bio-inspired flow sensors have been proposed for various applications (Pinto et al. 2011; Eberhardt et al. 2011; Tao et al. 2011).

A specific application in which bio-inspired flow sensors have been used for SHM is the case of bridge scour monitoring. Bridge scour is the disruption of marine structure's foundations (e.g., overwater bridge piers and abutments) due to rapid water flow, flooding, or severe weather events, among others (Whitehouse 1998). For instance, (Swartz et al. 2014) proposed a wireless smart scour sensing post that consisted of bio-inspired magnetostrictive flow sensors attached to the post surface (Fig. 11.7). When scour advances and exposes these Galfenol whiskers at different buried depths, the sensors would deflect due to fluid flow and drag. A giant magnetostrictive sensor mounted at the base of the whisker would detect such a change, and the response would be transmitted wirelessly to a base station. A similar concept was also proposed by Wang et al. (2012), Loh et al. (2014), and Azhari et al. (2014) but with piezoelectric sensors.

While hair-like structures that protrude from the surface of the skin could be used for sensing flow, it is well known that the skin itself is a highly effective distributed sensor. In fact, the skin has inspired the development of thin films and coatings that are sensitive to different damage features. One particular technique for creating artificial skin sensors is to incorporate randomly distributed carbon nanotubes (CNT) in a thin, flexible, polymer matrix. Numerous techniques have been developed, and they include examples such as evaporation (Dinh-Trong et al. 2009), spin coating (Yim et al. 2008), layer-by-layer (Loh et al. 2007), and spraying (Kang et al. 2006), just to name a few.

In particular, Loh et al. (2007, 2005) demonstrated strain sensing using layer-by-layer thin films assembled with single-walled carbon nanotubes (SWNT) and various polyelectrolyte species. It was observed that the film's electrical properties (e.g., resistivity) varied linearly with applied strains up to at least 1 % strains, and strain sensitivity could be controlled by modifying the concentration of the CNT solution used during layer-by-layer fabrication (Loh et al. 2008). Like skin, these materials could be applied onto the surfaces of structures for SHM. However, for detecting damage within structural materials, these SWNT-based films were also

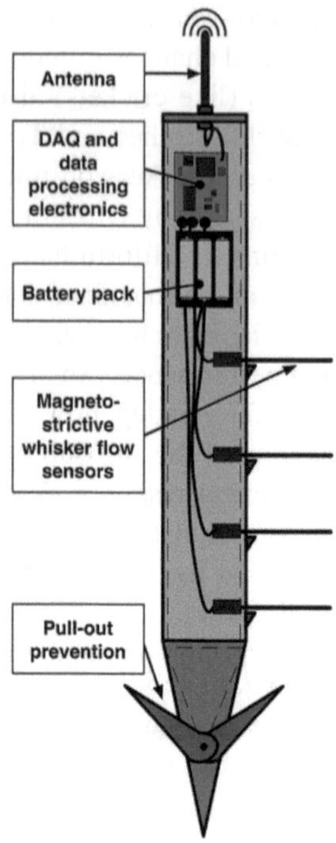

Fig. 11.7 A smart scour sensing post incorporates bio-inspired magnetostrictive whisker sensors for measuring fluid flow and bridge scour. (Image courtesy of Prof. R. Andrew Swartz, Michigan Technological University, USA)

embedded in structures like fiber-reinforced polymer (FRP) composites for embedded strain monitoring and damage detection (Fig. 11.8) (Loyola et al. 2010).

It should also be mentioned that another method for achieving embedded sensing was to modify the FRP composite's epoxy matrix, again, using conductive fillers like carbon nanotubes. Strain sensing, as well as monitoring delamination and micro-cracks, were validated by Thostenson and Chou (2006), Böger et al. (2008),

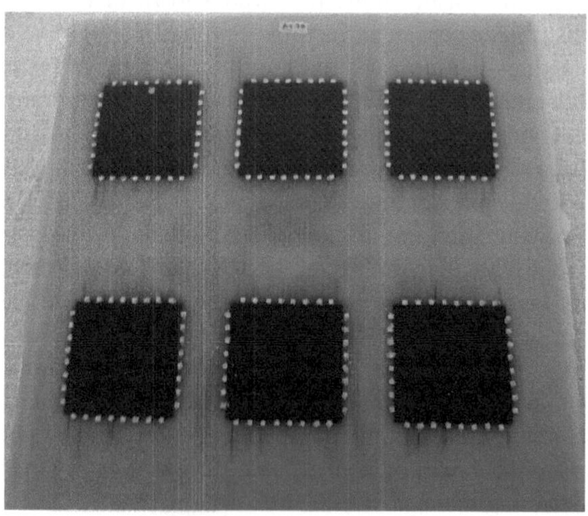

Fig. 11.8 Multiple carbon nanotube-based sensing skins are embedded underneath the surface of a large GFRP panel during the manufacturing process for achieving in situ damage detection and SHM

Nofar et al. (2009), and Kostopoulos et al. (2009). CNT concentrations as low as 0.1 w% was sufficient for producing electrically conductive, epoxy-based composites (Thostenson and Chou 2002; Thostenson and Chou 2006; Thostenson and Chou 2008).

Despite these advances in creating skin-like thin films that are piezoresistive, strain sensing could only be accomplished in an average sense. A resistivity measurement was directly correlated to strains being experienced by the underlying structure (i.e., at least on the surface). However, this strain measurement was averaged over the entire surface area or measurement area of the film. Yet, structural damage including cracks, corrosion, and impact were localized phenomena. An "averaged" strain measurement may not be able to detect such damage (i.e., depending on the size of damage and measurement area).

Instead, one could leverage the fact that the electrical resistivity (or equivalently conductivity) of every location in the film is calibrated to strain. Work by Hou et al. (2007b) and Loh et al. (2009) showed that an electrical impedance tomography (EIT) algorithm (Brown 2003) could be used for mapping the spatial conductivity distribution of CNT-based thin films. To produce this conductivity map without having to probe every location in the film, EIT relied on a set of instrumented electrodes along the boundaries of the thin film. This defined the sensing area. Then, EIT utilized boundary current input and voltage measurements for reconstructing the spatial conductivity distribution of the thin film (Borcea 2002; Brown 2003). By nature, the EIT inverse problem is an ill-posed problem, and one set of boundary voltage measurements would be insufficient for determining the spatial conductivity map of the sensing area (Holder 2005). Thus, electrical current was applied to numerous boundary electrodes, and sets of boundary voltage measurements were obtained for each current injection case. Since thin film conductivity (or resistivity) was correlated to applied strain, pH, or other damage phenomena (Loh et al. 2007, 2008; Loh 2008), the spatial conductivity maps were directly related to spatial damage in the underlying structure.

Previous works by Hou et al. (2007a, b) validated the EIT technique for spatial conductivity mapping of the aforementioned layer-by-layer CNT-based thin films. Square $25 \times 25\ \text{mm}^2$ SWNT-based thin films were instrumented with boundary electrodes, and EIT was used to obtain baseline conductivity maps of these films. Then, selected regions on the film surface were intentionally etched and removed so as to create straight, diagonal, and L-shaped cuts (Hou et al. 2007b). These cuts formed regions of zero conductivity in the film and were used for simulating some change or extreme damage. After mechanical etching, EIT was conducted again to obtain the relative change in thin film spatial conductivities. It was found that the spatial conductivity maps visually represented that of the actual films, and they could be used for identifying the locations of these cuts (Hou et al. 2007a, b). It was also shown that the EIT-estimated spatial conductivities were within 2 % error as compared to experimental measurements of the same film and sensing area.

With the EIT algorithm validated, these carbon nanotube *sensing skins* were then used for SHM and laboratory tests. For example, sensing skins were coated onto aluminum plates and mounted in an impact testing apparatus. It was shown

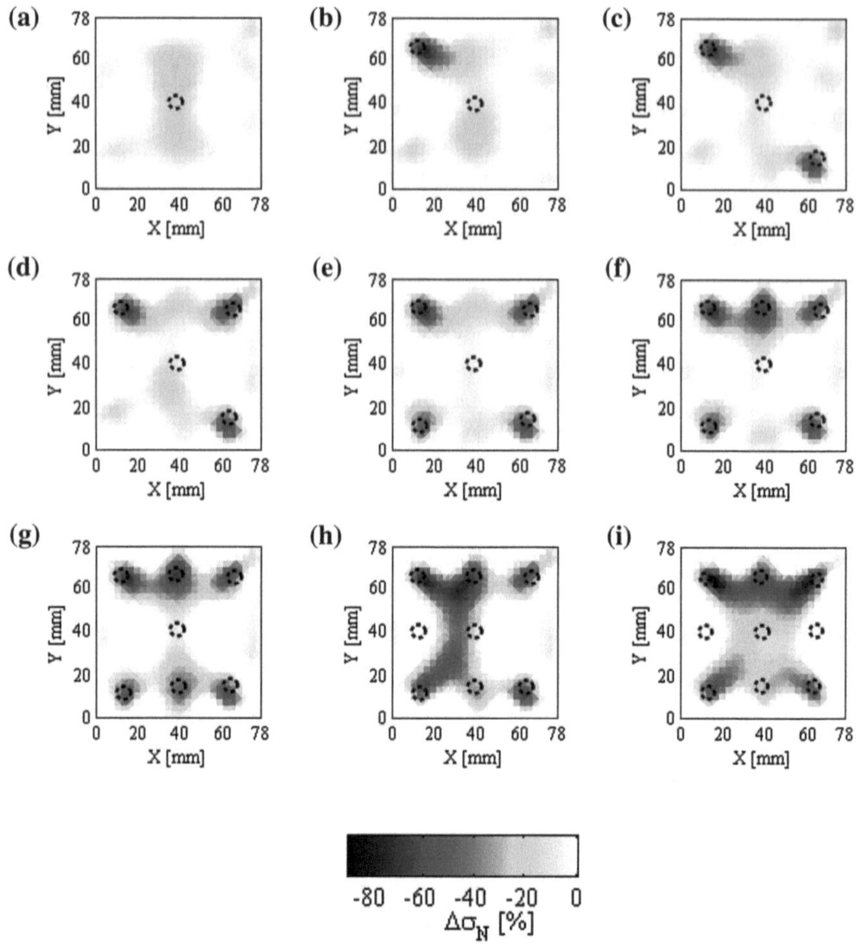

Fig. 11.9 The sensing skins and EIT algorithm are tested by verifying that this method could detect the location of drilled holes in sensing skin-enhanced GFRP panels. Each EIT image is obtained after a new hole is drilled in the **a** *center*, **b** *upper-left*, **c** *lower-right*, **d** *upper-right*, **e** *lower-left*, **f** *upper-center*, **g** *lower-center*, **h** *center-left*, and **i** *center-right* of the composite specimen (Loyola et al. 2013a). (Images provided courtesy of Sage publications)

that EIT was able to resolve the severity and location of four different magnitude impact damage sites (Loh et al. 2009). The monitoring of spatial variations in pH (i.e., as a precursor of corrosion) (Hou et al. 2007b) and changes in film conductivity due to corrosion/rust formation (Pyo et al. 2011) were also validated.

More recently, an airbrushing technique for spray deposition of these CNT-based sensing skins was developed for achieving rapid fabrication and field applications (Mortensen et al. 2013). In short, an ink solution was created by dispersing multi-walled carbon nanotubes (MWNT) in poly(sodium 4-styrene-sulfonate) (PSS) and the mixing it with a latex solution. The ink and latex mixture could then be sprayed onto virtually any substrate including metals, glass, concrete, and plastic. In particular, sensing skins were sprayed and embedded in glass fiber-reinforced polymer (FRP) composites during fabrication. Laboratory characterization tests were conducted, and the results verified the detection of spatially distributed impact damage (Loyola et al. 2013b) and drilled holes (Fig. 11.9) (Loyola et al. 2013a), again, using EIT. While these bio-inspired sensing skins demonstrated promise for SHM, more in-depth studies and larger-scale validation tests are needed for transitioning this technology to the commercial domain.

11.5 Summary

This chapter summarizes recent developments in bio-inspired computational tools and new technological developments for SHM. In particular, three specific areas have been discussed. First, computational methods inspired by the BIS have been shown to be applicable for distributed SHM sensing system. These computational tools can be used for detecting sensor faults in the network, optimizing sensor instrumentation strategy, or processing signals measured.

Second, mobile robotic sensors are inspired by different living being's ability to crawl or navigate around the world. These robotic devices offer the possibility of measuring densely distributed structural response by crawling around the host structure. Their small size and nimbleness allow them to navigate to inaccessible or dangerous places.

Finally, thin film-based sensors inspired by the human dermatological system (or skin) have also been proposed and validated in the laboratory. The skin is an extremely unique organ in which densely distributed and multi-modal sensing is accomplished in real-time. Recent developments in nanotechnology-enabled thin film sensors or coatings have permitted the design of artificial sensing skins. For instance, polymeric thin films that incorporate carbon nanotubes have been shown to be sensitive to strain. When combined with an electrical impedance tomographic spatial conductivity algorithm, spatial sensing and the detection of damage severity and location have been verified.

Despite numerous attempts of biomimicry, most artificial systems remain inferior to its natural counterparts and still require expensive and tedious materials processing. The authors expect that future trends in SHM will continue to see further developments in biologically inspired sensing systems. Not only will technological innovations mimic biological system behavior, but also, one could expect systems that are assembled in the same manner as in biology. This possibility is becoming more realistic, especially where the advent of nanotechnology has permitted the "bottom-up" assembly of molecules to form macro-scale devices. In addition, mechanisms ranging from actuation, healing, and energy transduction, among many others, would become integrated for advancing SHM and for achieving next-generation resilient infrastructure systems.

References

Abbasi M, Bin Abd Latiff MS, Chizari H (2014) Bioinspired evolutionary algorithm based for improving network coverage in wireless sensor networks. Sci World J 2014:8

Akiba T, Lee N, Mita A (2013) Sensor agent robot with servo-accelerometer for structural health monitoring. Key Eng Mater 558:289–296

Azhari F, Benassini J, Tom C, Loh KJ, Bombardelli FA (2014) Design and characterization of a piezoelectric sensor for monitoring scour hole evolution. In: SPIE—21st annual symposium on smart structures and materials and nondestructive evaluation and health monitoring, SPIE, San Diego, CA

Bai S, Li X, Xie Z, Zhou Z, Ou J (2014) A wireless fatigue monitoring system utilizing a bio-inspired tree ring data tracking technique. Sensors 14:4364–4383

Balaguer C, Gimenez A, Jardon A (2005) Climbing robots' mobility for inspection and maintenance of 3D complex environments. Auton Robots 18:157–169

Bar-Cohen Y (ed) (2006) Biomimetics: biologically inspired technologies. CRC Press, Boca Raton

Biezma MV, Schanack F (2007) Collapse of steel bridges. J Perform Constructed Facil 21:398–405

Boller C (2000) Next generation structural health monitoring and its integration into aircraft design. Int J Syst Sci 31:1333–1349

Borcea L (2002) Electrical impedance tomography. Inverse Problems, 18

Brown BH (2003) Electrical impedance tomography (EIT): a review. J Med Eng Technol 27:97–108

Böger L, Wichmann MHG, Meyer LO, Schulte K (2008) Load and health monitoring in glass fibre reinforced composites with an electrically conductive nanocomposite epoxy matrix. Compos Sci Technol 68:1886–1894

Celebi M (2002) Seismic instrumentation of buildings (with emphasis on federal buildings). United States Geological Survey, Menlo Park, CA

Celebi M (2006) Real-time seismic monitoring of the new Cape Giradeau bridge and preliminary analyses of recorded data: an overview. Earthq Spectra 22:609–630

Coombs S (2001) Smart skins: information processing by lateral line flow sensors. Auton Robots 11:255–261

Dasgupta D (2006) Advances in artificial immune systems. IEEE Comput Intell Mag 1:40–49

Dijkstra M, Barr JJV, Wiegerink RJ, Lammerink TSJ, Boer JHD, Krijnen GJM (2005) Artificial sensor hairs based on the flow sensitive receptor hairs of crickets. J Micromech Microeng 15:S132–S138

Dinh-Trong N, Steitz J, Lei B, Kanoun O (2009) Influence of the composition of MWCNTs layers on the properties of strain gauges. In: 9th IEEE conference on nanotechnology Genoa, Italy, pp 477–480

Drozda M, Schaust S, Schildt S, Szczerbicka H (2011b) Priming: making the reaction to intrusion or fault predictable. Nat Comput 10:243–274

Drozda M, Bate I, Timmis J (2011a) Bio-inspired error detection for complex systems. Dependable computing (PRDC), 2011. In: IEEE 17th Pacific rim international symposium, pp 154–163. 12–14 Dec 2011

Eberhardt WC, Shakhsheer YA, Calhoun BH, Paulus JR, Appleby MA (2011) Bio-inspired artificial whisker for fluid motion sensing with increased sensitivity and reliability. In: IEEE sensors, Limerick, pp 982–985

Enner F, Rollinson D, Choset H (2012) Simplified motion modeling for snake robots. In: IEEE International conference on robotics and automation, St. Paul, MN, pp 4216–4221

Enner F, Rollinson D, Choset H (2013) Motion estimation of snake robots in straight pipes. In: IEEE International conference on robotics and automation (ICRA), Karlsruhe, Germany, pp 5149–5153

Farrar CR, Worden K (2007) An introduction to structural health monitoring. Philos. Trans. R. Soc. A 365:(1851):303–315

Federal Highway Administration (FHWA) (2004) National bridge inspection standards
Fisher JW, Kaufmann EJ, Wright W, Xi Z, Tjiang H, Sivakumar B, Edberg W (2001) In: Hoan bridge forensic investigation failure analysis final report. Wisconsin Department of Transportation and the Federal Highway Administration, Madison, WI
Gandomi AH, Yun GJ, Yang X-S, Talatahari S (2013) Chaos-enhanced accelerated particle swarm optimization. Commun Nonlinear Sci Numer Simul 18:327–340
Hartle RA, Amrhein WJ, Iii KEW, Baughman DR, Tkacs JJ (1990) Bridge inspector's training manual/90. Federal Highway Administration, McLean
Holder DS (2005) Electrical impedance tomography—methods, history and applications. Institute of Physics Publishing, Bristol, UK
Horsley DA, Talin AA, Skinner JL (2008) Micromechanical and microfluidic devices incorporating metallic gratings fabricated using nanoimprint lithography. J Nanophotonics 2:021785/1-11
Hou T.-C, Loh KJ, Lynch JP (2007a) Electrical impedance tomography of carbon nanotube composite materials. SPIE, San Diego, CA 652926/1-10
Hou TC, Loh KJ, Lynch JP (2007b) Spatial conductivity mapping of carbon nanotube composite thin films by electrical impedance tomography for sensing applications. Nanotechnology 18:315501/1-9
Hunt J, Timmis J, Cooke E, Neal M, King C (1999) Jisys: the envelopment of an artificial immune system for real world applications. In: Dasgupta D (ed) Artificial immune systems and their applications. Springer, Berlin Heidelberg
Huston D, Esser B, Miller J, Wang X (2005) Robotic and mobile sensor systems for structural health monitoring. In: 5th International workshop on structural health monitoring, Stanford, CA, pp 1–8
Javadi AA, Farmani R, Tan TP (2005) A hybrid intelligent genetic algorithm. Adv Eng Inform 19:255–262
Kang I, Lee JW, Choi GR, Jung JY, Hwang S-H, Choi Y-S, Yoon KJ, Schulz MJ (2006) Structural health monitoring based on electrical impedance of a carbon nanotube neuron. Key Eng Mater 321–323:140–145
Kao I, Kumar A, Binder J (2007) Smart MEMS flow sensor: theoretical analysis and experimental characterization. IEEE Sens J 7:713–722
Kennedy J, Eberhart R (1995) Particle swarm optimization. In: Proceedings of IEEE international conference on neural networks, Perth, Australia, pp 1942–1948
Kersey AD (1996) A review of recent developments in fiber optic sensor technology. Opt Fiber Technol 2:291–317
Kim DR, Lee CH, Zheng X (2009) Probing flow velocity with silicon nanowire sensors. Nano Lett 9:1984 1988
Kirikera GR, Shinde V, Schulz MJ, Ghoshal A, Sundaresan MJ, Allemang RJ, Jong WL (2008) A structural neural system for real-time health monitoring of composite materials. Struct Health Monit 7:65–83
Kostopoulos V, Vavouliotis A, Karapappas P, Tsotra P, Paipetis A (2009) Damage monitoring of carbon fiber reinforced laminates using resistance measurements. improving sensitivity using carbon nanotube doped epoxy matrix system. J Intell Mater Syst Struct 20:1025–1034
Lee MH, Nicholls HR (1999) Review article tactile sensing for mechatronics—a state of the art survey. Mechatronics 9:1–31
Li Z, Dharap P, Nagarajaiah S, Barrera EV, Kim JD (2004) Carbon nanotube film sensors. Adv Mater 16:640–643
Lin T-K, Kiremidjian A, Lei C-Y (2010) A bio-inspired structural health monitoring system based on ambient vibration. Smart Mater Struct 19:115012
Lin Y, Sodano HA (2009) Fabrication and electromechanical characterization of a piezoelectric structural fiber for multifunctional composites. Adv Funct Mater 19:592–598
Liu C (2007) Micromachined biomimetic artificial haircell sensors. Bioinspiration Biomimetics 2:S162–S169

Liu W, Gao W-C, Sun Y, Xu M-J (2008) Optimal sensor placement for spatial lattice structure based on genetic algorithms. J Sound Vib 317:175–189
Loh KJ, Azhari F (2012) Recent advances in skin-inspired sensors enabled by nanotechnology. J Mater 64:793–801
Loh KJ, Hou T-C, Lynch JP, Kotov NA (2009) Carbon nanotube sensing skins for spatial strain and impact damage identification. J Nondestr Eval 28:9–25
Loh KJ, Kim JH, Lynch JP, Kam NWS, Kotov NA (2007) Multifunctional layer-by-layer carbon nanotube-polyelectrolyte thin films for strain and corrosion sensing. Smart Mater Struct 16:429–438
Loh KJ, Lynch JP, Shim BS, Kotov N (2008) Tailoring piezoresistive sensitivity of multilayer carbon nanotube composite strain sensors. J Intell Mater Syst Struct 19:747–764
Loh KJ, Tom C, Benassini J, Bombardelli F (2014) A distributed piezo-polymer scour net for bridge scour hole topography monitoring. Smart Struct Sys 1:183-195
Loh KJ (2008) Development of multifunctional carbon nanotube nanocomposite sensors for structural health monitoring. Ph.D., Department of Civil & Environmental Engineering, University of Michigan, Ann Arbor
Loh KJ, Lynch JP, Kotov NA (2005) Conformable single-walled carbon nanotube thin film strain sensors for structural monitoring. In: 5th International workshop on structural health monitoring, Stanford, CA, pp 686–694
Loyola BR, Briggs TM, Arronche L, Loh KJ, Saponara VL, O'Bryan G, Skinner JL (2013a) Detection of spatially distributed damage in fiber-reinforced polymer composites. Struct Health Monit 12:225–240
Loyola BR, la Saponara V, Loh KJ (2010) In situ strain monitoring of fiber-reinforced polymers using embedded piezoresistive nanocomposites. J Mater Sci 45:6786–6798
Loyola BR, la Saponara V, Loh KJ, Briggs TM, O'Bryan G, Skinner JL (2013b) Spatial sensing using electrical impedance tomography. IEEE Sens J 12:2357–2367
Lynch JP, Loh KJ (2006) A summary review of wireless sensors and sensor networks for structural health monitoring. Shock Vibr Digest 38:91–128
Mcgary PD, Tan L, Zou J, Stadler BJH, Downey PR, Flatau AB (2006) Magnetic nanowires for acoustic sensors. J Appl Phys 99:08B310/1-6
Moore M, Phares B, Graybeal B, Rolander D, Washer G (2001) Reliability of visual inspection for highway bridges. Federal Highway Administration, Washington
Mortensen LP, Ryu D, Zhao Y, Loh KJ (2013) Rapid assembly of multifunctional thin film sensors for wind turbine blade monitoring. Key Eng Mater 569–570:512–522
Mukai T, Onishi M, Odashima T, Hirano S, Luo Z (2008) Development of the tactile sensor system of a human-interactive robot "RI-MAN". IEEE Trans Rob 24:505–512
National Transportation Safety Board (NTSB) (2003) Natural gas pipeline rupture and fire near Carlsbad, New Mexico. In: Pipeline accident report, Washington DC
Nofar M, Hoa SV, Pugh MD (2009) Development of novel single-wall carbon nanotube–epoxy composite ply actuators. Compos Sci Technol 69:1599–1606
Oh J-K, Jang G, Oh S, Lee JH, Yi B-J, Moon YS, Lee JS, Choi Y (2009) Bridge inspection robot system with machine vision. Autom Constr 18:929–941
Peckens CA, Lynch JP (2013) Utilizing the cochlea as a bio-inspired compressive sensing technique. Smart Mater Struct 22:105027
Peekema RM (2013) Causes of natural gas pipeline explosive ruptures. ASCE J Pipeline Syst Eng Pract 4:74–80
Pfatteicher SKA, Tongue MP (2002) What drives diversity? Engineering education and underrepresentation of minority groups. In: 32nd Annual frontiers in education, Boston, MA, pp S1C-1-6
Philen MK, Shan Y, Prakash P, Wang KW, Rahn CD, Zydney AL, Bakis CE (2007) Fibrillar network adaptive structure with ion-transport actuation. J Intell Mater Syst Struct 18:323–334
Pinto, PA, Sarles SA, Leo DJ (2011) Bio-inspired flow sensors fabricated from carbon nanomaterials. In: ASME 2011 Conference on smart materials, adaptive structures and intelligent systems, Scottsdale, AZ, pp 725–732

Pyo S, Loh KJ, Hou T-C, Jarva E, Lynch JP (2011) A wireless impedance analzyer for automated tomographic mapping of a nanoengineered sensing skin. Smart Struct Syst 8:137–153

Rytter A (1993) Vibration based inspection of civil engineering structures. Aalborg University, Aalborg Ph.D. Ph.D

Salowitz N, Guo Z, Li Y-H, Kim K, Lanzara G, Chang F-K (2013) Bio-inspired stretchable network-based intelligent composites. J Compos Mater 47:97–105

Sarles SA, Madden JDW, Leo DJ (2011) Hair cell inspired mechanotransduction with a gel-supported, artificial lipid, membrane. Soft Matter 7:4644–4653

Schmidt RF (1986) Fundamentals of sensory physiology. Springer, New York

Spencer Jr BF, Ruiz-Sandoval ME, Kurata N (2004) Smart sensing technology: opportunities and challenges. J Struct Control Health Monit 11:349–368

Sumitro S, Kurokawa S, Shimano K, Wang ML (2005) Monitoring based maintenance utilizing actual stress sensory technology. Smart Mater Struct 14:S68–S78

Swartz RA, Flatau AB, Brooks CN, Barkdoll BD, NA S-M, Endsley KA (2014) Bio-inspired magnetostrictive whisker sensors for autonomous bridge scour sensing. In: Transportation research board 93rd annual meeting, Washington, DC, pp 1–12

Tadigadapa S, Mateti K (2009) Piezoelectric MEMS sensors: state-of-the-art and perspectives. Meas Sci Technol 20:1–30

Tao J, Yu XB, Berrilla J (2011) Bio-inspired flow and acoustic sensor. In: Proceedings of the Sensors, and command, control, communications, SPIE and intelligence (C3I) technologies for homeland security and homeland defense X, Orlando, FL, pp 80190R/1-10

Thostenson ET, Chou T-W (2002) Aligned multi-walled carbon nanotube-reinforced composites: processing and mechanical characterization. J Phys D 35:L77–L80

Thostenson ET, Chou T-W (2006) Carbon nanotube networks: sensing of distributed strain and damage for life prediction and self healing. Adv Mater 18:2837–2841

Thostenson ET, Chou T-W (2008) Real-time in situ sensing of damage evolution in advanced fiber composites using carbon nanotube networks. Nanotechnology 19:215713

Tomizuka M, Bergman LA, Shapiro B, Shoureshi R, Spencer BF, Taya M (2007) Bio-inspired autonomous engineered systems. Smart Struct Syst 3:495–505

Tsuda H, Takahashi J, Urabe K, Ikeguchi T (1999) Damage monitoring of carbon fiber-reinforced plastics with Michelson interferometric fiber-optic sensors. J Mater Sci 34:4163–4172

Twycross J, Aickelin U (2009) An Immune Inspired Approach to Anomaly Detection. arXiv preprint arXiv:0910.3117

Unver O, Uneri A, Aydemir A, Sitti M (2006) Geckobot: a gecko inspired climbing robot using elastomer adhesives. In: Proceedings of the IEEE International conference on robotics and automation, Orlando, FL. pp 2329–2335

Wang C-Y, Wang H-L, Ho C-C (2012) A piezoelectric film type scour monitoring system for bridge Pier. Adv Struct Eng 15:897–905

Whitehouse R (1998) Scour at Marine Structures. Thomas Telford, Heron Quay, London

Wiegerink RJ, Floris A, Jaganatharaja RK, Izadi N, Lammerink TSJ, Krijnen GJM (2007) Biomimetic flow-sensor arrays based on the filiform hairs on the cerci of crickets. In: IEEE sensors conference, pp 1073–1076

Worden K, Burrows AP (2001) Optimal sensor placement for fault detection. Eng Struct 23:885–901

Wu X, Wang D, Zhao A, Li D, Mei T (2013) A wall-climbing robot with biomimetic adhesive pedrail. In: Zhang D (ed) Advanced mechatronics and MEMS devices. Springer, New York

Yao L, Sethares WA, Kammer DC (1993) Sensor placement for on-orbit modal identification via a genetic algorithm. AIAA J 31:1922–1928

Yi T-H, Li H-N, Zhang X-D (2012) Sensor placement on canton tower for health monitoring using asynchronous-climb monkey algorithm. Smart Mater Struct 21:125023

Yim JH, Kim YS, Koh KH, Lee S (2008) Fabrication of transparent single wall carbon nanotube films with low sheet resistance. J Vac Sci Technol B 26:851–855

Yousef H, Boukallei M, Althoefer K (2011) Tactile sensing for dexterous in-hand manipulation in robotics—a review. Sens Actuators A 167:171–187

Yu L, Santoni-Bottai G, Xu B, Liu W, Giurgiutiu V (2008) Piezoelectric wafer active sensors for in situ ultrasonic-guided wave SHM. Fatigue Fract Eng Mater Struct 31:611–628

Yu X, Tao J, Berilla J (2010) A Bio-inspired flow sensor. In: Proceedings of SPIE—nanosensors, biosensors, and info-tech sensors and systems

Zhang G, Zhang Z, Ficher C (2007) Structural health monitoring of a long-span cable-stayed bridge. J Intell Mater Syst Struct 18:835–843

Zhao X, Qian T, Mei G, Kwan C, Zane R, Walsh C, Paing T, Popovic Z (2007) Active health monitoring of an aircraft wing with an embedded piezoelectric sensor/actuator network: II. wireless approaches. Smart Mater Struct 16:1218–1225

Zhu D, Yi X, Wang Y, Lee K-M, Guo J (2010) A mobile sensing system for structural health monitoring: design and validation. Smart Mater Struct 19:055011/1-11

Chapter 12
Bio-inspired, Flexible Structures and Materials

J. Lienhard, S. Schleicher and J. Knippers

Abstract This chapter discusses the potential of biomimetics in formfinding and the development of structural systems based on constant or reversible elastic deformation. The existence of high strength elastic materials are the preconditions for the technical realisation of such elastic structures. Therefore, this chapter will start by introducing elastic building materials and biomimetic abstraction techniques individually before bringing the two together by presenting case studies which successfully combined both aspects.

12.1 Introduction

The inseparable combination of form, material, and function is a criterion, which is typically more appropriate to describe natural than technical system. However, due to the increasing availability of high-performing materials on the one hand and the rising computational power of our simulation and planning tools on the other, it is now possible to explore structures whose form arise from their material behaviour instead of being the result of pre-defined typologies. The research presented in this chapter aims to explore exactly this relationship between form and material behaviour on the basis of biological principles and will discuss possible functions for architectural applications. Therefore, examples of static and kinetic structures whose form and adaptability are based on elastic deformation will be examined closely. From this point onwards, these systems will be referred to as 'bending-active' and 'elastic-kinetic' structures (Knippers et al. 2011a, b).

Bending-active structures are structural systems that include curved beam or shell elements, which base their geometry on the elastic deformation from an

J. Lienhard · S. Schleicher · J. Knippers (✉)
Institute of Building Structures and Structural Design (Itke),
University of Stuttgart, Stuttgart, Germany
e-mail: j.knippers@itke.uni-stuttgart.de

F. Pacheco Torgal et al. (eds.), *Biotechnologies and Biomimetics for Civil Engineering*,
DOI 10.1007/978-3-319-09287-4_12

initially straight or planar configuration. The category of elastic-kinetic structures takes this form-defining approach one step further by considering that the deformation process can also be reversible thus allowing the creation of adaptive systems and compliant mechanisms (Lienhard et al. 2010; Schleicher et al. 2011).

An initial review by Lienhard et al. (2013a, b) on bending-active structures has shown how these systems face two major challenges in the design process: Firstly, since bending-active structures are not based on known typologies, the central question is about the form and functional inspiration. Secondly, there is the difficulty of predicting the system's geometrical shape or structural performance without immediately consulting advanced digital simulations or physical tests.

Recent studies suggest that biomimetic abstraction processes can provide helpful scientific methods that may be used for the design of elastic systems entirely born out of a process based on material behaviour rather than typology based model thinking. (Knippers and Speck 2012; Knippers 2013). This trend is also supported by computational developments, which enable the simulation of nonlinear behaviour in large elastic deformations. In fact, today, the engineering of a structure is increasingly becoming an essential part in modern design processes and is no longer a downstream proof calculation that comes into play when the decisions about shape and form have already been made.

A constructional design approach based on flexibility rather than on stiff members and movable joints however has not yet established itself. In terms of today's building industry there is uncharted ground, in which experience and expertise is lacking and role models are scarce. Interestingly, however, soft and flexible structures are rampant in Nature and indicate the promising potential of this construction principle. It is therefore reasonable to combine research on flexible structures and materials with the topic of biomimetics; with the goal of understanding flexible structures in biology and to transferring, scale-up, and implement their underlying principles into bio-inspired technical devices and structures.

So far, however, most bio-inspired technical structures consider applications either purely theoretically and/or include very simple models on laboratory scale (Marder and Papanicolaou 2006; Kobayashi et al. 2000; Liang and Mahadevan 2009; de Focatiis and Guest 2002; Jenkins 2005). Even the most well known elastic kinetic structures, namely deployable space structures like foldable solar panels and retractable satellite antennas (Miura 1993; Furuya and Satou 2008), which are often discussed in relation to biologically inspired folding and unfolding techniques, are inspired by traditional Japanese folding techniques (e.g. Origami) rather than actual studies of flexible structures in plants and animals.

Only recently, some projects like the patented Fin Ray Effect® (Patent: Leif Kniese 2012) or the Flectofin® (Patent: Knippers et al. 2011a, b) have attracted attention and open up the potential for a successful implementation of bio-inspired flexible structures and materials.

12.2 The Design Potential of Elastic Construction Materials

A fundamental pre-requisite in the designing of bending-active structures is the characteristics of particular building materials which provide a beneficial ratio of high strength and low bending stiffness. Traditional building materials such as timber, for example, offer a suitable elastic range. Fibre Reinforced Polymers (FRP) embody an even greater potential for lightweight bending-active structures due to their low density and high breaking strain. Constant improvements in the development and composition of custom-made, high-performing FRPs suggest that there will be even more materials to choose from in the future.

12.2.1 Material Overview

Among the materials we can consider working with in building and construction, there is a wide range of strength, stiffness, and densities available. Some materials are more suitable for bending-active structures than others. Comparing steel with rubber, for example, one may expect that rubber is the more appropriate choice due to its compliant nature. However, the low tensile strength of rubber makes it unsuitable for structural applications. The much stronger steal, in comparison, offers high tensile strength but is also comparably stiff. As a result neither of these two materials can really be considered as a favorable option for the design of bending-active structures.

In the screening process for the most suitable materials it is therefore important not to focus on one material characteristic only. Instead, the combination of material properties is much more significant. Choosing the right material parameters necessitates an understanding of the mechanical relationships in large elastic bending deformations. From the Euler-Bernoulli law we know that the bending curvature is proportional to the bending moment *M* (1) (Fertis 2006, p. 9). For flat sections or plates the width has no influence on the bending stress which can therefore be expressed proportional to the thickness *t* and curvature *1/r* as shown in (2). This leads to the formulation of the minimal bending radius (3) which we can use to analyse different building materials for their potential use in bending-active structures. The most important variables to set into relation here are the Young's Modulus *E* and permissible bending stress $\sigma_{M,Rk}$. It can thus be concluded that an optimal material for bending-active structures would ideally combine low stiffness and high tensile strength.

$$\frac{1}{r(x_0)} = \frac{M(x_0)}{E \cdot I} \tag{12.1}$$

$$\sigma_M = \frac{E \cdot I}{r \cdot w} = \frac{E \cdot t}{2 \cdot r} \tag{12.2}$$

$$r_{\min} = \frac{E \cdot t}{2 \cdot \sigma_{M,Rk}} \tag{3}$$

Previous work by Ashby has provided very helpful tools in the selection of adequate materials for a given design task (Ashby 2005). The 'Ashby diagrams' open up a property-space by comparing multiple material classes to each other as well as identifying the property range within the individual classes. Providing design guidelines to define specific 'search regions' additionally facilitates the practical use of these diagrams. Figure 12.1 follows Ashby's approach and lists a selection of common building materials. Here, the materials are plotted on a graph with logarithmic scale and are brought into context based on their ratio of flexural strength to stiffness. The range of the axis is chosen to include the material classes investigated. Focusing on the context of building structures, the values in Table 12.1 are taken from the Eurocodes DIN EN 1993–1995 and 1999, DIN 1052:2004–08, and DIN 17221, as well as from Knippers et al. (2011a, b, p. 77) and Gas et al. (1985).

Based on this diagram, adequate materials for static bending-active structures should offer a ratio of $\sigma_{M,Rk}/E > 2.5$ (with $\sigma_{M,Rk}$ [MPa] and E [GPa]). When it comes to elastic-kinetic structures and large-scale compliant mechanisms, the additional requirements for fatigue control further limit the permissible permanent elastic stress; therefore, a ratio of $\sigma_{M,Rk}/E > 10$ is needed. This is indicated by the design guideline with the inclination $\sigma_{M,Rk}/E$. Moreover, the diagram shows clearly that Fibre-Reinforced Polymers (FRP) and certain types of timber and high strength metals are particularly appropriate materials for the use in bending-active structures. When it comes to elastic-kinetic structures and large-scale compliant mechanisms, FRPs mostly fulfil these requirements.

In addition to this short overview, it is critical to consider the material's long-term behaviour. For static bending-active structures, this means paying particular attention to time-dependent deformation (creep). The effects differ in the various materials and are significantly higher in timber than in FRP, for example. The creeping of a material needs to be considered because it can lower the pre-stress in the structure, which dependent on the design of the structure can be more or less relevant and may negatively affect the system's integrity. If the pre-stress is not playing a decisive role for the structure's stiffness, materials such as timber may be chosen. Regarding elastic-kinetic structures, the most important long-term behaviour that needs to be checked is a material's fatigue behaviour. Only if a chosen material has a sufficient fatigue life it can be guaranteed that the compliant mechanism can undergo high and cyclical loading. For the designer, however, it is difficult to assess if the mechanism can perform according to the prescribed functions because the motion of its deflecting members is limited by their strength, which is a property that is not easily readable from the outside. Therefore, precise knowledge of the material properties and the desired motion range is required when designing an elastic-kinetic structure.

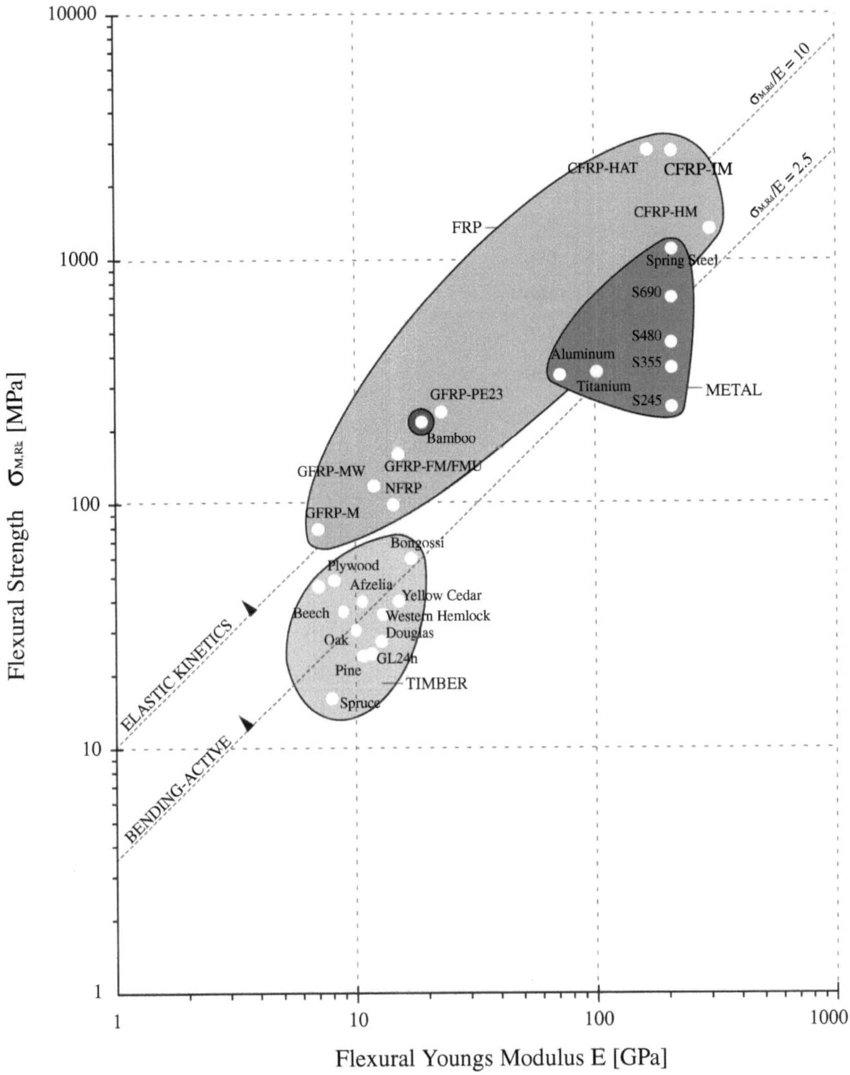

Fig. 12.1 Common building materials with ratio of strength $\sigma_{M,Rk}$ (MPa) to stiffness E (GPa). Adapted from Lienhard et al. (Lienhard et al. 2013a, b)

12.2.2 The Potential of Using Elastic Materials for Compliant Mechanisms

The reason why elastic materials are growing ever more popular, in particular for use in compliant mechanisms, is due to the various benefits that this construction principle possesses over conventional mechanics: By gaining motion through the

Table 12.1 Common building materials with ratio of strength $\sigma_{M,Rk}$ (MPa) to stiffness E (GPa)

	Flexural strength [MPa]	Flexural youngs modulus [GPa]	Ratio
Metals			
S245	245	210	1.17
S355	355	210	1.69
S450	450	210	2.14
S690	690	210	3.29
Spring steel	1100	210	5.24
Titanium	340	102	3.33
Aluminum	330	70	4.71
Timber			
Spruce	16	8	2
Pine	24	11	2.18
Douglas	30	12	2.5
Western hemlock	35	13	2.69
Yellow cedar	40	15	2.67
Oak	30	10	3
Beech	35	10	3.5
Afzelia	40	11	3.64
Bongossi	60	17	3.53
GL24 h	24	11.6	2.07
GL28 h	28	12.6	2.22
GL32 h	32	13.7	2.34
GL36 h	36	14.7	2.45
Birch plywood (6.4 mm)	50.9	12.737	4
Brich plywood (18 mm)	40.2	10.048	4
Combi-plywood (6.4 mm)	50.8	12.69	4
Combi-plywood (18 mm)	35.8	8.95	4
Softwood plywood (6.4 mm)	29.1	9.462	3.08
Softwood plywood (18 mm)	23	7.464	3.08
Bamboo	213	19.129	11.13
FRP			
Type			
CRFP-HAT	2800	165	16.97
CRFP-IM	2800	210	13.33
CRFP-HM	1350	300	4.5

(continued)

Table 12.1 (continued)

	Flexural strength [MPa]	Flexural youngs modulus [GPa]	Ratio
P E 23	300	23	13.04
GRFP-M	80	7	11.43
GRFP-MW	120	12	10
GRFP-FM/FMU	160	15	10.67
NRFP	101	14.38	7.02

flexibility of the construction, rather than linking multiple rigid parts together, one may cause a dramatic reduction in the total number of parts required to accomplish a specific mechanical task. Reducing the amount of parts may decrease assembly time and simplify manufacturing processes, which in turn may result in significant cost reduction.

Another important benefit is that compliant mechanisms often show an increased performance regarding reliability. This is due to the fact that they have few or no hinges, comparable to conventional revolute or sliding joints. Thus, the wear in these structures and the need for lubrication is very low. These are valuable characteristics, especially for applications where mechanisms are either difficult to access or operating in harsh environments, in which conventional joints would rust and require significant maintenance.

Furthermore, one of the main advantages of compliant mechanisms is their capacity to store energy in their deflected flexible members. If a compliant mechanism is based on material with little creeping behaviour, it can store strain energy over a longer time period and release it at wish at a later stage, or use it to reset the mechanism to its original state again. But beyond that, it is also possible to fine-tune the force-deflection relationships in the flexible structure to correlate energy and motion and vice versa. The transmission ratio therein can be tailored to suit specific functional needs with targeted amplification effects. For a given energy input the mechanical response of compliant mechanisms can be customised to achieve, for instance, either a maximum displacement (displacement multiplication), or a maximum resulting force (force multiplication).

12.3 Biomimetic Approach

In the last decade, using nature as an inspirational source to solve technical problems has become scientifically recognised. Through the evolutionary pressure of natural selection on organisms to survive under particular boundary conditions, highly-adapted systems have developed. In order to use such systems as role models for technical applications a 'top-down approach' can be used to solve technical problems by formulating a concise technical question which evolution may have already developed an answer for. Elastic-kinetic structures may be derived, for example, from flexible deployable systems found in nature. These can be observed especially in movements of plants or plant organs. Of particular interest are non-autonomous plant movements which usually show a clear interrelation of form, actuation, and kinetics. In the following section two strategies are presented which may be used to extract principles from nature for bending-active structures and compliant mechanisms.

12.3.1 Direct Methods in the Context of Compliant Mechanisms

In general, the work with biological role models can be initiated by two different biomimetic process sequences, which, in the further development of an iterative design process, are often mixed. The basic processes are:

> Bottom-up = biology push: In this approach new biomimetic research projects for technical implementation are born from new and promising results of fundamental biological research. The first step of the process (…) is to analyse the biomechanics and functional morphology of a biological system. In the next step quantitative analysis leads to a principal and detailed understanding of the biological structures, shapes and functions. On the abstraction level, which follows, separation of the principles discovered in the biological model takes place. Abstraction often proves to be one of the most important as well as most difficult steps in a biomimetic projects. (Speck and Speck 2008, p. 6)

> Top-down = technology pull: A biomimetic project following the top-down process typically starts with the work of an engineer. In this approach biomimetic innovations and improvements are sought for in already existing technical products. These products might either be in a final state of industrial development, or are often already successfully established on the market. For a successful top-down process well founded expertise is required from company representatives (engineers) as well as from fundamental researchers (biologists), and also readiness to talk with the parties on both sides. The improvement or further development of an existing product stands in the centre of the cooperation during a top-down process. (Speck and Speck 2008, p. 7)

So, the top down approach begins by defining the question. For instance, the overriding question in the development of compliant mechanisms is the optimisation of adaptability and energy efficiency of kinetic systems as well as the lowering of their weight and maintenance costs.

When aiming for biomimetic solutions, a proven first step can be a screening of natural concept generators with a high potential for translation into technical applications. Usually, the most evolved and robust answers can be found in organisms which developed under high selective pressure (Reith et al. 2007). Plant movements, for example, are particularly adequate for translation into kinetic architectural structures. Unlike animal locomotion, which is usually laid out for a variety of complex movements, the actuation systems of plants are evolutionary optimised to perform a single type of movement (Sitte et al. 1991). Furthermore, the structures of the mostly sessile plants are often adapted for similar boundary conditions such as those which affect the design of the architectural structures.

Once the objective is set the screening process can be carried out. In the particular cases analysed in this chapter the biological role models which best meet the requirements and demonstrate possible solutions for potential technical implementation are gathered.

In order to narrow down the search parameters some screening criteria should be further defined. For example, in the search for compliant mechanisms the focus was placed on reversible elastic or visco-elastic deformations in plants. Plant movements are based on many different motion principles, some of which only occur in highly specialised plant groups like trapping mechanisms in carnivorous plants or pollination mechanisms in the flowers of a specific plant family.

These optimised biological mechanical systems are promising concept generators for the design of elastic structures in architecture since they often combine sensors and actuators within one mechanical system.

For the further analysis of the selected specimens it is helpful to build up a phenomenological understanding of the underlying physical principles that are involved in the observed mechanisms. Therefore, it can be a good approach to apply even more stringent selection criteria in order to find examples with the greatest potential regarding cost efficiency, energy and material solutions. Again, in the context of compliant mechanisms, a precisely defined selection criterion can be the prioritisation of research for systems with large bending radii, small actuation and beneficial energy transmission, or an alternative freedom of movement that can usually not be found in conventional rigid body mechanics.

The final step in this process is the abstraction of the role model into a bio-inspired mechanism. This is probably the most difficult step in the transfer process since it requires both scientific rigor as well as a high level of creativity. In order to unveil the basic mechanical principles involved, one must study the functional-morphological relationships in the biological role model very systematically. The process of reduction and dissection are very helpful examples. Here, the functionality of a mechanism in a specimen is tested by progressively cutting off all the elements that seem to be unrelated to the mechanism. By following this approach, one can narrow down the constituent parts that play a key role in the mechanism. This is of particular importance because the knowledge about the basic building blocks that are needed for a mechanism opens the door for their creative use. Then, one can start to modify these building blocks, reconfigure them, or fine-tune and optimise them in order to address various tasks.

This insight significantly broadens the design freedom and allows for concepts beyond the direct mimicry of the natural system. A reinterpreting of the mechanical principle and the development of novel kinetic systems can be followed through, with the help of physical and digital models, for example. Upon further analysis of the abstracted mechanical principle, the working process may be opened up for reconfigurations and adaptations to meet the boundary conditions given by a technical implementation.

The technical implementation in the form of a reduced scale or full scale prototype lies not at the end, but rather amidst an iterative design process in which feedback from real construction can reach as far back as fostering the understanding of the biological role model.

12.3.2 Indirect Biomimetic Approach

While the approach described above concentrates on singular phenomena and their direct technical implementation, biomimetics can also support the design process when looking at natural design and construction principles from a more general perspective. Knippers and Speck (2012) highlight a number of design principles that lead to structures with multiple network functions based on a multi-layered, finely tuned and differentiated combination of basic components. Additionally, Gruber (2011) selects some design strategies found in natural organisms that may be of relevance for architectural constructions. From these sources, a list of natural principles can be summarised that may be particularly applicable for the integrated design approach pursued with bending-active structures and compliant mechanisms. (Based on Knippers and Speck (2012) and Gruber (2011):

- *Heterogeneity*: Natural constructions are characterised by geometric differentiation.
- *Anisotropy*: Many natural constructions consist of fibre reinforced composite materials.
- *Hierarchy*: Biological structures are characterised by a multilevel hierarchical structure.
- *Multifunctionality*: In botany, fibres simultaneously serve mechanical and diverse physiological functions.
- *Adaptability*: Many short- and long term adaptations are known from natural systems that adapt to their environment.

It should be noted at this point, however, that while all these principles already provide a helpful guidance and inspiration for the design of bending-active structures and compliant mechanisms, this list makes no claim of completeness. A detailed explanaition of these principles is given in VDI 6226 (2014). In fact, it could be easily expanded by many other key aspects (e.g. redundancy, self-healing), in order to address specific levels of perspective or to reach paramount objectives such as closed material loops, effective energy conversion or sparing use of resources.

Nevertheless, this first overview of natural principles already provides a useful starting point. With focus on bending-active structures in the context of lightweight structures, for example, it can be shown how these natural principles may also be linked to the threefold engineering classification of lightweight structures (Wiedenmann 1989):

- *Light Materials*: Materials of low density combined with high mechanical strength
- *Light Structures*: Following the flow of forces on minimal paths
- *Light Systems*: Integration of multiple functions into a single element

In Table 12.2 the engineering classifications of lightweight structure and natural design principals are compared, to identify natural construction principles which may be consulted to find new engineering solutions for lightweight structures. Such a comparison may be used as a starting point in architectural design and engineering when aiming to include biomimetic principles in the working process.

12.4 Case Study Projects

The following two case study projects exemplify a typical direct top-down approach (Flectofin®) and an indirect approach (M1).

12.4.1 Flectofin®

The first case study followed a direct biomimetic approach. In this project, the Bird-Of-Paradise flower, also known as *Strelitzia reginae* (Fig. 12.2), was chosen as a role model because it features a sophisticated externally actuated pollination mechanism. The motion is part of an elastic and reversible valvular pollination mechanism, which is driven by the direct application of an external force. This mechanism was intensively studied in order to reveal the underlying principles that are responsible for the plant's mechanical performance. The plant's distinct form-structure-function relationship was then further abstracted and inspired the development of an elastic flapping mechanism called Flectofin®. Exemplarily, it is used for the conceptualisation of an adaptive façade shading system (Fig. 12.8), in which the fins allow for opening angles ranging from −90° to +90°.

12.4.1.1 Biomimetic Approach

As an adjustment to its relatively large and heavy pollinators (mainly birds), *S. reginae* has developed a protruding perch of two adnate petals that act as a landing platform. When a bird lands on this structure to reach for nectar, its weight causes the perch to bend downwards, as shown in Fig. 12.2. This initial deformation triggers a secondary sideways flapping of two thick petal wings. As a result, the previously enclosed stamens are exposed and pollen is transferred to the bird's feet. When the bird flies away, the open perch resets to a protective closed state

Table 12.2 Comparison of engineering and natural design principles

Engineering principles	Natural principles
Light materials	Anisotropy
Light structures	Heterogeneity
	Hierarchy
Light systems	Multifunctionality
	Adaptability

again and the bird may transport the pollen to another Strelitzia flower where pollination takes place.

The flapping mechanism of the *S. reginae* is particularly suitable for a technological transfer because it can be triggered at any time by applying an external mechanical force at a specific location of the structure. When loaded this way it shows what is defined in engineering terms as a distinct tipping failure. This phenomenon permits a series of experiments with which the mechanism can be investigated precisely in order to learn more about its functions.

In this case the abstraction can take place via a reduction principle. While doing so, parts of the plant that may not have anything to do with the basic kinematics are gradually cut away. Hereby it becomes clear, that only individual halves of the leaves with directly connected reinforcing strip are responsible for the flapping mechanism (Fig. 12.3). Such a system can easily be imitated and simulated with physical models (Fig. 12.4).

This biological principle unveils an elastic flapping mechanism which is transferrable to technical fields of application. In contrast to common kinetic structures which work on the basis of locally arranged hinged, jointed and angular connections, the biomimetic approach works on the basis of pliability of large surfaces. This enables a wide range of adaptation and the wear of material at the joints and hinges can be avoided. This principle represents a paradigmatic shift in civil and structural engineering wherever it was previously recommended to avoid any form of structural failure. The actual mechanism behind this movement is known as lateral torsional buckling in engineering, a so-called 'failure mode' that engineers generally try to avoid by sizing structural members to adequate stiffness. The physical model in Fig. 12.4 shows the effect of lateral torsional buckling in a constellation as can be found in the plant's mechanism. In the *Strelitzia*, however, this failure mode has no negative connotation but simply utilises as fundamental motion principle for a highly effective compliant mechanism. In fact, *S. reginae* cleverly exploits the potential of an unsymmetrical bending motion as an integrative part within a reversible deformable structure with multiple deflected equilibrium positions.

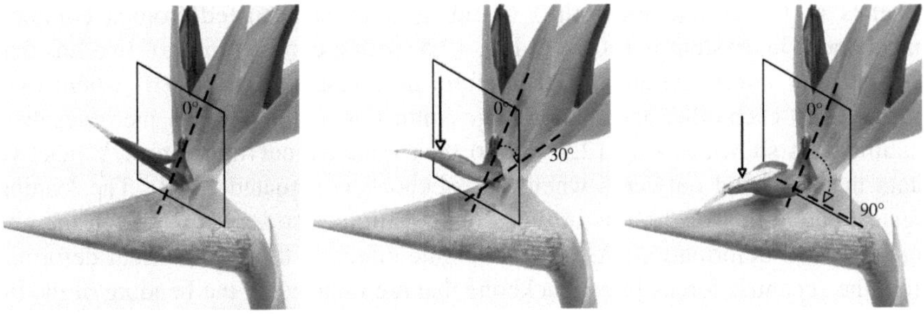

Fig. 12.2 The elastic deformation in the *Strelitzia reginae* flower. Adapted from Lienhard et al. (2011)

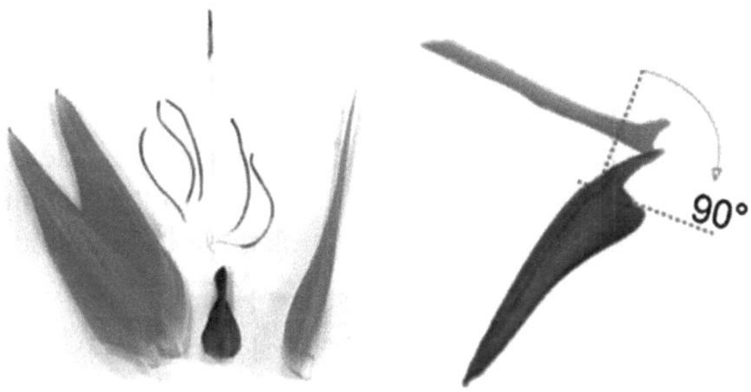

Fig. 12.3 Abstraction of the Flectofin® mechanism via a reduction principle. Adapted from Lienhard et al. (2009)

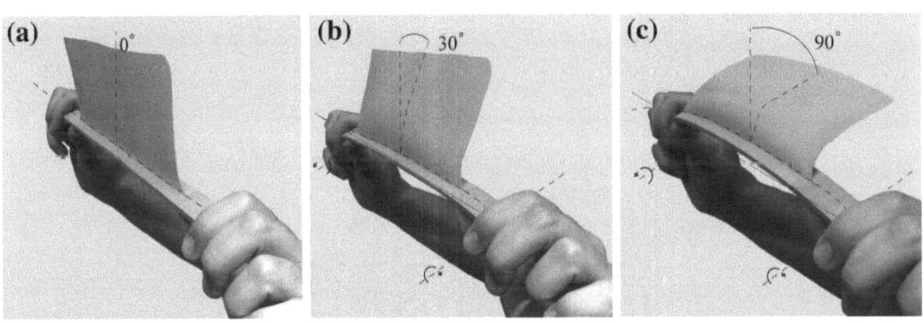

Fig. 12.4 Abstraction of the deformation principle in the flower of *S. reginae*, realised with a simple physical model. Adapted from Lienhard et al. (2011)

12.4.1.2 Optimisation of the Basic Flectofin® Principle

Further optimisation of the basic Flectofin® principle was aimed at folding two fins outward from one common element and thereby doubling the shaded area per element while hardly affecting the view. In contrast to the *S. reginae* the elastic kinetics of the double Flectofin® would need to be actuated from a common backbone. This system is shown in Fig. 12.5, with a configuration of two fins that theoretically interpenetrate (A). Therefore, they rest in position (B) where they push against each other and share a large contact area which highly increases their stability. As shown in Fig. 12.5, due to their concave curvature in their inactive state the fins bend outwards when the backbone is actuated (C–F). The leaning against each other of the fins in their 'rest position' also serves to stiffen the fins against wind deformation. As a positive side effect of the symmetrical deformation, the eccentric forces in the backbone that are induced by the bending of the fin counteract each other. This limits the torsion in the backbone and results in a more filigree profile.

It was found that the elastic kinetic system relies on perfect symmetry which is difficult to produce on a larger scale. Therefore additional geometric changes were studied to make the elastic kinetics more robust. Here, a study of the *Aldrovanda* snap-trapping mechanism (Fig. 12.6) led to the idea of using curved line folding logics to force the fins into a consistently outward folding motion. The *Aldrovanda* traps consist of two sickle-shaped lobes that are connected to a lens-shaped central portion by a curved living hinge. In the symmetry axis of the trap and in the middle of the central portion there is another distinct element to be found—a midrib that acts as the "driver" for the closing movement. When a prey (e.g., small crustaceans like water fleas) stimulates the trap by touching its sensory hairs, a sudden

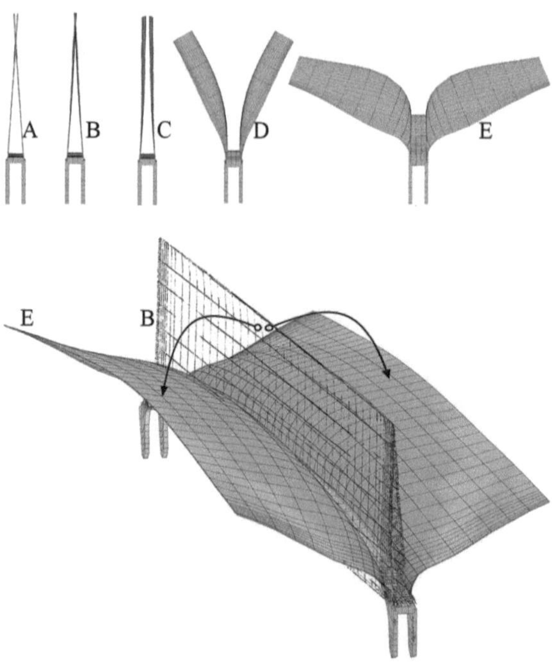

Fig. 12.5 Simulation of a double Flectofin®, **B** position of the planar fins, **B** real position of the fins pushing against each other, **E** opened fins due to bending of the backbone (Lienhard et al. 2011)

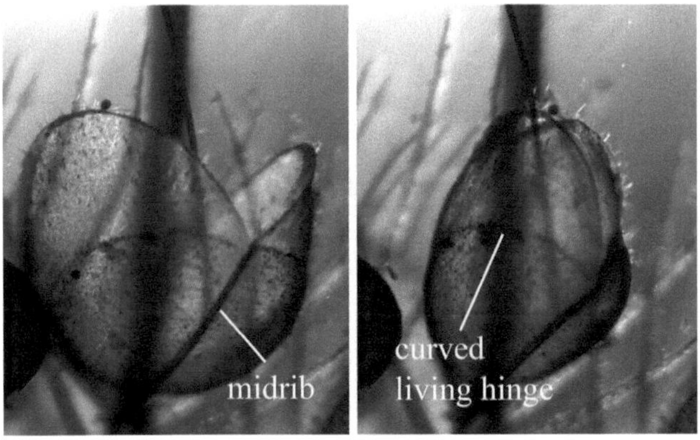

Fig. 12.6 The trap leaves of *Aldrovanda vesiculosa* in open and closed configurations

bending of the midrib leads to an instant closure of the trap lobes. From a mechanical perspective, this rapid motion results from a controlled sequence of multiple interconnected deformations. At first, the midrib starts to bend, driven by hydraulic actuation. This causes bending of the central portion, which transmits the deformation further and triggers an out-of-plane bending of the adjacent softer lobes. Similar to the Flectofin®, the mechanism is actuated by an externally induced support displacement. This translation movement causes bending in the centre surface and triggers the flapping motion of the lobes (Fig. 12.7, left)(Schleicher et al. (2010). Typically, for curved-line folding, this principle couples convex and concave surface bending, which increases in curvature as the folding proceeds. Due to the orientation of the curved-line fold, however, the lobes can rotate in the opposite direction of the Flectofin® and thus can be understood as an inverse mechanism to it. Turning the curvature around and applying it to the line between the Flectofin® fin and its backbone will consequentially force it into an outward bending motion (Fig. 12.7, right).

This Double Flectofin® is a further development of the initially proposed façade component, which has an increased shading efficiency and higher wind stability, shown in Fig. 12.8.

12.4.1.3 Architectural Application

A significant expansion for possible future applications is provided by the fact that the system functions without a straight turning axis. The geometrical flexibility of these bio-inspired compliant mechanisms renders the unique possibility to enter, for instance, the neglected market niche of shading double curved façades. Compared to traditional sun protections (e.g. blinds and louvers) whose mechanics are designed for planar and standardised facades, elastic-kinetic structures like the Flectofin® preserve their functionality even when distorted or scaled. In particular

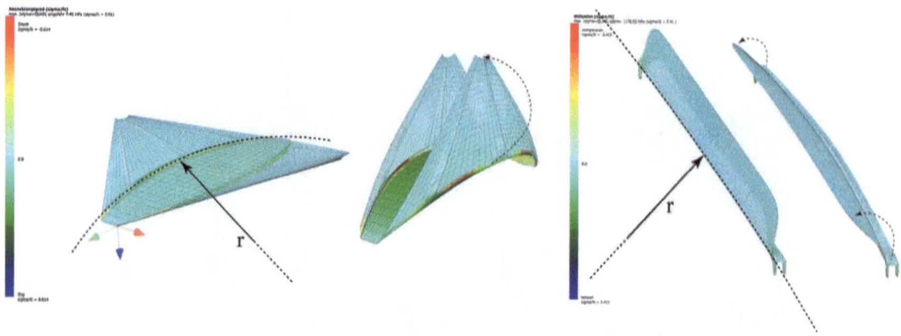

Fig. 12.7 *Left* Kinetic model of Aldrovanda's snap-trapping mechanism in FEM. Adapted from Schleicher et al. (2014). *Right curve* attachment line of the a single Flectofi® fin to its backbone

Fig. 12.8 Facade Mock-up with three Flectofin® fins with various degrees of opening

in the context of cladding double curved facades this becomes a very important and game-changing characteristic. For although the meshes of double curved façades may seem to have a smooth global appearance, they consist in fact of thousands of individual panels, which render the use of standardised shading devices practically impossible. Thus, applying a transformable shading system to this kind of facade would require the individual scaling and distortion of its components—a feature that elastic-kinetic structures can offer. Furthermore, the Flectofin® principle suits a wider range of applications, from small-scale micro-systems to large-scale architectural building components.

This concept was put to the test through the Flectofin® being the triggering inspiration for the façade of the Thematic Pavilion at the EXPO 2012 trade-fair in Yeosu, Korea, by soma-architecture and Knippers Helbig Advanced engineering. The engineering company was then commissioned with the planning and constructional design of this kinetic façade. The mechanism that should move the 108 lamellas was still unclear at the start; the central question being how one could also guarantee their functional suitability according to the expected case of a typhoon. In a first investigation, it was checked whether the Flectofin® principle could be magnified to the large scale of 108 lamellas with heights varying between 3 m and 14 m. This proved that up-scaling of the basic principle was possible, yet could not entirely fulfil the architectural intentions of the façade. Inspired by the Flectofin®, an alternative elastic kinetic mechanism was developed which is based similarly on structural failure (buckling), but does not correspond to the direct abstraction of the *S. reginae*. These further developments show the potential of such basic discoveries, in this case, the instrumentalisation of failure and deformation. In the context of architectural design, it also became clear that targeted adaptations are needed in order to be able to handle the profile of requirements in the intended field of use.

12.4.2 Textile Hybrid: M1

The Textile Hybrid M1 at La Tour de l'Architecte, France showcases the research on hybrid form- and bending-active structure systems by the Institute of Computational Design (ICD) and Institute of Building Structures and Structural Design (ITKE) with students of the University of Stuttgart and ABK Stuttgart. The scientific goal of the project was the exploration of formal and functional possibilities in highly integrated equilibrium systems of bending-active elements and multi-dimensional form-active membranes. The resulting multi-layered membrane surfaces allow not only structural integration, but also serve as a functional integration by differentiating the geometry and orientation of the membrane surfaces. The site selected for the design is a historical and structurally sensitive tower in Monthoiron, France. The tower is based on a design by Leonardo Da Vinci from the 16th century, which brought the owners to the idea of making the tower usable for exhibitions. On the basis of a spatial program, a Textile Hybrid system was developed where short-cutting of forces produced a minimisation of the loading on the tower. In the context of this project, the M1 was developed as a representative temporary pavilion.

The Textile Hybrid system was constructed with GFRP rods of diameters ranging from 3 to 24 mm in combination with textile membranes as continuous surfaces and open-weave meshes. The elastic rods gain their stiffness from active bending into curved leaf-shaped modules which are networked into a global structural system. Stress stiffening effects are activated by further deformation of the system through the integration of a pre-stressed membrane surface, and therefore create a fully Textile Hybrid system (Lienhard et al. 2013b).

12.4.2.1 Form-Finding and Construction

The system and geometry was developed by students from the University of Stuttgart through a long process of physical model building, which was later optimised through digital simulation. Aim of this process was to break free from traded structural typologies in the search for novel formal and functional possibilities in highly integrated structural systems. This process was accompanied by the consideration of natural construction principles as introduced above.

For generative studies, the spring-based modelling environment SpringFrom (Ahlquist et al. 2014) was utilised alongside exhaustive physical form-finding experiments. The computational modelling allowed for complex topologies to be developed and altered, quickly registering feedback from the prototypical physical studies. This approach was utilised for the form-finding of the secondary textile hybrid system, in particular; a series of differentiated cells providing additional structure to the primary envelope and variation to the illumination qualities of the space (Fig. 12.10). As both a design avenue and method for material specification, FEM was utilised. Here the parameters of the complex equilibrium system were explored to determine the exact geometry and evaluate the structural viability. Custom programmed methods inside the general purpose FE-Software Sofistik® allowed for great degrees of displacement to be calculated in order to form-find the beam positions. The beams were initialised as straight elements and gradually deformed into interconnected curved geometries, finally being reshaped by the inclusion of pre-stressed membrane surfaces (Fig. 12.9, right). The geometric data therein was determined initially by the physical form-finding models in defining the lengths and association points on the rods for the topology of FE beam elements (Fig. 12.9, left). Given the unrolled geometry and connection points of the rods it was possible to simulate the erection process and thereby the residual stress in a finite element based form-finding process (Fig. 12.11).

12.4.2.2 Biomimetic Approach

For the M1 project the aspect of Biomimetics is not used to look for a biological role model to find a technical solution, but define a set of strategies that influenced the entire design process. The key construction principles that influenced the design process of the M1 were:

Anisotropy: Textile materials where used on all scales; their anisotropy was used as a driving force in the design and form-finding process of the material system. For the bending rods made of pultruded GFRPs this meant that maximum material strength could be accessed, as fibres ran in the main stress direction. On a detailed level the connection points needed reinforcement by laminating fibres in the circumferential direction of the rods.

Heterogeneity: An important feature in the design was the structural integration and heterogeneity, leaving the limits of strictly categorised building structures by accumulating different load bearing strategies in an associative system. The M1

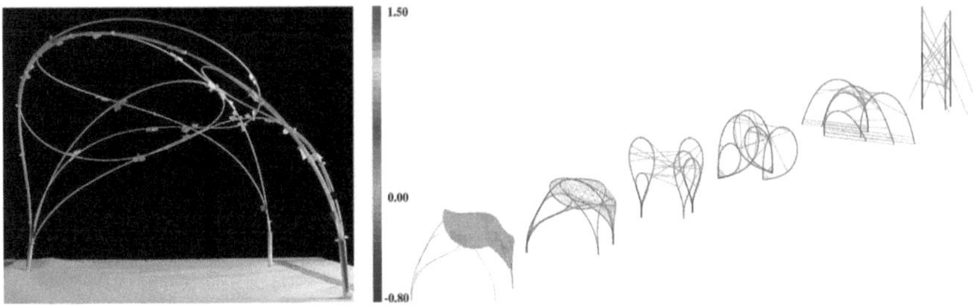

Fig. 12.9 Physical and numerical form-finding of the M1. Adapted from Lienhard et al. (2013b)

Fig. 12.10 Inner cell structure of the M1

Fig. 12.11 The textile hybrid structure M1

shows this in the way its structure freely yet continuously passes through several types of load bearing mechanisms. Thereby the system is able to adapt to changing environmental requirements without the need of changing its materiality.

Hierarchy: This system is featured on two hierarchical levels: a macro-system of interwoven bending rods that form leaf-like shapes and a meso-scale differentiated cell logic. At both scales the system logic is based upon a Textile Hybrid system in which pre-stressed membranes stabilised elastically bent rods. Repeating the macro system on a meso-scale offered the prospect of functional integration next to additional structural rigidity.

Integration: Even though, factually, the cells of the M1 do not yet serve additional functions on a physical building level, they could still be seen as placeholders for integrating such functions into small scale yet structurally relevant parts of the system. Their spatial separation of the two main membrane layers particularly offers the integration of thermal insulation and sound damping. On another level, also present in the actual M1 project, they can serve light diffusion.

This Biomimetic approach in the M1 project may show how, even though abstract in their implementation, design and construction principles found in nature can complement architectural and engineering design processes. Further examples of implementation using this approach are given in VDI Guideline 6226 (2014).

12.5 Conclusions

This chapter on bio-inspired, flexible structures introduced a general overview of elastic building materials and biomimetic abstraction techniques and showed their promising application in two case study structures.

With the Flectofin® project it was shown how direct abstraction methods may be used to extract mechanical principles from natural systems to develop novel technical engineering solutions. This example shows how nature and engineering differ in problem solving and highlights that the structures and principles identified in biological concept generators can provide innovative impulses for the development of elastic-kinetic structures beyond traditional preconceptions. The analysis of the anatomy, functional morphology and biomechanics of plants may thereby be seen as a promising approach for concept generation and optimisation in the field of architecture, building and construction.

Through the M1 project a more indirect biomimetic approach was shown. Here, the approach of studying general phenomena found in natural construction principles was successfully used to develop new structural solutions.

In general both projects show how working with biomimetics in an interdisciplinary team can help finding novel engineering solutions which are more than an optimisation of existing typologies, potentially even represent a paradigmatic change.

References

Ahlquist A, Kampowski T, Oliyan Torghabehi O, Menges A, Speck T (2014) Development of a digital framework for the computation of complex material and morphological behavior of biological and technological systems. Computer-Aided Design, special issue: material ecologies, available online. http://dx.doi.org/10.1016/j.cad.2014.01.013

Ashby MF (2005) Materials selection in mechanical design. MRS Bulletin

De Focatiis DSA, Guest SD (2002) Deployable membranes designed from folding tree leaves. Philos Trans R Soc Lond, Ser A: Math, Phy Eng Sci 360(1791):227–238

Fertis D (2006) Nonlinear Structural Engineering. With Unique Theories and Methods to Solve Effectively Complex Nonlinear Problems

Furuya H, Satou Y (2008) Deployment and retraction mechanisms for spinning solar sail membrane. In: AIAA structures, structural dynamics, and materials conference, AIAA-2008-2051, pp 1–10

Gas S, Druesedau H, Hennike J (1985) IL 31 Bambus—Bamboo. Karl Krämer Verlag, Stuttgart

Gruber P (2011) Biomimetics in architecture [architekturbionik] architecture of life and buildings. Springer, Berlin

Jenkins CH (2005) Compliant structures in nature and engineering. Wit Pr/Computational Mechanics. Southampton, Boston

Kniese L (2012) Vorrichtung zur Aufnahme von Kräften, mit einer flexiblen Aussenhaut (Device for taking up forces, with a flexible outer skin). European Patent Office, 02.05.2012, EP 000001316651 A2

Knippers J (2013) From model thinking to process design. AD Architectural Design 02(2013): 74–81

Knippers J, Speck T (2012) Design and construction principles in nature and architecture. Bioinspir Biomim 7:1–10

Knippers J, Cremers J, Gabler M, Lienhard J (2011a) Construction manual for polymers + membranes. Institut für internationale Architektur-Dokumentation, München

Knippers J, Lienhard J, Masselter T, Poppinga S, Schleicher S, Speck T (2011b) Gelenkloser, stufenlos verformbarer Klappmechanismus (Hingeless, infinitely deformable folding mechanism). European Patent Office, 11.05.2011, EP 2320015 A2

Kobayashi H, Daimaruya M, Vincent JFV (2000) Folding/unfolding manner of tree leaves as deployable structures. In: Pellegrino S, Guest SD (eds) IUTAM-IASS symposium on deployable structure : theory and applications. Kluwer Academic Pub., London, pp 211–220

Liang H, Mahadevan L (2009) The shape of a long leaf. Proc Natl Acad Sci 106(52): 22049–22054

Lienhard J, Poppinga S, Schleicher S, Masselter T, Speck T, Knippers J (2009) Abstraction of plant movements for deployable structures in architecture. In: Thibaut B (ed) Proceedings of the 6th plant biomechanics international conference, Ecofog, Cayenne, French Guyana, pp 389–397

Lienhard J, Poppinga S, Schleicher S, Speck T, Knippers J (2010) Elastic architecture: nature inspired pliable structures. In: Brebbia CA (ed) Design & Nature V: Comparing Design in Nature With Science and Engineering. WIT Press, Southampton, Boston, pp 469–477

Lienhard J, Schleicher S, Poppinga S, Masselter T, Milwich M, Speck T, Knippers J (2011) Flectofin: a hinge-less flapping mechanism inspired by nature. Bioinspir Biomim 6:045001

Lienhard J, Alpermann H, Gengnagel C, Knippers J (2013a) Active Bending, a Review on structures where bending is used as a self-formation process. Int J Space Struct 28(3&4): 187–196

Lienhard J, Ahlquist S, Menges A, Knippers J (2013b) Finite element modelling in integral design strategies of form- and bending-active hybrid structures. In: Bletzinger K-U, Kröplin B, Onate E (eds) Structural Membranes—Proceedings of the VI international conference on textile composites and inflatable structures

Marder M, Papanicolaou N (2006) Geometry and elasticity of strips and flowers. J Stat Phys 125(5–6):1065–1092
Miura K (1993) Concepts of deployable space structures. Int J Space Struct 8: 3–16
Reith M, Baumann G, Claßen-Bockhoff R, Speck T (2007) New insights in the functional morphology of the lever mechanism of Salvia pratensis. Ann Bot 100:393–400
Schleicher S, Lienhard J, Poppinga S, Speck T, Knippers J (2010) Abstraction of bio-inspired curved-line folding patterns for elastic foils and membranes in architecture. In: Brebbia CA (ed) Design & Nature V: Comparing Design in Nature With Science and Engineering. WIT Press, Southampton, Boston, pp 479–489
Schleicher S, Lienhard J, Knippers J, Poppinga S, Masselter T, Speck T (2011) Bio-inspired kinematics of adaptive shading systems for free form facades. In: Nethercot D et al. (eds) In: Proceedings of the 35th annual symposium of iabse / 52nd annual symposium of iass / 6th international conference on space structures. Taller Longer Lighter—Meeting growing demand with limited resources, London,UK
Schleicher S, Lienhard J, Poppinga S, Speck T, Knippers J (2014) A methodology for transferring principles of plant movements to elastic systems in architecture, Computer-Aided Design, special issue: material ecologies, available online 29 January 2014, ISSN 0010-4485. http://dx.doi.org/10.1016/j.cad.2014.01.005
Sitte P, Ziegler H, Ehrendorfer F, Bresinsky A (1991) Strasburger. Lehrbuch der Botanik. Gustav Fischer Verlag (33. Auflage), p 456–469
Speck T, Speck O (2008) Process sequences in biomimetic research. In: Brebbia CA (ed) Design and Nature IV. WIT Press, Southampton, pp 3–11
Sitte P, Ziegler H, Ehrendorfer F, Bresinsky A (1991) Strasburger. Lehrbuch der Botanik. Gustav Fischer Verlag (33. Auflage), p 456–469
VDI 6226 (2014) Biomimetics—Architecture, civil engineering, industrial design. VDI-Gesellschaft Technologies of Life Sciences, Düsseldorf 2014
Wiedenmann J (I989) Leichtbau. Bd. 1 +2. Springer, Berlin, Heidelberg

Chapter 13
Bioinspired Concrete

Brent R. Constantz, Mark A. Bewernitz, Christopher L. Camiré, Seung-Hee Kang, Jacob Schneider and Richard R. Wade II

Abstract New bioinspired cement can be used to form artificial limestone aggregate for concrete, as well as cement for the binding phase of concrete. Amorphous calcium carbonate precursors were first used in combination with unstable calcium phosphates to form high strength, rapid setting carbonated calcium phosphate cements, similar to the mineral phase of bone. Later, these amorphous calcium carbonate precursors and other unstable polymorphs of calcium carbonate were used in combination to form calcium carbonate cements with high strength and other advantageous properties. In the last decade, mechanisms to use the carbon dioxide from the combustion of fossil fuels were developed, allowing very large quantities of calcium carbonate cementing precursor materials to be formed, making it a foreseeable reality that new concrete mixes comprising calcium carbonate, both as the aggregate component and the cementing phase of concrete can be establish broadly on a worldwide basis. Calcium carbonate concrete compositions enable a sustainable pathway for concrete as a construction material.

13.1 Introduction

Earth's most massive biological structures are coral reefs that may extend for hundreds of miles and are composed mainly of calcium carbonate minerals. The biological formation of minerals, biomineralization, plays a dominant role as a reservoir of carbon on Earth by converting carbon dioxide to carbonate minerals. In fact, most of Earth's accessable carbon is in the form of carbonate rocks in Earth's lithospheric crust as limestone, comprised of the fossil remains of ancient

B.R. Constantz (✉) · M.A. Bewernitz · C.L. Camiré · S.-H. Kang · J. Schneider · R.R. Wade II
Blue Planet Ltd, 100 Cooper Ct, Los Gatos, CA 95037, USA
e-mail: brentc@stanford.edu

F. Pacheco Torgal et al. (eds.), *Biotechnologies and Biomimetics for Civil Engineering*,
DOI 10.1007/978-3-319-09287-4_13

calcium carbonate forming organisms. Man-made structures of similar extent include highways, dams, bridges, and building that are usually constructed from concrete.

Concrete is the most used construction material on Earth. It is the most traded material other than water. Last year over 15 gigatons of concrete was placed worldwide. An agglomeration of rock aggregates, concrete is held together by Portland cement, binding the rock particles to form a solid. In today's construction concretes, the aggregate component comprises most of the concrete mix and the Portland cement binder comprises a relatively small component of the concrete mix.

In other significant hydraulic cement applications outside construction, such as orthopedic surgery, where new bioinspired cements have found application, cements are used primarily alone without aggregate. These cements see about 5 million surgical procedures every year, and address significant human pain and suffering. Some of these cements are also significantly stronger than Portland cement or concretes formed with Portland cement. These new bioinspired cements show near-perfect dimensional stability, low heat evolution, and high chemical stability during their operational life.

"Bioinspired" means that the composition and crystallographic nature of the cement or concrete is seen in biological materials, usually mineralized skeletons. This relationship is often termed 'biomimetic'. Biomimetic contrasts with 'bio-duplication' which refers to the cement's method of synthesis where it mimics a formation process from biological systems, such as a skeletal formation process.

Modern concrete mix designs are in part constrained by the cement-binding phase, which today is Portland cement. Due to its dimensional instability, large proportions of aggregate are needed to prevent substantial dimensional changes during the curing of the concrete as well. Aggregates are also applied because of the cost of the cement. Further, the heat evolution from the setting and curing reactions of Portland cement also requires the use of large fraction of aggregates in order to prevent excessive heat accumulation. More recently, the carbon footprint of Portland cement, nearly the entire carbon footprint of concrete, has given further incentive to reduce the use of Portland cement in concrete.

Due to the importance of concrete in society and the global community, there is a large and pressing need to address the continued use of Portland cement. Last year over 3.6 gigatons of Portland cement were manufactured, releasing over 3 gigatons of carbon dioxide into Earth's atmosphere—the third largest source of anthropogenic greenhouse gas emissions. The fastest way to mitigate the impacts of Portland cement production of climate change is to capture and store the carbon dioxide emitted from Portland cement plants, making the cement a carbon neutral product. This does not solve the dimensional stability or heat evolution problems associated with Portland cement, or secondary problems such as alkali-aggregate reactions, etc. It also does not address the cost problem associated with Portland cement, as most alternatives appear not only environmentally favorable and functionally more sustainable, but also less expensive and more economically sustainable.

Therefore, it is evident that a replacement for Portland cement as the binding component of concrete would benefit mankind in a number of different ways. Large concrete structures pre-date the invention of Portland cement by millennia, including Egyptian and Roman concretes that are known for their durability in large structures still standing today. Taking a biologically inspired or biomimetic approach to understanding how nature builds structure of mineral with cement and concrete-like properties may provide insight about what other possibilities may be used in concrete.

13.2 Overview

Earth's solar system is composed principally of a few anionic complexes—silicates, carbonates, phosphates, sulfates, and nitrates. While many other anionic complexes are also important, they are not in abundance. Silicates are present in abundance in high temperature forms such as seen in igneous and metamorphic rocks and are the source rock for some sedimentary rocks which are all used for aggregate in concrete. At the same time Portland cement is also a high temperature silicate. Silicates are also present as well as low temperature hydrated forms, such as the skeletons of sponges, and are the reaction products of Portland cement.

Elements scavenged by organisms are often found in the ocean at depth, where light does not penetrate to drive photosynthesis and form the basis of a food chain. In surface oceanic waters, versus deep ocean waters, silica, nitrates, and phosphates are biolimiting elements in this photic zone of the ocean because they are taken up by marine phytoplankton, so they are not readily available in low temperature systems. By contrast, carbonates are by far the most broadly employed and abundant materials seen in massive biological structures such as the skeleton of massive mineralizing organisms like coral comprising coral reefs. Carbonate forms from the interaction of carbon dioxide gas with water, in a carbonate buffer system, such as seawater. Carbonate rocks comprise about 10 % of the lithospheric crust of Earth, and are mined for aggregates used in concrete as well as for the raw materials for the formation of Portland cement via high temperature processing that drives off carbon dioxide into the atmosphere. Phosphates and nitrates are biolimiting elements, used as fertilizer, and tend to be expensive since they are relatively rare. However, there appear to be some applications of phosphate-based cement in concrete construction stemming from their initial development in bone cement applications. Sulfates and nitrates are more soluble than carbonates, silicates, and phosphates and do not appear to have broad applications in concrete construction as a primary component.

We define a cement as the result of combining ingredients that when mixed form a moldable paste that will set to a hardened mass and cure to a structurally stable solid. The process of cementation can occur in an open system, where new elements are introduced, from molding to setting to curing, but in general most common cements will be relatively closed systems where most of the components in the final cured cement were also present in the starting materials. Many examples exist, mostly sedimentology studies focused on beachstone, a term used to describe the lithification of sands or sediments by carbonate mineral formation , resulting in carbonate binding the sediments into a hardened mass. This is also commonly termed cemenation. For the discussion herein, we will focus on aqueous or "hydraulic" cements, where mineral particles, are combined in an aqueous solution to form a paste that will set to a solid mass, then cure to acquire structural integrity.

13.3 Bioinspired Cements

Cement can be "bioinspired" in various ways. The first is the process of cement formation and secondly is the resulting cemented composition and structure, in some cases termed concrete and mortar. In the former case, organisms form mineralized structures in ambient temperature environments by making use of unstable precursor materials (commonly ionic solutions) to engage a mineralizing reaction. Almost ubiquitously, organisms begin with amorphous precursors that attain their final structure and composition through crystallization to more stable forms that involve an aqueous cementing reaction.

13.3.1 Unstable Calcium Carbonate Cement Components

Calcium carbonate formation is well known to occur in aqueous solution through the initial formation of amorphous phases, including precursors to metastable phases which also convert to more stable phase polymorphs through a continuous pathway. Lowenstam first described amorphous calcium carbonate precursors at the California Institute of Technology in the 1950s, and summarized them in the classic paper on biomineralization published in SCIENCE in 1981 (Lowenstam 1981), and further in his book, 'On Biomineralization' published in 1987 (Lowenstam and Weiner 1989). Several amorphous calcium carbonate and calcium phosphate precursors where described: including ACC-calcite precursor, ACC-vaterite precursor, ACC, as well as ACP-brushite precursor, ACP-octacalcium phosphate precursor, and ACP-dahlite precursor. All of these compounds were found to be synthesized by organisms, existing in there metastable states in vivo, and to transform to their eventual more stable crystalline polymorph when explanted (Gunasekaran et al. 1994).

Metastable precursors such as proto calcite ACC and proto vaterite ACC have been investigated as highlighted in Gebauer et al. (2010). Elucidating the development of the precursor chemistry, resulting in final carbonate polymorphs, has been largely limited by the resolution of spectrographic and imaging techniques employed to observe the long and short-range order of the crystal lattice. In depth analyses of the binding chemistry of the carbonate chemical components along

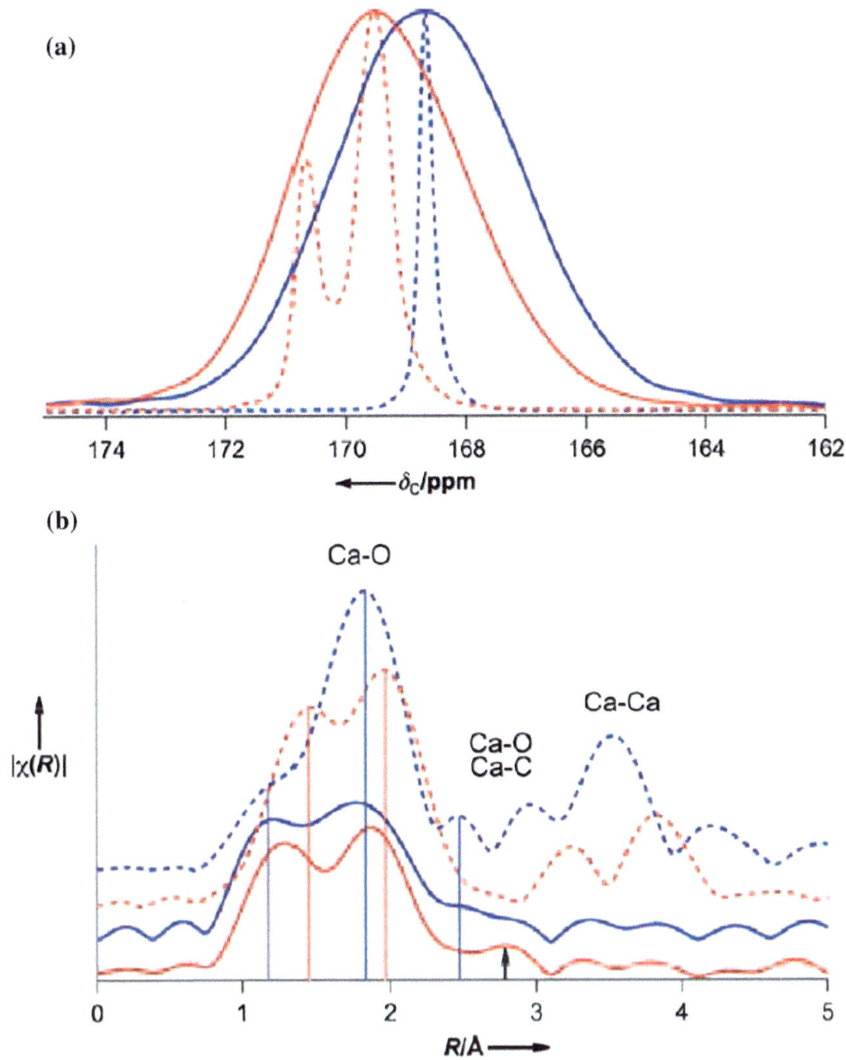

Fig. 13.1 Spectra of calcite (*dashed blue lines*), vaterite (*dashed red lines*), pc-ACC (*blue lines*), and pv ACC (*red lines*). **a** ^{13}C solid-state NMR spectra recorded by single pulses at a magnetic field of 9.4 T and a MAS rate of 8.0 kHz. **b** Fourier Transform of calcium K-edge EXAFS plotted in the R space (*lines* as in (**a**)). The expected interactions of the first three coordination shells are indicated. The *black arrow* marks a peak, which may relate to the coordination of structural water. The *vertical lines* are a guide for the eye (Gebaur et al. 2010)

with NMR and advanced X-ray techniques (EXAFS) (Fig. 13.1) has led the field in a new direction of understanding these materials more accurately as they arise from ions in liquids, gathering together and eventually taking the form of what is currently identified as carbonate structures. Cartwright et al. (2012) further discusses these metastable precursors as identified by the preceding techniques and identifies them on the continuum of two axes, synthetic to biogenic and stabilized to transient. Within this arena of amorphous precursor carbonate chemistry, also identified as nascent phase chemistry, various precursors are differentiated. The various precursors are commonly identifiable by advanced technique however are identified as proto-aragonite ACC (pv-ACC), proto-vaterite ACC (pv-ACC), proto-calcite ACC (pc-ACC), Liquid Condensed Phase (LCP), Polymer Induced Liquid Precursor (PILP), Dynamically ordered liquidlike oxyanion polymer (DOLLOP), structured, hydrated, anhydrous and monohydrocalcite-like ACC forms. The ordered forms of crystallographically identifiable carbonate are fewer and easily identified by common spectrographic technique.

The first use of unstable calcium carbonate compositions of vaterite and various forms of ACC in a cement composition were reported in SCIENCE in 1995 (Constantz et al. 1995), by a team at Norian Corporation. Calcium carbonate, that had been precipitated was employed as a component of the first carbonated cement. The cement set fast and attained compressive strengths in excess of 35 MPa. This cement is now used on a worldwide basis for orthopedic surgery. Precursor phases of calcium carbonate, including ACC-vaterite precursor, ACC-aragonite precursor, vaterite and aragonite are used in the cement. The carbonated cement is an example of a biomimetic design used to mimic the mineral phase of bone. While very useful clinically, these cement are produced in relatively small volume and are very expensive, and not scalable for use in civil engineering applications.

In 2002, General Motors developed the method to use carbon dioxide from fossil fuel combustion to form calcium carbonates through an aqueous precipitation process that converts carbon dioxide in flue gases to aqueous carbonate, and subsequently to calcium carbonate, sequestering the carbon dioxide in mineralized form. This allowed the possibility of forming very large volumes of metastable calcium carbonate precursors, for application in concrete, both for synthetic limestone aggregate, as well as the cementing phase as a calcium carbonate cement (Dziedzic et al. 2006, 2010, 2012). These carbonates that originate from the carbon dioxide released from the combustion of fossil fuels, and are enriched with regard to the light stable isotope of carbon, ^{12}C, and depleted with respect to the heavy stable isotope of carbon, ^{13}C, and can be readily identified as having negative carbon isotope values relative to PDB and SMOW isotopic scales.

Also with regard to point source power production, carbon capture technologies are being developed and have distinct synergies with the industrial processes that produce large amounts of CO_2. Refineries, power plants and cement production release a high % CO_2 in their flu gas. Current carbon capture technologies such as amine capture have been researched for the past 80 years; however, they are still largely energy intensive. Newer technologies that use carbonate mineralization techniques have been of interest since 2007. Below (Fig. 13.2) we see an example of one such plant that has been demonstrated at the 10 MW scale.

Fig. 13.2 First Generation carbonate mineralization was demonstrated at a pilot and demonstration scale between 2008 and 2010 in Moss Landing, California, using flue gas from the Moss Landing Power Station, the largest operating power plant on the west coats of the United States

13.3.2 Calcium Carbonate Cements

A diverse variety of calcium carbonate cements were tested in the 1990s and found to have different properties, including very high strengths and rates of biological incorporation (Constantz et al. 1998). Calcium carbonate cements were further developed in France (Fontaine et al. 2001) at the same time that General Motors was developing the large-scale carbon mineralization processes, around 2002. Since Norian, other examples of calcium carbonate cementing systems have been observed in biological cement applications (Combes et al. 2006). Coombs found that by combining various metastable calcium carbonate, both ACCs and various crystalline polymorphs such as vaterite and aragonite in combination, and optionally with calcium phosphates as the Norian team had done, a workable cement resulted. Coombs formulated a series of calcium carbonate cement with a variety of properties, including high strength and engineering porosity (Combes et al. 2006).

In this work, various types of synthetic limestone are precipitated from aqueous solution, namely two types of amorphous calcium carbonate, specifically strontium containing and magnesium-stabilized amorphous carbonates are precipitated and dried as powders. These amorphous carbonate powders were blended with a secondary metastable calcium carbonate polymorph, namely vaterite, powder and hydrated with liquid creating cement pastes. The liquid phase used in these cements was saline, mimicking the surgical environment, with blood contact, that the cement would be setting and curing in. The cement paste was then applied in vitro by allowing setting and curing in a 100 % relative humid environment at physiological body temperature, 37 °C. Setting and curing reactions were observed using both mechanical and analytical techniques in an effort to qualitatively and quantitatively track cement reaction progress. Had other sample curing methods been used, the relatively good compressive strength of their sample would be expected to be an order of magnitude more than what was reported for the clinically relevant condition of blood emersion simulated in their testing methodology. During communications with the research team it was understood that relatively high strengths were attained in their testing that did not simulate immersion in blood, but rather conventional concrete and cement sample preparation techniques. The results with the same formulations using standard sample preparation techniques exceeded 35 MPa in compressive strength routinely.

The two principle chemistries applied in Coomb's work resulted in two final cement mineralogies. The strontium inclusive amorphous carbonate when blended with vaterite, results in a calcite final structure. The amorphous magnesium carbonate when blended with vaterite will generate aragonite structure after hydration, due to the presence of magnesium. The presence of additional ions during formation of carbonates has been well understood (Bischoff and Fyfe 1968; Bischoff 1968a, b).

Further, published in 2007, a series of cements were investigated (Tas et al. 2007). In this work, the author precipitates and characterizes two calcium carbonate polymorphs; calcite and vaterite. The materials were precipitated. The carbonates were dried and applied as dried 100 % carbonate powers, including 100 % vaterite cement powder being submersed in a phosphate bearing setting solution mimicking a common biological interaction. Time-related study shows the materials eventually incorporating some phosphate, mimicking human bone.

Forming mechanically stable carbonate materials has been investigated significantly and it was written by K. Hosoi; "Marble and coral, which are formed in nature consist mainly of calcium carbonate. However, it is difficult to solidify calcium carbonate powder by ordinary sintering methods due to the characteristic of thermal decomposition at high temperature (700 °C). It is obvious that a low temperature chemically-bonded ceramic, or hydraulic cementing reaction is the most beneficial method to form calcium carbonate solids, akin to natural limestone. Therefore, there is almost no man made processing route to sinter only calcium carbonate powder." (Hosoi et al. 1996) Hosoi continues to comment that if this were possible it would open a host of new possibilities for building panels and other building materials. Other research groups have investigated carbonates in the framework of low temperature solidification, mainly from a ceramics perspective. In later works, calcium carbonate solids formed from unstable precursors, including amorphous calcium carbonates showed high apparent densities and overall strengths of 133 MPa (Yamasaki et al. 1993).

13.4 Environmental Challenges with Cements and Concrete

Portland cement contains mineral compounds that are formed during high temperature processing. Portland cement is produced, in the simplest of terms, by heating limestone with clay and calcium sulfate compounds to high temperatures (1,100–1,600 °C). During the production of Portland cement 0.8 ton of carbon dioxide is produced for each ton of cement produced. This carbon dioxide is resulting from the fuel burned as well as the CO_2 that is produced by limestone ($CaCO_3$) reducing to $CaO + CO_2$. Current methods of off-setting these exhausts and pollutants are by focusing on lower Portland cement containing mix designs.

Carbon sensitivity and awareness are growing, however; and industries are changing. The carbon sensitivity is mainly associated with Portland cement due to its high volume use. One method of reducing the amount of Portland cement in a mix design, without sacrificing performance, is by the addition of limestone to the cement during the final stages of milling. This final cementitious product is then termed a limestone interground cement and is commonly specified and applied in Europe, North America, and other countries. Newer technologies exist applying membrane methods to CO_2 capture and mineralization techniques to form 'synthetic limestone'. These synthetic limestone materials generated from these processes can be used interchangeably with natural calcium carbonate limestone.

Unless specifically introduced in the form of a limestone interground cement, or as limestone aggregate, carbonate is only found in concrete after long periods in small quantities. This arises from the aging of concrete in environments that contain gaseous carbon dioxide and is only introduced to the hydraulic composition in small amounts via the aqueous transport of carbon dioxide in the carbonate equilibria established with atmospheric carbon dioxide and the aqueous solution used to hydrate the cement. For instance, lime cements and later Portland cement.

13.4.1 Carbon Sequestration in Concrete

The global cement (3.6 billion tons) and aggregate (37.4 billion tons) markets are the only product-related markets that have the ability to sequester the mass and volume characteristic of the global carbon epidemic. With ever increasing CO_2 levels (Francey 2013) it is important to usefully store the carbon in a societal relevant form. Cement and aggregate are typically mixed with ground limestone or pozzolans, which in some cases are inert and occasionally reactive. Understanding the reaction of ordinary portland cement (OPC) with various carbon negative aggregates, pozzolons and additives are increasingly important to the successful inclusion of carbonated anthropogenic CO_2 in the human built environment.

Much can be leaned by astutely observing the biological world around us in an ever more accurate fashion. To solve the impending carbon question of the

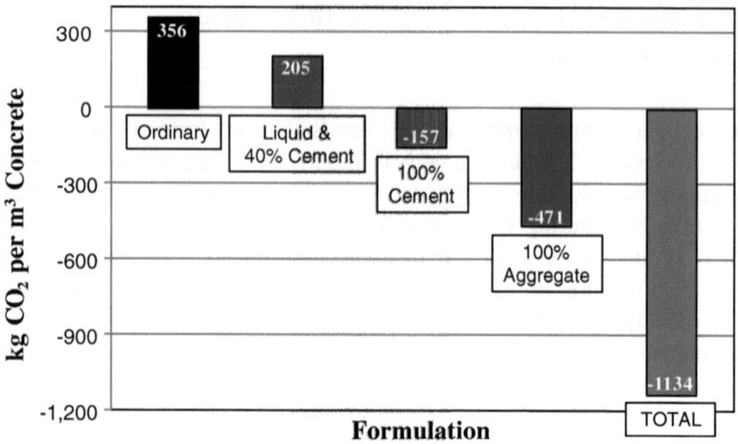

Fig. 13.3 GHG analyses of different concrete formulations where traditional components are replaced by novel carbon-reducing carbon sequestered limestone components. The formulations are based on a generic "moderate-strength" mix design that includes, per m^3 concrete, 356 kg cement, 178 kg water and 1,880 kg aggregate (848 kg fine, 1,032 kg coarse)

twenty-first century we cannot simply assess the situation from the similar viewpoint as the nineteenth and twentieth century. Considering the civil engineered world around us is a core component to an overall sustainable future. Reducing cement production by offsetting cement needs in concrete, using concrete as a sink to receive captured carbon materials are solutions leading to a direction that can solve our challenges of today and tomorrow.

Below we see the green house gas (GHG) analysis for ordinary concrete as compared to advanced mix designs that not only reduce the amount of carbon per yard of ordinarily mix designed concrete compared to low cement mix designs as well as synthetic limestone inclusive mix designs (Fig. 13.3).

A green house gas analysis of kilograms of carbon dioxide per cubic meter (kg CO_2/m^3) concrete in Fig. 13.1 exemplifies that replacement of traditional components in a concrete formulation by novel, carbon-reducing components, can lead to a significant reduction in the kg CO_2/m^3 concrete. If we consider that an average of 927 kg of CO_2 is emitted for every 1,000 kg of OPC produced in the U.S., depending on fuel type, raw ingredients and the energy efficiency of the cement plant (Source: "Concrete CO_2 Fact Sheet", National Ready-Mix Concrete Association, Feb 2012, based on the most recent survey of Portland Cement Association members), an ordinary moderate-strength mix design has 356 kg CO_2/m^3 concrete; this also assumes no CO_2 contributions from the water and aggregate components of the concrete mix design. If, for example, a carbon-reducing liquid that contains 5 % by weight CO_2 is used as a complete water replacement, the kg CO_2/m^3 concrete is reduced by 9 kg CO_2. When the carbon-reducing liquid is used in combination with a 40 % replacement of OPC by interground limestone, the kg CO_2/m^3 concrete is reduced to a mere 205 kg CO_2/m^3 (from 356 kg CO_2/m^3).

This is due to the 5 % by weight CO_2 in the liquid and the 40 % offset of CO_2 that would have otherwise come from the OPC. Other examples include a complete replacement of OPC by, for example, $CaCO_3$ cement, or complete replacement of aggregate by $CaCO_3$ aggregate. For these examples it is noteworthy to assume the $CaCO_3$ is produced by the following stoichiometric chemical reaction: $CaCl_2 + 2\ NaHCO_3 = CaCO_3 + 2\ NaCl + CO_2 + H_2O$, i.e., each equivalent of $CaCO_3$ sequesters one equivalent of CO_2. If a concrete formulation were to replace traditional components with novel carbon-reducing liquid, cement and aggregate, the total kg CO_2/m^3 decreases by more than −1,100 kg.

13.5 Conclusions

Concrete may provide a significant sink for anthropogenic carbonate dioxide in the form of aggregate and calcium carbonate cement. These carbonate concretes hold promise specifically in regard to the greater carbon problem and may one day provide an energy efficient platform for the reuse of carbon dioxide in our built environment on a very large scale and at a very low energy demand. Low temperature cement precursors may usher in a new era of possibilities in concrete compositions.

References

Bentz DP (1997) Three-dimensional computer simulation of Portland cement hydration and microstructure development. J Am Ceram Soc 80(1):3–21

Bischoff JL, Fyfe WS (1968) Catalysis, inhibition, and the calcite-aragonite problem: I. The aragonite-calcite transformation. Am J Sci 266:65–79

Bischoff JL (1968a) Catalysis, inhibition, and the calcite-aragonite problem: II. The vaterite-aragonite transformation. Am J Sci 266:80–90

Bischoff JL (1968b) Kinetics of calcite nucleation: Magnesium ion inhibition and ionic strength catalysis. J Geophys Res 73(10):3315–3322

Cartwright JHE, Checa AG, Gale JD, Gebauer D, Sainz-Díaz CI (2012) Calcium carbonate polyamorphism and its role in biomineralization: how many amorphous calcium carbonates are there? Angew Chem Int Ed 51:11960–11970. doi:10.1002/anie.201203125

Combes C, Miao B, Bareille R, Rey C (2006) Preparation, physical-chemical characterization and cytocompatibility of calcium carbonate cements. Biomaterials 27:1945–1954

Constantz BR, Ison IC, Fulmer MT, Poser RD, Smith ST, Van Wagoner M, Ross J, Goldstein SA, Jupiter JB, Rosenthal DI (1995) Skeletal repair by in situ formation of the mineral phase of bone. Science 267:1796–1799

Constantz BR (2004) Crossing the chasm: adoption of new medical technology. In: Greco R (ed) Nanotechnology. CRC Press, Boca Ranton, Ann Arbor, London, Tokyo.

Constantz BR, Barr BM, Ison IC, Fulmer MT, Baker J., McKinney LA, Goodman SB, Gunasekaren S, Delaney DC, Ross J, Poser RD (1998) Histological, chemical, and crystallographic analysis of four calcium phosphate cements in different rabbit osseous sites. J Biomed Mat Res 43(4):451–461

Dziedzic D, Gross KB, Gorski RA, Johnson JT (2006) Sequestration of carbon dioxide. United States Patent Number: 7132090
Dziedzic, D.; Johnson, J. T.; Gross, K. B. (2010) Process for immobilization of protein catalysts, products, and use. United States Patent Number: 7642076
Dziedzic D, Johnson JT, Gross KB (2012) Process for immobilization of protein catalysts, products, and use. United States Patent Number: 20100068784
Francey RJ (2013) Atmospheric verification of anthropogenic CO_2 emission trends. Nat Clim Change 3:520–524
Fontaine ML, Combes C, Mounic S, Rey C (2001) Composition De Ciment Hydraulique A Base De Carbonates De Calcium. French Patent Number: 20010012696
Gebauer D, Gunawidjaja PN, Ko JYP, Bacsik Z, Aziz B, Liu L, Hu Y, Bergström L, Tai C-W, Sham T-K, Edén M, Hedin N (2010) Proto-calcite and proto-vaterite in amorphous calcium carbonates. Angew Chem 122:9073–9075. doi:10.1002/ange.201003220
Gunasekaran S, Gaspodorwitz D, Barr P, Constantz BR (1994) Mineralized collagen. In: Brown PW, Constantz BR (eds) Hydroxyapatite and related materials. CRC Press, Boca Raton, Ann Arbor, London, Tokyo, pp 215–223
Hosoi K, Hashida T, Takahashi H, Yamasaki N, Korenaga T (1996) Low temperature solidification of calcium carbonate through vaterite-calcite wet transformation. J Mater Sci Lett 15:812–814
Lowenstam HA (1981) Minerals formed by organisms. Science 211(4487):1126–1131
Lowenstam HA, Weiner S (1989) On biomineralization. Oxford Press, Oxford
Mehta PK (2006) Concrete: microstructure, properties and materials, 3rd edn. The McGraw-Hill Companies Inc, New York, USA
Tas AC (2007) Use of vaterite and calcite in forming calcium phosphate cement scaffolds. In: Brito M, Case E, Kriven WM, Salem J, Zhu D (eds) Developments in porous, biological and geopolymer ceramics: ceramic engineering and science Proceedings, vol 28, Issue 9. Wiley, Hoboken, NJ, USA. doi:10.1002/9780470339749.ch14
USGS. http://minerals.usgs.gov/minerals/pubs/commodity/cement/myb1-2011-cemen.pdf
Yamasaki N, Weiping T, Yanagisawa K (1993) Solidification of $CaCO_3$ containing $SrCO_3$ by hydrothermal hot-pressing. J Mater Res 8:1972–1976

Chapter 14
Production of Bacteria for Structural Concrete

Varenyam Achal

Abstract This chapter reviews a novel, green and economical concrete based on microbially induced calcium carbonate precipitation (MICP). Microbial or bacterial concrete is product of MICP, produced by ureolytic bacteria, requires much less energy to produce. Such bacteria are abundant in nature in almost every environment and can be reproduced at fast rate at low cost. Calcium carbonate precipitated during the process of MICP might help building materials and structures by improving compressive strength and impermeability, and ultimately their durability. Harnessing this novel process of biogeochemistry may bring in enormous economical benefits to construction industries and will open a new door to the research in the arena of geotechnical and structural engineering. This chapter critically reviews the production and mechanism of MICP. Further, a thorough understanding of the research in the area of microbial-based cementitious materials, which lead to improving the durability of building materials and structures, has been discussed.

14.1 Introduction

Everything comes with a price, an old saying; however, true even when we talk about our modern civilization. Thanks (in one way) to cement, which builds modern civilization, however at the cost of massive pollution to the health and environment. On the other hand, the cement production is energy consuming and environmentally unfriendly process as contributes about 7 % of global anthropogenic CO_2 emissions (Worrell et al. 2001). It is true that we cannot replace cement completely with other building material; however, there is high scope to reduce

V. Achal (✉)
Shanghai Key Lab for Urban Ecological Processes and Eco-Restoration,
East China Normal University, Shanghai, China
e-mail: varenyam@re.ecnu.edu.cn

F. Pacheco Torgal et al. (eds.), *Biotechnologies and Biomimetics for Civil Engineering*,
DOI 10.1007/978-3-319-09287-4_14

this substance in construction to lower the environmental hazard caused by cement. Again question rises, what about the quality of such structure? We need to find a novel way to do this without compromising the quality of building structures. Finding such a sustainable material means, we could reduce the environmental impact of cement.

Concrete is the most widely used building material and most of the building structures are made up of it. However, natural processes including earthquake or weathering or land subsidence and human activities play enough roles to degrade or reduce the durability of concrete structures. Durability of concrete is the ability of a concrete to resist deterioration, particularly deterioration due to weather exposure, chemical exposure or surface abrasion (Reddy et al. 2012).

Though ignored for centuries with respect to their role in construction industry, bacteria have enormous potential in carbonate formation leading to increment in compressive strength, a key parameter while designing buildings structures. Moreover, bacteria are omnipresent, especially in soil, regardless of normal to harsh environmental conditions. It can be used perfectly in construction as live building materials, rather than inert one, as it can precipitate carbonate when required, even once construction is over. Together with building components, nutrient solutions; bacteria in cementitious materials will form "concrete ecosystem." Concrete ecosystem (perhaps a term coined first time here) looks simpler as will contain only microbes as living component in conjunction with cement, sand, aggregates and some other building materials; however, the ecosystem process under it is highly complex due to harsh environment. It provides favorable condition for Microbially induced Calcium Carbonate Precipitation (MICP).

The importance of MICP has been reported in several applications including remediation of heavy metals (Achal et al. 2011a, 2012a), soil strengthening/improvement (Whiffin et al. 2007), restoration of calcareous stone materials (Tiano 1995; Castanier et al. 1999; Stocks-Fisher et al. 1999; Rodriguez-Navarro et al. 2003), wastewater treatment (Hammes et al. 2003), sand consolidation (Achal et al. 2009a), strengthening of concrete (Ramchandran et al. 2001), and durability of building materials (Achal et al. 2010a). The present chapter outlines, based on the reports from researchers worldwide, the mechanism driving MICP, concrete construction using bacteria (thus known as *microbial concrete*, a term coined by Achal et al. (2011b)) and how microbial concrete is effective to enhance the durability of building structures.

14.2 Microbially Induced Carbonate Precipitation

Microbially induced calcite precipitation is resultant of complex biochemical reaction often governed by an enzyme urease (urea amidohydrolase; EC 3.5.1.5) produced by microbes. This reaction/precipitation requires urea as substrate while calcium source as chief agent for calcite production. During microbial urease activity, 1 mol of urea is hydrolyzed intracellularly to 1 mol of ammonia and

1 mol of carbamate (Eq. 14.1), which spontaneously hydrolyses to form an additional 1 mol of ammonia and carbonic acid (Eq. 14.2) (Burne and Chen 2000). These products subsequently equilibrate in water to form bicarbonate and 2 mol of ammonium and hydroxide ions (Eqs. 14.3 and 14.4) that give rise to an increase in pH and ultimately shift the bicarbonate equilibrium, resulting in the formation of carbonate ions (Eq. 14.5). High pH condition favors the formation of CO_3^{2-} from HCO_3^- (Knoll 2003). Finally, the carbonate concentration will increase, inducing an increase in supersaturation level leading to $CaCO_3$ precipitation around the cell in the presence of soluble calcium ions (Eqs. 14.6 and 14.7).

$$CO(NH_2)_2 + H_2O \rightarrow NH_2COOH + NH_3 \tag{14.1}$$

$$NH_2COOH + H_2O \rightarrow NH_3 + H_2CO_3 \tag{14.2}$$

$$H_2CO_3 \leftrightarrow HCO_3^- + H^+ \tag{14.3}$$

$$2NH_3 + 2H_2O \leftrightarrow 2NH_4^+ + 2OH^- \tag{14.4}$$

$$HCO_3^- + H^+ + 2NH_4^+ + 2OH^- \leftrightarrow CO_3^{2-} + 2NH_4^+ + 2H_2O \tag{14.5}$$

$$Ca^{2+} + Cell \rightarrow Cell - Ca^{2+} \tag{14.6}$$

$$Cell - Ca^{2+} + CO_3^{2-} \rightarrow Cell - CaCO_3^- \tag{14.7}$$

Based on the above equations, it may be said that calcium carbonate precipitation is a biochemical process governed mainly by five key factors: (1) the urease, (2) the calcium concentration, (3) the concentration of dissolved inorganic carbon (DIC), (4) the pH, and (5) the availability of nucleation sites. The availability of nucleation site is very important for continuous and stable calcium carbonate formation. It is isolated from the environment by a delimiting geometry by limiting the diffusion in and out of the system (Sarayu et al. 2014). The ion movement is enabled by active pumping with organelles or passive diffusion to enable the microorganisms to use a great variety of anatomical arrangements (Perry 2003), as shown in Fig. 14.1. The production of CO_3^{2-} from bicarbonate (HCO_3^-) in water is strongly pH dependent, an increase in CO_3^{2-} concentration occurs under alkaline conditions that lead to deprotonation of the functional groups like carboxyl, hydroxyl, and phosphate of the bacterial cell wall and creates a strong electrostatic affinity to attract cations and enables the accumulation of calcium ions on the surface of the cell wall. On the other hand, the calcium ions in the solution are attracted to the bacterial cell wall due to the negative charge of the later. Upon addition of urea to the bacteria, dissolved inorganic carbon and ammonium are released in the microenvironment of the bacteria. In the presence of calcium ions, this can result in a local supersaturation and hence heterogeneous precipitation of calcium carbonate occurs on the bacterial cell wall (Reddy et al. 2012).

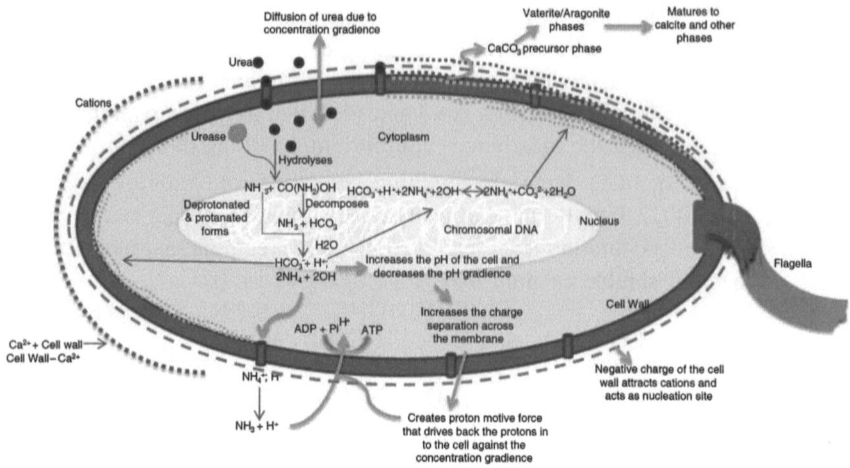

Fig. 14.1 Pathway of biomineral secretion and precipitation in a bacterial cell (Reprinted with permission from Sarayu et al. 2014)

The concrete ecosystem, rich in calcium source, provides favorable condition for MICP as calcium carbonate precipitation readily occurs in alkaline environments abundant of the calcium ($Ca^{+2)}$ and carbonate (CO_3^{2-}) ions (Stocks-Fischer et al. 1999; Ramachandran et al. 2001; Qian et al. 2010a).

14.3 Why MICP and Microbial Concrete?

Natural processes, such as weathering, faults, land subsidence, earthquakes, and human activities create fractures and fissures in concrete structures or monuments. These fractures and fissures are detrimental since they can reduce the service life of the structure (Achal et al. 2011b). Such processes often weaken strength, induce porosity and also give an unattractive appearance, and cracks lead to easy passage for aggressive environment to reach the reinforcement and initiate corrosion. There are synthetic agents (epoxy, hydroxyl-epoxy) or latex binding agents (such as acrylic, polyvinyl acetate, butadiene styrene) or silanes or organic–inorganic products available in the market to protect or repair damaged concrete structures; however, suffer from being expensive, and problems associated with different thermal expansion, degradation with age and the need for constant maintenance. Appearance of cracks and fissures is an inevitable phenomenon during the aging process of concrete structures when exposed to weather changes. Such cracking leads to easy passage for aggressive environment to reach the reinforcement and initiate corrosion. Moreover, sometimes repair is carried out in the areas where it is not possible to shut down the plant or it is hazardous for human beings. Hence, in such situations, a way should be found out to self healing materials that seal the cracks automatically. Recently, a novel technique has been reported that utilizes microorganisms in remediation of cracks and fissures in natural and man-made structures by precipitation of calcium carbonate.

On the other hand, the cement industry has for some time been seeking procedures that would effectively reduce the high energy requirements and environmental costs of cement manufacture (Rong and Qian 2012). The answer very much depends up on "microbial concrete" that is based on MICP process consists of three materials, namely, alkalophilic microbes, substrate solution and calcium ion solution. The great promise of MICP-based microbial concrete has been demonstrated to enhance the durability of building materials, consolidation of sand columns, and repair of limestone monuments and concrete (Gollapudi et al. 1995; Tiano et al. 1999a; Ramachandran et al. 2001; De Muynck et al. 2008; Qian et al. 2009; Achal et al. 2011b; Rong et al. 2012). Microbial concrete can improve the strength and durability of structures, which are considered to be the requirements for concrete or any other building materials. A major goal of using microbial concrete is to ensure the quality parameters or durability of building structures.

14.4 Quality Parameters for Concrete Structures

The quality of any building material depends on three major parameters, (i) strength, (ii) permeability, and (iii) corrosion. For an efficient microbial concrete it should produce more compressive strength, less permeability and should not affect corrosion of any reinforcement (Reddy et al. 2012). When MICP came in practice, the first building material to test was sand where the process of MICP was well established. Sand is the common material used to make most of the building materials and structures, thus research to confine a novel methodology with sand is warranted.

14.4.1 Biosandstone

The process of MICP was proposed as a novel method for cementing loose sands to produce structural materials, termed as biosandstone (Fig. 14.2). It consists only of alkalophilic urease producing bacteria, substrate solution (urea), calcium source and sand. A typical set-up for the sand consolidation experiment to develop biosandstone was simplified in Reddy et al. (2012), where sand (either mixed with bacterial culture or later injected directly in the column) was plugged through a plastic column and the cementation fluid (consists of nutrient media with urea and calcium source) was injected or dropped at specific rate in the column in gravity free flow direction.

The cementation fluid also contains nutrients that are necessary for the bacterial growth and metabolism to perform MICP and can be transported through

Fig. 14.2 Biosandstone (Reprinted with permission from Achal et al. 2011a)

sandstone cores (Jenneman et al. 1984). The bacterial sand consolidation was resulted into porosity reduction from sand, as reported by Kantzas et al. (1992) when they found up to 50 and 90 % reduction in porosity and permeability, respectively in sand consolidated by *Bacillus pasteurii*. Such reduction might be due to high deposition of calcite in column, as Achal et al. (2009a) found 40 % calcite deposition in the sand column consolidated by a mutant of Sporosarcina pasteurii. The sand column of size 32.10 and 18.40 mm showed good amount of compressive strength, measured up to 2 MPa (Mega Pascal) when $CaCl_2$ was used as calcium source for biosandstone (Qian et al. 2010b). The microbial induced precipitated substance in bio-sandstone was checked using X-ray diffraction (XRD) and energy dispersion spectroscopy (EDS), and calcite as main microbial induced substance was seen in biosandstone (Fig. 14.3).

The positive results of MICP on biosandstone lead researchers to think beyond this building material i.e., sand. Researchers started focus on more complex building material such as cement and concrete, and studied effect of MICP on the improvement of strength, permeability and other durability parameters.

14.4.2 Microbial Concrete and Compressive Strength

Compressive strength is the first and foremost prerequisite to judge building materials or structures, expressed as the ultimate compression load per cross sectional area, usually in psi, Pa or Kg/cm^2. It is required to determine whether the concrete mixture used in construction meets the requirements of the specified strength in the job specification or not. There are several researches on mortars, where microbial concrete was used to enhance its compressive strength. Mortar or cement mortar refers to a workable paste to use during construction, consisting of cement, sand and water, while the term concrete consists of similar constituents in addition to aggregates.

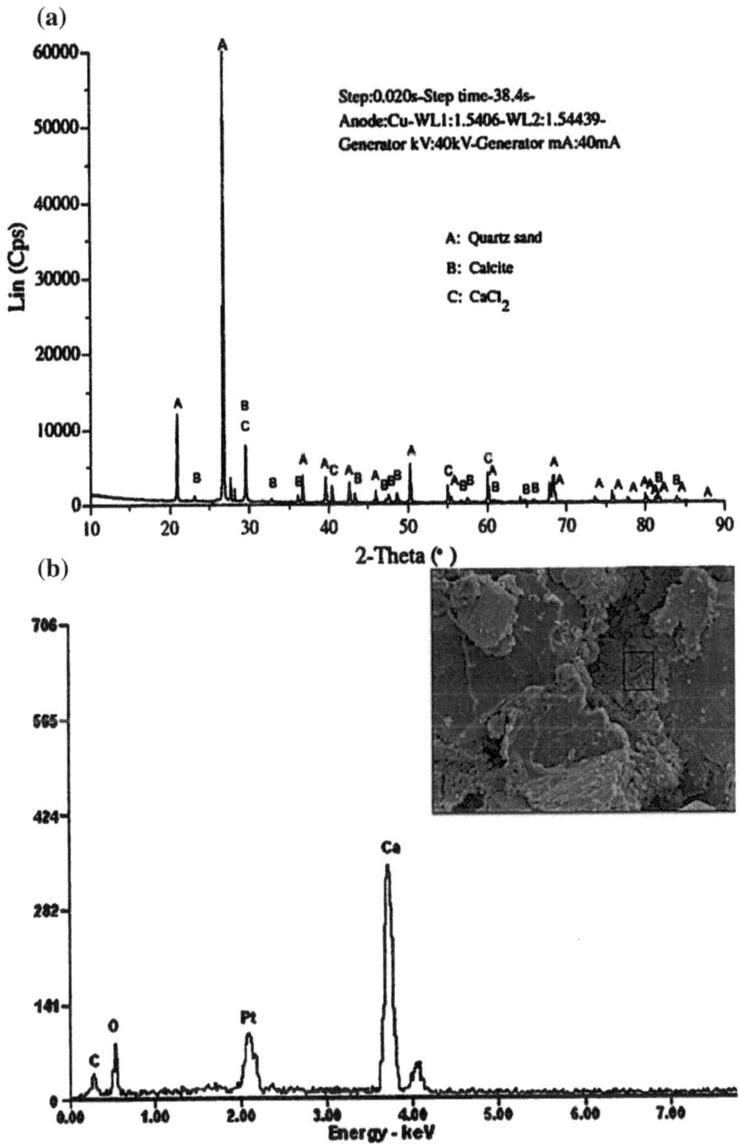

Fig. 14.3 **a** X-ray diffraction, and **b** Energy dispersion spectroscopy of mineral phases microbial induced precipitated biosandstones (Reprinted with permission from Rong et al. 2012)

Perhaps, the improvement of the compressive strength of Portland cement mortar cubes based on MICP was initiated by Ramachandran et al. (2001). They experimented on two different bacteria, namely, *Bacillus pasteurii* (later renamed as *Sporosarcina pasteurii*) and *Pseudomonas aeruginosa* and mixed bacterial cells in form of pellets to mortars. The mortars were cured in solution containing urea and calcium chloride for 7 days. When tested for compressive strength, it was recorded about 65 MPa in the presence of *S. pasteurii*, which was relatively higher than control mortars (55 MPa). As *P. aeruginosa* is not reported to induce calcite precipitation, it couldn't improve the compressive strength of mortars.

Other than bacteria of genus *Bacillus*, *Shewanella* sp. was used to enhance the compressive strength of mortars by 17 and 25 % at 7 and 28 days, respectively; however, they cured mortar specimens in air (Ghosh et al. 2005). Just like *P. aeruginosa*, there was no increment in the compressive strength with *Escherichia coli* (non-urease producing bacterium). Later, while mixing spores of *Bacillus pseudofi rmus* and *Bacillus cohnii,* an increment of 10 % mortar compressive strength was recorded (Jonkers and Schlangen 2007).

One of problems associated, while studying compressive strength, with microbial concrete is cost factor used in growing bacterial cells or curing in nutrient media. To overcome this, Achal et al. (2009b) replaced the commercially available nutrients with some industrial by products such as lactose mother liquor (LML) and corn steep liquor (CSL). An improvement of 17 % in the compressive strength of mortar cubes was noticed with *S. pasteurii* grown and cured with LML medium compared to control (23.2 MPa) at the end of 28 days. Later, Achal et al. (2010a) reported 35 % improvement in the compressive strength of mortar at 28 days with *S. pasteurii* prepared mortar cubes with CSL-urea medium. The similar experiment resulted into 36 % improvement in the compressive strength, when CSL was replaced with commercial nutrient medium (Achal et al. 2011b). The effect of different media on the compressive strength of cement mortar using *S. pasteurii* has been summarized in Fig. 14.4. Their experiments were of great potential with respect to economization of microbial concrete preparation.

The overall trend of an increase in the compressive strength was very much dependent on calcium carbonate precipitation induced by bacterial cells and the behavior of bacterial cells within the cement mortar matrix. As cement mortar remains still porous during the initial curing period, though bacterial cells get good nourishment; but growth might not be proper due to the completely new environment for bacteria, especially high cement pH (Achal et al. 2011b). As the curing period proceed, bacterial cells started growing and start precipitating calcium carbonate within mortar matrix. The bacterial growth and curing period led to plugging of pores in the matrix and the flow of the nutrients and oxygen to the bacterial cells stops that causes the cells either died or turned into endospores and acts as an organic fiber that may enhance the compressive strength of the mortar cubes (Ramachandran et al. 2001).

Further, to confirm MICP process in the improvement of compressive strength, researchers analyzed mortar specimens with techniques such as X-ray diffraction (XRD) and visualized with scanning electron microscope (SEM).

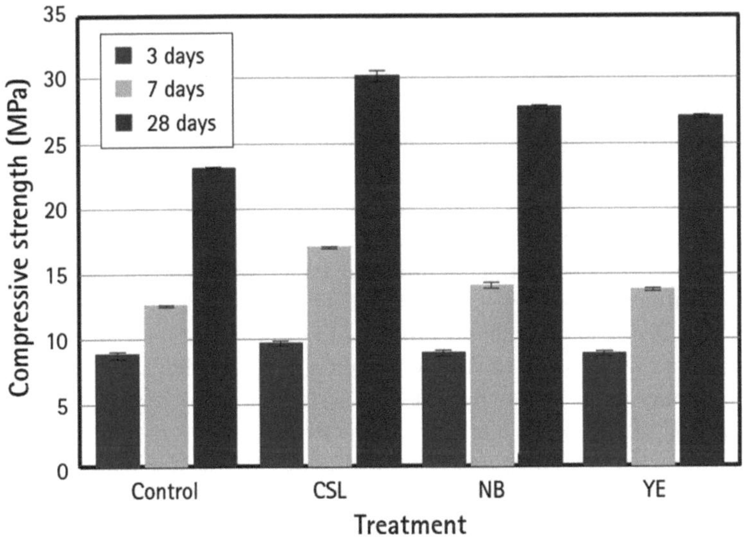

Fig. 14.4 Effect of *S. pasteurii* on the compressive strength of cement mortar cubes grown in different media at 3, 7 and 28 days (Reprinted with permission from Achal et al. 2010a)

14.4.3 Microbial Concrete and Permeability

Permeability is an important factor on which concrete durability depends, known to be property that governs the rate of flow of fluid into porous mortar or concrete. Such property controls the ingress of moisture, gas or harmful substances to the concrete structures. Any adverse condition affecting building materials or structures targets permeation properties easily. Some of commercially available substances, which can be used to make such surface impermeable, are not successfully used due to disadvantages such as, an incompatibility of the protective layer and the underlying layer due to differences in their thermal expansion coefficient or disintegration of the protective layer over time and a need for constant maintenance (Reddy et al. 2012).

The ability of MICP to improve impermeability on building material surface was first observed by Tiano et al. (1992) when they successfully used organic matrix macromolecules extracted from *Mytilus californianus* shells to induce the precipitation of calcium carbonate within the pores of the stone. The calcite precipitation resulted in a slight decrease in porosity and water absorption by capillarity (Tiano 1995), which was reduced about 60 % from the limestone (Tiano et al. 1999b). Further, Le Metayer-Levrel et al. (1999) confirmed bacterial carbonatogenesis/biocalcification on the stone surface resulted into permeability reduction without affecting its aesthetic appearance, with conclusion that biological mortars or cement could be used to affix small pieces broken from statues and to fill small cavities on limestone surfaces. Later Dick et al. (2006) reported calcite induced by *Bacillus* sp. was effective in reducing the water absorption rate of limestone. Such researches provided way to choose other building materials. An increase in the resistance of concrete toward alkali, sulfate, freeze thaw attack and drying shrinkage was observed with calcite precipitating bacteria (Ramakrishnan et al. 1998). While studying the durable effect of *Bacillus sphaericus* on mortars, De Muynck et al. (2008) found a significant decrease of the water uptake compared to untreated specimens (a reduction of 45, 43, and 24 % with increasing w/c) and 19 % decrease of the chloride migration coefficient. Later, they (De Muynck et al. 2008) concluded that the carbonate precipitation was mainly a surface phenomenon due to the limited penetration of the bacteria in the porous matrix, resulted in a decrease of water absorption and gas permeability from mortars.

Bacillus sp. CT-5, isolated from commercially available cement, was used to prepare mortars and a sorptivity test was performed on it (Achal et al. 2011b). Over a period of 168 h, the mortars with bacterial cells absorbed nearly six times less water than the control cubes. The presence of bacteria resulted in a significant decrease of the water uptake compared to untreated mortars. The deposition of a layer of calcium carbonate crystals on the surface resulted in a decrease of the sorptivity.

Achal et al. (2011c) performed the water impermeability test on the concrete cubes of dimension 150 mm (M20 grade), prepared with mutant *S. pasteurii* grown in commercially available nutrient broth (NB) and economic corn steep liquor (CSL) media with urea as substrate and calcium chloride as calcium source. The results indicated that the permeability of the concrete cubes prepared with bacterial cells was lower than that of the control irrespective of media used. The penetration at the sides of concrete was higher than that at the top due to better compaction and closing of pores at the top by calcite precipitated by bacterial cells. Further they reported that the resistance of concrete to chloride penetration increased with MICP. The permeability class type was recorded ''moderate'' for control concrete specimens, while the class changed to ''low'' type of concrete with bacterial cells as per ASTM C1202-05. For control samples, the average charge passed was 3,177 C, whereas for samples prepared with bacterial cells in NB and CSL media it was 1,019 and 1,185 C, respectively.

14.4.4 Microbial Concrete and Corrosion

The corrosion of steel and reinforcing bar is a predominant factor causing widespread premature deterioration of concrete constructions worldwide (Raupach and Schiebl 2001). Corrosion and permeability goes together, higher the permeability, more would be corrosion and vice versa. The ingress of moisture, chloride ions, and carbon dioxide initiates corrosion through the concrete to the steel surface. Chloride-induced corrosion of reinforcing steel is one of the most pressing problems worldwide that the construction industry is facing today. The corrosion products of iron oxides/hydroxides expose the reinforcement to direct environmental attack that results in accelerated deterioration of the structure (Neville 1995). The solution to prevent corrosion of such structure can be achieved by sealing the paths of ingress to improve the life of the reinforced concrete (RC) structures.

As MICP promises to alleviate permeability and transport of pollutants inside concrete, it can be effective in reducing corrosion in RC by making protective layer or carbonate followed by calcite precipitation. However, there is scarce research on the role of MICP in corrosion prevention of RC structure. Such research was mainly reported by Achal et al. (2012b) where they performed detailed investigation leading to positive impact of MICP in the RC corrosion prevention.

To determine the effect of MICP, Achal et al. (2012b) prepared the RC specimens with bacterial cells (*Bacillus* sp. CT-5) and induced corrosion by applying a constant anodic potential of 40 V for 7 days. There was visible calcite precipitation on bacterially treated RC specimens. After 7 days of accelerated corrosion, numerous (at least seven) cracks with widths nearly 0.2 mm (0.008 in.) were observed on control specimens with one longitudinal localized crack of width 0.3 mm within 36 h, whereas in bacterially treated samples, a crack of that width appeared not before 168 h. The control specimens had significantly higher *Icorr* (60.83 mA/m^2 [39.25 $mA/in.^2$]) compared to MICP samples (14.78 mA/m^2 [9.53 $mA/in.^2$]) in nutrient and 20.03 mA/m^2 (12.92 $mA/in.^2$) in CSL media. An approximate four-fold reduction in *Icorr* by *Bacillus* sp. CT-5 suggests that the calcite precipitation has the effect of greatly reducing corrosion. Achal et al. (2012b) concluded that the formation of calcite might facilitate the protective passive film around the steel and act as a corrosion inhibitor by interrupting the transport process in such samples. Further, they also found that pullout strength was enhanced and mass loss of the reinforcing bar was reduced due to MICP.

Based on calcium carbonate induced by *B. pasteurii*, Qian et al. (2010a) showed improvement in the surface impermeability of cement mortars, resulted in resistance to the acid ($pH > 1.5$). They concluded MICP ability in the prevention of corrosion of building materials and structures. The results of various researchers on microbial concrete with their target materials have been summarized in Table 14.1.

14.5 Cost Analysis of Microbial Concrete?

As microbial concrete is novel product, which can be used to enhance the durability of building structures, many researchers or engineers doubt on its production cost. The costs of microbial concrete depend very much on the price of bacteria and nutrients. Further, the price of bacteria varies country to country; however, one standard bacterial strain, if bought from ATCC, costs US $500, and from MTCC, costs US $10, while CGMCC sells at US $200. De Muynck et al. (2010) reported the cost analysis of microbial concrete based on personal communication by an employee of the Belgian company FTB remmers 2008; http://www.ftbremmers.com/). The price of 1 kg lyophilized bacteria is about US $1,500 (1,100 €) and

Table 14.1 Overview of BioCement applications with respect to target, microorganism used and origin of country

Target	Microorganism	Origin	References
Limestone protection	Biocalcin producing bacteria	France	Adolphe et al. (1990), Le Metayer-Levrel et al. (1999)
Monumental stone protection	*Mytilus californianus*	Italy	Tiano et al. (1992), Tiano (1995)
Monumental stone porosity	*Micrococcus* sp.	Italy	Tiano et al. (1999a)
	Bacillus subtilis		
Biochemical properties of MICP	*B. pasteurii*	USA	Stocks-Fischer et al. (1999)
Concrete crack remediation	*B. pasteurii*	USA	Ramachandran et al. (2001)
Ornamental stone conservation	*Myxococcus xanthus*	Spain	Rodriguez-Navarro et al. (2003)
Concrete durability	*B. sphaericus*	Belgium	De Muynck et al. (2008)
Concrete crack repair	*B. sphaericus*	Belgium	Van Tittelboom et al. (2009)
Compressive strength of cement mortars	*Shewanella* sp.	India	Ghosh et al. (2005)
	Bacillus sp. CT-5		Achal et al. (2011b)
Greener media in BioCement	*S. pasteurii*	India	Achal et al. (2009b, 2010a)
Corrosion prevention	*Bacillus* sp. CT-5	India	Achal et al. (2012b)
Bio-sandstone	Alkalophilic microbes, *Bacillus* spp.	China	Qian et al. (2010a), Rong et al. (2012), Rong and Qian (2012)

2–3 g/m^2 is applied in concrete, which costs about US \$4 (3 €)/m^2. The cost of nutrients is estimated to be about US \$250 (180 €)/kg. Based on this analysis, the dosage for microbial concrete application will generally range between 0.04 and 0. 08 kg/m^2, bringing the cost of nutrients to US \$7–15 (5–10 €)/m^2 and the total product cost is estimated around US \$31–39 (23–28 €)/m^2.

The additional cost during the preparation of biological mortar or microbial concrete will be that of bacteria and nutrient; however, the cost of nutrients can be reduced significantly by replacing standard or commercially available nutrients with such industrial by products, rich in carbohydrate/protein/energy sources. Achal et al. (2011c) successfully reduced the cost of microbial concrete production by replacing standard nutrient with corn steep liquor (CSL). Corn steep liquor can typically be available locally with a price of nearly US \$2 (1.5 €)/L, which is very economic compared with standard nutrient medium and this brings the biodeposition cost to US \$0.5–1.0 (0.3–0.7 €)/m^2. The performance of CSL was significantly better than standard laboratory nutrients in terms of microbial concrete production. Hence, CSL offers an economic advantage over the standard nutrient medium and the overall process cost reduces dramatically, and finding such other economic nutrient solution is of need.

14.6 Conclusions

The quality of building structures is very much dependent upon its strength and impermeability. However, many natural processes causes damage to such structures. In such a situation, it is very important to find novel additive to use during construction, which can improve the durability of structures as well, can be used in the remediation of damaged affects. Hence, a sustainable building material is need of time. Microbially induced calcium carbonate precipitation has great potential not only in the area of microbiology or environmental biotechnology, but also in civil and geotechnical engineering. The introduction of MICP-based microbial concrete offers a novel additive to cement-based materials with adequate impermeability, compressive strength, and reduced reinforced corrosion. The laboratory-based researches provide enough evidence for the successful use of microbial concrete; however, the real challenges are to use it in field studies and in construction of new structures. More research requires converting results achieved in the laboratory into practical applications. The microbial concrete can also be utilized in rehabilitation of heritage stone and lime mortar structures. This MICP-based process can also be carefully used in the remediation of structures that contain hazardous materials such as nuclear fill buildings. The production of bacteria for structural concrete will provide the basis for an alternative and high quality concrete sealant that is highly economic and environmentally safe, leading to the enhancement in the durability of building materials and structures.

References

Achal V, Mukherjee A, Reddy MS (2011a) Effect of calcifying bacteria on permeation properties of concrete structures. J Ind Microbiol Biotechnol 38:1229–1234

Achal V, Mukherjee A, Reddy MS (2011b) Microbial concrete- a way to enhance the durability of building structures. J Mater Civ Eng 23:730–734

Achal V, Pan X, Zhang D (2011c) Remediation of copper contaminated soil by *Kocuria flava* CR1, based on microbially induced calcite precipitation. Ecol Eng 37:1601–1605

Achal V, Mukherjee A, Basu PC, Reddy MS (2009a) Lactose mother liquor as an alternative nutrient source for microbial concrete production by *Sporosarcina pasteurii*. J Ind Microbiol Biotechnol 36:433–438

Achal V, Mukherjee A, Basu PC, Reddy MS (2009b) Strain improvement of *Sporosarcina pasteurii* for enhanced urease and calcite production. J Ind Microbiol Biotechnol 36:981–988

Achal V, Mukherjee A, Goyal S, Reddy MS (2012a) Corrosion prevention of reinforced concrete with microbial calcite precipitation. ACI Mater J 109:157–164

Achal V, Mukherjee A, Reddy MS (2010) Biocalcification by *Sporosarcina pasteurii* using Corn steep liquor as nutrient source. Industrial Biotechnology 6:170–174

Achal V, Pan X, Zhang D (2012b) Bioremediation of strontium (Sr) contaminated aquifer quartz sand based on calcite precipitation induced by Sr resistant *Halomonas* sp. Chemosphere 89:764–768

Adolphe J-P, Loubière J-F, Paradas J, Soleilhavoup F (1990). Procédé de traitement biologique d'une surface artificielle. European Patent 90400G97.0

Burne RA, Chen RE (2000) Bacterial ureases in infectious diseases. Microbes Infect 2:533–542

Castanier S, Le Metayer-Levrel G, Perthuisot J-P (1999) Ca-carbonates precipitation and limestone genesis—the microbiogeologist point of view. Sed Geol 126(1–4):9–23

De Muynck W, De Belie N, Verstraete W (2010) Microbial carbonate precipitation in construction materials: a review. Ecol Eng 36:118–136

De Muynck W, Debrouwer D, De Belie N, Verstraete W (2008) Bacterial carbonate precipitation improves the durability of cementitious materials. Cem Concr Res 38:1005–1014

Dick J, De Windt W, De Graef B, Saveyn H, Van der Meeren P, De Belie N, Verstraete W (2006) Bio-deposition of a calcium carbonate layer on degraded limestone by *Bacillus* species. Biodegradation 17:357–367

Ghosh P, Mandal S, Chattopadhyay BD, Pal S (2005) Use of microorganism to improve the strength of cement mortar. Cem Concr Res 35:1980–1983

Gollapudi UK, Knutson CL, Bang SS, Islam MR (1995) A new method for controlling leaching through permeable channels. Chemosphere 30:695–705

Hammes F, Boon N, Clement G, de Villiers J, Siciliano SD, Verstraete W (2003) Molecular, biochemical and ecological characterisation of a bio-catalytic calcification reactor. Appl Microbiol Biotechnol 62:191–201

Jenneman GE, Knapp RM, McInerney MJ, Menzie DE, Revus DE (1984) Experimental studies of in situ microbial enhanced oil recovery. Soc Petrol Eng J 24:33–37

Jonkers HM, Schlangen E (2007) Crack repair by concrete-immobilized bacteria. In: Schmets AJM, Van der Zwaag S (eds) Proceedings of first international conference on self healing materials. noordwijk, the netherlands, p 7

Kantzas A, Ferris FG, Stehmeier L, Marentette DF, Jha KN, Mourits FM (1992) A novel method of sand consolidation through bacteriogenic mineral plugging (CIM 92-46). In: Proceedings of the CIM 1992 annual technical conference, Vol 2, pp 1–15. Petroleum Society of CIM, Calgary, Canada

Knoll AH (2003) Biomineralization and evolutionary history. Rev Mineral Geochem 54:329–356

Le Metayer-Levrel G, Castanier S, Orial G, Loubiere JF, Perthuisot JP (1999) Applications of bacterial carbonatogenesis to the protection and regeneration of limestones in buildings and historic patrimony. Sed Geol 126:25–34

Neville A (1995) Chloride attack of reinforced concrete: an overview. Mater Struct 28(2):63–70

Perry CC (2003) Silicification. Rev Mineral Geochem 54:291–327

Qian CX, Wang RX, Cheng L, Wang J (2010a) Theory of microbial carbonate precipitation and its application in restoration of cement-based materials defects. Chin J Chem 28(5):847–857. doi:10.1002/cjoc.201090156

Qian CX, Pan QF, Wang RX (2010b) Cementation of sand grains based on carbonate precipitation induced by microorganism. Sci China Technol Sci 53:2198–21206

Qian CX, Wang JY, Wang RX, Cheng L (2009) Corrosion protection of cement-based building materials by surface deposition of $CaCO_3$ by *Bacillus pasteurii*. Mater Sci Eng, C 29:1273–1280

Ramachandran SK, Ramakrishnan V, Bang SS (2001) Remediation of concrete using microorganisms. ACI Mater J 98:3–9

Ramakrishnan V, Bang SS, Deo KS (1998) A novel technique for repairing cracks in high performance concrete using bacteria. In: Proceedings of international conference on high performance high strength concrete, Perth, Australia, pp 597–618

Raupach M, Schiebl P (2001) Macrocell sensor systems for monitoring of the corrosion risk of the reinforcement in concrete structures. NDT and E Int 34:435–442

Reddy MS, Achal V, Mukherjee A (2012) Microbial concrete, a wonder metabolic product that remediates the defects in building structures. In: Satyanarayana T, Johry BN, Prakash A (eds) Microorganisms in Environmental Management: Microbes and Environment. Springer Publishers, USA, pp 547–568

Rodriguez-Navarro C, Rodriguez-Gallego M, Ben Chekroun K, Gonzalez-Muñoz MT (2003) Conservation of ornamental stone by *Myxococcus xanthus*-induced carbonate biomineralization. Appl Environ Microbiol 69:2182–2193

Rong H, Qian CX, Li LZ (2012) Study on microstructure and properties of sandstone cemented by microbe cement. Constr Build Mater 36:687–694
Rong H, Qian CX (2012) Development of microbe cementitious material in China. J Shanghai Jiaotong Univ (Sci) 17:350–355. doi:10.1007/s12204-012-1285-x
Sarayu K, Iyer NR, Ramachandra Murthy A (2014) Exploration on the biotechnological aspect of the ureolytic bacteria for the production of the cementitious materials—a review. Appl Bio Biotechnol 172:2308–2323. doi:10.1007/s12010-013-0686-0
Stocks-Fischer S, Galinat JK, Bang SS (1999) Microbiological precipitation of CaCO3. Soil Biol Biochem 31:1563–1571
Tiano P, Addadi L, Weiner S (1992). Stone reinforcement by induction of calcite crystals using organic matrix macromolecules: feasibility study. In 7th International Congress on Deterioration and Conservation of Stone, Lisbon. pp 1317–1326
Tiano P, Biagiotti L, Mastromei G (1999a) Bacterial bio-mediated calcite precipitation for monumental stones conservation: methods of evaluation. J Microbiol Methods 36:139–145
Tiano P, Biagiotti L, Mastromei G (1999b) Bacterial bio-mediated calcite precipitation for monumental stones conservation: methods of evaluation. J Microbiol Methods 36:138–145
Tiano P (1995) Stone reinforcement by calcite crystal precipitation induced by organic matrix macromolecules. Stud Conserv 40:171–176
Van Tittelboom K, De Belie N, De Muynck W, Verstraete W (2009) Use of bacteria to repair cracks in concrete. Cem Concr Res 40:157–166
Whiffin VS, Van Paassen LA, Harkes MP (2007) Microbial carbonate precipitation as a soil improvement technique. Geomicrobiol J 24:417–423
Worrell E, Price L, Martin N, Hendriks C, Meida LO (2001) Carbon dioxide emissions from the global cement industry. Annu Rev Energy Env 26:303–329

Chapter 15
Bacteria for Concrete Surface Treatment

Peihao Li and Wenjun Qu

Abstract Bacterial induced calcium carbonate deposition, i.e., biodeposition is a widespread natural process, occurring under different conditions in the biosphere. For the moment, biodeposition has been investigated extensively both in natural processes and under laboratory conditions. Biodeposition has led to the exploration in the field of construction materials and has been studied in detail with numerous applications in civil engineering. Various mechanisms of bacterial induced deposition have been proposed. Biodeposition can be influenced by the environmental physicochemical conditions, and it is correlated with both the metabolic activity and the cell surface structures of bacteria. Surface treatment of concrete materials and structures by means of biodeposition, i.e., a bacterially deposited carbonate layer presents a promising novel biotechnology for the enhancement or improvement of durability of concrete materials and structures. Biodeposition make bacterial concrete, a novel most important metabolic byproduct, can remediate concrete structures. This chapter reviews the main mechanisms of the process and literature on biodeposition carbonates as surface treatment agents for the decrease in permeability of concrete materials and structures, bacterial induced carbonates as a binder material, i.e., biocementation, have been added to concrete for the improvement of compressive strength and the remediation of concrete surface cracks. The chapter suggests potential applications of biodeposition as an ecological and novel alternative to traditional techniques in subsurface remediation of concrete structures and accordingly enhancement in their service life.

P. Li (✉)
College of Civil Engineering and Architecture, Zhejiang University of Technology, Hangzhou, China
e-mail: pumalph@163.com; pumalph@tongji.edu.cn

W. Qu
College of Civil Engineering, Tongji University, Shanghai, China
e-mail: Quwenjun@tongji.edu.cn

F. Pacheco Torgal et al. (eds.), *Biotechnologies and Biomimetics for Civil Engineering*,
DOI 10.1007/978-3-319-09287-4_15

15.1 Introduction

Concrete material is the most widely used in building all over the world. Current remediation of degraded concrete structures has drawn the attention to the methods to slow down or even to eliminate concrete degradation. Nowadays a broad range of organic and inorganic products is applied to the protection of concrete surfaces, such as a variety of coatings, water repellents, and pore blockers. Synthetic agents such as epoxies and surface treatments with water repellents such as silanes or siloxanes or with pore blockers are applied for remediation of these structures. These conventional surface treatments show, however, beside their pro influences also a number of disadvantageous aspects, e.g., different thermal expansion coefficient of the treated layers, degradation over time and the need for constant maintenance, and environmental pollution (Moon et al. 2007; De Muynck et al. 2008a, b). Organic treatments commonly result in the formation of incompatible and often harmful surface films. Additionally, because large quantities of organic solvents are used, they contribute to pollution (Camaiti et al. 1988). Inorganic consolidation may be preferable since stone materials and protective or consolidating materials share some physicochemical affinity (Rodriguez-Navarro et al. 2003). Surface cracks and fissures is an inevitable phenomenon during the course of concrete structures when exposed to weather changes. Shortcomings have drawn the attention to alternative techniques. It is necessary to develop some new eco-friendly self-healing techniques to concrete structures to meet various demands to enhance their durability. Bacterial induced carbonate deposition (i.e., biodeposition) treatment was similar as conventional treatments. Promising results of techniques based on microbial mineralization have lead to several investigations on the use of bacteria in concrete. Alkali-resistant spore-forming bacteria represent promising candidates for application in concrete and probably other cement-based materials (Jonkers et al. 2010). The use of stone consolidants aims at re-establishing the cohesion between grains of deteriorated stone. However, both conservation treatments are subject to frequent controversy due to their nonreversible action and their limited long-term performance. Because of problems related to incompatibility with the stone, both water repellents and consolidants have often been reported to accelerate stone decay (Moropoulou et al. 2003). Biodeposition has been proposed as an eco-friendly method to protect decayed building stone. The method relies on biodeposition formation of a compatible carbonate precipitate on limestone, and unlike the lime-water treatment, the carbonate cement appears to be highly coherent (Le Metayer-Levrel et al. 1999).

Bacteria are small, prokaryotic, microorganism that are ubiquitous in terrestrial and aquatic habitats. Some of them can cause deterioration to construction materials such as stone and concrete by the weathering action of various physical, chemical, and biological damage factors at the object site (Warscheid and Braams 2000; Gaylarde et al. 2003). Actually, some of them are also capable of remediation of buildings materials. Microbial activity can have strong impact on the durability of building materials. Some can cause deterioration to construction materials by the weathering action of various physical, chemical, and biological damage factors. The controlled use of bacteria offers new approaches for conservators to help preserve, protect, and restore building materials. Techniques based on biodeposition have lead to several investigations on the use of bacteria in concrete and the exploration in the field of construction materials. Biodeposition consolidating and/or protecting efficacy are of great worth, because bacteria induce carbonate cementation to a depth of several hundred micrometers ($\geq$500 μm). Furthermore, no plugging or blocking of pores takes place during this cementation (Rodriguez-Navarro et al. 2003). Bacterially induced carbonate mineralization has been proposed as a novel method on corrosion protection of construction materials (Qian et al. 2009; De Muynck et al. 2010). Bacterial induced/mediated calcium carbonate deposition, i.e., biodeposition is a widespread natural process, occurring under different conditions in the biosphere. For the moment, biodeposition has been investigated extensively both in natural processes and under laboratory conditions. Biodeposition has led to the exploration in the field of construction materials and has been studied in detail with numerous applications in civil engineering.

Bacterial induced carbonate deposition is a widespread natural process among bacteria, occurring under different conditions in the biosphere (Boquet et al. 1973) and common in different environments such as terrestrial and aquatic habitats (Castanier et al. 1999; Ehrlich 1998). Biodeposition applications include biomimetic processes and materials and examples of bioremediation (and stabilization) in several fields ranging from applied environmental microbiology, for example, leaching and bioremediation of inorganic contaminants to civil and environmental engineering, for example, bioplugging, sediment dikes, biogrouting, and remediation of concrete and limestone structures (Rodriguez-Navarro et al. 2003; Barabesi et al. 2007; De Muynck et al. 2010).

Some studies of biodeposition have proposed different mechanisms (Ehrlich 1996) and have pointed out the complexity of deposition that can be influenced by the environmental physicochemical conditions and its process is correlated with both the metabolic activity and the cell surface structures of bacteria (Fortin et al. 1997; Castanier et al. 1999). Although biodeposition is a widespread occurring natural process and has been investigated extensively both in natural environments and under defined laboratory conditions, the key role played by bacteria in the course of deposition is still worthy of further discussing (Von Knorre and Krumbein 2000; Zavarzin 2002).

Biodeposition has been proposed as a novel method on corrosion protection of building materials. Surface treatment of concrete materials and structures by means of biodeposition presents a promising novel biotechnology for the enhancement or improvement of durability of concrete materials and structures. This chapter starts with a brief overview of the main commonly recognized features of biodeposition (Sect. 7.2), followed by the description of concrete surface treatment based on biodeposition and literature on biodeposition carbonates as surface treatment agents for the decrease of permeability in concrete materials and structures (Sect. 7.3) and bacterial self-healing concrete and bacterial induced carbonates as a binder material, i.e., biocementation, have been added to concrete for the improvement of compressive strength and the remediation of concrete surface cracks (Sect. 7.4). The chapter suggests potential applications of biodeposition as an eco-novel alternative to traditional techniques in remediation of defects in concrete structures and accordingly enhancement in their service life.

15.2 Bacterial Induced Deposition

15.2.1 Bacterial Induced/Mediated Deposition

Bacterial induced carbonate deposition is a widespread natural process, occurring by most soil bacterial isolates on a solid medium with added calcium and concluded that under suitable conditions (Boquet et al. 1973), biodeposition as a process which bacteria conduct both in bacterially induced (active) and/or in bacterially influenced or mediated (passive) way (Dupraz et al. 2009). Different bacterial strains can induce deposition of different amounts, shapes, and types of carbonate crystals from exactly the same synthetic medium, with an apparent occurring of environment- and species-specific biodeposition (Hammes and Verstraete 2002). Some authors emphasize that carbonate deposition by bacteria is an unwanted byproduct of bacterial physiological activities under special environmental conditions, a simple physiological accident. Bacteria would not precipitate carbonate particles by a specific mechanism and the supply of a structure by bacteria would not be necessary (von Knorre and Krumbein 2000). Other authors emphasize that the role of bacteria in carbonate deposition can be specific with ecological benefits for the precipitating organisms (McConnaughey and Whelan 1997; Castanier et al. 1999; Barabesi et al. 2007). Castanier et al. (1999) distinguish between passive and active deposition mechanisms which may occur, often concurrently, in heterotrophic bacteria. Passive deposition (or passive carbonatogenesis) operates by producing carbonate and bicarbonate ions and inducing chemical modifications in the medium through metabolic pathways (e.g., linked to nitrogen and sulfur cycles). In active deposition (or active carbonatogenesis), the carbonate particles would be produced by ionic exchanges through the cell membrane by activation of calcium and/or magnesium ionic pumps or channels, probably coupled with carbonate ion production. Active deposition would be independent of specific metabolic pathways (von Knorre and Krumbein 2000). Accordingly, the role of bacteria in $CaCO_3$ deposition and different mechanisms of biodeposition were proposed by different author (Ehrlich 1996), but the key role played by bacteria in the processes is still debated and worthy of further discussing. They include calcium concentration of the medium by microbial binding, metabolic alteration of the medium that results in changes in bicarbonate concentration and pH, and bacterial bodies acting as crystal nucleation sites (Little et al. 1997). It seems that different bacteria as well as abiotic factors seem to contribute in a different ways to carbonate deposition in a wide range of environments. Biodeposition is a specific type of the biologically induced mineralization, referring to deposition that results from interactions between bacterial metabolic activities and the natural environment (Weiner and Dove 2003; Dupraz et al. 2009). Bacterial metabolic activities and cell surface structures and their interactions with environmental physicochemical parameters are commonly recognized as the key factors in carbonate deposition (Fortin et al. 1997; Douglas and Beveridge 1998; Rodriguez-Navarro et al. 2003). Essential condition of carbonate deposition is a carbonate alkalinity and the availability of free calcium ions (the two are combined as a saturation index, i.e., SI) and concentrations of both free carbonate $CO_3{}^{2-}$ and Ca^{2+} ions must exceed saturation (Dupraz et al. 2009; Decho 2010), carbonate deposition is a rather straightforward chemical process governed by four key factors (Hammes and Verstraete 2002):

(1) the calcium (Ca^{2+}) concentration,
(2) the concentration of dissolved inorganic carbon (DIC),
(3) the pH, and
(4) the availability of nucleation sites.

In the deposition process, bacteria create an alkaline environment by an increase in pH to 8.0 and higher and a DIC increase through their metabolic activities (Castanier et al. 1999; Douglas and Beveridge 1998).

Bacterial surface cells are able to induce pH variations in the medium, which result in variations of the pH in the surrounding microenvironment (Fortin et al. 1997). Accordingly, bacterial surface cells also play an important role in calcium deposition (Fortin et al. 1997). Surface bacterial (macro) molecules can induce carbonate deposition, providing a template for carbonate nucleation, both as part of bacterial cell and as cell-free, when released in the environment; in the latter case, primarily as polymeric substances. Bacterial surfaces, and consequently bacterial cells, can act as important sites for the absorption of cations and constitute particularly favorable templates for heterogeneous nucleation and crystal growth (Fortin et al. 1997). Due to the presence of several negatively charged groups, at neutral pH, positively charged metal ions could be bound on bacterial surfaces, favoring heterogeneous nucleation (Fortin et al. 1997; Douglas and Beveridge 1998). Among surface structures, bacterial cell walls have been studied for their ability to complex metals (Jiang et al. 2004). Negative charges result predominantly from deprotonation of carboxyl, phosphate, and hydroxyl functional groups exposed on the outer surface of the cell wall (Fein et al. 1997). The first step is a stoichiometric interaction of metal with reactive chemical groups, which reside primarily in the peptidoglycan. After complexation, these sites nucleate the deposition of more metal as a chemical precipitate. Surface reactivity of bacterial cells depends on the metabolic state of the cells (Jiang et al. 2004). In the absence of metabolic activity, passive interactions may occur in which microbial cells (inactive living or dead) behave as solid-phase sorbents of dissolved metals, and heterogeneous nucleation templates for authigenic mineral deposition (Beveridge 1989). *B. subtilis* dead cells, as well as a cell fraction comprising the cell wall, were demonstrated able to induce calcite formation in a laboratory test, acting as heterogeneous crystallization nuclei, with an applicative potential in the reinforcement of monumental calcareous stones (Barabesi et al. 2003). Functional groups on the *B. subtilis* cell wall were considered able to bind dissolved Ca^{2+} in a calcite dissolution study with dead cells (Friis et al. 2003). Like other deposition processes, calcium carbonate ($CaCO_3$) deposition can occur by two different mechanisms: biologically controlled or induced. In biologically controlled mechanisms, the organism controls the process, i.e., nucleation and growth of the mineral particles, to a high degree. The organism synthesizes crystals in a form that is unique to that species, independently of environmental conditions. Examples of controlled mineralization are magnetite formation in magnetotactic bacteria and silica deposition in the unicellular algae coccolithophores and diatoms, respectively (Bazylinski et al. 2007; Barabesi et al. 2007). Different types of bacteria, as well as abiotic factors (salinity and composition of the synthetic medium), seem to contribute in a variety of ways to calcium carbonate deposition in a wide range of different environments (Von Knorre and Krumbein 2000; Rivadeneyra et al. 2004). The type of calcium carbonate deposition is largely dependent on the environmental conditions; accordingly, biodeposition has been generally regarded as induced mechanisms (Rivadeneyra et al. 1994; Perito and Mastromei 2011).

15.2.2 Role of Bacteria on Calcium Carbonate Deposition

According to calcium carbonate deposition governed by four key factors (Hammes and Verstraete 2002), $CaCO_3$ deposition requires sufficient calcium and carbonate ions so that the ion activity product (IAP) exceeds the solubility constant (Ksp) (Eqs. (15.1) and (15.1)). From the comparison of the IAP with the Ksp, the saturation state (Ω) of the system can be defined; if $\Omega > 1$ the system is oversaturated and deposition is likely (Morse 1983):

$$Ca^{2+} + CO_3^{2-} \leftrightarrow CaCO_3 \tag{15.1}$$

$$\Omega = \left(Ca^{2+}\right)\left(CO_3^{2-}\right)/K_{sp} \text{ with } K_{spcalcite,25^\circ} = 4.8 \times 10^{-9} \tag{15.2}$$

The concentration of carbonate ions is related to the concentration of DIC and the pH of a given aquatic system. In addition, the concentration of DIC depends on several environmental parameters such as temperature and the partial pressure of carbon dioxide (for systems exposed to the atmosphere). The equilibrium reactions and constants governing the dissolution of CO_2 in aqueous media (25 °C and 1 atm) are given in Eqs. (15.3)–(15.6) (Stumm and Morgan 1981):

$$CO_{2(g)} \leftrightarrow CO_{2(aq)} \ (pK_h = 1.468) \tag{15.3}$$

$$CO_{2(aq)} + H_2O \leftrightarrow H_2CO_3 \ (pK = 2.84) \tag{15.4}$$

$$H_2CO_3^* \leftrightarrow H^+ + HCO_3^- (pK1 = 6.352) \tag{15.5}$$

$$HCO_3^- \leftrightarrow CO_3^{2-} + H^+ \ (pK_2 = 10.329) \tag{15.6}$$

Several authors recognized specific metabolic pathways to be involved in carbonate deposition, with the pH change of the medium as the main mechanism favoring deposition. In general, metabolic pathways able to shift the environmental pH toward alkalinity can, in the presence of calcium ions, foster carbonate deposition when a state of oversaturation develops (Fortin et al. 1997). Bacteria can induce deposition by altering almost any of the deposition parameters described above, either separately or in various combinations with one another. Both autotrophic and heterotrophic pathways create an alkaline environment. While the environmental conditions of heterotrophic pathways are diverse (aerobiosis, anaerobiosis, and microaerophily), carbonate deposition always appears to be a response of the heterotrophic bacterial communities to an enrichment of the environment in organic matter (Castanier et al. 1999). Besides changes induced in the macro-environment, bacteria have also been reported to influence calcium carbonate deposition by acting as sites of nucleation or calcium enrichment (Morita 1980). Due to the presence of several negatively charged groups on the cell wall, at a neutral pH, positively charged metal ions can be bound on bacterial surfaces (Douglas and Beveridge 1998). Such bound metal ions (e.g., calcium) may subsequently react with anions (e.g., carbonate) to form an insoluble salt (e.g., calcium carbonate). In the case of a sufficient excess of the required cations and anions, the metal salt on the cell surface initiates mineral formation by acting as a nucleation site. The anion (e.g., carbonate) in this reaction may be a product of the bacterial metabolism, or it may have an abiotic origin (Ehrlich 1998). Furthermore, it has been demonstrated that specific bacterial outer structures (glycocalyx and parietal polymers) consisting of exopolysaccharides and amino acids play an essential role in the morphology and mineralogy of bacterially induced carbonate deposition (Braissant et al. 2003; Ercole et al. 2007).

An alkaline barrier formed as a result of bacterial metabolic activities, responsible for carbonate deposition. The alkaline barrier is primarily due to decomposition of anions rather than to production of alkali. Organisms that increase the medium pH by the elimination of anions leave the scene ready for calcium deposition. Rather than activities of single species or groups, activities of microbial communities should be considered in natural environments (Zavarzin 2002). Metabolic pathways involved in bicarbonate deposition include autotrophic as well as heterotrophic pathways (both in aerobiosis and in anaerobiosis) with a different contribute. Three main kinds of bacteria are involved in autotrophic production methanogenic archaebacteria, sulfurous or non-sulfurous, purple and green bacteria, and cyanobacteria. All obtain carbon from gaseous or dissolved CO_2, the origin of which is complex and use it as carbon source to produce organic matter. These pathways induce local CO_2 depletion either of the medium or of the immediate environment of bacteria (Castanier et al. 1999). When calcium ions are present in well-buffered alkaline or neutral media, such depletion favors carbonate deposition according to the overall reaction (Eq. 15.7):

$$Ca^{2+} + 2HCO_3^- \leftrightarrow CaCO_3 + CO_2 + H_2O \quad (15.7)$$

During consumption of bicarbonate in photosynthesis, cyanobacteria generate carbonate and thereby create alkaline surroundings (Eqs. 15.8 and 15.10), which induce the deposition of carbonate by Ca^{2+} dissolved in water. The process is based on the metabolic utilization of dissolved CO_2, which exists in chemical equilibrium with HCO_3^- and CO_3^{2-} (Eq. 15.9) in the medium surrounding the bacteria. This would induce a shift in the bicarbonate equilibrium and a subsequent pH rise in the bulk medium, deposition could occur if soluble calcium ions are present (Eqs. 15.9 and 15.10). So the most common form of bicarbonate deposition is in excuse of carbonate deposition by photosynthetic cyanobacteria in aqueous environments such as marine and/or freshwater ones (McConnaughey and Whelan 1997; Ehrlich 1998).

$$2HCO_3^- \leftrightarrow CO_2 + CO_3^{2+} + H_2O \quad (15.8)$$

$$CO_2 + H_2O \leftrightarrow (CH_2O) + O_2 \quad (15.9)$$

$$CO_3^{2-} + H_2O \leftrightarrow HCO_3^- + OH^- \quad (15.10)$$

$$HCO_3^- + H_2O \leftrightarrow (CH_2O) + O_2 + OH^- \quad (15.11)$$

$$HCO_3^- + OH^- \rightarrow CO_3^{2+} + H_2O \quad (15.12)$$

The cyanobacterium can convert intracellular HCO3$^-$ photosynthetically into reduced carbon (CH2O), extracellular HCO_3^- is exchanged for intracellular OH^- across the cell membrane (Eq. 15.11). The extracellular OH^- generates an alkaline peri-cellular region where CO_3^{2-} is generated from HCO_3^- (Eq. 15.12). The resulting CO_3^{2-} immediately reacts with Ca^{2+} at the cell surface to form $CaCO_3$. The carbonate deposits can be in the form of marl sediment and massive bioherms (Ehrlich 1998).

Biodeposition generally results from a series of complex biochemical reactions (Stocks-Fischer et al. 1999) involving urease (urea amidohydrolase). Urea or uric acid hydrolysis by the urease enzyme is a simple model, and it has often been used in technological application of bicarbonate deposition. During microbial urease activity, 1 mol of urea is hydrolyzed intracellularly to 1 mol of ammonia and 1 mol of carbamate (Eq. 15.13), which spontaneously hydrolyses to form an additional 1 mol of ammonia and carbonic acid (Eq. 15.14) (Burne and Chen 2000).

$$CO(NH_2)_2 + H_2O \rightarrow NH_2COOH + NH_3 \quad (5.13)$$

$$NH_2COOH + H_2O \rightarrow NH_3 + H_2CO_3 \quad (5.14)$$

These products subsequently equilibrate in water to form bicarbonate and 2 mol of ammonium and hydroxide ions (Eqs. 15.15 and 15.16). The two reactions result in a pH increase, which in turn shifts the bicarbonate equilibrium, resulting in the formation of carbonate ions (Eq. 15.17). The pH increase takes place initially in the local microenvironment around the bacterial cell and propagates in the bulk solution of the bacterial cell suspension.

$$H_2CO_3 \leftrightarrow HCO3^- + H^+ \quad (5.15)$$

$$NH_3 + 2H_2O \leftrightarrow 2NH_4^+ + 2OH^- \quad (5.16)$$

$$HCO_3^- + H^+ + 2NH_4^+ + 2OH^- \leftrightarrow CO_3^{2-} + 2NH_4^+ + 2H_2O \quad (5.17)$$

Thus, the carbonate concentration will increase, inducing an increase in Ω (according to Eq. 15.2) and resulting in $CaCO_3$ deposition around the cell in the presence of soluble calcium ions as shown in Eqs. 15.18 and 15.19.

$$Ca^{2+} + Cell \rightarrow Cell - Ca^{2+} \quad (5.18)$$

$$Cell - Ca^{2+} + CO_3^{2-} \rightarrow Cell - CaCO_3 \downarrow \quad (5.19)$$

The pH of solution occurs have strongly influence on the production of $CO_3{}^{2-}$ from bicarbonate ($HCO_3{}^-$), an increase in $CO_3{}^{2-}$ concentration with increase of alkaline conditions. Therefore, calcium carbonate deposition readily occurs in alkaline environments abundant of the calcium (Ca^{2+}) and carbonate ($CO_3{}^{2-}$) ions (Stocks-Fischer et al. 1999; Ramachandran et al. 2001; Qian et al. 2010a). Urease activity promotes deposition outside the cells. Calcium ions in the solution are attracted to the bacterial cell wall with the negative charge. In the presence of calcium ions, a local supersaturation result in heterogeneous deposition of calcium carbonate occurs on the bacterial cell wall. Furthermore, it has been demonstrated that specific bacterial outer structures (glycocalyx and parietal polymers), a variety of organic polymers outside the cell wall (Lappin-Scott et al. 1988; MacLeod et al. 1988) or exopolymeric substances consisting of exopolysaccharides, proteins, and amino acids play an essential role in the morphology and mineralogy of bacterially induced carbonate deposition (Braissant et al. 2003; Ercole et al. 2007). The carbonate deposition has subsequently led to the exploration of this process in the field of bioremediation. The use of carbonate deposition has also been proposed for the removal of heavy metals (Warren et al. 2001) and biodegradation of pollutants (Simon et al. 2004; Chaturvedi et al. 2006). Urease is the key enzyme involved in the process of calcite deposition induced by bacteria. Bacteria are known to hydrolyze urea by urease for the purposes of (1) increasing the ambient pH (Burne and Marquis 2000), (2) utilizing it as a nitrogen source (Burne and Chen 2000), and (3) using it as a source of energy. There are many species of Bacillus reported that produce large amount of urease, further helps in carbonate deposition and biocementation (De Muynck et al. 2008a, b; Achal et al. 2009). Sporosarcina pasteurii is another specialist organism that has a different use for urease, other than nitrogen assimilation. *S. pasteurii* is a moderately alkaliphilic organism with a growth optimum at pH 9.25. Urea hydrolysis is the most easily controlled of the carbonate generating reactions, with the potential to produce high concentrations of carbonate within a short time. Beside conventional bioremediation measurement, a number of applications involving bacterial deposition have been attempted in the area of construction industry.

Bacterial induced carbonate deposition has subsequently led to the exploration of this process in a variety of fields. A first series of applications is situated in the field of bioremediation. In addition to conventional bioremediation strategies which rely on the biodegradation of organic pollutants (Chaturvedi et al. 2006; Simon et al. 2004). Applications include the treatment of groundwater contaminated with heavy metals (Warren et al. 2001) and radionucleotides (Fujita et al. 2004), the removal of calcium from wastewater (Hammes et al. 2003a, b). Another series of applications aims at modifying the properties of soil, i.e., for the enhancement of oil recovery from oil reservoirs (Nemati and Voordouw 2003; Nemati et al. 2005), plugging (Ferris and Stehmeier 1992) and strengthening of sand columns (DeJong et al. 2006; Whiffin et al. 2007). In recent years, biodeposition has been investigated for its potential to improve the durability of concrete materials. The deposition can be divided into processes for the deposition of a protective surface layer with consolidating and/or waterproofing properties, i.e., biodeposition for surface treatment, and processes for the generation of a biologically induced binder, i.e., biocementation for bacterial concrete. Following section deals with detailed review regarding parameters affecting the durability of concrete materials and structures.

15.3 Concrete Surface Treatment by Biodeposition

15.3.1 Bioremediation of Concrete Surface Cracks

Concrete materials have been used extensively as worldwide construction materials. They have been typically characterized by a high-compressive strength and a relative low-tensile strength. The latter can be offsetted by the application of steel or other material reinforcements taking over tensile forces. High-tensile stresses may be result from external loads, imposed deformations (due to confined shrinkage, temperature gradients, or differential settlement), plastic settlement, plastic shrinkage, and expansive reactions (e.g., due to alkali silica reaction, reinforcement corrosion, or sulfate attack). Microcrack formation is a commonly phenomenon in concrete structures due to various natural or human activities, execrably affects concrete structural properties, increases permeability, and substantially reduces the durability of concrete structures ingress of detrimental components in moist environments. Once microcracks form a continuous network, they may provide an easy path for the transport of potentially contain detrimental substances and substantially contribute to the permeability of cementitious materials, thereby reducing the resistance against attack of detrimental substances. If microcracks grow further and reach the reinforcement, not only the concrete itself may be attacked, but also the reinforcement will be corroded when it is exposed to water and oxygen, and possibly carbon dioxide and chlorides. Microcracks may reduce or impair the durability of concrete structures and are therefore precursors to structural failure. Some important measures were taken to diminish the permeability of detrimental components or to slow down or even to eliminate concrete degradation. Many of techniques are available but traditional repair systems have a number of disadvantageous aspects such as different thermal expansion coefficient compared to concrete and environmental and health hazards and so on. Biodeposition plays an important role in limiting the infiltration of detrimental components into concrete and lead to the exploration of remediation technique in the field of cementitious materials (Peihao et al. 2012). Cracks were plugged and healed with a mixture of bacteria, nutrients, and a filler material. Among the different materials that were mixed with *S. pasteurii*, the silica fume and sand mixture lead to the highest compressive strength and lowest permeability. Different bacteria have been used to increase the compressive strength of cement mortar and for the remediation of cracks in concrete. Ghosh et al. (2005) demonstrated the positive effect of the addition of Shewanella on the compressive strength of mortar specimens. An increase of 25 % of the 28 days compressive strength was obtained for a cell concentration of about 105 cellsmL^{-1}. Microbial cells also prevent ingress of water effectively in different concentrations of fly ash-amended concrete. It is therefore more advisable and economical to restrict the development of early age small cracks the moment they appear, than to repair them after they have developed to large cracks.

The use of a microbial mineral plugging system based on the deposition of carbonates was suggested (Ferris and Stehmeier 1992; Zhong and Islam 1995). While initial research on biodeposition in sand columns was mainly focused on the decrease in porosity and permeability as a result of the physical presence of the newly formed carbonates (Ferris and Stehmeier 1992), subsequently investigations focus on the improvement of strength as a result of the cementation of sand particles due to the particle binding properties of the bacterially deposition carbonates (Kucharski et al. 2006). The hydrolysis of urea was selected as a very suitable pathway for the production of carbonate ions due to its ability to alkalinize the environment. Furthermore, urea is an important organic nitrogen carrier in natural environments and is commonly used as an agricultural fertilizer (Nielsen et al. 1998). Furthermore, the ability to hydrolyze urea is widely distributed among indigenous bacteria in soils and groundwater systems (Fujita et al. 2000). Urea-utilizing bacteria such as *Sporosarcina pasteurii* and *Sporosarcina ureae* are commonly isolated from soil, water, sewage, and incrustations on urinals. The participation of *S. pasteurii* in sand consolidation has been demonstrated by Kantzas et al. (1992). Gollapudi et al. (1995) further investigated the use of *S. pasteurii* for the plugging of sand columns. Although the bacteria were mixed with the sand slurry, consolidation mainly occurred near the surface.

The use of bacteria products as a long-term remediation tool has promised high potential for crack healing of various structural formations such as granite and concrete (Gollapudi et al. 1995). Stocks-Fischer et al. (1999) showed that bacteria were directly involved in deposition by providing a nucleation site and by creating an alkaline environment which favored carbonate deposition. Zhong and Islam (1995) used the consolidation of sand mixtures for the remediation of cracks in granite. Cracks in granite were packed with a mixture of bacteria, nutrients, and a filler material. Among the different materials that were mixed with *B. pasteurii*, the silica fume (10 %) and sand (90 %) mixture lead to the highest compressive strength and lowest permeability. Bacterially enhanced crack remediation has been reported by Bang et al. (2001). They used Bacillus pasteurii to induce $CaCO_3$ deposition. Scanning electron micrography (SEM) and X-ray diffraction (XRD) analysis has shown the direct involvement of bacteria in the process of deposition. As a further extension to this research, Ramachandran et al. (2001) investigated the microbiological remediation of cracks in concrete and proposed carbonate deposition as an effective way to seal cracks. The appearance of cracks and fissures is an inevitable phenomenon during the aging process of concrete structures upon exposure to weather changes. If left untreated, cracks tend to expand further and eventually lead to costly repair. Specimens with cracks filled with bacteria, nutrients, and sand demonstrated a significant increase in compressive strength and stiffness values when compared with those without cells. The presence of calcite was, however, limited to the surface areas of the crack. *B. pasteurii* grows more actively in the presence of oxygen. Still, the highly alkaline pH (12.5) of concrete was a major hindering factor to the growth of the moderate alkaliphile *B. pasteurii*, whose growth optimum is around a pH of 9. To retain high metabolic activities of bacterial cells at such a high pH, immobilization technology (where microbial cells are encapsulated in polymers) can be applied. Bang et al. (2001) found that physicochemically versatile polyurethane is an effective enhancement tool in microbiologically induced calcite deposition in concrete cracks. In order to protect the cells from high pH, Day et al. (2003) investigated the effect of different filler materials on the effectiveness of the crack remediation. Beams treated with bacteria and polyurethane showed a higher improvement in stiffness compared to filler materials such as lime, silica, fly ash, and sand. The porous nature of the polyurethane minimizes transfer limitations to substrates and supports the growth of bacteria more efficiently than other filling materials, enabling an accumulation of calcite in deeper areas of the crack. No differences could be observed between the overall performances of free or polyurethane-immobilized cells in the deposition of carbonate. As an extension to their research on biodeposition on cementitious materials, De Muynck et al. (2008a, b) further investigated the use of bacterially induced carbonate deposition for the repair of concrete cracks. Bacteria were immobilized with a silica sol to protect *B. sphaericus* from the alkaline pH conditions. Upon the addition of a salt, a bioceramic material (biocer) was formed, which was able to bridge the crack. Subsequent addition of a urea and calcium chloride solution resulted in the deposition of carbonate crystals inside the pores of the biocer and concomitantly sealing of the crack. As a result, a decrease in the water permeability, similar to that obtained with traditional epoxy injections, was observed. Cracks filled with a mixture of *S. pasteurii* and sand showed a significant increase in compressive strength and stiffness, compared to cracks without cells. Microscopy confirmed the presence of calcite crystals and cells near the surface of the cracks. Ramachandran et al. (2001) studied the effect of microbiological calcite deposition of cracks of various depths on the compressive strength values in Portland cement mortar cubes and found an increase of the strength in the presence of *B. pasteurii* in the cubes prepared with the deepest cracks (25.4 mm), whose microbial remediation increased the compressive strength by approximately 61 % of that of the control concrete. Recently, Qian et al. (2010a) reported that the compressive strength of the treated specimens could be restored to 84 % and demonstrated that this biorestoration method is effective in repairing surface defects of cementitious materials. Sand particle surfaces were covered by $CaCO_3$ deposition which could cement loose sand and supply conjunction between individual particles. SEM results indicated that $CaCO_3$ crystals were precipitated on crack surface, whereas sand was consolidated and cemented by $CaCO_3$ crystals resulting in compressive strength recovery.

Some form of enhanced crack repair might be obtained through biodeposition treatment with *B. sphaericus* culture, which is incorporated in a gel matrix and a calcium source is provided, and silica gel was used to protect the bacteria against the high pH in concrete. Protection of the bacteria by means of this gel matrix seemed to be effective as $CaCO_3$ crystals were precipitated inside the matrix which was not the case if only bacteria were used, without immobilization in the silica gel (see Fig. 15.1). Crack sealing by means of this biological treatment resulted in a decrease in water permeability. However, it was seen that the decrease in water flow was also obtained if autoclaved bacteria were used instead of active bacteria. This corroborates that the greater part of the decrease in water permeability is attributed to crack filling by the sol–gel matrix. TGA analysis on the crack repair material showed only in the case of active bacteria the presence of $CaCO_3$ crystals. Precipitation of these crystals inside the gel matrix may enhance the durability of this repair material. Efficiency of this biological treatment was also evaluated by means of ultrasonic transmission measurements and visual examination. Crack treatment with *B. sphaericus*, immobilized in silica gel, resulted in an increase in ultrasonic pulse velocity, indicating that crack bridging was obtained. Visual examination of the cracks proved that this technique resulted in complete filling of the cracks. The use of this biological repair technique is highly desirable because the mineral precipitation induced as a result of microbial activities is pollution-free and natural. However, further experiments have to be done to examine the durability of this crack repair technique (Van Tittelboom et al. 2010). In Fig. 15.1, some top views of the specimens and some cross-sections of the treated cracks are shown. The top views were taken from realistic cracks after the specimens were subjected to the water permeability test. Cross-sections were obtained by sawing samples with repaired standardized cracks after ultrasonic measurements were taken. Figure 15.1a shows a non-treated crack. At both crack faces, crystal deposition can be observed, showing that the untreated specimens had undergone a certain extent of autogenous healing during the water permeability test. For the specimens with treated cracks, no crystals were observed after performance of the water permeability test. This can be explained as follows: for the untreated samples, the water flow was so fast that the upper compartment, of the test setup, became completely empty between two successive readings. This brought the concrete surface into contact with the atmosphere and led to carbonation of Ca(OH)2 into CaCO3 crystals. In Fig. 15.1b, a crack treated with BS + CaCl2 is shown. No CaCO3 crystals were detected by means of the microscope used, probably, because bacteria were not protected against the high pH in concrete. However, when the bacteria are immobilized in sol–gel, complete filling of the cracks occurs as shown in Fig. 15.1c. Figure 15.1d, e shows a cross-section of a standard crack filled with grout and epoxy, respectively. Cement grout only covers the surface of the samples and does not fill the cracks because of the big grain size of the grout. Epoxy treatment, by contrast, resulted in complete filling of cracks of both 10 and 20 mm deep. Treatment with only sol–gel or with sol–gel + BS + $CaCl_2$ resulted in cracking of the gel matrix as shown in Fig. 15.1g, h. When the gel hardens, it shrinks and this gives rise to cracking. Samples treated with BS in sol–gel + CaCl2 or $Ca(NO_3)_2$ or Ca $(CH_3COO)_2$ were placed in a urea–calcium solution immediately after filling of the cracks with silica gel and bacteria. During immersion, bacteria started to precipitate CaCO3 resulting in a compete filling of the cracks (see Fig. 15.1c). However, complete filling was only feasible for 10-mm deep cracks (see Fig. 15.1f). As shown in Fig. 15.1i, these treatments were not able to fill 20-mm deep cracks. This was also observed from the ultrasonic measurements. The 10-mm deep cracks treated with BS in sol–gel + CaCl2 or Ca(NO3)2 or Ca(CH3COO)2 performed almost as good as cracks treated with epoxy, which was no longer the case for 20-mm deep cracks, which were only completely filled when epoxy was used.

Microbe cement as a biogrouting could consolidate loose particles to improve mechanical properties. Microbe cement has drawn much attention because of the ever increasing awareness of environmental protection. This paper confirms the feasibility of binding loose sand particles using microbe cement and details the cementation mechanism of microbe cement. In addition, the microstructure and properties of representative bio-sandstones have been analyzed by X-ray computed tomography (XRCT), scanning electron microscopy (SEM), and mercury intrusion porosimetry (MIP). The experimental results indicate that the compressive strength of bio-sandstone could be up to 6.1 MPa and the bottom region microstructure in bio-sandstone is denser and less fragile than the top region due to more calcite precipitated in the former one (Rong et al. 2012). Kim et al. (2013) placed the

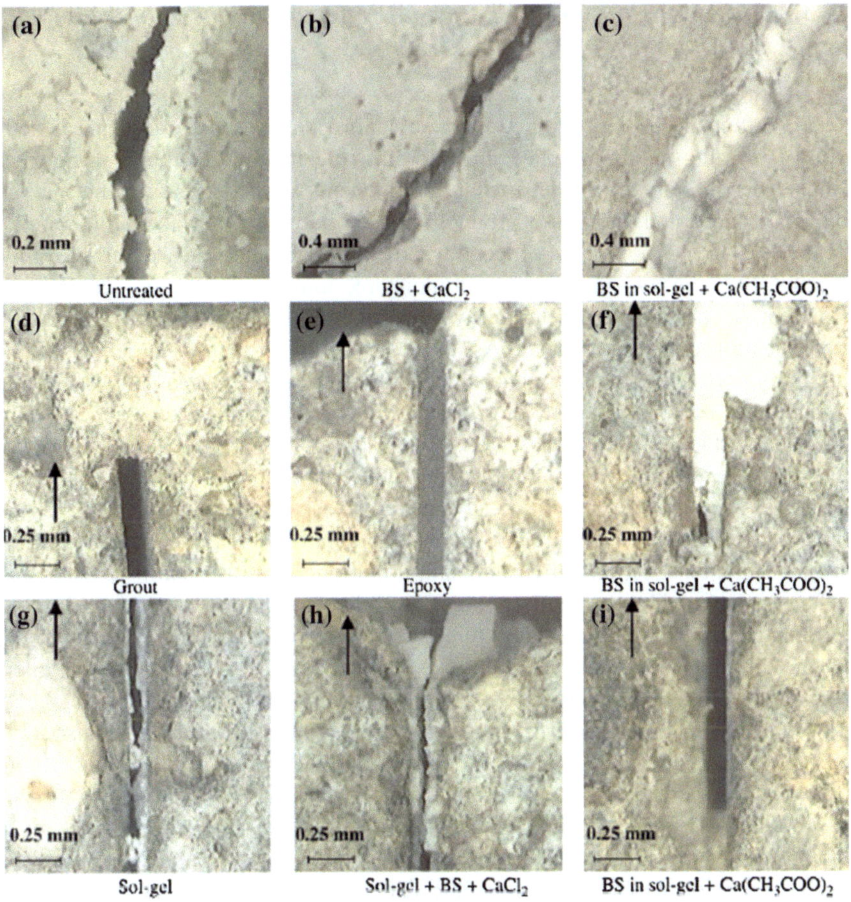

Fig. 15.1 *Top view* of an untreated crack (**a**), crack repaired with BS + $CaCl_2$ (**b**) or BS in sol–gel + $Ca(CH_3COO)_2$ (**c**), cross-section of cracks repaired with grout (**d**), epoxy (**e**), sol–gel (**g**), Sol–gel + BS + $CaCl_2$ (**h**) and cross-section of the tip of a 10 mm (**f**) and 20 mm (**i**) deep crack repaired with BS in sol–gel + $Ca(CH_3COO)_2$ (*upper surface* of sample is indicated with an *arrow*) (Van Tittelboom et al. 2010)

emphasis of the studies on the comparison of the characteristics of calcium carbonate precipitation on concrete by two types of ureolytic bacteria, *S. pasteurii* and *B. sphaericus*. The present treatment methods can also be applied on horizontal and vertical surfaces. Both the normal and lightweight concrete specimens treated by the medium with *B. sphaericus* showed the lowest weight increases per unit area. It was considered that the denseness of the calcium carbonate crystals of the medium with *B. sphaericus* was higher than that of the cases with other treatments (Kim et al. 2013).

15.3.2 Influence on CO_2, Cl^-, H_2O Penetrativity of Concrete by Biodeposition

An important measure to protect concrete against damage is diminishing penetrativity. Surface treatments play an important role in limiting the infiltration of water and consequently of detrimental components into concrete. In the market, lots of organic and inorganic products are available for surface protection of concrete, such as a variety of coatings, water repellents, and pores blockers. Penetrativity controls the ingress of moisture, ionic and gaseous species into concrete; accordingly, the durability of concrete depends on the permeation properties of concrete surface. As the permeation of concrete decreases, its durability performance resisting physicochemical corrosion increases. Nowadays a broad range of organic and inorganic products is applied to the protection of concrete surfaces, such as a variety of coatings, water repellents, and pore blockers. These conventional surface treatments show, however, beside their pro influences also a number of disadvantageous aspects, e.g., different thermal expansion coefficient of the treated layers, degradation over time and the need for constant maintenance, and environmental pollution. There are some conventional techniques available to improve impermeability to enhance the durability of building materials. Chemical admixtures such as plasticizers, superplasticisers, and water reducing agents help to improve the workability by reducing the intergranular friction ultimately affecting the porosity and distribution of pores. However, these conventional methods of protection have a number of disadvantages, such as (1) an incompatibility of the protective layer and the underlying layer due to differences in their thermal expansion coefficient; (2) disintegration of the protective layer over time; and (3) a need for constant maintenance by a treatment that is reversible and repeatable. Such chemical treatments commonly result in the formation of incompatible and often harmful surface films. Additionally, because large quantities of chemical solvents are used, they contribute to pollution (Camaiti et al. 1988; Rodriguez-Navarro et al. 2003) and also such techniques are not long lasting.

Shortcomings have drawn the attention to alternative techniques to reduce permeation properties for the improvement of the durability of concrete. Tiano (1995) proposed the use of organic matrix macromolecules extracted from *Mytilus californianus* shells to induce the deposition of calcium carbonate within the pores of the stone. The organic matrix was shown to produce a more relevant and durable carbonate deposition compared to the single use of calcium chloride or hydroxide. This deposition resulted in a slight decrease in porosity and water absorption by capillarity. Tiano et al. (1999) also observed a reduction of about 60 % from the limestone samples treated with biodeposition. Le Metayer-Levrel et al. (1999) studied the bacterial carbonatogenesis for the protection and regeneration of limestone in buildings, monuments, and statues. The biodeposition of protective surface on the stone was confirmed to reduce its permeability for gas without affecting its esthetic appearance. They further state that biodeposition layer material is remarkably similar to the limestone substrate. Moreover, biodeposition follows the same natural process that formed many limestones. Finally, they concluded that bacterial mortars or cement could be used to affix small pieces broken from statues and to fill small cavities on limestone surfaces. Nemati and Voordouw (2003) noticed a decrease in the permeability of sandstone cores after injecting $CaCO_3$ forming reactants. Dick et al. (2006) studied such deposition on degraded limestone by Bacillus species. They concluded that such deposition reduces the water absorption rate of limestone. They studied the capillary water absorption and absorption under vacuum. A decrease in the ability to absorb water will result in a deceleration of the weathering process. The positive results of carbonate deposition on the sandstone and limestone attracted many researchers to apply this technique on mortar and concrete to reduce permeability. Microbial carbonate deposition (biodeposition) has also been reported to decrease the permeation properties of mortar and concrete. De Muynck et al. (2008a, b) investigated the effects of biodeposition on the permeability of concrete and mortar. To determine the increase in resistance toward water penetration, they carried out a sorptivity test. They coated the mortar specimens with Bacillus sphaericus, oven dried then dipped into 10 $\pm$ 1 mm of water. They found that the presence of only bacteria resulted in a significant decrease of the water uptake compared to untreated specimens (a reduction of 45, 43, and 24 % with increasing w/c). When a calcium source was added to the medium an additional significant decrease of the water absorption coefficient was noticed. They also studied the durability of the treated surface by measuring the resistance to carbonation and chloride ingress, as chloride ions are one of the corrosion-causing agents. They obtained a 19 % decrease of the chloride migration coefficient, by biotreatment, compared to untreated cubes. The addition of bacterial biomass resulted in a significant smaller carbonation rate compared to untreated cubes. The deposition of a layer of calcium carbonate induced by *Bacillus* sphaericus on the surface of the mortar specimens resulted in a decrease of water absorption and gas permeability (De Muynck et al. 2008a, b). The biodeposition treatment on the surface of concrete materials should be regarded as a coating system. This could be attributed to the fact that the carbonate deposition was mainly a surface phenomenon due to the limited penetration of the bacteria in the porous matrix. Ramakrishnan et al. (1998) investigated the effect of this technique on the durability of concrete. The presence of bacteria was observed to increase the resistance of concrete toward alkali, sulfate, and freeze thaw attack and drying shrinkage; the effect being more pronounced with increasing concentrations of bacterial cells. The authors attributed this to the presence of a calcite layer on the surface, as confirmed by XRD analysis, lowering the permeability of the specimens. Recently, Achal et al. (2011a, b, c) reported that the mortar cubes treated with *Bacillus sp.* a layer of biodeposition crystals on the surface of the mortar resulted in a decrease of the permeation properties. As a consequence, the ingress of harmful substances may be limited. From these reports, it is clear that the presence of a layer of carbonate crystals on the surface has the potential to improve the resistance of concrete materials toward degradation processes. Biodeposition treatment was similar as conventional treatments.

Promising results of techniques based on microbial mineralization have lead to several investigations on the use of bacteria in concrete. Alkali-resistant spore-forming bacteria represent promising candidates for application in concrete and probably other cement-based materials. The controlled use of bacteria offers new approaches for conservators to help preserve, protect, and restore building materials.

Permeability of concrete is believed to be the most important characteristic of concrete that affects its durability. The principal result of the intrusion of chloride (i.e., salt water) into concrete is the corrosion of the reinforcing steel. Once this occurs, the structure will no longer maintain its structural integrity; the lifespan is reduced, and the general safety of the public is severely degraded. It is increasingly apparent that for many concrete members, the ability of the concrete to resist chloride penetration is an essential factor in determining its successful performance over an extended period. The decrease in gas permeability due to the biodeposition treatments resulted in an increased resistance toward carbonation (Siddique and Kaur Chahal 2011).

Sorptivity coefficient K for different grades of mortar applied with different types of surface treatments helped to determine the increase in resistance toward the water penetration which is best depicted (Fig. 15.2a). So, absorption of bacteria and precipitation of carbonate crystals resulted in weight increase of the mortar specimens. So, absorption of biodeposition resulted in weight increase of the mortar specimens. The most pronounced reduction in water absorption compared to untreated samples was reached for the most porous mixture w/c (0.6) and where urea, nutrient broth, and external calcium source were provided, the most of water absorbed by the mortar samples after 2 h was the decreased by factor 5 (Fig. 15.2b). Except for the water repellents, similar tendencies were observed between the gas permeability and carbonation rate results. The rate of carbonation and the performance of the surface treatment were correlated with the water–cement ratio. Carbonation was shown to be related to the nature and connectivity of the pores, with larger pores giving rise to higher carbonation depths. Significant differences in carbonation depth between treated and untreated specimens were already noticeable after 2 weeks of accelerated carbonation. The protective effect of the biodeposition treatment toward carbonation could be improved by additional treatments with bacteria and a calcium source or an increased concentration of calcium ions. Film forming coatings and sealants to be effective against carbonation, the thickness of the treatment should be at least 200 μm (Basheer et al. 2001). The mean thickness of deposition layer was about 30–50 μm; nevertheless, an improved resistance toward carbonation was already noticed in Fig. 15.2c.

Resistance toward chloride penetration of biodeposition treated samples was measured with the use of an accelerated migration test, chloride migration coefficient. The increased resistance toward the migration of chlorides of cubes treated with biodeposition was similar to that of the acrylic coating and the water repellent silanes and silicones and larger than in the case of the silanes/siloxanes mixture, which were all reported to be effective in decreasing the rate of reinforcement corrosion (Fig. 15.2d). Based on the studied properties, the conventional methods

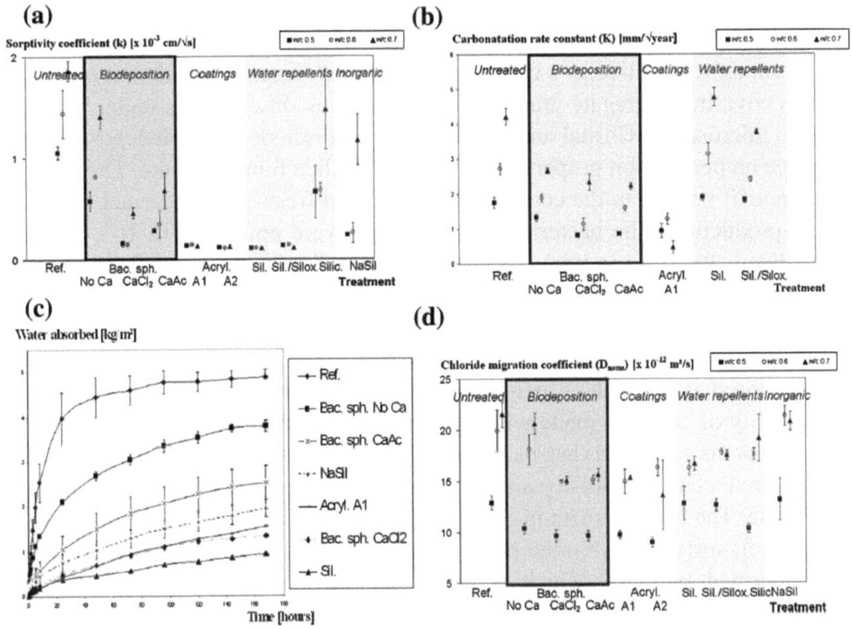

Fig. 15.2 Effect on different grades of mortar applied with different types of surface treatments. (De Muynck et al. 2008b; Siddique and Kaur Chahal 2011). **a** Sorptivity coefficients, *k*. **b** The rate of water absorption versus time for w/c 0.6. **c** Carbonation rate constants, K. **d** Chloride migration coefficients, D_{nssm}

protect concrete from degradation proves to be the advantageous property. Biodeposition appears to be a promising technique. The effect tests of the ureolytic bacteria (*Sporosarcina pasteurii*) on the compressive strength, water absorption, and rapid chloride permeability of concrete made with silica fume were performed at the age of 28 and 91 days. Test results indicated that inclusion of *S. pasteurii* in concrete enhanced the compressive strength, reduced the porosity and permeability of the concrete. Maximum increase of 38.2 and 44 MPa in compressive strength at 28 and 91 days was observed. Moreover, reduction in water absorption was observed with 105 cells/ml of bacteria. This improvement in compressive strength was possibly due to deposition on the bacteria cell surfaces within the pores. Scanning electron microscopy revealed the direct involvement of *S. pasteurii* in calcium carbonate precipitation which was further confirmed by XRD. Due to calcite deposition in concrete, it was observed that reduction in chloride permeability of concrete occurred (Chahal et al. 2012). Grabiec et al. (2012) explored surface modification of recycled aggregate concrete using biodeposition involving a method employing *Sporosarcina pasteurii* (*Bacillus pasteurii*) bacteria. It was possible to obtain reduction in water absorption of aggregate; the effect was more visible in case of finer fractions and for aggregates originating from inferior quality concrete. Calcium chloride was used for precipitation of calcium carbonate, while culture medium consisting of beef extract, peptone, and urea was used for cultivation of bacteria. Whey, ecologically dangerous byproduct from dairy industry, was found to be effective as a culture medium. The presence of calcium carbonate crystals covering aggregate grains was confirmed by observations under scanning electron microscope. Chahal and Siddique (2013) discussed that biodeposition had influence on permeation properties of fly ash and silica fume concrete. The bacteria (*S. pasteurii*) present in the concrete rapidly sealed freshly formed cracks through calcite production. The bacterial concentrations were optimized to 10^3, 10^5, and 10^7 cells/ml. In concrete mix, cement was replaced with fly ash and silica fume. The percentage replacement of fly ash and silica fume was by weight of cement. The percentage use of fly ash was 0, 10, 20, and 30 %, and that silica fume was 0, 5, and 10 %. The experiments were carried out to evaluate the effect of *S. pasteurii* on the compressive strength, water absorption, water porosity, and rapid chloride permeability of concrete made with fly ash and silica fume up to the age 91 days. The test results indicated that inclusion of *S. pasteurii* enhanced the compressive strength, reduced the porosity and permeability of the concrete with fly ash and silica fume. The improvement in compressive strength was due to deposition on the bacteria cell surfaces within the pores which was scanned by electron microscopy and confirmed by XRD which revealed calcium carbonate precipitation. This precipitation reduced the chloride permeability in concrete with fly ash and silica fume. The bacteria improve the permeability of concrete by improving its pore structure and thereby enhancing the life of concrete structures.

15.3.3 Influence on Strength of Concrete by Biodeposition

The compressive strength of the concrete is considered as an index to assess the overall quality of concrete, generally an improvement in the compressive strength results in improvement of all other properties. Bacterial induced carbonates as a binder material, i.e., biocementation, have been added to concrete for the improvement of compressive strength. Several studies have shown that biodeposition can be used to improve the compressive strength of mortar (Bang et al. 2001; Ghosh et al. 2005; Achal et al. 2009, 2011a, b, c; Park et al. 2010) where bacteria, salinity, and composition of the synthetic medium have also been applied in the concrete mixture. According to Ramachandran et al. (2001), the use of bacteria in concrete remediation was considered unorthodox. Biodeposition is highly desirable because the biodeposition carbonate crystal is pollution-free, natural, and eco-friendly. Ramachandran et al. (2001) investigated the use of microbiologically induced mineral deposition for the improvement of the compressive strength of Portland cement mortar cubes at the age of 7 and 28 days. They used live and killed cells of different concentrations of *Bacillus pasteurii* and found that the live cells, at lower concentrations, increase the compressive strength of cement mortar with a longer incubation period. The overall increase of strength, therefore, resulted from the presence of an adequate amount of organic substances in the matrix due to the microbial biomass. To achieve greatest improvement in the compressive strength, cell concentrations/microbial doses need to be optimized. They concluded that biodeposition enhanced the compressive strength of cement mortar cubes. Biodeposition has unique features and performance that have encouraged innovative high-performance composite applications in construction materials. The application of biodeposition in concrete is a potential field for further research efforts. Compressive strength of concrete with *B. pasteurii* as microbes had marginal (5–10 %) increase in strength (in case of BU and BP bacterial suspended in urea–$CaCl_2$, bacteria suspended in phosphate buffer when conc. with BW (Bacteria suspended in water) had marginal decrease in strength (10 %) when compared to controlled concrete samples. Park et al. (2010) used four different species of calcite-forming bacteria (*Sporosarcina soli*, *Bacillus massiliensis*, *Arthrobacter crystallopoietes*, and *Lysinibacillus fusiformis*) for compressive strength improvement in mortar. The 28-day strength test for consolidated mortar showed that the cube treated with Arthrobacter crystallopoietes had the greatest improvement in compressive strength, and the observed change in compressive strength from 7 to 28 days was 22 %. Ghosh et al. (2005) demonstrated the positive effect of biodeposition on the compressive strength of mortar specimens. They found that the strength of mortar cubes increased at all levels of anaerobic microbe addition. For these samples, the presence of a fibrous material inside the pores could be noticed. As a result, a modification of the pore size distribution was observed. The positive effect of biodeposition improved with increasing curing times. The greatest improvement in compressive strength was reported to occur at cell concentrations of 10^5 cells/ml for all ages (3, 7, 14, and 28 days). For a concentration of 10^5 cells/ml, an increase of the compressive strength of 17 and 25 % was observed after 7 and 28 days, respectively. However, no increase of the compressive strength was observed with additions of *Escherichia coli* (non-urease-producing microbe) to the mortar mixture. This led the authors to suggest that the choice of the bacteria plays an important role in the improvement of the compressive strength. Jonkers and Schlangen (2007) investigated that the addition of a high number of bacterial spores (10^8/cm ^{3}) by spore-forming bacteria (*Bacillus pseudofirmus* and *Bacillus cohnii*) resulted in 10 % increase of strength. Achal et al. (2009) showed that nutrients to grow bacteria also play an important role when used with construction materials. To economize the overall process, they replaced standard nutrient with some industrial byproducts such as lactose mother liquor (LML) and corn steep liquor (CSL). They used *Sporosarcina pasteurii* in mortar cubes with LML medium and found 17 % improvement in compressive strength at 28 days (26.3 MPa) compared to control (23.2 MPa). There was no significant difference when LML medium was replaced by standard nutrient medium. Further, medium containing CSL resulted in significantly higher compressive strength even when compared with commercially available medium. All the reports suggest significant increase in the compressive strength at the age of 28 days compared to early ages (3 and 7 days). The overall trend of an increase in compressive strength up to 28 days might be attributed to the behavior of microbial cells within the cement mortar matrix. During the initial curing period, microbial cells obtained good nourishment, because the cement mortar was still porous; but growth might not be proper due to the completely new environment for microbes (Achal et al. 2011a, b, c). It may also be possible that as the pH of the cement remained high, cells were in inactive condition and as curing period was increased, it started growing slowly. Upon cell growth, calcite would have precipitated on the cell surface as well as within the cement mortar matrix. Once many of the pores in the matrix were plugged, the flow of the nutrients and oxygen to the bacterial cells stopped, eventually the cells either died or turned into endospores and acted as an organic fiber, that may enhance the compressive strength of the mortar cubes (Ramachandran et al. 2001). Despite the importance of urease-producing bacteria in remediation of cracks and fissures toward enhancement in the durability of building structures, very few bacteria have been exploited. Bacteria inhabit all possible locations including extremes and exhibit growth and reproduction in such environments. There is a need to explore extreme alkaline environments to isolate indigenous bacteria that can survive in concrete structures for effective biodeposition.

Recently, Achal et al. (2011a, b, c) isolated bacteria from cement, Bacillus sp. CT-5 and used to study compressive strength. They reported 36 % increase in compressive strength of cement mortar with the addition of bacterial cells. Further, it has suggested that due to the ability to tolerate high pH, Bacillus sp. CT-5 enhanced the compressive strength of cement mortar cubes significantly. To determine whether the increase in compressive strength of the specimens prepared with bacteria could be attributed to the microbial calcite deposition, biodeposition was quantified by X-ray diffraction (XRD) analysis and visualized by SEM (De Muynck et al. 2008a, b; Achal et al. 2009, 2011a, b, c)

From these findings, it can be concluded that compressive strength of cement mortar increases with an addition of urease-producing microbes such as Bacillus, Shewanella, and Arthrobacter species. This improvement in compressive strength might be due to deposition on the microorganism cell surfaces and within the pores of cement–sand matrix, which plug the pores within the mortar as a result of biodeposition (Ramakrishnan et al. 1998). The strength of the bacteria modified mortar increases due to the deposition of the new material gehlenite by the bacterial activity, which in turn increases mainly the uniformity of SiO_2 concentration of the mortar. The high Ca/Si ratio at 10^5 cells/ml corresponds to the optimum strength of bacteria modified mortar at such cell concentration. Also the protein secreted by the bacterium leaches silica and helps in formation of new silicate phases that fill the micropores. This protein increases the strength of mortar when it is added separately (Ghosh et al. 2009). Pei et al. (2013) explored the role of bacterial cell walls of Bacillus subtilis as a concrete admixture to improve the mechanical performance of concrete. The bacterial cell walls are known to mediate microbialy induced carbonate precipitation, a process in which CaCO3 is formed from Ca2 + ions and dissolved CO2. Consistent with such knowledge, incorporation of bacterial cell walls increased carbonation of Ca(OH)2 and formation of CaCO3 in concrete. Furthermore, the bacterial cell walls significantly increased compressive strengths of concrete by 15 % while also decreased porosity at 28 days of curing. Assay for CaCO3 precipitation in vitro indicated that bacterial cell walls, but not dead cells, accelerated carbonation of Ca2 + ions in Ca(OH)2 solution. Since CaCO3 formed can fill up the void, decrease the porosity and increase the compressive strength in concrete, bacterial cell walls could act as a promising concrete admixture with benefits in enhancing mechanical performance and improving other carbonation-related properties (Pei et al. 2013).

15.3.4 Improvement in the Durability of Concrete or Mortar Surface Treatment by Biodeposition

Nowadays, performance-based durability criteria have drawn the attention in specifications for building materials. A major goal of any performance-based specification is to ensure the durability of building materials and structures. Durability is regarded as the ability of the materials and structure to perform satisfactorily with minimum maintenance over the anticipated service life. The quality of any building material depends on three major parameters, such as (i) compressive strength, (ii) permeability, and (iii) corrosion. For an efficient bacterial concrete, it should produce more compressive strength, less permeability and should not cause corrosion of any reinforcement. To improve the durability of concrete structures, engineers started using mineral additives in the form of waste materials from industries such as fly ash, ground granulated blast furnace slag. In spite of so much development in the ingredients of concrete final product, concrete was not adapt well to environment and has the ability of self-healing. Most of these additives were passive in nature. More research and development is required to take some measures to enhance the resistance of permeability, ingress of chloride and carbonation to protect the reinforcing material added in concrete from deterioration. Several tests are available to the engineer for diagnosing the condition of the concrete. The alkalinity of the concrete at the surface and at various depths down to the position of the reinforcement will give an indication of how much and how soon the reinforcement may be at risk. This is known as the depth of the carbonation front. The presence of chlorides can be determined, again at varying depths. The permeability of the concrete will give an indication of its susceptibility to absorption of air and water. Particularly cracking of the surface layer of concrete reduces material durability as ingress water and detrimental chemicals cause a range of matrix degradation processes as well as corrosion of the embedded steel reinforcement (Neville 1996). Durability problems such as crack formation are typically tackled by manual inspection and repair, i.e., by impregnation of cracks with cement or epoxy-based or other synthetic fillers (Neville 1996). There are so many synthetic agents or latex binding agents (such as acrylic, polyvinyl acetate, butadiene styrene) which are used to avoid any kind of fractures and fissures in the concrete structures or used in repair applications such as the bonding of fresh concrete, sprayed concrete, or sand/cement repair mortar to hardened concrete.

There is reluctance by some repair product manufacturers to adequately address the problem, with a very limited range of appearances available from their materials in different countries. There is a propensity for specifiers to recommend overcoating of the whole building, such as with a high-build acrylic paint in order to achieve a uniform appearance, which is also easily matched in future if further repairs become necessary.

As a general rule, these traditional repair systems have a number of disadvantageous aspects such as different thermal expansion coefficient compared to concrete, weak bonding, disappearance, environmental and health hazards, and even costly. Therefore, many researchers proposed biodeposition as an alternative and eco-friendly technique to enhance the durability of building materials and structures. So far, most of studies have focused on improvement in the durability of concrete or mortar surface treatment by biodeposition (Reddy, et al. 2012). De Muynck et al. (2008a, b) compared the durability (concerning capillary water uptake and gas permeability) of concrete when its surface was treated with pure and mixed cultures of ureolytic bacteria. They concluded that the type of bacterial culture and the medium composition had a profound impact on $CaCO_3$ crystal morphology, being that the use of pure cultures resulted in a more pronounced decrease in the uptake of water. They also concluded that the durability performance obtained with cultures of the species *B. sphaericus* was comparable to the ones obtained with conventional water repellents (silanes, siloxanes). De Muynck et al. (2008a, b) studied different durability parameters (carbonation, chloride penetration, and freezing and thawing) confirming that the biodeposition treatment showed a similar protection toward degradation processes when compared to some of the conventional surface treatments under investigation. They also mention the need for investigations regarding the durability of the treatment under acidic media. They further mentioned that biological generated calcite is less soluble than the one inorganically precipitated, thus suggesting a higher performance. Achal et al. (2011a, b, c) mention a six times reduction in water absorption due to the microbial calcite deposition. In a different study, Achal used a phenotypic mutant of *S. pasteurii* (BpM-3) with improved urease activity also reporting a significant reduction in water absorption, permeability, and chloride permeability. Okwadha and Li (2011) used bacterium *S. pasteurii* strain ATCC 11859 to create a biosealant on a PCB-contaminated concrete surface reporting a reduction on water permeability by 1–5 orders of magnitude. They also state that the treated concrete had a high resistance to carbonation. Li and Qu (2012) confirm that bacterially mediated carbonate precipitation on concrete surface reduces capillary water uptake, leading to the carbonation rate constant to be decreased by 25–40 %. Biomaterials can lead concrete to a more sustainable in civil engineering. Much research has drawn the attention in the field of concrete surface treatments by biodeposition; however, it is still far from being a proved and reliable technique capable of replacing current common concrete surface treatments based on organic polymers sealers. Biodeposition helps to fill micropores and cracks, thus reducing its permeability. However, the highly alkaline pH of concrete reduces the activities of the bacteria. To overcome this problem, different authors have suggested the use of different immobilization technology (clay capsules, silica gel, or polyurethane encapsulation). The use of bacteria in concrete mix also needs further research efforts. Results from practical applications in which there is exposure to environmental conditions are still needed in order to confirm the importance of this new approach (Pacheco-Torgal and Labrincha 2013).

Microbially induced calcium carbonate precipitation is a naturally occurring biological process that has various applications in remediation and restoration of range of building materials. In the present study, the role of bacteria Bacillus sp. on the durability properties and remediation of cracks in cementitious structures were studied. Biocement induced by a Bacillus sp. lead to more than 50 % reduction in the porosity of mortar specimens, while chloride permeability of concrete changed from moderate to very low as indicated by rapid chloride permeability test. The bacteria successfully healed the simulated cracks of depths including 27.2 mm in cement mortars with increase in the compressive strength as high as 40 % of that of control. The results clearly showed that microbially induced calcium carbonate precipitation can be applied for various building materials for remediation of cracks and enhancement of durability (Achal et al. 2013).

15.4 Bacterial Self-healing Concrete

Besides external application of bacteria in the case of remediation of cracks, bacteria have also been applied in the concrete mixture. Until now, research has mainly focused on the consequences of this addition on the material properties of concrete, i.e., strength and durability. Both properties depend on the microstructure of the concrete. However, the effects of the presence of the bacteria and/or biodeposition on the microstructure still need to be elucidated, especially the interaction between the biomass and the cement matrix. Ramachandran et al. (2001) investigated the use of microbiologically induced mineral deposition for the improvement of the compressive strength of Portland cement mortar cubes. This study identified the effect of the buffer solution and type and amount of *S. pasteurii* and *P. aeruginosa* used. Furthermore, in order to study the effect of the biomass, the influence of both living and dead cells was investigated. Before addition to the mortar mixture, bacteria were centrifuged and washed twice. The final pellets were then suspended in either saline or phosphate buffer, which was subsequently added to the mixture. After demolding, the mortar specimens were stored in a solution containing urea and calcium chloride for 7 days. Subsequently, the specimens were cured in air until the measurement of the compressive strength. At lower concentrations, the presence of *S. pasteurii* was shown to increase the compressive strength of mortar cubes. While the 28-day compressive strength of the control cubes amounted to about 55 $\pm$ 1 MPa, specimens treated with 103 cells cm^{-3} had a compressive strength of about 65 $\pm$ 1 MPa. The contribution of *P. aeruginosa* to the strength was found to be insignificant. From the X-ray diffraction (XRD) analysis, no significant increased amounts of calcite could be found in mortar specimens treated with bacteria. This could be attributed to the inhibition of the bacteria by the high pH and the lack of oxygen inside the mortar mixture. The overall increase of strength, therefore, resulted from the presence of an adequate amount of organic substances in the matrix due to the microbial biomass. However, an increase of the biomass, as dead cells in particular, resulted in a decreased strength. According to the authors, this could be attributed to the disintegration of the organic matter with time, making the matrix more porous (Ramachandran et al. 2001).

Ramakrishnan et al. (2001) investigated the effect of this technique on the durability of concrete. The presence of bacteria was observed to increase the resistance of concrete toward alkali, sulfate, freeze–thaw attack, and drying shrinkage; the effect being more pronounced with increasing concentrations of bacterial cells. The authors attributed this to the presence of a calcite layer on the surface, as confirmed by XRD analysis, lowering the permeability of the specimens. The best results were obtained with the phosphate buffer. Ghosh et al. (2005) demonstrated the positive effect of the addition of Shewanella on the compressive strength of mortar specimens. Contrary to the aforementioned research, these authors did not intend mineral deposition, as these specimens were cured in air and not in a nutrient containing medium. An increase of 25 % of the 28 days compressive strength was obtained for a cell concentration of about 10^5 cells mL^{-1} and a water to cement ratio of 0.4. For these samples, the presence of a fibrous material inside the pores could be noticed. As a result, a modification of the pore size distribution was observed. The positive effect of the addition of Shewanella improved with increasing curing times. For a concentration of 105 cells mL^{-1}, an increase of the compressive strength of 17 and 25 % was observed after 7 and 28 days, respectively. However, no increase of the compressive strength was observed with additions of Escherichia coli to the mortar mixture. This led the authors to suggest that the choice of the microorganism plays an important role in the improvement of the compressive strength. More specifically, the production of EPS by the bacteria seemed to be of importance.

Figure 15.3 shows direct stereomicroscopic observation of cracks from control and bacteria-based specimens before and after 100 days of immersion in tap water. Width of completely healed cracks was significantly larger in bacteria-based specimens (0.46 mm) compared to control specimens (0.18 mm).

Two-component biochemical self-healing agent, consisting of a mixture of bacterial spores and calcium lactate, can be successfully applied to promote and enhance the self-healing capacity of concrete as the maximum healable crack width more than doubled. Moreover, oxygen measurements provided evidence that concrete incorporating bacterial spores embedded in expanded clay particles and derived active bacteria remain viable and functional several months after concrete casting. The microbial enhanced crack-healing ability is presumably due to combined direct and indirect calcium carbonate formation: (i) direct $CaCO_3$ precipitation through metabolic conversion of calcium lactate and (ii) indirect formation due to reaction of metabolically produced CO_2 molecules with Ca(OH)2 minerals present in the concrete matrix leading to additional $CaCO_3$ precipitation.

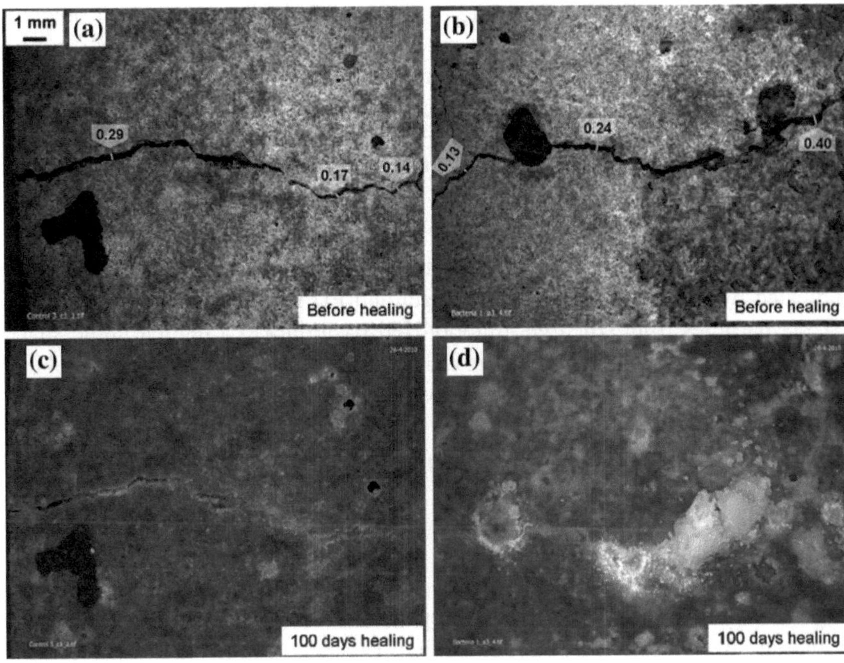

Fig. 15.3 **a** Stereomicroscopic images of crack-healing process in control mortar specimen before and **b** in biochemical agent-based specimen before, and **c** after 100 days healing, and **d** after 100 days healing (Wiktor and Jonkers 2011)

In addition, as the metabolically active bacteria consume oxygen, the healing agent may act as an oxygen diffusion barrier protecting the steel reinforcement against corrosion. So far, bacteria have never been used to remove oxygen from the concrete matrix to inhibit reinforcement corrosion, and further studies are needed to quantify this potentially additional beneficial process. While in this study the enhanced self-healing capacity of bacteria-based concrete has been quantified, several other characteristics such as long-term (years) durability and cost efficiency of this novel type of concrete need to be resolved before practical application can be considered. Anticipated potential advantages of this bacteria-based concrete are presumably primarily in reduction of maintenance and repair costs and extension of the service life of concrete constructions (Wiktor and Jonkers 2011).

Bacteria and polyurethane (PU) prepolymer were applied internally (to heal cracks from the inside). PU foam should form in the crack automatically when cracking occurs and bacteria are incorporated inside the foam at the same time. This study was to investigate the potential use of silica gel or polyurethane-immobilized bacteria to bring about self-healing concrete. Bacteria can still conserve ureolytic activity and carbonatogenesis activity after being immobilized into silica gel and polyurethane foam. TGA results proved the formation of biodeposition in a mimic self-healing process which indicated the great potential use of immobilized bacteria in self-healing concrete cracks. More self-healing efficiency (higher strength regain and more pronounced decrease of water permeability) was obtained from the specimens with incorporated polyurethane-immobilized bacteria. Therefore, it is promising to use polyurethane-immobilized bacteria to self-heal early concrete microcracks. Further research is needed to increase the amount of biodeposition to obtain more self-healing (Wang et al. 2012a, b).

Figure 15.4 shows the scanning electron micrograph of CaCO3 precipitation (indicated by the black arrow) in silica gel and PU foam. It can be seen that the CaCO3 particles from silica gel-immobilized bacteria had a regular cubic shape, and the size was about 100 μm. CaCO3 in PU foam showed an irregular block shape with a similar particle size (also about 100 μm). It can be seen from Fig. 15.4a, c that CaCO3 particles were distributed more homogenously in silica gel than in PU foam. The reason is that polyurethane prepolymer is more viscous than silica sol (a state before gel). So it is easier for bacteria to distribute homogeneously in silica sol than in polyurethane before silica sol became gel or polyurethane foam formed.

Microcapsules were applied to encapsulate bacterial spores for self-healing concrete. The viability of encapsulated spores and the influence of microcapsules on mortar specimens were investigated first. Breakage of the microcapsules upon cracking was verified by scanning electron microscopy. Self-healing capacity was evaluated by crack-healing ratio and the water permeability. The results indicated that the healing ratio in the specimens with biomicrocapsules was higher (48–80 %) than in those without bacteria (18–50 %). The maximum crack width healed in the specimens of the bacteria series was 970 μm, about 4 times that of the non-bacteria series (max 250 μm). The overall water permeability in the bacteria series was about 10 times lower than that in non-bacteria series. Wet–dry cycles were found to stimulate self-healing in mortar specimens with encapsulated bacteria. No self-healing was observed in all specimens stored at 95 %RH, indicating that the presence of liquid water is an essential component for self-healing (Wang et al. 2014).

Mechanical properties, self-healing capacity, and bonding behavior of a sustainable bio-based mortar repair system for concrete are discussed. Two different mixes of strain-hardening cement-based composites (SHCC) have been used. The bio-based agent added to the SHCC consists of both bacteria and food for the bacteria. The metabolic activity of bacteria was monitored by oxygen profile measurements, which reveals O_2 consumption by bacteria-based samples, but not by control samples. Biodeposition results in improved mechanical properties and bonding strength of a concrete-compatible fiber-reinforced repair system. When applied as a repair material the mortar with bio-based agent shows reduced delamination from the concrete substrate compared to mortar without the bio-based agent. Furthermore, after cracking and healing, the mixtures with bio-based healing agent show a slightly better recovery of both flexural strength and deflection capacity from control mixtures without bio-based healing agent.

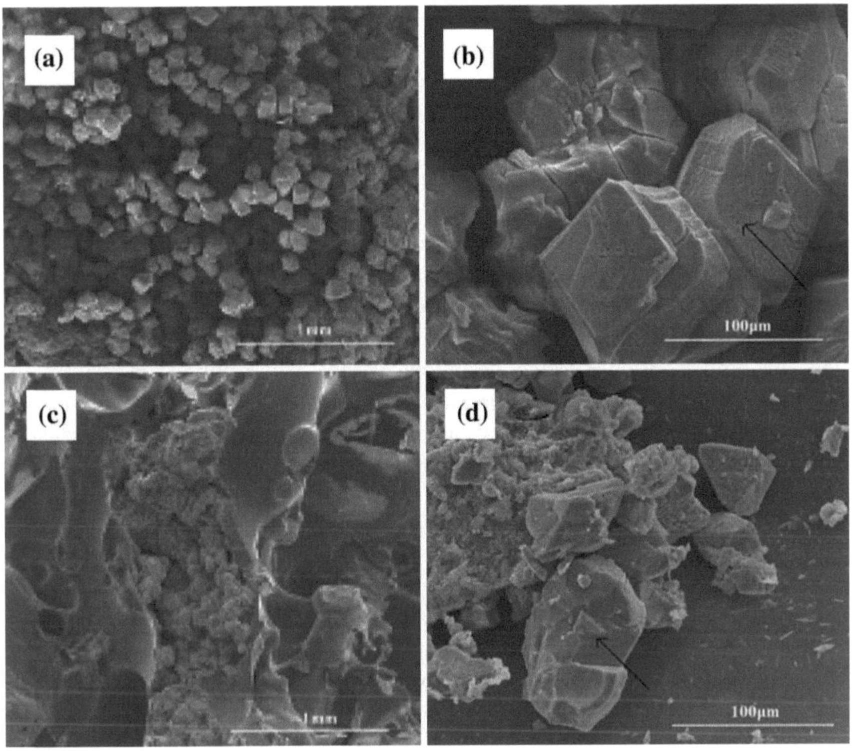

Fig. 15.4 Scanning electron micrographs of CaCO3 precipitation in the silica gel and PU foam. **a** CaCO3 in silica gel (100x), **b** CaCO3 in silica gel (1000x), **c** CaCO3 in PU foam (100x), and **d** CaCO3 in PU foam (1000x) (Wang et al. 2012a)

Although oxygen measurements indicate that bacteria were metabolically active in the bio-based specimens, observed amounts of calcium carbonate precipitate did not appear to substantially differ from control specimens. The reason for the apparent uncoupling between metabolic activity and lack of enhanced calcium carbonate precipitation can possibly be attributed to limited amounts of feed applied and these remains to be clarified in pending studies (Guadalupe et al. 2014).

15.5 Conclusions

Some specific bacteria metabolic activities and deposition in concrete are responsible to improve overall behaviors of concrete. Generally, it has been hypothesized that almost all bacteria are capable of $CaCO_3$ production because precipitation occurs as a byproduct of common metabolic processes such as photosynthesis, sulfate reduction, and urea hydrolysis. Even the effect of bacteria on various parameters in concrete proves to be beneficial development. Based on the studied properties like compressive strength, permeability, water absorption, chloride ingression, the microbial mineral precipitation appears to be a promising technique at this state of development.

These promising results of surface treatment by biodeposition for the improvement of the durability of concrete materials and structures have drawn the attention of research groups all over the world. However, so far, no data have described a sufficient degree toward enhancement in the durability of building structures by biodeposition. Hence, a lot of research is still necessary before such technology is ready for applications in civil engineering. Moreover, long-term effect of biodeposition surface treatment is not yet reported. However, the important parameter such as permeation properties was not dealt with. As biodeposition fills the subsurface pores, the extent of permeability reduction must be studied in detail. The efficiency of biodeposition in resisting carbonation and reduction in permeability to improve the corrosion resistance must be studied in detail. Most of the studies based on biodeposition were conducted to evaluate compressive strength, water absorption, and crack remediation of mortars and concretes. The effects of biodeposition on reducing the corrosion rate of reinforced concretes have not been studied in detail and needed to further research efforts.

Acknowledgment Financial supports from Natural Science Foundation of China (No. 51008281), the China postdoctoral science Foundation (No. 2012M510119), and National Basic Research Program of China (No. 2009CB623200) are gratefully appreciated.

References

Achal V, Mukherjee A, Basu P, Reddy MS (2009) Strain improvement of Sporosarcina pasteurii for enhanced urease and calcite production. J Ind Microbiol Biotechnol 36:981–988

Achal V, Mukherjee A, Reddy M (2011a) Microbial concrete: way to enhance the durability of building structures. J Mater Civil Eng 23:730–734

Achal V, Mukherjee A, Reddy M (2011b) Effect of calcifying bacteria on permeation properties of concrete structures. J Ind Microbiol Biotechnol 38:1229–1234

Achal V, Pan X, Ozyurt N (2011c) Improved strength and durability of fly ash-amended concrete by microbial calcite precipitation. Ecol Eng 37:554–559

Achal V, Mukerjee A, Reddy MS (2013) Biogenic treatment improves the durability and remediates the cracks of concrete structures. Constr Build Mater 48:1–5

Afifudin H, Hamidah M, Hana H, Kartini K (2011) Microorganism precipitation in enhancing concrete properties. Appl Mech Mater 99–100:1157

Bang SS, Galinat JK, Ramakrishnan V (2001) Calcite precipitation induced by polyurethane-immobilized Sporosarcina pasteurii. Enzyme Microb Technol 28(4–5):404–409

Barabesi C, Salvianti F, Mastromei G, Perito B (2003) Microbial calcium carbonate precipitation for reinforcement of monumental stones. In: Saiz-Jimenez C (ed) Molecular biology and cultural heritage. AA Balkema Publishers, Lisse, pp 209–212

Barabesi C, Galizzi A, Mastromei G, Rossi M, Tamburini E, Perito B (2007) Bacillus subtilis gene cluster involved in calcium carbonate biomineralization. J Bacteriol 189(1):228–235

Basheer L, Kropp J, Cleland DJ (2001) Assessment of the durability of concrete from its permeation properties a review. Constr Build Mater 15(2–3):93–103
Bazylinski DA, Frankel RB, Konhauser KO (2007) Modes of biomineralization of magnetite by microbes. Geomicrobiol J 24:465–475
Boquet E, Boronat A, Ramos-Cormenzana A (1973) Production of calcite (calcium carbonate) crystals by soil bacteria is a general phenomenon. Nature 246(5434):527–529
Braissant O, Cailleau G, Dupraz C, Verrecchia E (2003) Bacterially induced mineralization of calcium carbonate in terrestrial environments: the role of exopolysaccharides and amino acids. J Sediment Res 73(3):485–490
Burne RA, Chen RE (2000) Microbes Infect 2:533–542
Camaiti M, Borselli G, Matteol U (1988) Prodotti consalidanti impiegati nelle operazioni di restauro. Edilizia 10:125–134
Castanier S, Métayer-Levrel L, Perthuisot J-P (1999) Ca-carbonates precipitation and limestone genesis-the microbiologist point of view. Sediment Geol 126:9–23
Chahal N, Siddique R (2013) Permeation properties of concrete made with fly ash and silica fume: influence of Ureolytic bacteria. Constr Build Mater 49:161–174
Chahal N, Siddique R, Rajor A (2012) Influence of bacterial treated cement kiln dust on the properties of concrete. Constr Build Mater 52:42–51
Day JL, Ramakrishnan V, Bang SS (2003) Microbiologically induced sealant for concrete crack remediation. In: Proceedings of 16th engineering mechanics conference, Seattle
De Muynck W, Cox K, De Belie N, Verstraete W (2008a) Bacterial carbonate precipitation as an alternative surface treatment for concrete. Constr Build Mater 22(5):875–885
De Muynck W, Debrouwer D, De Belie N, Verstraete W (2008b) Bacterial carbonate precipitation improves the durability of cementitious materials. Cem Concr Res 38(7):1005–1014
De Muynck W, De Belie N, Verstraete W (2010) Microbial carbonate precipitation in construction materials: a review. Ecol Eng 36:118–136
Decho AW (2010) Overview of biopolymer-induced mineralization: what goes on in biofilms? Ecol Eng 36:137–144
DeJong JT, Fritzges MB, Nusslein K (2006) Microbially induced cementation to control sand response to undrained shear. J Geotech Geoenviron 132(11):1381–1392
Dick J, De Windt W, De Graef B, Saveyn H, Van der Meeren P, De Belie N, Verstraete W (2006) Bio-deposition of a calcium carbonate layer on degraded limestone by Bacillus species. Biodegradation 17(4):357–367
Douglas S, Beveridge TJ (1998) Mineral formation by bacteria in natural microbial communities. FEMS Microbiol Ecol 26(2):79–88
Dupraz C, Reid RP, Braissant O, Decho AW, Norman RS, Visscher PT (2009) Process of carbonate precipitation in modern microbial mats. Earth Sci Rev 96:141–162
Ehrlich HL (1998) Geomicrobiology: its significance for geology. Earth Sci Rev 45(1–2):45–60
Ercole C, Cacchio P, Botta AL, Centi V, Lepidi A (2007) Bacterially induced mineralization of calcium carbonate: the role of exopolysaccharides and capsular polysaccharides. Microsc Microanal 13:42–50
Fein JB, Daughney CJ, Yee N, Davis TA (1997) A chemical equilibrium model for metal adsorption onto bacterial surfaces. Geochim Cosmochim Acta 61:3319–3328
Ferris FG, Stehmeier LG (1992) Bacteriogenic mineral plugging. USA Patent US5143155
Fortin D, Ferris FG, Beveridge TJ (1997) Surface-mediated mineral development by bacteria. Rev Mineral 35:161–180
Friis AK, Davis TA, Figueira MM, Paquette J, Mucci A (2003) Influence of bacillus subtilis cell walls and EDTA on calcite dissolution rates and crystal surface features. Environ Sci Technol 37:2376–2382
Fujita Y, Ferris FG, Lawson RD, Colwell FS, Smith RW (2000) Calcium carbonate precipitation by ureolytic subsurface bacteria. Geomicrobiol J 17(4):305–318

Fujita Y, Redden GD, Ingram JC, Cortez MM, Ferris FG, Smith RW (2004) Strontium incorporation into calcite generated by bacterial ureolysis. Geochim Cosmochim Acta 68(15):3261–3270
Gaylarde C, Ribas SM, Warscheid T (2003) Microbial impact on building materials: an overview. Mater Struct 36(3):342–352
Ghosh P, Mandal S, Chattopadhyay B, Pal S (2005) Use of microorganism to improve the strength of cement mortar. Cem Concr Res 35:1980–1983
Ghosh P, Biswas M, Chattopadhyay B, Mandal S (2009) Microbial activity on the microstructure of bacteria modified mortar. Cem Concr Compos 31:93–98
Gollapudi UK, Knutson CL, Bang SS, Islam MR (1995) A new methodbfor controlling leaching through permeable channels. Chemosphere 30(4):695–705
Grabiec A, Klama J, Zawal D, Krupa D (2012) Modification of recycled concrete aggregate by calcium carbonate biodeposition. Constr Build Mater 34:145–150
Guadalupe M, Sierra-Beltran, Jonkers HM, Schlangen E (2014) Characterization of sustainable bio-based mortar for concrete repair. Constr Build Mater. http://dx.doi.org/10.1016/j.conbuildmat.2014.01.012
Hammes F, Verstraete W (2002) Key roles of pH and calcium metabolism in microbial carbonate precipitation. Rev Environ Sci Biotechnol 1:3–7
Hammes F, Boon N, de Villiers J, Verstraete W, Siciliano SD (2003a) Strainspecific ureolytic microbial calcium carbonate precipitation. Appl Environ Microbiol 69(8):4901–4909
Hammes F, Seka A, de Knijf S, Verstraete W (2003b) A novel approach to calcium removal from calcium-rich industrial wastewater. Water Res 37(3):699–704
Jiang W, Saxena A, Bongkeun S, Ward BB, Beveridge TJ, Myneni CB (2004) Elucidation of functional groups on Gram-positive and Gram-negative bacterial surfaces using infrared spectroscopy. Langmuir 20:11433–11442
Jonkers H (2007) Self healing concrete: a biological approach. In: van der Zwaag S (ed) Self healing materials: an alternative approach to 20 centuries of materials science. Springer, Netherlands, pp 195–204
Jonkers HM, Thijssen A, Muyzer G (2010) Application of bacteria as self-healing agent for the development of sustainable concrete. Ecol Eng 36(2):230–235
Kantzas A, Ferris FG, Jha KN, Mourits FM (1992) A novelmethod of sand consolidation through bacteriogenic mineral plugging. In: Proceedings of CIM annual technical conference, Calgary
Kim HK, Park SJ, Han JI, Lee HK (2013) Microbially mediated calcium carbonate precipitation on normal and lightweight concrete. Constr Build Mater 38:1073–1082
Kucharski ES, Cord-Ruwisch R, Whiffin V, Al-Thawadi SMJ (2006) Microbial biocementation. World intellectual property organization, WO2006/066326 A1
Lappin-Scott HM, Cusack F, Costerton JW (1988) Nutrient resuscitation and growth of starved cells in sandstone cores: a novel approach to enhanced oil recovery. Appl Environ Microbiol 54(6):1373–1382
Le Metayer-Levrel G, Castanier S, Orial G, Loubiere JF, Perthuisot JP (1999) Applications of bacterial carbonatogenesis to the protection and regeneration of limestones in buildings and historic patrimony. Sediment Geol 126(1–4):25–34
Li PH, Jin B (2012) Healing of cracked concrete by *Sporosarcina pasteurii* mediated carbonate deposition. Appl Mech Mater 164:103–106
Li P, Qu W (2012) Microbial carbonate mineralization as an improvement method for durability of concrete structures. Adv Mater Res 365:280–286
Little BJ, Wagner PA, Lewandowski Z (1997) Spatial relationship between bacteria and mineral surfaces. Rev Mineral 35:123–159
MacLeod A, Lappin-Scott HM, Costerton JW (1988) Plugging of a model rock system by using starved bacteria. Appl Environ Microbiol 54:1365–1372
McConnaughey TA, Whelan JF (1997) Calcification generates protons for nutrient and bicarbonate uptake. Earth Sci Rev 42(1–2):95–117

Moon HY, Shin DG, Choi DS (2007) Evaluation of the durability of mortar and concrete applied with inorganic coating material and surface treatment system. Constr Build Mater 21(2):362–369

Morita R (1980) Calcite precipitation by marine bacteria. Geomicrobiol J 2:63–82

Moropoulou A, Kouloumbi N, Haralampopoulos G, Konstanti A, Michailidis P (2003) Criteria and methodology for the evaluation of conservation interventions on treated porous stone susceptible to salt decay. Prog Org Coat 48:259–270

Morse JW (1983) The kinetics of calcium carbonate dissolution and precipitation. In: Reeder RJ (ed.) Carbonates: mineralogy and chemistry, vol. 11. Mineralogic Society of America, Washington,, pp 227–264

Muynck D, De Belie W, Verstraete NW (2010) Microbial carbonate precipitation in construction materials: a review. Ecol Eng 36(2):118–136

Nemati M, Voordouw G (2003) Modification of porous media permeability, using calcium carbonate produced enzymatically in situ. Enzyme Microb Technol 33(5):635–642

Nemati M, Greene EA, Voordouw G (2005) Permeability profile modification using bacterially formed calcium carbonate: comparison with enzymic option. Process Biochem 40(2):925–933

Nielsen TH, Bonde TA, Sorensen J (1998) Significance of microbial urea turnover in Ncycling of three Danish agricultural soils. FEMS Microbiol Ecol 25(2):147–157

Okwadha G, Li J (2011) Biocontainment of polychlorinated biphenyls (PCBs) on flat concrete surfaces by microbial carbonate precipitation. J Environ Manage 92:2860–2864

Pacheco-Torgal F, Labrincha JA (2013) Biotech cementitious materials: some aspects of an innovative approach for concrete with enhanced durability. Constr Build Mater 40:1136–1141

Park SJ, Park YM, Chun WY, Kim WJ, Ghim SY (2010) Calcite-forming bacteria for compressive strength improvement in mortar. J Microbiol Biotechnol 20:782–788

Pei Ruoting, Liu Jun, Wang Shuangshuang, Yang Mijia (2013) Use of bacterial cell walls to improve the mechanical performance of concrete. Cem Concr Compos 39:122–130

Perito B, Mastromei G (2011) Molecular basis of bacterial calcium carbonate precipitation W.E.G. In: Muller (ed.) Molecular biomineralization, progress in molecular and subcellular Biology 52. doi:10.1007/978-3-642-21230-7_5

Qian CX, Wang JY, Wang RX (2009) Corrosion protection of cement-based building materials by surface deposition of CaCO3 by Bacillus pasteurii. Mater Sci Eng 29(4):1273–1280

Qian C, Wang R, Cheng L, Wang J (2010a) Theory of microbial carbonate precipitation and its application in restoration of cement-based materials defects. Chin J Chem 28:847–857

Qian CX, Pan QF, Wang RX (2010b) China Tech Sci 53:2198–2206

Ramachandran SK, Ramakrishnan V, Bang SS (2001) Remediation of concrete using micro-organisms. ACI Mater J 98:3–9

Ramakrishnan SK, Panchalan RK, Bang SS (2001) Improvement of concrete durability by bacterial mineral precipitation. In: 11th International conference on Fracture, Turin, Italy

Reddy S, Rao M, Aparna P, Sasikala C (2010) Performance of standard grade bacterial (bacillus subtilis) concrete. Asian J Civ Eng (Build Housing) 11:43–55

Reddy MS, Achal V, Mukherjee A (2012) Microbial concrete, a wonder metabolic product that remediates the defects in building structures. In: Satyanarayana T, et al (eds.) Bacteria in environmental management: microbes and environment. doi:10.1007/978-94-007-2229-3_24

Rivadeneyra MA, Delgado R, del Moral A, Ferrer MR, Ramos-Cormenzana A (1994) Precipitation of calcium carbonate by Vibrio spp. from an inland saltern. FEMS Microbiol Ecol 13(3):197–204

Rivadeneyra MA, Parraga J, Delgado R, Ramos-Cormenzana A, Delgado G (2004) Biomineralization of carbonates by Halobacillus trueperi in solid and liquidmedia with different salinities. FEMS Microbiol Ecol 48:39–46

Rodriguez-Navarro C, Rodriguez-Gallego M, Ben CK (2003) Conservation of ornamental stone by Myxococcus xanthus-induced carbonate biomineralization. Appl Environ Microb 69(4):2182-2193

Rong H, Qian CX, Li LZ (2012) Study on microstructure and properties of sandstone cemented by microbe cement. Constr Build Mater 36:687–694

Siddique R, Kaur Chahal N (2011) Effect of ureolytic bacteria on concrete properties. Constr Build Mater 25:3791–3801

Stocks-Fischer S, Galinat JK, Bang SS (1999) Microbiological precipitation of CaCO3. Soil Biol Biochem 31(11):1563–1571

Tiano P (1995) Stone reinforcement by calcite crystal precipitation induced by organic matrix macromolecules. Stud Conserv 40(3):171–176

Tiano P, Biagiotti L, Mastromei G (1999) Bacterial bio-mediated calcite precipitation for monumental stones conservation: methods of evaluation. J Microbiol Methods 36(1–2):139–145

Tiano P, Cantisani E, Sutherland I, Paget JM (2006) Biomediated reinforcement of weathered calcareous stones. J Cult Herit 7(1):49–55

Van Tittelboom K, De Belie N, De Muynck W, Verstraete W (2010) Use of bacteria to repair cracks in concrete. Cem Concr Res 40:157–166

Von Knorre H, Krumbein WE (2000) Bacterial calcification. In: Riding RE, Awramik SM (eds) Microbial sediments. Springer, Berlin, pp 25–31

Wang J, De Belie N, Verstraete W (2012a) Diatomaceous earth as a protective vehicle for bacteria applied for self-healing concrete. J Ind Microbiol Biotechnol 39:567–577

Wang J, Van Tittelboom K, De Belie N, Verstraete W (2012b) Use of silica gel or polyurethane immobilized bacteria for self-healing concrete. Constr Build Mater 26:532–540

Wang JY, Soens H, Verstraete W, De Belie N (2014) Self-healing concrete by use of microencapsulated bacterial spores. Cem Concr Res 56:139–152

Warren LA, Maurice PA, Parmar N, Ferris FG (2001) Microbially mediated calcium carbonate precipitation: implications for interpreting calcite precipitation and for solid-phase capture of inorganic contaminants. Geomicrobiol J 18:93–125

Warscheid T, Braams J (2000) Biodeterioration of stone: a review. Int Biodeterior Biodegrad 46:343–368

Weiner S, Dove PM (2003) An overview of biomineralization and the problem of the vital effect. Am Rev Mineral Geochem 54:1–31

Whiffin VS, van Paassen L, Harkes MP (2007) Microbial carbonate precipitation as a soil improvement technique. Geomicrobiol J 24:417–423

Wiktor V, Jonkers H (2011) Quantification of crack-healing in novel bacteria-based self-healing concrete. Cem Concr Compos 33:763–770

Zavarzin GA (2002) Microbial geochemical calcium cycle. Microbiol Transl Mikrobiologiya 71:1–17

Zhong L, Islam MR (1995) A newmicrobial plugging process and its impact on fracture remediation. In: Proceedings of Society of Petroleum Engineers. Annual Technical Conference, Dallas, Texas, pp 703–715

Chapter 16
A Case Study: Bacterial Surface Treatment of Normal and Lightweight Concrete

H.K. Kim and H.K. Lee

Abstract Bacterial surface treatments for concrete have become increasingly popular due to their strong potential to improve the durability of concrete structures for practical usage. Compared to bacteria inoculated into the cement matrix, it is easier to provide the proper environment for bacteria when the bacteria are applied onto the surface of the concrete. Moreover, the bacterial surface treatment method has a number of advantages in comparison with conventional surface treatment methods using polymer-based coating materials, including water repellents or pore-blockers. Three advantages are as follows: (1) a similar thermal expansion property between the microbially precipitated calcium carbonate and the concrete surface, (2) environmentally friendly characteristics, and (3) the potential for self-healing. A surface treatment, especially for lightweight concrete, is very important to ensure good durability, as the durability of lightweight concrete is generally lower than that of normal concrete. In the present chapter, a previous work of the authors, a study of the bacterial surface treatment of normal and lightweight concrete (Kim et al. 2013) is reviewed and summarized. The surfaces of normal and lightweight concrete specimens were treated with a liquid medium containing bacteria. Macro- and micrographic assessments were done to analyze the shapes and distribution of the calcium carbonate crystals. The capillary water absorption of the concrete specimens was measured to evaluate the effects of the bacterial precipitation of calcium carbonate on the moisture transport properties, as these properties were thought to affect the durability of the concrete.

H.K. Kim
School of Architecture, Chosun University, Gwangju, South Korea

H.K. Lee (✉)
Department of Civil and Environmental Engineering,
Korea Advanced Institute of Science and Technology (KAIST),
Daejeon, South Korea
e-mail: leeh@kaist.ac.kr

F. Pacheco Torgal et al. (eds.), *Biotechnologies and Biomimetics for Civil Engineering*,
DOI 10.1007/978-3-319-09287-4_16

16.1 Introduction

There has been considerable effort to apply special types of bacteria which produce calcium carbonate particles in various fields of civil engineering (Bellucci et al. 1999; Ramachandran et al. 2001; Beklemishev and Kozliak 2003; de Graef et al. 2005; Phutane et al. 2007; Van Tittelboom et al. 2010). Among these efforts, bacterial surface treatment for concrete has been of particular interest given its strong potential to improve the durability of concrete structures (De Muynck et al. 2008a, b). By inoculating bacteria onto the surface of concrete, a thin and stable layer of calcium carbonate crystals can be formed via the metabolism of the bacteria. The layer of calcium carbonate crystals may diminish the uptake of water or protect the concrete from chemical deterioration, leading to an improvement of the durability of the concrete structures (Stocks-Fischer et al. 1999; De Muynck et al. 2008a, b; Kim et al. 2013).

Compared to the application of bacteria inside the cement matrix, bacterial surface treatment is also considered as a promising method for practical reasons. Regarding this approach, it is relatively easy to supply nutrients, oxygen, and water to the bacteria, as the bacteria are applied onto the surface of the concrete, compared to when the bacteria are inoculated into the cement matrix (Ghosh et al. 2005, 2009; Qian et al. 2009; Jonkers et al. 2010; Chahal et al. 2012). Moreover, the pH of concrete surfaces is generally neutral due to carbonation by CO_2 gas in the atmosphere, making the surface a suitable environment in which to survive, whereas the pH of the inner part of concrete is too basic for the bacteria to survive (Wiktor and Jonkers 2011).

Compared to conventional surface treatment methods using polymer-based coating materials, including water repellents or pore-blockers, the bacterial surface treatment method has a number of advantages. Most of all, the problem of the delamination of polymer-based coating materials from the concrete surface is mitigated (De Muynck et al. 2008a, b). The delamination of coating materials stems from the different thermal expansion coefficients between the polymer and the concrete (De Muynck et al. 2008b). However, calcium carbonate crystals precipitated by bacteria have a thermal expansion coefficient similar to that of the cement matrix (Kim et al. 2013). Moreover, the risk of environmental pollution induced by polymer-based coating materials can be eliminated. For example, Bisphenol A, one of the components resulting from the manufacturing of epoxy resin, can act as an endocrine disruptor in the human body (Maffini et al. 2006; Kimura et al. 1998). However, the bacteria used in concrete surface treatments only have limited health effects, such as only isolated cases of mild eye and skin irritation (US EPA 1998). In addition, the bacterial surface treatment may have self-healing characteristics if the bacteria survive after they are supplied with nutrients, whereas polymer-based coating materials become ineffective after delamination from the concrete surface (c.f. Wiktor and Jonkers 2011).

In terms of energy and time consumption for surface treatments, there are no significant differences between bacterial treatments and the use of a conventional polymer coating. For suitable hardening of an epoxy resin, the curing temperature should exceed 15 °C and the curing time can require anywhere from two to seven days (Issa and Debs 2007; Ha et al. 2010; Lee et al. 2011). Similarly, for a bacterial surface treatment, a sufficient temperature is one that exceeds 20 °C, and the curing time is again about 4–7 days (De Muynck et al. 2008a, b; Qian et al. 2009).

In the present chapter, a previous work of the authors, a case study involving the bacterial surface treatment of normal and lightweight concrete (Kim et al. 2013) is reviewed and summarized. In this work, both normal and lightweight concrete were subjected to bacterial surface treatment. It is well known that the durability of lightweight concrete is lower than that of normal concrete (Haque et al. 2004). For lightweight concrete, water, gas, and harmful chemicals can easily penetrate through the pore networks of the lightweight aggregates in the concrete (Vaysburd 1996; Haque et al. 2004). This leads to lower resistance of the concrete against the chloride penetration, carbonation, and freeze-thaw damage (Vaysburd 1996; Haque et al. 2004). As environmental attacks on concrete structures start on the surface, a surface protective layer is the first line of defense for both the concrete itself and the reinforcements embedded in it (Vaysburd 1996). Therefore, a surface treatment, especially for lightweight concrete, is very important to ensure its durability.

In the present work, the surfaces of normal and lightweight concrete specimens were treated with a liquid medium containing bacteria. Macro- and micrographic assessments were carried out to analyze the shapes and distribution of the calcium carbonate crystals. The capillary water absorption of the concrete specimens was measured to evaluate the effects of the bacterial precipitation of calcium carbonate on the moisture transport properties, which can affect the durability of the concrete.

16.2 Experimental Program

In this work, two types of ureolytic bacteria, *S. pasteurii* (ATCC 11859) and *B. sphaericus* (ATCC 13805), which can precipitate calcium carbonate crystals, were selected. The details of the preparation of the liquid media are available in Kim et al. (2013). As a reference, a cell-free medium was also prepared. The XRD patterns for the calcium carbonate crystals precipitated in media with and without bacteria are presented in Fig. 16.1. Both calcite and vaterite phases were found in all types of the media, but the intensity ratio between the calcite and vaterite phases differed in each case. The ratios between the peak intensities of 29.3° (calcite phase) and 27.0° (vaterite phase) were 1:0.72, 1:2.66, and 1:1.05, respectively, for the calcium carbonate crystals formed in a cell-free medium, the medium with *S. pasteurii*, and the medium with *B. sphaericus*. This indicates that both types of bacteria tended to form vaterite crystals compared to the medium without bacteria, where calcium carbonate crystals were precipitated by the natural reaction between the urea and the calcium acetate in that sample medium.

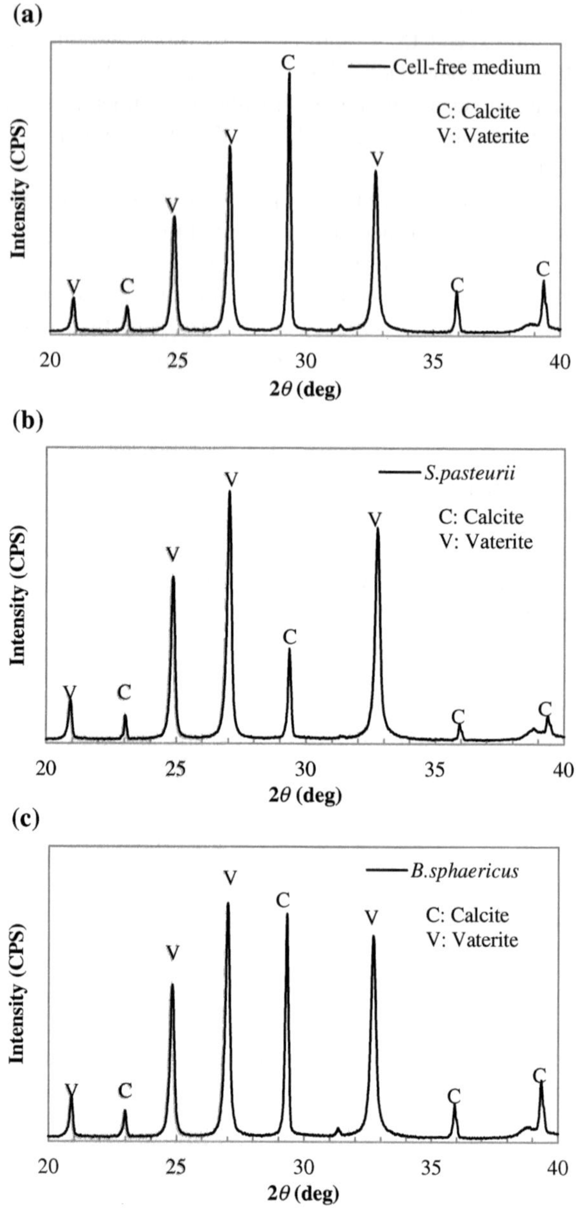

Fig. 16.1 XRD pattern of calcium carbonate crystals formed in the cell-free medium (**a**), the medium with *S. pasteurii* (**b**), and the medium with *B. sphaericus* (**c**). Reprinted from Kim et al. (2013), Copyright (2013), with permission from Elsevier

As concrete specimens for the bacterial surface treatment, normal and lightweight concrete were prepared. Coal bottom ash was applied as a lightweight aggregate material. The details of the preparation method of the concrete specimens and the procedures of the surface treatment are available in Kim et al. (2013).

The distributions of the calcium carbonate precipitation on the surfaces of the concrete specimens were photographed using a digital camera and a scanning electron microscope (SEM). The energy dispersive spectroscopy (EDS) spectra were evaluated to verify the chemical compositions of the crystals on the surfaces of the concrete specimens. Moreover, the amount of capillary water absorption of the concrete specimens was measured to evaluate the effects of the bacterial coating on the water transport properties. The experimental details of the present work can be found in Kim et al. (2013).

16.3 Results and Discussion

16.3.1 Precipitation of Calcium Carbonate on Concrete

Figure 16.2 shows the surfaces of normal and lightweight concrete specimens with various surface treatments.

As shown in Fig. 16.2, calcium carbonate was formed on the surfaces of the concrete specimens treated with the cell-free medium, similar to the cases of concrete specimens treated with a medium containing bacteria. However, it was found that small, light-brown spots covered the top surface of the lightweight concrete specimens treated with the cell-free medium. These were not found on the surfaces of lightweight concrete specimens treated with the medium containing bacteria. As marked by the red circle in Fig. 16.2d, these small spots have a hollow dome-shaped structure.

The assumed development process of the small spots by the cell-free medium on the surfaces of the lightweight concrete specimens is illustrated in Fig. 16.3a. The lightweight concrete had more pores due to the porous lightweight aggregate material used. Thus, some of the liquid medium was absorbed in the pores when the medium was poured on the surface of the lightweight concrete; consequently, a corresponding volume of air was simultaneously released from the pores. Although some amount of air also rose to the surface of the liquid medium, some remained on the substrate and formed air bubbles at the openings of the pores due to the surface tension which arose. Calcium carbonate then formed at the interface between the liquid medium and the air bubbles. After the medium on the concrete specimens was removed, dome-shaped calcium carbonate structures formed and the remaining medium under these structures dried out.

These dome-shaped structures, however, were easily broken due to their thin layer. However, the respiration characteristics of the bacteria were assumed to be hindered during the development of the dome-shaped structures on the lightweight concrete specimens. As shown in Fig. 16.3b, the air bubbles remaining at the openings of the pores due to the surface tension may have been consumed by the respiration of the bacteria. Therefore, calcium carbonate crystals could precipitate inside the pores.

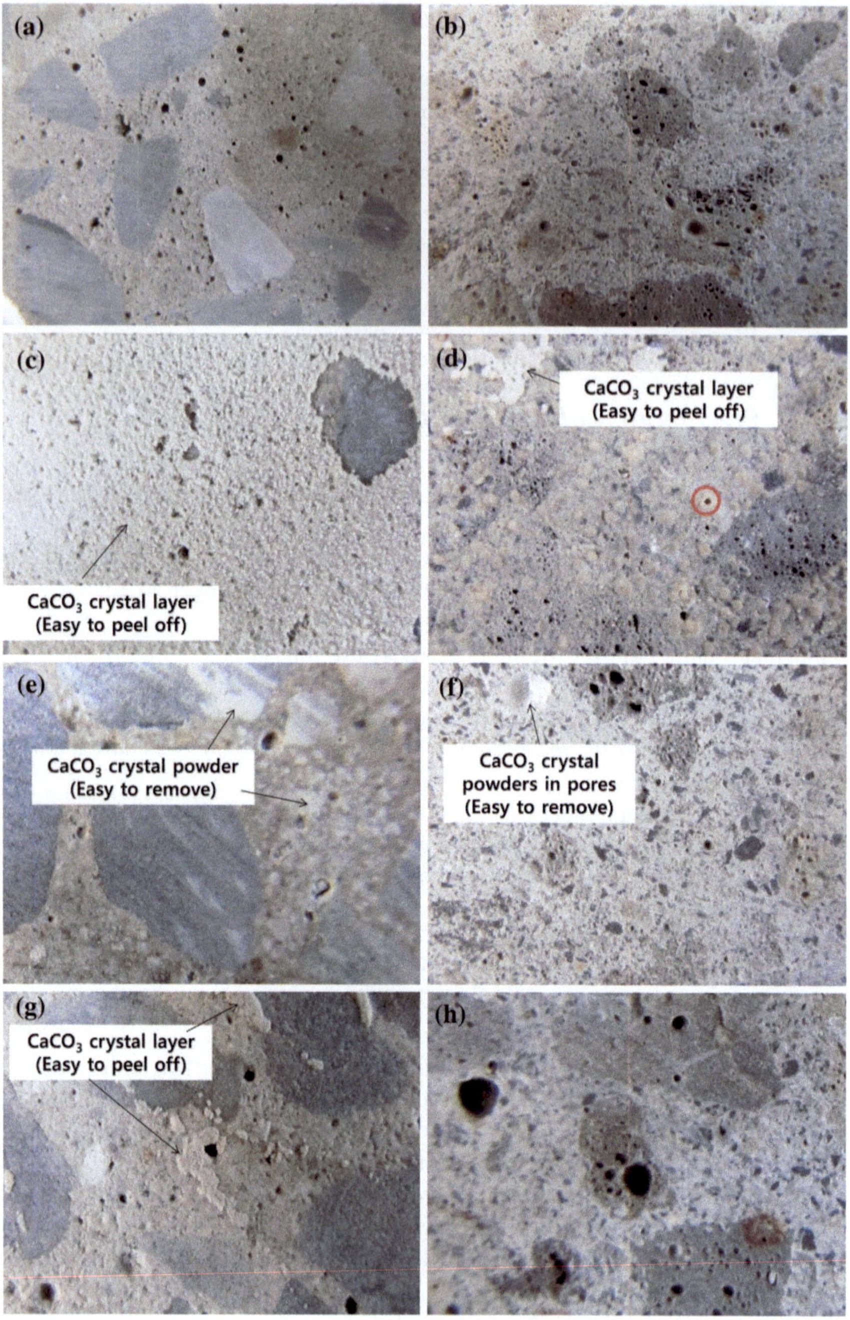
(a)
(b)
(c)
CaCO3 crystal layer
(Easy to peel off)
(d)
CaCO3 crystal layer
(Easy to peel off)
(e)
CaCO3 crystal powder
(Easy to remove)
(f)
CaCO3 crystal
powders in pores
(Easy to remove)
(g)
CaCO3 crystal layer
(Easy to peel off)
(h)

◀ **Fig. 16.2** *Top surfaces* of concrete specimens treated by water: **a** normal concrete and **b** lightweight concrete; *top surfaces* treated by cell-free medium: **c** normal concrete and **d** lightweight concrete; *top surfaces* of concrete specimens treated by medium with *S. pasteurii*: **e** normal concrete and **f** lightweight concrete; *top surfaces* of concrete specimens treated by medium with *B. sphaericus*: **g** normal concrete and **h** lightweight concrete photographed by using a digital camera (*height* 23 mm, *width* 30 mm). Reprinted from Kim et al. (2013), Copyright (2013), with permission from Elsevier

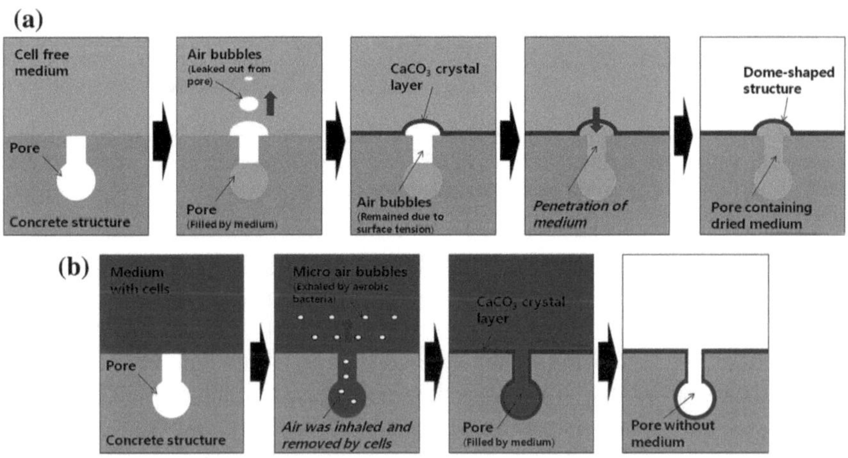

Fig. 16.3 **a** *Dome-shape* structure formation when being treated with cell-free medium, **b** crystal layer formation when being treated with cells. Reprinted from Kim et al. (2013), Copyright (2013), with permission from Elsevier

In addition, the fact that fewer bubbles arose in medium with bacteria was also attributable to the fact that the bacteria metabolized the proteins. For instance, *B. sphaericus* metabolized glucose via a metabolic process (Massie et al. 1985), possibly acting as a surfactant (Bam et al. 1995), which could then hinder the formation of air bubbles.

Meanwhile, the precipitations of calcium carbonate differed slightly depending on the type of bacteria. The medium with *S. pasteurii* showed the precipitation of very fine calcium carbonate powders (Fig. 16.3e and f). On the other hand, the medium with *B. sphaericus* precipitated thick calcium carbonate layers (Fig. 16.3g, h).

16.3.2 Microstructures of Precipitated Calcium Carbonate

Figures 16.4, 16.5, 16.6, 16.7, 16.8, 16.9 and 16.10 show SEM images and the EDS spectra of the surfaces of normal and lightweight concrete specimens treated with liquid media with and without bacteria. In the case of the cell-free medium,

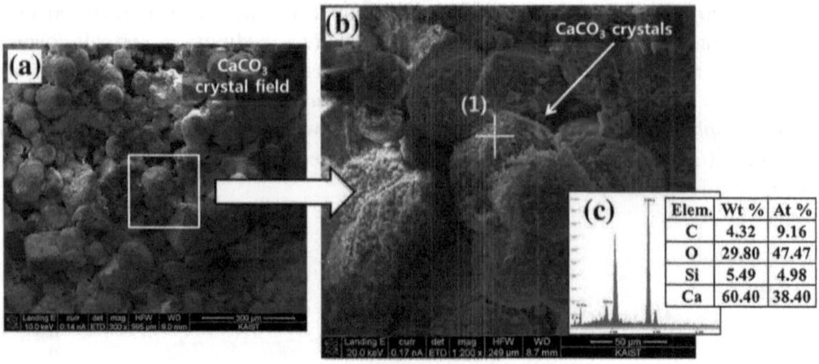

Elem.	Wt %	At %
C	4.32	9.16
O	29.80	47.47
Si	5.49	4.98
Ca	60.40	38.40

Fig. 16.4 SEM images of *top surface* of normal concrete treated by the cell-free medium measured in 300 μm (**a**) and 50 μm (**b**), and EDS spectra at point 1 (**c**). Reprinted from Kim et al. (2013), Copyright (2013), with permission from Elsevier

round-shaped calcium carbonate crystals were formed and were stacked upon each other on the normal and lightweight concrete specimens (Figs. 16.4 and 16.5). The diameters of these crystals were in the range of 10–80 μm. There were no significant differences between the normal concrete and lightweight concrete samples. It was noted that, when bacteria were not used, a single pure bulk crystal consisting of many plate-shaped, small crystals formed (Fig. 16.5b). Moreover, as shown in Fig. 16.6, the calcium carbonate crystals filled the pores of the lightweight aggregates.

Compared to the cell-free medium, the sizes of the calcium carbonate crystals were smaller and their distribution was denser in the medium with bacteria (Figs. 16.7, 16.8, 16.9 and 16.10). Moreover, the distribution of calcium carbonate on the lightweight concrete was slightly different from that on the normal concrete. As shown in Fig. 16.7, in the normal concrete, small calcium carbonate crystals with a size of approximately 500 nm were stacked like gravel on the surface of the

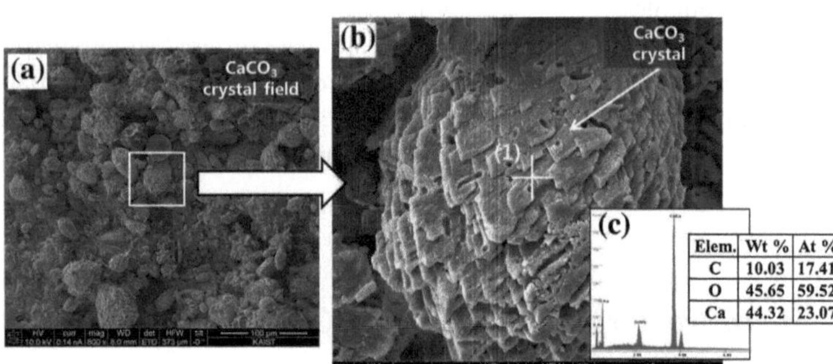

Elem.	Wt %	At %
C	10.03	17.41
O	45.65	59.52
Ca	44.32	23.07

Fig. 16.5 SEM images of *top surface* of lightweight concrete treated by cell-free medium measured in 100 μm (**a**), 10 μm (**b**), and EDS spectra at point 1 (**c**). Reprinted from Kim et al. (2013), Copyright (2013), with permission from Elsevier

Fig. 16.6 CaCO$_3$ crystal filled in pore of aggregate on *top surface* of lightweight concrete treated by cell-free medium. Reprinted from Kim et al. (2013), Copyright (2013), with permission from Elsevier

specimens and were loosely connected to each other. On the other hand, in the lightweight concrete, the assembly of many small-sized calcium carbonate crystals of a size of about 10 μm was noted (point 1 in Fig. 16.8b); these were connected to each other by web-shaped crystals (Point 2 in Fig. 16.8b). The chemical composition of these web-shaped crystals differed slightly from that of bulk calcium carbonate crystals (Fig. 16.8c, d). It is important to note that the EDS spectra of the pure calcium carbonate crystals should show 20 % of calcium ions, 20 % of carbon ions, and 60 % of oxygen ions. It is assumed that the web-shaped crystals

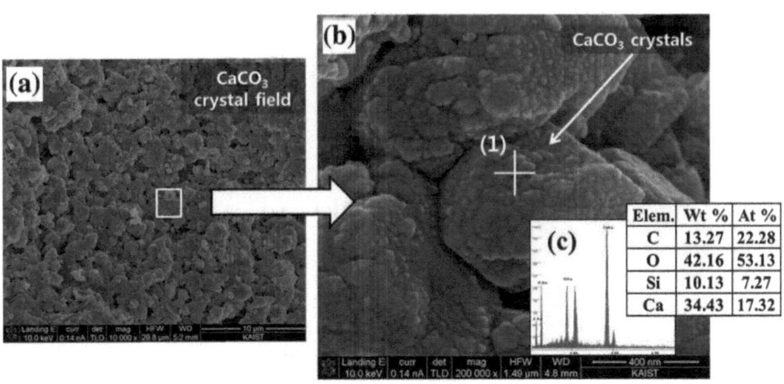

Fig. 16.7 SEM images of *top surface* of normal concrete treated by the medium with *S. pasteurii* measured in 10 μm (**a**), 400 nm (**b**), and EDS spectra at point 1 (**c**). Reprinted from Kim et al. (2013), Copyright (2013), with permission from Elsevier

Fig. 16.8 SEM images of *top surface* of lightweight concrete treated by the medium with *S. pasteurii* measured in 100 μm (**a**), 10 μm (**b**), and EDS spectra at point 1 (**c**) and at point 2 (**d**). Reprinted from Kim et al. (2013), Copyright (2013), with permission from Elsevier

formed after the bulk calcium carbonate crystals were developed, but the crystallization process is not yet known. Moreover, why the different shapes of the calcium carbonate crystals of the lightweight concrete and of the normal concrete formed is not known. However, it is considered that changes of pH and calcium ion concentration on the surface of the concrete by a pozzolanic reaction in the bottom ash aggregate may have led to this phenomenon (c.f. Zhang and Gjørv 1990; Mehta and Monteiro 2006; Chindaprasirt et al. 2009).

In addition, the calcium carbonate formed by the medium with *B. sphaericus* crystallized with pockmark shapes regardless of the type of concrete specimen, whereas the calcium carbonate crystals formed by the cell-free medium and the medium with *S. pasteurii* showed spherical or semi-spherical particle shapes (Figs. 16.9 and 16.10). The calcium carbonate developed by *S. pasteurii* rapidly

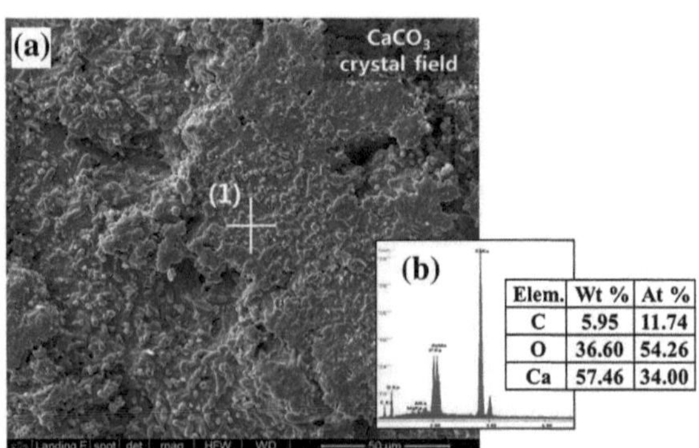

Fig. 16.9 SEM images of *top surface* of normal concrete treated by the medium with *B. sphaericus* measured in 50 μm (**a**), and EDS spectra at point 1 (**b**). Reprinted from Kim et al. (2013), Copyright (2013), with permission from Elsevier

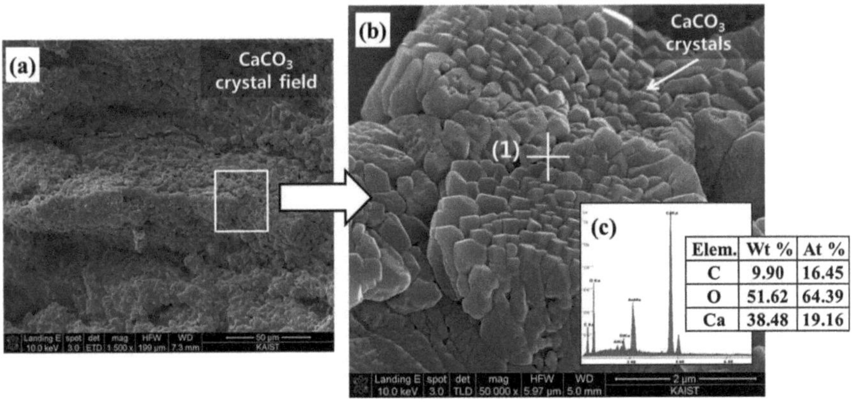

Fig. 16.10 SEM images of *top surface* of lightweight concrete treated by the medium with *B. sphaericus* measured in 50 μm (**a**), 2 μm (**b**), and EDS spectra at point 1 (**c**). Reprinted from Kim et al. (2013), Copyright (2013), with permission from Elsevier

was precipitated in the medium and thus formed spherical particle shapes and was deposited on the surface of the concrete (Parks 2009). On the other hand, for *B. sphaericus*, the calcium carbonate crystals were precipitated from the interface between the medium and the concrete specimen due to the formation of a biofilm (Kolari et al. 2001; Parks 2009). These crystals were precipitated with radial shapes and were developed between other crystals without a gap, covering the top surface densely (Fig. 16.10). The different shapes of the calcium carbonate crystals shown in the SEM images (Figs. 16.4, 16.5, 16.6, 16.7, 16.8, 16.9 and 16.10) may be affected by the different morphologies of the calcite and vaterite phases (de Leeuw and Parker 1998).

16.3.3 Effect of Precipitated Calcium Carbonate on Capillary Water Absorption of Concrete

Figure 16.11 presents the relative values of the capillary water absorption per unit area of concrete. It should be noted that the relative values of both the normal and lightweight concrete were similar, while the absolute value of lightweight concrete (1.4–2.6 kg/m^2) was higher than that of the normal concrete (4.7–9.3 kg/m^2).

As shown in Fig. 16.11, the capillary water absorption per unit area of concrete was decreased in both the medium with and the medium without bacteria. However, in the medium without bacteria, the relative values of water absorption of the concrete specimens were not stable in comparison with the concrete specimens treated with the medium with bacteria. The relative value of the normal concrete treated with the cell-free medium was higher at 6 h than that of the concrete treated with the medium with bacteria, whereas it was lower at 128 h. Moreover, the relative value of water absorption of the lightweight concrete at 6 h was very

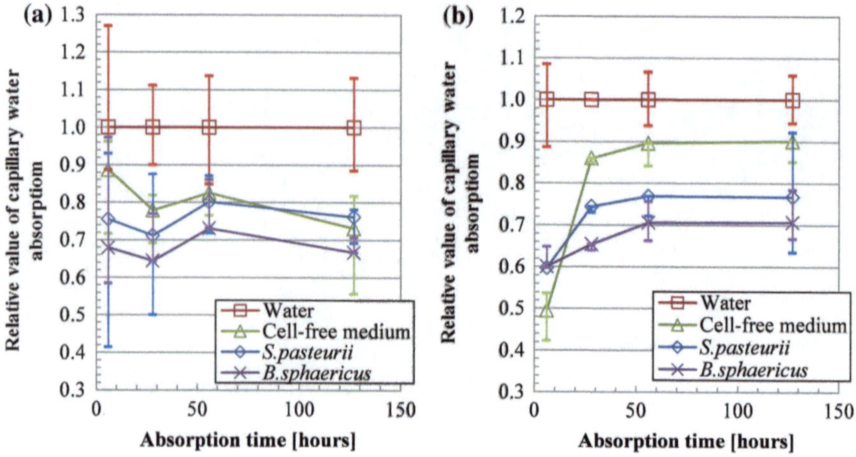

Fig. 16.11 Relative value of capillary water absorption: **a** normal concrete and **b** lightweight concrete. Reprinted from Kim et al. (2013), Copyright (2013), with permission from Elsevier

low, though this value increased at 128 h. This was considered to be attributable to the distribution of the calcium carbonate precipitation on the concrete. In the cell-free medium, calcium carbonate crystals formed with a stacked particle morphology on the surface of the concrete (Figs. 16.4 and 16.5); thus, these particles could easily become detached during the capillary water absorption test. On the other hand, although the calcium carbonate crystals precipitated by *S. pasteurii* had shapes similar to those that formed in the cell-free medium, their sizes were much smaller and the crystals were connected to each other (Figs. 16.7 and 16.8). Therefore, the relative value of the concrete treated with the medium with *S. pasteurii* was more stable than that of the concrete treated with the cell-free medium.

In addition, regarding the decrease in the capillary water absorption of the concrete, the efficiency of the bacterial treatment in the present work appears to be similar to experimental results obtained in similar studies. The surface treatment by *B. sphaericus* could reduce the capillary water absorption of cement mortar by about 50 % (De Muynck et al. 2008b). Moreover, the surface treatment by *B. pasteurii* decreased the capillary water absorption of cement paste by about 50–70 % (Qian et al. 2009).

16.4 Conclusion

In this chapter, a case study involving the bacterial surface treatment of normal and lightweight concrete samples was presented. The surfaces of concrete specimens were treated with a liquid medium with and without bacteria. Macro- and micrographic studies were conducted to analyze the shapes and distribution of the calcium carbonate crystals. The amounts of capillary water absorption of the concrete specimens were measured to evaluate the effects of the bacterial precipitation of calcium carbonate on the moisture transport properties, which may affect the durability of the concrete.

The size of the calcium carbonate crystals precipitated by the medium with bacteria was larger than that precipitated by the medium without bacteria. Moreover, the medium with *B. sphaericus* formed calcium carbonate crystals with nano-sized pockmark shapes, and the crystals covered the concrete surface densely. Thus, both the normal and lightweight concrete specimens treated with the medium containing *B. sphaericus* showed lower capillary water absorption levels per unit area than that of the concrete specimens treated with the cell-free medium. In addition, the respiration characteristics of the bacteria may hinder the development of the dome-shaped calcium carbonate structures noted on the lightweight concrete specimens.

Considering the water absorption rates from the surfaces of the concrete samples, the efficiency of bacterial treatments is still lower than that of conventional water-repellent materials, including epoxy or silicone resins (De Muynck et al. 2008b; Qian et al. 2009). Moreover, investigations of the long-term performance of a bacterial surface treatment and life-cycle assessments based on these results have not been reported, both of which are important for practical applications of this method. Therefore, further studies will be carried out along this line.

Acknowledgments This research was sponsored by a grant from the National Research Foundation of Korea (NRF) funded by the Korean government (2013028443).

References

Bam NB, Randolph TW, Cleland JL (1995) Stability of protein formulations: investigation of surfactant effects by a novel EPR spectroscopic technique. Pharm Res 12(1):2–11

Beklemishev MK, Kozliak EI (2003) Bioremediation of concrete contaminated with n-hexadecane and naphthalene. Acta Biotechnol 23(2–3):197–210

Bellucci R, Cremonesi P, Pignagnoli G (1999) A preliminary note on the use of enzymes in conservation: the removal of aged acrylic resin coatings with lipase. Stud Conserv 44(4):278–281

Chahal N, Siddique R, Rajor A (2012) Influence of bacteria on the compressive strength, water absorption and rapid chloride permeability of fly ash concrete. Constr Build Mater 28(1):351–356

Chindaprasirt P, Jaturapitakkul C, Chalee W, Rattanasak U (2009) Comparative study on the characteristics of fly ash and bottom ash geopolymers. Waste Manag 29(2):539–543

de Graef B, de Windt W, Dick J, Verstraete W, de Belie N (2005) Cleaning of concrete fouled by lichens with the aid of *Thiobacilli*. Mater Struct 38(10):875–882

de Leeuw NH, Parker SC (1998) Surface structure and morphology of calcium carbonate polymorphs calcite, aragonite, and vaterite: an atomistic approach. J Phys Chem B 102(16):2914–2922

De Muynck W, Cox K, De Belie N, Verstraete W (2008a) Bacterial carbonate precipitation as an alternative surface treatment for concrete. Constr Build Mater 22(5):875–885

De Muynck W, Debrouwer D, De Belie N, Verstraete W (2008b) Bacterial carbonate precipitation improves the durability of cementitious materials. Cem Concr Res 38(7):1005–1014
Ghosh P, Mandal S, Chattopadhyay BD, Pal S (2005) Use of microorganism to improve the strength of cement mortar. Cem Concr Res 35(10):1980–1983
Ghosh S, Biswas M, Chattopadhyay BD, Mandal S (2009) Microbial activity on the microstructure of bacteria modified mortar. Cement Concr Compos 31(2):93–98
Ha SK, Lee HK, Kang IS (2010) Shear behavior and performance of RC beams with polymer mortar systems under cyclic loading. J Reinf Plast Compos 29(17):2604–2620
Haque MN, Al-Khaiat H, Kayali O (2004) Strength and durability of lightweight concrete. Cement Concr Compos 26(4):307–314
Issa CA, Debs P (2007) Experimental study of epoxy repairing of cracks in concrete. Constr Build Mater 21(1):157–163
Jonkers HM, Thijssen A, Muyzer G, Copuroglu O, Schlangen E (2010) Application of bacteria as self-healing agent for the development of sustainable concrete. Ecol Eng 36(2):230–235
Kim HK, Park SJ, Han JI, Lee HK (2013) Microbially mediated calcium carbonate precipitation on normal and lightweight concrete. Constr Build Mater 38:1073–1082
Kimura H, Matsumoto A, Hasegawa K, Ohtsuka K, Fukuda A (1998) Epoxy resin cured by bisphenol A based benzoxazine. J Appl Polym Sci 68(12):1903–1910
Kolari M, Nuutinen J, Salkinoja-Salonen MS (2001) Mechanisms of biofilm formation in paper machine by Bacillus species: the role of *Deinococcus geothermalis*. J Ind Microbiol Biotechnol 27(6):343–351
Lee HK, Cheong SH, Ha SK, Lee CG (2011) Behavior and performance of RC T-section deep beams externally strengthened with CFRP sheets. Compos Struct 93:911–922
Maffini MV, Rubin BS, Sonnenschein C, Soto AM (2006) Endocrine disruptors and reproductive health: the case of bisphenol-A. Mol Cell Endocrinol 254–255:179–186
Massie J, Roberts G, White PJ (1985) Selective isolation of *Bacillus sphaericus* from soil by use of acetate as the only major source of carbon. Am Soc Microbiol 49(6):1478–1481
Mehta PK, Monteiro PJM (2006) Concrete: microstructure, properties, and materials, 3rd edn. McGraw Hill, Maidenhead
Parks SL (2009) Kinetics of calcite precipitation by ureolytic bacteria under aerobic and anaerobic conditions. Master degree thesis, Montana State University
Phutane SR, Renner JN, Nelson SL, Seames WS, Páca J, Sundstrom TJ, Kozliak EI (2007) Removal of 2,4-dinitrotoluene from concrete using bioremediation, agar extraction, and photocatalysis. Folia Microbiol 52(3):253–260
Qian C, Wang J, Wang R, Cheng L (2009) Corrosion protection of cement-based building materials by surface deposition of $CaCO_3$ by *Bacillus pasteurii*. Mater Sci Eng C 29(4):1273–1280
Ramachandran SK, Ramakrishnan V, Bang SS (2001) Remediation of concrete using microorganisms. ACI Mater J 98(1):3–9
Stocks-Fischer S, Galinat JK, Bang SS (1999) Microbiological precipitation of $CaCO_3$. Soil Biol Biochem 31(11):1563–1571
US Environmental protection agency (EPA) (1998) *Bacillus sphaericus*; exemption from the requirement of a tolerance, http://www.epa.gov/fedrgstr/EPA-PEST/1998/September/Day-11/p24469.htm
Van Tittelboom K, De Belie N, De Muynck W, Verstraete W (2010) Use of bacteria to repair cracks in concrete. Cem Concr Res 40(1):157–166
Vaysburd AM (1996) Durability of lightweight concrete bridge in severe environments. Concr Int 18(7):33–38
Wiktor V, Jonkers HM (2011) Quantification of crack-healing in novel bacteria-based self-healing concrete. Cement Concr Compos 33(7):763–770
Zhang MH, Gjørv OE (1990) Pozzolanic reactivity of lightweight aggregates. Cem Concr Res 20(6):884–890

Chapter 17
Biotechnological Aspects of Soil Decontamination

V. Sheoran and A. Sheoran

Abstract Soils have been subjected to several contaminants that vary in concentration and composition. Soil pollution causes significant damage to the environment and human health as a result of their mobility and solubility. Significant progress has been made in regulating soil pollution, with a parallel development of methodologies for soil assessment and remediation. The selection of most appropriate soil and sediment remediation method depends on the site characteristics, concentration, type of pollutants to be removed, and the end use of the contaminated medium. This chapter provides the developing biotechnological aspects of soil decontamination. The study also reviews other available remediation options, which includes physical, chemical, and thermal technologies. All these technologies may be used in conjunction with one another to reduce the contamination to an acceptable level, and may offer potential technical solution to most soil pollution.

17.1 Introduction

Globally, the increasing human population, industrial revolution, and the number of anthropogenic inefficiencies with unplanned growth of urban system have been putting an intense pressure on the consumption of natural resources, thus threatening human health and the environment. Over time, the quantities of these

V. Sheoran (✉)
Department of Zoology, Jai Narain Vyas University, Jodhpur 342011, India
e-mail: vimi_sheoran@yahoo.com

A. Sheoran
Department of Civil Engineering, Birla Institute of Technology & Science, Pilani 333031, Rajasthan, India
e-mail: arushai_sheoran@yahoo.com

F. Pacheco Torgal et al. (eds.), *Biotechnologies and Biomimetics for Civil Engineering*,
DOI 10.1007/978-3-319-09287-4_17

nonrecyclable toxic by-products are at the levels that present an unacceptable risk to the sensitive biosphere (Sheoran et al. 2008; Sikdar et al. 1998)

Historically, soil pollution has been of very little concern, and its contamination is a relatively recent issue when, actually, it began a long time ago, particularly after the industrial revolution, even though only recently has mankind become aware of its dimension, persistence, and harmful effects (Castelo-Grande et al. 2010). Pollutants can be built up in the soil directly or indirectly from several sources, such as industrial emissions, mining and smelting of metalliferous ore, electroplating energy and fuel production, and sludge dumping. A wide range of inorganic and organic compounds cause contamination, and these include heavy metals, combustible and putrescible substances, hazardous wastes, explosives, and petroleum products. There has been increasing concern over the last few decades regarding organic pollutants, which includes PCBs such as dioxin, polycyclic aromatic hydrocarbons (PAHs) such as benzoapyrene, nitroaromatics such as trinitrotoluene (TNT), and linear halogenated hydrocarbons such as trichloroethylene (TCE). Major components of inorganic contaminants are heavy metals (Ghosh and Singh 2005). Heavy metals cannot be destroyed biologically and are present in soils as free metal ions, soluble metal complexes (sequestered to ligands), exchangeable metal ions, organically bound metals, precipitated or insoluble compounds such as oxides, carbonates, and hydroxides, or they may form part of the structure of silicate materials (indigenous soil content) (Davies et al. 2001).

Contaminated soils are the target of several thousands of pollutants that vary in their compositions and concentration. Elevated concentration of these pollutants impair the functioning of human population and pose significant risk to the development of flora and fauna (Scullion 2006). Contaminated soil fails to support crop growth and negatively affects its development because of interference of phytotoxic contaminants with metabolic processes and sometimes also leading to plant death (Hoffmann 1983). Besides this, soil toxicity disrupts biological cycling of nutrients and also affects the hydrosphere compromising with the quality of drinking water resources, and threatening the aquatic ecosystem (Bilek 2004). Human beings are also at risk from polluted soils, thus the magnitude of pollution in our soils calls for immediate action (Friberg et al. 1986; Knasmuller et al. 1998; Nathanail and Earl 2001). When contaminant concentrations in soil are too high for natural biodegradation to occur, cleanup action is warranted (Sikdar et al. 1998). Thus, in response to a growing need to address this environmental contamination, many remediation technologies have been developed to treat soil contaminated by various pollutants, including in situ (treating relatively undisturbed soils) and ex situ (treating excavated soils) methods either onsite or in designated soil treatment facilities. In situ has the advantage of minimal disruption to activities on site or on adjacent land. Ex situ approaches generally offer greater scope for managing conditions to optimize treatment efficiency and for controlling potential spread of pollutants. In situ methods are favored over the ex situ techniques due to their low cost and reduced impact on the ecosystem (Reed et al. 1992).

Biological, physical, chemical, and other technologies can be used in conjunction with one another to reduce the contamination to a safe and acceptable level (RAAG 2000). Physical methods employs soil washing, encapsulation, and solidification; precipitation and ion exchange are chemical treatments, and for the biological treatment plants are used. Even though many technologies are available for the decontamination of polluted sites, the selection depends on contaminant and site-by-site basis, regulatory requirements, costs, and time constraints. Since most remediation techniques are site-specific, the selection of appropriate technology is often a difficult, but extremely important step in the successful remediation of a contaminated site. Therefore, the successful decontamination of a contaminated site depends on proper selection of the methodology, its design, and adjustment of the remediation technology's operations based on properties of the contaminants and soils, and on the performance of the whole system (USEPA 1998; Khan et al. 2004 and Pazos et al. 2010).

This chapter provides the developing biotechnological aspects of soil decontamination, and also reviews various other physical, chemical, electro-remediation or electro-reclamation techniques. It also outlines the types of waste and media in which the technology could be successfully applied so that wide-scale implementation and commercialization of the technique may be recommended on a global basis.

17.2 Physical Techniques

Decontamination of soil relies on an understanding of the physical behavior of the pollutants in the site specific environment. Physical remediation methods are most effective in coarser textured soils, although fracturing of finer textured soils may extend their applicability, and for pollutants that are more soluble or volatile. This treatment can enhance the effectiveness of biological degradation of contaminates or indirectly cause their destruction (Scullion 2006).

17.2.1 Off-Site Management

The most common traditional remediation technique is off-site management. The contaminated soil is taken for burial at land fill sites. This method of remediation merely shifts the contamination problem else where. Additionally, there are hazards associated with the transport of contaminated soil and migration of contaminant from landfill into adjacent environment (Williams 1988).

17.2.2 Isolation and Containment

Contaminants can be isolated and contained, to prevent further movement, to reduce the permeability of the waste to less than 1×10^{-7}m/s (as required by the USEPA), and to increase the strength or bearing capacity of the waste (USEPA 1994). This technique consists of the use of barriers that inhibit the migration of contaminants to the neighboring uncontaminated site. Physical barriers made of steel, cement, bentonite, and grout walls can be used for capping, vertical, and horizontal containment. Capping is a site specific proven technology, which uses synthetic membranes to reduce water infiltration. Horizontal barriers restrict the downward movement of metal contaminant within the soil, whereas vertical barrier reduces the migration from one site to another. These barriers are made of slurry walls, grout, or geomembrane curtains, and sheet pile walls. It is the least expensive approach, but leaves the contaminant in place without treatment. The selection of each technology is site-specific. They are beneficial where the area of contaminant is shallow but large. More research is required to match reactive media with contaminants, model life time performance, optimize retention times, and develop methods for regeneration of reactive media.

In terms of risk management, these above mentioned approaches aimed to control the pathway linking hazard and receptor without treating the source of the hazard. Remediation practices emphasized containment rather than treatment.

17.2.3 Solidification/Stabilization

Solidification process is a nondestructive physical method to immobilize the contaminants by encapsulating them in a solid of high structural integrity, while stabilization includes chemical reactions to reduce contaminant mobility. It is also known as waste fixation through both physical and chemical means. Some variants like liquid monomers that polymerize, pozzolans, bitumen, fly-ash, asphalt, and cement are injected to encapsulate the soils. Capping or jacketing or complete coating of the contaminated sediment with sandy material, such as clean sediment, sand, or gravel, which decreases the direct contact area between the water and the contaminated sediment, is what is done under encapsulation (Peng et al. 2009). Two ways of encapsulation are (i) microencapsulation (ii) macroencapsulation. Port land cement, pozzolans, or lime/hydrated lime, and organic polymers may be used for microencapsulation, whereas concrete, organic materials (polythene, polyesters, etc.), sulfur cement, etc. can be used for macroencapsulation.

Some researchers have reported that a good cap thickness was approximately 50 cm; and through capping the sediment by sands materials, the heavy metal concentration in water could reduce to 80 %. The cost of implementing this technology is dependent on the lithology of the site and the depth of the contaminant. As the depth of contamination increases, so does the cost (Khan et al. 2004). The stabilization converts the contaminants into less soluble immobilized and less toxic form by mixing soil or waste with chemical binders, such as cement, sulfide, and phosphate binders, polyester resins, or polysiloxane compounds to create a slurry, paste or other semi-liquid state, and is allowed time to cure into solid form (Wang et al. 2012). Additives through which solidification is achieved are either cement-based, pozzolon-based, the thermoplastic methods, the organic polymerization methods, the encapsulation method, and organophillic clay-based (Wang et al. 2012). Among these methods, cement-based solidification/stabilization is of increasing importance as option for remediating contaminated sites because of its low material and equipment cost.

This technique is suitable for contaminant in shallow depths and of large volume, and is not suitable for metals, which are not highly soluble, and do not form hydroxides, such as arsenic, chromium, and mercury. In situ solidification/stabilization techniques are preferred since labor and energy costs are lower, but site conditions, such as bedrock, large boulders, clay, and oily patches may cause mixing problems (Mulligan et al. 2001).

17.2.4 Vapor Extraction and Air Sparging

Vapor extraction and air sparging techniques are based on the manipulation of pollutant distribution between liquid and vapor phases. These treatments promote the volatilization of pollutants (e.g., benzene, toluene, ethyl benzene, xylene, and chlorobenzenes) in unsaturated and saturated zones. Extracted gases or vapors may be adsorbed onto activated carbon or treated (e.g., by oxidation). The effectiveness of vapor extraction systems may be extended to semi-volatile pollutants by injection of heated air or heating by microwave/radiowave, and rates of extraction may be improved by increasing air flow rates to a point when mass transfer limits volatilization (George et al. 1992; Park et al. 2005). Air sparging also known as in situ air stripping or in situ volatilization induces partitioning of dissolved and free phase contaminants into the vapor phase, and increases in dissolved oxygen can stimulate aerobic degradation. Benzene removal by air sparging has been shown by Adams and Reddy (2003). Both these approaches are less well suited to fine textured soils because of restricted rates of movement in the mobile phases and increased distances over which volatile organic contaminants have to diffuse through an aqueous phase. Also treatment rates are slower in soils with higher organic contents (Gomez-Lahoz et al. 1995).

17.2.5 Vitrification

Vitrification of molten glass is another method of solidification/stabilization process requiring thermal energy. It uses heat of up to > 1,000 > 1,600 °C - > 2,000 °C to destroy organic pollutants by pyrolysis and immobilize most of the pollutants in organic (Gavrilesseu et al. 2009). It is a two stage process in which pollutants are desorbed at lower (<600 °C) temperature and then combusted. Vitrification is mainly used to remediate soils contaminated with heavy metals mixed with radioactive elements (Wang et al. 2012). It involves insertion of electrodes into the soil, which must be able to carry a current, and then to solidify, as it cools. Full scale applications exist for arsenic, lead, and chromium contaminated soils. Vitrification is expensive and suitable for shallow contamination and also toxic gases can be produced during the process, but is applicable to mixed wastes where few technologies are available (Mulligan et al. 2001).

The in situ or ex situ (where the soil is excavated and treated) vitrification consists in the insertion of graphite electrodes into the soil creating a high electric current, such that the released heat provokes the fusion of the soil matrix (Castelo-Grande et al. 2005). This leads to the formation of vitrified end product into which the contaminants are incorporated and subsequently immobilized. During this process, the majority of contaminants initially present in the soil are volatilized reducing their concentration in the soil and the waste, while the remainder are converted into a chemically inert, stable glass, and crystalline product. The vitrification can be performed by three different processes namely electrical process, thermal process, and plasma process. Precisely, electrical process makes use of application of electrical energy through graphite electrodes inserted into ground, whereas thermal process requires an external heat source and a typical reactor (Dermatas and Meng 2003); and moreover in plasma process, electrical discharges are used to achieve temperatures up to 5,000 °C .The advantage of this method is that the volume of waste can be reduced with long-term stability but it is a costly method (Suthersan 1997).

17.2.6 Mechanical Separation

This involves the size selection process to remove larger, cleaner particles from the smaller, and more polluted ones. Characterization in terms of particle size and contaminant level in each fraction is the most important parameter in determining the suitability of this process. They include hydrocyclones, which separate the larger particles greater than 10–20 μm by centrifugal force from the smaller particles, fluidized separation removes smaller particles at the top (Less than 50 μm) in countercurrent overflow in a vertical column, by gravimetric settling and flotation, which is based on different surface characteristics of contaminated particles, magnetic separation, used to separate these from ferrous metals. These methods have been used in mineral ore processing.

17.2.7 Pyrometallurgical Separation

Pyrometallurgical processes use high temperature furnace to volatilize metals in contaminated soil. Temperatures of 200–700 °C are used to evaporate the contaminant. After volatilization, metals are then recovered or immobilized. These methods are most applicable to mercury since it is easily converted to its metallic form at high temperature. Other valuable metals such as gold and platinum can also be recovered from low soil concentration. This type of treatment is usually performed off-site due to lack of mobile units, and is applicable to highly contaminated soils (5–20 %) where metal recovery is profitable. Prior to pyrometallurgical separation the soil must be concentrated by physical or soil washing.

17.2.8 Soil Washing

Soil decontamination can also be carried out by in situ washing of soil. It is a physical separation technique, which consists of extraction of contaminants by suspending them in watery solutions, i.e., by dissolution. The main principle of soil washing is a selective classification of highly contaminated pollutants followed by the solid or liquid phase separation of the remaining suspension (Bradl and Xenidis 2005). It may consist of excavation, fragmentation, separation in different grain sizes, washing of the different fraction, and their disposal (Castelo-Grande et al. 2010). This technique is often considered as a pretreatment for the reduction of the toxic contents of the contaminated soil, and to be treated by another technology. The physical process of soil washing involves two principal steps:

Mechanical/Physical sorting Physical separation may include screening followed by density or gravity separation. Mechanical screens and hydrocyclones are often used to separate the soils into various size fractions. The bulk oversize material consists of clean or slightly contaminated cobbles and stones, and may undergo a water rinse before being returned to the site as fill. The slit and clay fraction generally contains the highest concentration of the contaminants, and is usually treated by the solidification/stabilization techniques to immobilize the contaminants prior to land filling. The remaining fine and coarse sands can be further treated using density/gravity separation process to separate high density aggregates and metal fragments. Magnetic contents are removed manually with the help of magnets.

Wash water treatment Fresh water and cleaned process water are added to the soil. From the suspension, cleaned soils fractions are separated and contaminants are further proceeded for further treatment.

For the process of soil washing the knowledge of particle size dependent pollutant distribution is of vital significance. It is one of the few permanent treatment alternatives to remove metal contaminants from the soils. This technology is particularly cost-effective and well established in mineral industry.

17.3 Chemical Techniques

A range of chemical processes have been applied to soil to destroy or convert pollutants into less toxic forms, to extract them, or to immobilize them. Wood (2001) suggested that chemical treatments can be highly specific for some pollutants, e.g., PCBs and halogenated alkanes.

17.3.1 Oxidation-Reduction Reaction

Chemical treatment by reductive as well as oxidative mechanisms may be used to detoxify or decrease the mobility of metal contaminants (Evanko and Dzombak 1997). This is commonly used for waste water treatment. Oxidation reactions detoxify, precipitate, or solubilize metals and involve addition of potassium permanganate, hydrogen peroxide, and hypochlorite, or chlorine gas. Neutralization reactions are performed to adjust the pH of acidic or basic soils (lime). Reduction reactions are induced through the addition of alkali metals, such as sodium, sulfur dioxide, sulfite salts, and ferrous sulfate. Sometimes chemical treatment is used to pretreat the soil for solidification or other treatments. These reactions are, however, not specific and, therefore, there is a risk of converting other metals into more toxic or mobile forms. Arsenic is most applicable for chemical oxidation since As (V) is less toxic than As (III). Hg, Pb, Se, and Ag are also applicable for reduction. These chemical treatments can be performed in situ by injection into ground water, but have the potential to introduce further contamination.

Over the last two decades, Fenton treatment has emerged as a viable remediation technology for PAH-contaminated soils. Several reviews on various Fenton-based treatments for contaminated soils have been published (Cravotto et al. 2005). In these works, PAHs have been grouped as hydrophobic or semi-volatile contaminants. PAHs are known to be toxic, mutagenic, carcinogenic, and teratrogenic, most commonly found at sites contaminated with coal tar and creosote, especially as the heritage from the manufactured gas plants (MGP) and wood treatment facilities of the last few centuries. Effective decontamination of the soil can be achieved by using advanced oxidation process (AOPs), which is based on Fenton's reaction (hydrogen peroxide catalysed by iron), involving solely one or a combination of physical, chemical, biological, and thermal processes .Faster and more efficient degradation of recalcitrant compounds such as PAHs can be achieved using AOPs (Cravotto et al. 2007).

17.3.2 Immobilization

Immobilization approaches are favored where pollution covers an extensive area, and where the main targets to be protected are water resources and plants. Immobilization can be achieved by complexing the contaminants or through increasing the soil pH by various amendments including addition of liming materials, phosphate compounds, and biosoilds (Alloway and Jackson 1991). Increased pH decreases the solubility of heavy metals like Cd, Cu, Ni, and Zn in soil. Although the risk of potential exposure to plants is reduced, their concentration remains unchanged. Mechanisms include increasing metal adsorption through higher surface charge, formation of insoluble metal complexes, precipitation, and redox reactions leading to immobile valency form. Basta et al. (2001) found that a range of soil amendments, especially alkaline biosolids, reduced the extractability and phytotoxicity of smelter waste polluted with Cd, Pb, or Zn.

17.3.3 Soil Washing (with Solvents)

Soil contaminated with metal pollutants can be decontaminated by two treatment methodologies:

- That leaves the metal in the soil such as solidification/stabilization and vitrification, which immobilizes the contaminants, thus limiting their movement.
- That removes the heavy metals from the soils. Technology such as soil washing, in situ soil flushing transfer the contaminants to liquid phase by desorption and solubilization.

Soil washing for metals after physical treatment processes is then washed with solvents on the basis of their ability to solubilize specific contaminants and to transform them into nonhazardous material and also on their environmental impacts (Feng et al. 2001; Chu and Chan 2003; Khan et al. 2004). Soil washing usually employs wash solutions, such as acids, bases, chelating agents, reducing agents, or other additives as the extracting agents. Thus, heavy metals can be removed from soils using various agents added to the soil. This can be done in reactors or as heap leaching. These agents are inorganic acids such as sulfuric and hydrochloric acids (pH less than 2), organic acids including acetic and citric acids (pH not less than 4), chelating agents such as ethylenediaminetetracetic acid (EDTA) and nitrilotriacetic acid (NTA), and the various combinations of the abovementioned .The cleaned soil can then be returned to the original site. Both organics and metals are removed. The effectiveness of this treatment approach can be high for hydrophilic pollutants such as aniline and phenols (Rajput et al. 1994). Metal removing efficiencies during soil washing depends on the soil particle size, metal characteristics, extractant chemistry, and processing conditions. pH plays a very important role in metal extraction from soils (Peters 1999). Limited

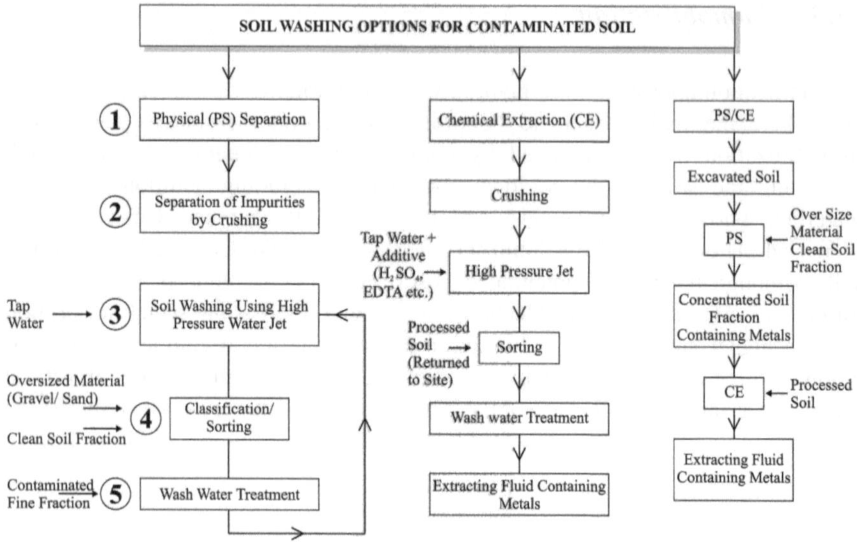

Fig. 17.1 Schematic diagram of soil washing

experience with such technique has demonstrated that they create new problems, e.g., increase in mobility (and bioavailable) of the remaining metals in the decontaminated substratum, redistribution of pollution to other remaining fractions of the process (resins, sludges, etc.) (Woelders 1998; Dermont et al. 2008) (Fig. 17.1).

17.3.4 Soil Flushing

This is in situ innovative remediation technology in which water is used with or without additives as flushing solutions to solubilize contaminants in soil to an area where they can be removed, with various methods of infiltration, such as infiltration basins, injection wells, and an infiltration trench. Various additives include organic or inorganic acids, sodium hydroxide, which can dissolve organic soil matter, water soluble solvents such as methanol, displacement of toxic cations with nontoxic cations, complexing agents such as EDTA, acids in combination with complexing agents or oxidizing/reducing agents. Once the water is pumped from soil, it must be extracted and then treated to remove the metals in wastewater treatment facilities or reused in the flushing process (Fig.17.2). Significant removal of chromium was achieved. Levels of chromium were reduced to 18 from 2,000 mg/l (USEPA 1996). Urlings (1990) decrease the Cd content 90 % of the soil from 10 to less than 1 mg/kg with dilute hydrochloric acid (pH 3). Since

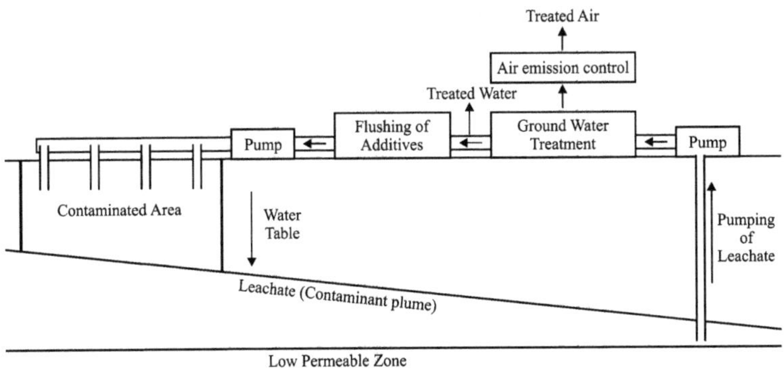

Fig. 17.2 Schematic representation of soil flushing

flushing is conducted in situ, it reduces the need for excavation, handling or transportation of hazardous substances (Khan et al. 2004). More demonstrations are needed in this methodology, in addition to developing more understanding into the mechanisms for solution, metal recovery, and use of nontoxic additives.

17.3.5 Dechlorination

Sometimes, reduction reaction can contribute to the decontamination of polluted soils. Chemical-reductive dechlorination is one such technique. Dechlorination, also known as dehalogenation, is a chemical technique based on the loss of halogen atoms (i.e., atoms of chlorine, fluorine, bromine, and iodine) from the halogenated organic molecules. Thus, converting toxic compounds into less toxic substances. These substances are frequently soluble in water, thus enhancing their separation from the soil. This technique applicable to polychlorinated biphenyls (PCBs), halogenated semivolatile volatile organic compounds, and pesticides (Castelo-Grande et al. 2010).

17.4 Biological Techniques

These techniques use living organisms in order to transform or degrade contaminants into less toxic form or remove the toxic contaminants of the soil (Kavamura and Esposito 2010). Microorganisms, soil invertebrates, and plants have all been exploited as potential agents of soil bioremediation, although most treatments have been based on microbial activity. The strategies can be applied in both in situ or ex situ ways depending on the contaminant and soil characteristics. Although biological solutions to inorganic pollution exist, the main emphasis has been on the treatment of organic compounds.

17.4.1 Biodegradation of Soil Pollutants

Degradation of soil pollutants mostly involves consortia of microorganisms, and may be achieved using in situ, on-site, or bioreactor approaches. On-site processes or bioreactors are likely to involve solid phase approaches such as landfarming (spreading of excavated contaminated soils in a thin layer usually combined with cultivation and/or nutrient inputs) or more "engineered" solutions compositing or the use of bio-piles (piles of polluted soils constructed to facilitate aeration and addition of nutrients). It can be applied to soils contaminated by crude oil, and also efficient in degradation of polycyclic hydrocarbons (PAHs). The management of the treatment environment is easiest with on site process bioreactors and difficult with in situ approaches, with bio-piles and compost techniques intermediate in this respect. Biodegradation technique is effective on a range of organic constituents. Bioaugmentation (addition of cultured microorganisms with the capacity to degrade target contaminants) or biostimulation (addition of nutrients to increase indigenous biomass or of substrates to promote cometabolism) of soil microbial populations may provide a means of accelerating pollutant degradation (Singer et al. 2005). There are many environmental factors that limit microbial biodegradation of soil pollutants including low temperature, restricted activity under anaerobic conditions, low levels of available nutrients or co-substrates, and limited bioavailability of pollutants. Although soil animals are not thought to have a significant direct role in the biological degradation of contaminants, their activities may stimulate microorganisms and improve the soil environment for microbial degradation (Haimi 2000).

Soils on many polluted sites are physically degraded and macro-fauna such as earthworms can improve these conditions (Scullion and Malik 2000). Recently, interest in the role of earthworm as "bioreactors" for degradation of contaminant such as trinitrotoluene have been reported. Biodegradation rate is controlled by microbial catabolic capacity, (Renoux et al. 2000). The presence of high concentration of heavy metals may inhibit the microbial growth and also the volatile components tend to evaporate rather than biodegrade. Thus, in most of these treatments, there will be some loss to atmosphere through volatilization and some physical or chemical stabilization of pollutants. Bioreactor treatments are increasingly favored as the microbes may also lead to formation of biofilm, which might accumulate organic pollutants for subsequent degradation of high molecular weight PAHs and heavy metals (Sheoran et al. 2010).

17.4.2 Bioleaching

It is a process of recovery of metals by some microorganisms capable of dissolving them from the environment. It is an effective alternative to chemical extraction processes. Microorganisms like *Thiobacillus ferrooxidans* and *T. thiooxidans* bacteria under aerobic and acidic conditions (pH 4) at temperature between 15 and 55 °C, depending on strain promote the microbiological leaching of metals such as copper, silver, uranium, and zinc by the oxidation followed by electron transfer to oxygen inducing the metal solubilization (Gadd 2004; Kavamura and Esposito 2010). Leaching can be performed directly by oxidation of metal sulfides to produce sulfuric acid, which then can desorb the metals on the soil by substitution of protons. Indirect leaching involves conversion of Fe^{2+} to Fe^{3+}, which in turn oxidizes sulfur minerals to Fe^{2+} producing acidity. Several options are available for bioleaching including heap leaching, bioslurry reactors, and in situ processes. Anoxic sediments are more suitable for treatment since the bacteria can solubilize the metal compounds without substantially decreasing the pH. Copper, zinc, uranium, and gold have been removed by *Thiobacillus* species in biohydrometallurgical processes (Karavaiko et al. 1988). Another fungus *Aspergillus niger*, which can produce citric and gluconic acids has a potential for remediation of metal contaminated soil. They can act as acids (pH 3.5) and chelating agents for the removal of metals such as copper from oxide mining residues (Mulligan et al. 1999b). Mercury and cadmium can be oxidized, while arsenic and iron can be reduced by microorganism. Cr (VI) can be oxidized to Cr (III) that is less mobile and toxic. Bacteria such as *Bacillus subtilis* and sulfate reducing bacteria in the presence of sulfur can perform this reaction.

17.4.3 Biosorption

Biosorption is the process of binding of metals to cell surfaces (Vidali 2001). It is a biological treatment method, which involves the adsorption of metals into biomass of algal or bacterial cells that can be dead or alive. With the use of *Trichoderma reesei* adsorption and desorption of cadmium and copper were analyzed by Kim et al. (2003). Pb(II) and Cd (II) were recovered employing biomass of *Amanita rubescens* from aqueous solution using the ability of the macrofungus by Sari and Tuzen (2009). If large scale, inexpensive production techniques for the biomass are developed, this heavy metal treatment is promising (Hazardous waste consultant 1996).This method is only applicable for low concentrations of metals in water. Therefore, the cells could potentially be placed in permeable barriers for adsorption of metals in ground water.

17.4.4 Biodegradable Biosurfactants

Surfactants are a class of natural chemicals, which are amphiphilic in nature, that promote the solubilization and emulsification of various types of organic and inorganic contaminants. Bio-surfactants are produced by bacteria and yeast. Biodegradable biosurfactants (surfactin, rhamnolipid and sophorolipid) are used to remove heavy metals from an oil contaminated soil (Mulligan et al. 1999a). The first two agents are produced by bacteria, while the last is produced by yeast. Biosurfactants are also able to remove metals from the sediments. Caustic surfactant could be used for removing the organically associated metals, while foam surfactant can be employed for extracting metals bound to carbonates and oxides (Dermont et al. 2008; Wang and Catherine 2004). Since these agents are biodegradable, they are able to enhance hydrocarbon removal, and can potentially be produced in situ, as they have a great potential for soil washing and soil flushing. This technology yet needs to be performed on commercial scale.

17.4.5 Bioventing

This technique can be applied in situ with a source of oxygen necessary to initiate the processes. It injects air into the contaminated media necessary to run the process at a rate designed to maximize in situ biodegradation. It eliminates the off-gassing of volatilized contaminants to the atmosphere (Khan et al. 2004). The simulation of decontamination of several organic pollutants including variety of petroleum refinery products by bioventing was performed by Sui et al. (2007). Unlike biosparging, which involves pumping air and nutrients into the saturated zone, bioventing pumps the air only into the unsaturated or vadose zone (USEPA 1998). Bioventing also degrades less volatile organic contaminants and, because a reduced volume of air is required, it allows for the treatment of less permeable soils. During the application of bioventing in volatile solute transportation in soils, volatilization has an important role in the first day followed by biodegradation after this period (Suko et al. 2006). Baker and Moore (2000) have reported the optimized performance and effectiveness of in situ bioventing. Diele et al. (2002) have discussed numerical models and their applications in bioventing system design and operation. Any aerobic degradable substance can be treated by bioventing and ultimately leading to biodegradation. Bioventing is most successful on mid-weight petroleum products like diesel since lighter products tend to volatilize quickly, while the heavier products generally take longer time to biodegrade. If the contaminant has to be cleaned to a level lower than 0.1 ppm or if total petroleum hydrocarbon (TPH) has to be reduced to greater than 95 %, bioventing at that site may not be effective, and some other method may be needed for this saturated zone (USEPA 1998).

17.4.6 Phytoremediation

In the past few years, green plants have shown several response patterns to the presence of potentially toxic concentrations of heavy metal ions. Most are sensitive even to very low concentrations; others have developed resistance and tolerance and accumulate toxic metals within roots and above-ground tissues, such as shoot, flower, stem, and leaves, etc. (Barcelo and Poschenrieder 2003). Such extraordinary ability of plants to accumulate heavy metals is described as hyperaccumulators. Hyperaccumulators are able to accumulate Zn concentration higher than 1 % and Cu, Pb, and Ni higher than 0.1 % of the tissue weight. The current criterion used to define a hyperaccumulator is a plant that can accumulate metal to a concentration that is 100 times greater than "normal" plants growing in the same environment (Brooks 1977; Reeves and Brooks 1983). This particular capacity to accumulate and tolerate large metal concentrations has opened up the possibility to use for remediation of polluted soils and waters. The use of hyperaccumulator plants to remove, destroy, or sequester hazardous toxic heavy metals is termed phytoremediation (Schnoor 1997).

Hyperaccumulator plants such as *Thlaspi*, *Utrica*, *Chenopodium*, *Polygonum sachalase,* and *Alyssim* have the capability to accumulate cadmium, lead, nickel, and zinc (Baker et al. 1991). In addition to hyperaccumulator, plants such as trees (Poplar) and grasses (Vetiver) are now being actively evaluated though, their metal bioconcentrating capability is well below that of hyperaccumulator plants (Burken and Schnoor 1998; Sebastiani et al. 2004). Decontamination of soil from radioactive contaminants by phytoremediation has been reported (Van denhove 2013). Two novel approaches, the use of agrobacterium transformed plant roots and mycelia cultures of fungi have been reported as research tool in the study of remediation of contaminated soil by Wenzel et al. (1999).

Phytoremediation is often also referred as botanical bioremediation or green remediation (Chaney et al. 1997). It involves the repeated cropping of plants on heavy metal contaminated soils until the soils metal concentrations have reached acceptable levels. After each cropping, the plant biomass is removed from the area and may be washed to reduce its volume where upon it can be stored in an appropriate area that does not pose a risk to the environment (Raskin et al. 1994; Chaney et al. 1997). Currently phytoremediation is used for treating many classes of contaminants including elemental (heavy metals and radionuclides) as well as organic pollutants (PCBs, PAHs, nitroaromatics) (Cunningham et al. 1996; Dushenkov 2003). This is most applicable to shallow soils with low levels of contamination. The main disadvantage is that longer times are required compared to other methods. Phytoremediation can be classified according to the mechanism and nature of contaminant (Table 17.1).

17.4.6.1 Phytoextraction

It is the biological approach to remove the contamination primarily from soil and isolate it, without destroying the soil structure and fertility. It is also called phytoaccumulation. It is the process that involves the uptake and translocation of heavy metals by roots into the above ground portions of "hyperaccumulator" plants (Brown et al. 1994) (Table 17.2).

Table 17.1 Phytoremediation—on the basis of mechanism and nature of contaminant (Sheoran et al. 2012)

No.	Process	Mechanism	Contaminant
1.	Phytostabilisation	Complexation	Inorganics
2.	Phytoextraction	Hyperaccumulation	Inorganics
3.	Phytovolatilization	Volatilization by leaves	Organics/inorganics
4.	Phytodegradation/ phytotransformation	Degradation in plant	Organics
5.	Rhizofilteration	Rhizosphere accumulation	Organics/inorganics
6.	Rhizodegradation	Rhizosphere degradation	Organics

In phytoextraction practice, metal accumulating plants are seeded or transplanted into metal polluted soil, and are cultivated using established agricultural practices. The roots of established plants absorb metal elements from the soil and translocate them to the above-ground shoots where they accumulate. After sufficient plant growth and metal accumulation, the above-ground portions of the plant are harvested and removed, resulting in the permanent removal of metal from the site. Following harvesting of pollutant-enriched plants the weight and volume of contaminated material can be further reduced by washing or compositing. Metal enriched plants can be disposed of as hazardous material or, if economically feasible, used for metal recovery (Cunningham and Ow 1996). Phytoextraction of metals for commercial gain is called phytomining (Au, Tl, Ni) (Chaney et al. 1998). This approach is suitable to remove most metals (such as Pb, Cd, Ni, Cu, Cr) and excess nutrient from contaminated soils. Examples of plant species (Table 17.2) used are plants belonging to Brassicaceae family, such as *Thlaspi* sp., *Brassica* sp., (Kumar et al. 1995a, b), and *Alyssum* sp. (Kramer et al. 1996). Radioactive contaminants have also been decontaminated by phytoextraction (Van denhove 2013) (Fig. 17.3).

There are two basic strategies of phytoextraction namely continuous phytoextraction and chelate assisted or induced phytoextraction. Continuous phytoextraction is the removal of metals, which depends on the natural ability of the plant to extract extraordinarily metal concentration from metal contaminated soil. Natural hyperaccumulators have the ability to solubilize readily available metals from the soil matrix, efficiently absorb them into the root, and translocate them to the shoot and storage in a non-phytotoxic form in the aerial portions (Pollard et al. 2002). Some of the natural metal accumulating plants secrete metal chelating compounds such as "phytosiderophores" (mugenic acid) to the rhizosphere and some secrete organic acids (citric, malic, and oxalic acid), which act as metal chelators and decrease the rhizosphere pH, thus increase the bioavailability of metals that are tightly bound to the soil and help to carry them into plant tissues (Kinnersely 1993; Nascimento and Xing 2006). The mycorrhizal fungi associated to rhizosphere also play important role in metal (Chen et al. 2005; Khan 2005; Abou-Shanab et al. 2006).

Table 17.2 Some of the plants with potential for phytoextraction of various metals (Sheoran et al. 2011, 2012)

Metal	Plant species	References
Cadmium	*Chamomilla recutita*	Kral'ova and Masarovicova (2003)
	Helianthus annus	Fenus and MacNeil (2003)
	Arabidopsis halleri	Macnair (2002)
	Brassica juncea	Kumar et al. (1995a), Salt et al. 1995b, Ebbs and Kochian (1998), Huang et al. (1997)
	Thlaspi caerulescens	Escarre et al. (2000), Lombi et al. (2001), Basic et al. (2006), Keller et al. (2006)
	Salsola kali	De la Rosa et al. (2004)
	Hypericum perforatum	Kral'ova and Masarovicova (2003)
	Medicago sativa	Drazic et al. (2006)
	Zea mays	
Copper	*Commelina communis*	Tang et al. (1997)
	B. juncea	Ebbs and Kochian (1998)
	Ipomea alpina	Malaisse et al. (1979), Baker and Walker (1990)
	Erica andevalensis	Asensi et al. (1999)
	Elsholtzia splendens	Jiang et al. (2002)
	Pelargonium species	Krishnaraj et al. (1999)
	Silene vulgaris	Song et al. (2004)
	Hirschfeldia incana	
	Haumaniastrum katangense, Crepidorhopalon perennis, Acalypha cupricola	Faucon et al. (2007)
Chromium	*Helianthus annus*	Davis et al. (2001)
	Brassica juncea	Kumar et al. (1995a), Huang et al. (1997), Han et al. (2004)
	Convolvulus arvensis	Gardea-Torresdey et al. (2004a, b)
	Pelaronium species	Krishnaraj et al. (1999)
	Prosopis species	Aldrich et al. (2003)
	Salsola kali	Gardea-Torresdey et al. (2005)
	Sutera fodina, Dicoma niccolifera, Leptospermum scoparium, Genipa americana	
	Typha spp.	Barbosa et al. (2007)
	Amaranthus viridis	Dong et al. (2007)
	Miscanthus	Zou et al. (2006)
	Oryza sativa	Arduini et al. (2006),
	Convonvulus arvensis	Bhattacharyya et al. (2005)
	Leucaea leucocephalla	Gardea-Torresdey et al. (2004a, b)
	Willows (Salix sp.)	Rout et al. (1999)
	Loilium perenne	Yu and Gu (2008)

(continued)

Table 17.2 (continued)

Metal	Plant species	References
		Vernay et al. (2007)
Mercury	*Eichhornia crassipes*	Riddle et al. (2002)
Nickel	*Psyshotria douarrei*	Davis et al. (2001)
	Brassicae juncea	Ebbs and Kochian (1998)
	Thlaspi goesingense	Reeves and Baker (1984)
	Streptanthus polygaloides	Reeves et al. (1981)
	Alyssum bertoloni	Minguzzi and Vergnano (1948)
	Berkheya codii	Robinson et al. (1997a)
	Alyssum murale	Robinson et al. (1997b)
		Bani et al. (2007), Chaney et al. (2008)
	Alyssum narkgrafii	Vinterhalter and Vinterhalter (2005)
		Perrier et al. (2004)
	Sebertia acuminate Phyllanthus species, Euphorbia helenae, Leucocroton flavicans, L. linearifolius	Berazain et al. (2007a, b)
Lead	*Dittrichia viscose*	Melendo et al. (2002)
	B. pekinesis, B.campetris, B. carinata, B. juncea, B. napus, B. nigra, Helianthus annus, Pisum sativum	
	Thlaspi rotundifolium Zea mays	Blaylock et al. (1997), Ebbs and Kochian (1998)
	Sesbania drummondii	Reeves and Brooks (1983)
	Pelargonium species	Huang and Cunningham (1996)
	Vetiveria zizaniodes	Sahi et al. (2002)
	Pelargonium crispum	Krishnaraj et al. (1999)
	Helianthus annus, Triticum aestivum L., Trifolium respens L.	Chen et al. (2000), Boonyapookana et al. (2005), Krishnaraj et al. (2000)
	Vicia faba	Yang et al. (1996)
	B. juncea	Srivastava et al. (2005)
	Hemidesmus indicus	Liu et al. (2000), Clemente et al. (2005)
		Chandra Sekhar et al. (2005)
Zinc	*Sedum alfredii*	Long et al. (2002), Yang et al. (2002)
	B. juncea, B.napus, B.rapa, Hordeum vulgare, Avena sativa,	Ebbs and Kochian (1997), (1998),
	Arabidopsis halleri	Dahmani-Muller et al. (2000),
	Viola calaminaria, Thlaspi calaminare	McGrath et al. (2006)
		Baumann (1885)
	Thlaspi careulescens	Baker and Walker (1990), Li et al. (2006)
	Polygonium aviculare	
		Gonzalez and Gonzalez-Chavez (2006)

(continued)

Table 17.2 (continued)

Metal	Plant species	References
Selenium	*B.napus, Festuca arundianacea, Hibiscus cannabinus*	
	Astragalus racemose	Banuelos et al. (1997)
	Sinapis arvensis	Rosenfeld and Beath (1964)
	Astragallus bisulcatus	Hambuckers et al. (2008)
	Grindelia squarosa, Stangeria pinnata	Smrkolj et al. (2007)
	Larrea tridentate, Salvia roemeriana	Goodson et al. (2003)
	Dryopteris fern, Pteris genera	Cruiz-Jimenez et al. (2005)
	Typha spp.	Srivastava et al. (2005)
		Pollard et al. (2007)
Uranium	*B. chinensis, B. juncea, B. narinosa, Amaranthus species*	Huang et al. (1998)
	Uncinia leptostachya, Coprosma arborea	Whiting et al. (2004), Chang et al. (2005)
	Picea mariana	Babula et al. (2008)
Thallium	*Iberis intermedia, Biscutela laevigata*	Anderson et al. (1999), Leblanc et al. (1999)
	Zea mays, B. napus	
	Hirschfeldia incana, Diplotaxis catholica	Madejon et al. (2007)
	Lolium perenne, B. napus, Phaseolus vulgaris	
	B. oleracea acephala,	Makridis et al. (1996)
	Iberis intermedia	Al-Nazar et al. (2005)
Cobalt	*Haumaniastrum roberti*	Brooks (1977)
	Haumaniastrum katangense, Crepidorhopalon perennis, Acalypha cupricola	Faucon et al. (2007)
	Anisopapus chinesis	
Arsenic	*B. junceae*	Pickering et al. (2000)
	Pteris vittata(Brake fern)	Ma et al. (2001), Caille et al. (2004)
	Eleocharis spp.,	Flores-Tavizon et al. (2003)
	Pityrogramma calomelanos	Francesconi et al. (2002)
	Pteris cretica, Pteris longifolia, Pteris umbrosa	Zhao et al. (2002)
Gold	*B. junceae,*	Anderson et al. (1999)
	B. codii, Chicory	Lamb et al. (2001), Msuya et al. (2000)
	C. linearis	
		Gardea-Torresdey et al. (2005)

(continued)

Table 17.2 (continued)

Metal	Plant species	References
Silver	*B. juncea, Medicago sativa*	Harris and Bali (2008)
		Borovicka et al. (2007)
	Amanita strobiliformis	
Manganese	*Macadamia neurophylla*	Brooks (1997)
	Phytolacca acinosa	Xue et al. (2004)
Platinum	*Sinapis alba*	Alt et al. (1998), (Kologziej et al. 2007), Babula et al. (2007)
	Lolium perenne	

Table 17.3 Some of the plants with potential for phytostabilization of various metals

Plant species	Metals	References
Agrostis tenuis, Festuca rubra L.	Pb, Zn, Cu	Smith and Bradshaw (1992)
Sesbania rostrata	Pb, Zn	Yang et al. (1997)
Cynodon dactylon and *Festuca rubra, Typha latifolia, Phragmites australis*	Pb, Zn, Cu	Wong, (1982), Ye et al. (1997, 1998)
Paspalum notatum, C. dactylcon, Imperata cyclindrica	Pb, Zn	Shu et al. (2000)
Lolium italicum, Festuca arundinaceae	Pb, Zn	Rizzi et al. (2004)
Hyparrhenia hirta, Zygophyllum fabago	Pb, Zn, Cu	Conesa et al. (2006)
Horedeum vulgare, Lupinus angustifolius, Secale cereale	As	Mains et al. (2006a, b)
B. juncea	Cd Zn, Cu, Mn, Fe, Pb, Cd	Bolan et al. (2003), Clemente et al. (2003, 2006)
Anthyllss vulneraria, Festuca arvernensis, Koeleria vallesiana, Armeria arenaria	Zn, Cd, Pb	Frerot et al. (2006)
H. hirta, Z. fabago	Pb, Zn, Cu	Conesa et al. (2006)

17.4.6.2 Phytostabilization

Phytostabilization, also known as phytorestoration, is a plant-based innovative remediation technique that stabilizes wastes and prevents exposure pathways via wind and water erosion; provides hydraulic control, which suppresses the vertical migration of contaminants into ground water, and physically and chemically immobilizes contaminants by root sorption and by chemical fixation with various soil amendments (Cunningham et al. 1995; Salt et al. 1995a; Berti and Cunningham 2000). It may also serve as an interim strategy to reduce risk at sites where complications delay the selection of the most appropriate technique for the site (Fig. 17.4 and Table 17.3).

Table 17.4 Some of the plants with potential for phytovolatilization of Hg and Se

Plant species	Metals	References
Arabidopsis thaliana	Hg	Rugh et al. (1996)
Liriondendron tulipifera		Rugh et al. (1998)
Nicotiana tabacum		
Astragalus racemosus	Se	Evans et al. (1968)
Brassica juncea		Pilon-Smits (2005)
Salicornia bigelowii		Pilon-Smits et al. (1999)
Typha latifolia		

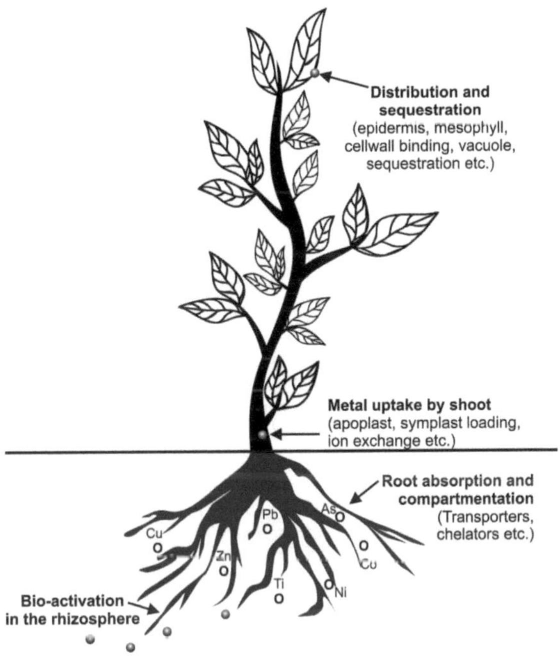

Fig. 17.3 Schematic diagram showing mechanism of phytoextraction

For phytostabilization, the normal practice is to choose drought-resistant, fast growing crops or fodder, which can grow in metal contaminated and nutrient deficient soils. Plant should also be poor translocator of metal contaminants to above-ground plant tissues that could be consumed by humans or animals, the lack of appreciable metals in shoot tissues also eliminates the necessity of treating harvested shoot residue as hazardous waste. Metal tolerant plant species (Table 17.4) immobilize heavy metals through adsorption and accumulation by roots, absorption on to roots, or precipitation within the rhizosphere (Flathman and Lanza 1998). Phytostabilization also involves soil amendments (organic matter) to promote the formation of insoluble metal complexes that reduce biological availability and plant uptake, thus preventing metals from entering the food chain (Berti and Cunningham 2000).

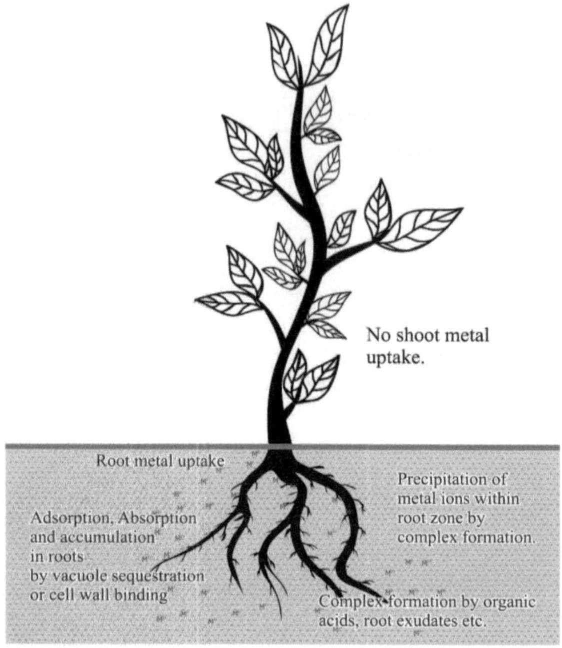

Fig. 17.4 Schematic diagram showing mechanism of phytostabilization

Table 17.5 Some of the plants with potential for rhizofilteration of various metals

Metals	Plant species	References
As, Cd, Cr, Ni, Pb, Zn	*Eichnornia crassipes Populus sp., Thlaspi* sp.	Zhu et al. (1999), Mangabeira et al. (2004), Salt et al. (1995a, b)
Fe, Cu, Cr	*Lemna minor, Azolla pinnata*	Jain et al. (1989)
Cr	*Hydrocotyle umballata*	Yong-pisanphop et al. (2005)
Cr	*Bacopa monnieri*	Mangabeira et al. (2004),
Pb, Cu, U, Sr, Cs, Co Zn	*Helianthus annus*	Dushenkov et al. (1995)
Cr	*Spirodela polyrhiza*	Appenroth et al. (2000)
Pb, Ni, Cd, Zn, Cd, Cr, Cu	*Brassica juncea*	Raskin et al. (1997)
Pb	*Hemidesmus indicus*	Sekhar et al. (2004)

17.4.6.3 Phytovolatilization

This involves use of naturally occurring or genetically modified plants that are capable of absorbing elemental forms of metal contaminants of subgroups II, V, and VI of the periodic table, such as As, Hg, and Se from the soil, and biologically converting them to gaseous species in the plant and releasing them into the

Table 17.6 Plant enzymes that have a role in transforming organic compounds (Susarla et al.2002)

Enzymes	Plants known to produce enzymatic activity	Application
Dehalogenases	Hybrid poplar (*populus* spp.), algae (various spp.), Parrot feather (*Myriophyllum aquaticum)*	Dehalogenates chlorinated solvents
Laccase	Stonewort (*Nitella spp.),* parrot-feather (*Myriophyllum aquaticum*)	Cleaves aromatic ring after TNT is reduced to triaminotoluene
Nitrilase	Willow (*Salix spp.*)	Cleaves cyanide groups from aromatic rings
Nitroreductase	Hybrid poplar (*Populus spp.*), Stonewort (*Nitella spp.*), parrot feather (*Myriophyllum aquaticum*)	Reduces nitro groups on explosives and other nitroaromatic compounds, and removes nitrogen from ring structures
Peroxidase	Horseradish (*Armoracia rusticana* P.Gaertner, Meyer & Scherb)	Degradation of phenols (mainly used in wastewater treatment)
Phosphatase	Giant duckweed (*Spirodela polyrhiza*)	Cleaves phosphate groups from large organophosphate pesticides

atmosphere (LeDuc et al. 2004). The well-known example of genetic manipulation is the transfer and expression of a modified *E. coli* Hg^{2+} reductase gene (*merA9pe*) in transgenic *Arabidopsis thaliana* plants (Rugh et al. 1996). Bacteria possessing *merA* are capable of converting highly toxic, Hg^{2+} to less toxic elemental Hg. Thus, expression of *merA* in transgenic plants helps the removal of elemental Hg as vapors through natural mechanisms of respiration. Rugh et al. (1998) also examined the ability of yellow poplar (L*iriondendron tulipifera*) tissue cultures and plantlets to express modified mercuric reductase (*merA*) gene constructs (Table 17.4).

Phytovolatization potentially offers a low cost alternative for Se removal from soil and water. During the process of Se volatilization, plants metabolize various inorganic species of Se [e.g., selenate, selenite, and Se-Met (Met)] into a gaseous form dimethyl selenide, the major volatile form of Se, is more than 600 times less toxic than inorganic forms (Evans et al. 1968; Berken et al. 2002; Neumann et al. 2003). Terry et al. (1992) reported that members of Brassicaceae are capable of releasing up to 40 g Se/ha/day as various gaseous compounds. Indian mustard (*Brassica juncea*) has a high rate of Se accumulation and volatilization, and a fast growth rate, making it a promising species for Se remediation (Pilon-Smits 2005). Some aquatic plants, such as cattail (*Typha latifolia L.*), have potential for Se phytoremediation (Pilon-Smits et al. 1999). Volatilization of arsenic (As) as dimethylarsenite has also been postulated as a resistance mechanism in marine algae (Salt et al. 1995a, b). Phytovolatilization has been successful in tritium (3H), a radioisotope of hydrogen; it is decayed to stable helium with a half-life of about 12 years reported by Dushenkov (2003). This remediation method has the added benefits of minimal site disturbance, less erosion, and no need to dispose of contaminated plant material (Heaton et al. 1998). However, phytovolatilization should be avoided for sites near population centers and at places with unique meteorological conditions that promote the rapid deposition of volatile compounds. Hence, the consequences of releasing the metals to the atmosphere need to be considered carefully before adopting this method as a remediation tool (Suko et al. 2006; Padmavathiamma and Loretta 2007).

17.4.6.4 Rhizofilteration/Phytofilteration

Plant assisted technique, which involves the use of both terrestrial and aquatic plants, to absorb, concentrate, and precipitate contaminants in the aqueous system has low contaminant concentration in their roots or seedlings (blastofilteration) (Dushenkov et al. 1995; Prasad and Frietas 2003). Mechanisms involved in rhizofilteration include chemisorption, complexation, ion exchange, micro precipitation, hydroxide condensation onto the biosurface, and surface adsorption (Gardea-Torresdey et al. 2004a, b). Root exudates and changes in rhizosphere pH also may cause metals to precipitate onto root surfaces. As they become saturated with metal contaminants, roots or whole plants are harvested for disposal (Flathman and Lanza 1998). Rhizofilteration can partially treat industrial discharge, agricultural runoff, or acid mine drainage. It can be used for Pb, Cr, Cd, Ca, Cu, Ni, and Zn, excess nutrients, and radionuclides (U, Cs, Sr), which are primarily retained with the roots (Ensley 2000) (Table 17.5).

17.4.6.5 Phytodegradation

It involves the breakdown of organics to simpler molecules that are incorporated into the plant tissues. Plant contains enzymes or enzyme cofactors that can breakdown and convert ammunition wastes, chlorinated solvents such as trichloroethylene and other herbicides (Newman and Reynolds 2004). Various plant species that can degrade aromatic rings in the absence of microorganisms have been described by Dec and Bollag (1994) and Singh and Jain (2003). Polychlorinated biphenyls (PCBs) have been metabolized by sterile plant tissues. Phenols have been degraded by plants, such as horseradish, potato (*Solanum tuberosum*), and white radish (*Raphanus sativus*) that contain peroxidases (Roper et al. 1996). Poplar trees (*populus* species) are capable of transforming trichloroethylene in soil and ground water (Newman et al. 1997). The enzymes are usually dehalogenases (transformation of chlorinated compounds), peroxidases (transformation of phenolic compounds), nitroreductases (transformation of explosives and other nitrated compounds), nitrilase (transformation of cyanated aromatic compounds), and phosphatases (transformation of organophosphates pesticides (Boyajian and Carriera 1997) (Table 17.6).

17.4.6.6 Rhizodegradation

Rhizodegradation is a biological treatment of a contaminant by enhanced bacterial and fungal activity in the rhizosphere of certain vascular plants. The rhizosphere is a zone of increased microbial density and activity at the root surface, and was described originally for legumes. Plants and microorganisms often have symbiotic relationships making the root zone or rhizosphere an area of very active microbial activity (bacteria and fungi) (Kirk et al. 2005). Plant litter and root exudates provide nutrients such as nitrate and phosphate that reduce or eliminate the need for costly fertilizer additives. Plant roots penetrate the soil, providing zones of aeration and stimulate aerobic biodegradation (Anderson et al. 1993).

Many plant molecules released by root dies and this exudation resembles common contaminants chemically and can be used as co-substrates. For example, phenolic substances released by plants have been found to stimulate the growth of PCB degrading bacteria (Fletcher and Hegde 1995). Recent studies have described enhanced degradation of pentachlorophenol in the rhizosphere of wheat grass (*Agropyron cristatum*), increased initial mineralization of surfactants in soil-plant cores, and enhanced degradation of TCE in soils collected from the rhizospheres (Knabel and Vestal 1992; Ferro et al. 1994).

17.4.7 Biochar

It is environmental friendly, carbon rich, fine grained, and porous substance, which is produced by thermal decomposition of several kinds of biomass under oxygen-limited conditions and at a relatively low temperatures, and have the capability of moisture and nutrients retention (Tang et al. 2013). It can also mitigate climate change by sequestrating C from atmosphere into the soil (Marris 2006) and also improve soil properties and enhance recycle of agricultural and forestry waste (Luo et al. 2011), and also microbial activity (Lehman et al. 2011). Amendment by adding biochar in the soil with poor fertility can improve the crop yield. It has been reported by various studies that biochar acts as an efficient sorbent of various organic and inorganic contaminants because of its increased surface area and special structure. Biochar can also be used for heavy metal removal from contaminated soils. Several kinds of organic waste like animal manure, woodchips, and crop waste can serve as source materials of biochar, thus showing a relationship among biochar, waste recycle, and soil decontamination.

Surface adsorbtion and partition of pollutant molecules in the micropores of biochar enhances the bioremediation process .However, from different point of view, further studies on the safety uses of biochar is needed to be carried out (Beesley et al. 2011).

17.5 Special Techniques

17.5.1 Electrokinetic Enhanced Phytoremediation

Electro-kinetic remediation is an environmental restoration technique, which involves energy application of a low DC current or a low potential gradient, in the order of mA/cm^2, to the two electrodes that are inserted into the sediments and encompass the contaminated zone. The application of the electric potential causes the ions to move to their respective electrodes designated as cathode and anode for subsequent removal out of the contaminated soil. It is especially designed for the in situ remediation of the contaminated soils (Cameselle et al. 2013). The electro-kinetic technology has been researched for over last two decades for the decontamination of soil. It helps in removal of heavy metals, recalcitrant, and hydrophobic organic contaminants.

Cameselle et al. (2013) reported the development of new coupled technology of electro-kinetic enhanced phytoremediation. This technology may lead to more effective and efficient remedial strategy as compared to the sequential use of these individual technologies (Fig. 17.5).

Basically, the coupled phytoremediation–electrokinetic technology consists of the application of a low intensity electric field to the contaminated soil in the vicinity of growing plants. The electric field may enhance the removal of the contaminants by increasing the bioavailability of the contaminants by desorption and transport of contaminants, even over short distances (Cameselle et al. 2013).

Variables that affect the coupled technology are: the use of AC or DC current, voltage level and mode of voltage application (continuous or periodic), soil pH evolution, and the addition of facilitating agents to enhance the mobility and bioavailability of the contaminants.

In the coupled phytoremediation-EK technology, the removal or degradation of the contaminants is performed by the plants; where as the electric field enhances the plant activity by increasing the bioavailability of the contaminants. Since the

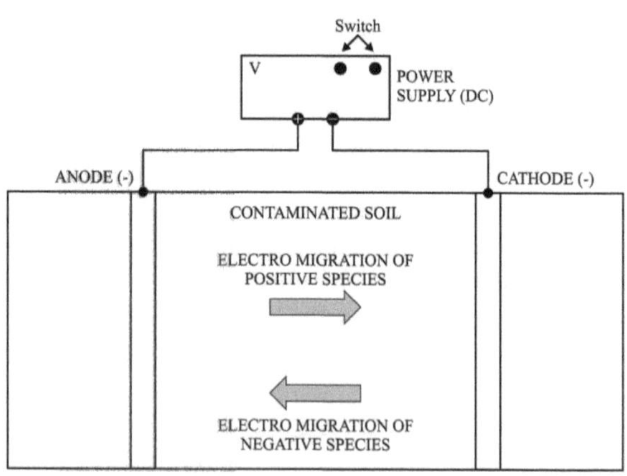

Fig. 17.5 Schematic principle of electro-kinetic soil remediation

electric field efficiency drives increased amount of soluble heavy metals toward plant roots, which results in stress conditions for the plants, hyperaccumulator plant with a rapid growth period are considered the best candidates for use in combination with EK technique .Phytoremediation can be applied after EKR to remove residual concentration of contaminants, and to achieve cleaner soil (Wan et al. 2012). The coupled EK–phytoremediation technology has showed very promising results for the restoration of heavy metal contaminated soils, and may lead to more effective and efficient remedial strategy as compared to the sequential use of these technologies (Cameselle et al. 2013).

17.6 Conclusion and Future Scope

Soil contamination is a global concern and disrupts the health of the biosphere in numerous ways; as a result reduces the capacity of the soil to meet the needs of future generation. There is an urgent need to develop an effective and affordable technological solution. There are physical, chemical, and biological methods available for such remediation, but effective remediation of polluted soils requires accurate information on the distribution and behavior of contaminants as they interact with soil and broader environment. The selection of each technique is site specific. On site with a range of organic and inorganic pollutants present, combination of different treatment approaches may offer the best prospect for effective remediation. Physical treatment process is an inexpensive comparison to chemical treatment, but most methods of physical treatments remove pollutants from the complex polluted form for further treatment or disposal. Chemical process converts the pollutant into less toxic form, or to extract them, or to immobilize them. This is highly scientific and technical process, and requires expert manpower with technological resources. Chemical additives increase the remediation cost and in situ application also increases the chances of leaching of pollutant to the other uncontaminated area and ground water. More field demonstrations are required to match reactive media with contaminants, model lifetime performance, optimize retention times, and develop methods for regeneration of reactive media. Biological methods include microorganisms (bacteria), soil invertebrates, and plant. Various field of research are needed to optimize the efficiency of biological methodology including identification of microorganisms capable of promoting their degradation, and better systems for delivering microbes, and nutrients to pollutants. Phytoremediation is environment friendly and cost-effective emerging new technology for remediation of low to moderate area of contamination as well as have important role in ecology restoration. The majority of the research for phytoremediation has been conducted in laboratories under relatively controlled conditions for short period of time.

Finally, to optimize the ecologic and economic efficiencies, it must be recognized that we need to focus more on ecological engineering approach, which is more sustainable. Assessing remediation progress and efficiency is also important, especially because pollutants are rarely completely destroyed or removed from polluted soils and evaluation of remediation has focused on the extent to which they achieve acceptable reductions in the risks posed by pollutants. Although it is clear that remediation of environment contamination is important, the need of the hour is to shift the focus from remediation to prevention for sustainable future.

References

Abou-Shanab RAI, Angle JS, Chaney RL (2006) Bacterial inoculants affecting nickel uptake by *Alyssum murale* from low, moderate and high Ni soils. Soil Biol Biochem 38:2882–2889

Adams JA, Reddy KR (2003) Extent of benzene biodegradation in saturated soil column during air sparging. Ground Water Moint Remediat 23:85–94

Aldrich MV, Gardea-Torresdey JL, Peralta-Videa JR, Parsons JG (2003) Uptake and reduction of Cr(VI) to Cr(III) by mesquite (*Prosopis* spp.): Chromate-plant interaction in hydroponics and solid media studied using XAS. Environ Sci Technol 37:1859

Alloway BJ, Jackson AP (1991) The behaviour of heavy metals in sewage-sludge amended soils. Sci Total Environ 100:151–176

Al-Nazar H, Kaschl A, Schulz R, Romheld V (2005) Effects of thallium fractions in the soil and pollution origin in thallium uptake by hyperaccumulator plants: a key factor for assessment of phytoextraction. Int J Phytorem 7:55–67

Alt F, Messerschmidt J, Weber G (1998) Investigation of low molecular weight platinum species in grass. Anal Chim Acta 359:65–70

Anderson CWN, Brooks RR, Chiarucci A, LaCoste CJ, Leblanc M, Robinson BH, Simcock R, Stewart RB (1999) Phytomining for nickel, thallium and gold. J Geochem Explor 67:407–415

Anderson TA, Guthrie EA, Walton BT (1993) Bioremediation in the rhizosphere. Environ Sci Technol 27:2630–2636

Appenroth KJ, Bischoff M, Gabrys H, Stoeckel J, Walckzak T (2000) Kinetics of chromium (V) formation and reduction in fronds of duckweed *Spirodela polyrhiza*-a low frequency EPR study. J Inorg Biochem 78:235–242

Arduini I, Masoni A, Ercoli L (2006) Effects of high chromium applications on *Miscanthus* during the period of maximum growth. Environ Exp Bot 58:234–243

Asensi A, Bennet F, Brooks R, Robinson B, Stewart R (1999) Copper uptake studies on *Erica andevalensis,* a metal-tolerant plant from Southwestern Spain. Commun Soil Sci Plant Anal 30(11, 12):1615–1624

Babula P, Supalkova V, Adam V, Havel L, Beklova M, Sladky Z, Kizek R (2007) An influence of cisplatin on the cell culture of *Nicotiana tabacum* BY-2. Plant Soil Environ 53:350–354

Babula P, Vojtech A, Radka O, Zehnalek J, Havel L, Kizek R (2008) Uncommon heavy metals, metalloids and their plant toxicity: a review. Environ Chem Lett 6:189–213

Baker AJM, Reeves RD, McGrath SP (1991) In situ decontamination of heavy metal polluted soils using crops of metal-accumulating plants—a feasibility study. In: Hinchee RE, Olfenbuttel RF (eds) In situ bioreclamation. Butterworth-Heinemann Publishers, Stoneham, pp 539–544

Baker AJM, Walker PL (1990) Ecophysiology of metal uptake by tolerant plants. In: Shaw AJ (ed) Heavy metal tolerance in plants: evolutionary aspects. CRC Press, Boca Raton, pp 155–177

Baker RS, Moore AT (2000) Optimizing the effectiveness of in situ bioventing. Pollut Eng 32(7):44–47

Bani A, Echevarria G, Sulce S, Morel JL, Mullai A (2007) In situ phytoextraction of Ni by a native population of *Alyssum murale* on an ultramafic site (Albania). Plant Soil 293:79–89

Banuelos GS, Airua HA, WU L, Guo X, Akohouy S, Zambrowski S (1997) Selenium induced growth reduction in *Brassicae landraces* considered for phytoremediation. Ecotoxicol Environ Saf 36:282–287

Barbosa RMT, deAlmeida A-AF, Mielke MS, Longuercio LL, mangabeira PAO, Gomes FP (2007) A physiological analysis of *Genipa americana* L.: a potential phytoremediator tree for chromium polluted watersheds. Environ Exp Bot 61:264–271

Barcelo J, Poschenrieder C (2003) Phytoremediation: principles and perspectives. Contrib Sci 2(3):333–344

Basic N, Salamin C, Keller N, Galland N, Besnard G (2006) Cadmium hyperaccumulation and genetic differentiation of *Thlaspi caerulescens* populations. Biochem Syst Ecol 34:667–677

Basta NT, Gradwohl R, Sneyhen KL, Schroder JL (2001) Chemical iimmobilization of lead, zinc, and cadmium in smelter-contaminated soils using biosolids and rock phosphate. J Environ Qual 30:1222–1230

Baumann A (1885) Das verhalten von zinksalzen gegen pflanzen und im boden. Landwirtscha Verss 3:1–53

Beesley L, Moreno-Jimenez E, Gomez-Eyles JL, Harris E, Robinson B, Sizmur T (2011) A review of biochars' potential role in the remediation, revegetation and restoration of contaminated soils. Environ Pollut 159:3269–3282

Berazain R, de la Fuente V, Rufo L, Rodriguez Nuria, Amils R, Diez-Garretas B, Sanchez-Mata D, Asensi A (2007a) Nickel localization in tissues of different hyperaccumulator species of euphorbiaceae from ultramafic areas of Cuba. Plant Soil 293:99–106

Berazain R, de la Fuente V, Sanchez-Mata D, Rufo L, Rodriguez N, Amils R (2007b) Nickel localization on tissues of hyperaccumulator species of *Phyllanthus* L. (Euphorbiaceae) from ultramafic areas of Cuba. Biol Trace Elem Res 115:67–86

Berken A, Mulholland MM, LeDuc DL, Terry N (2002) Genetic engineering of plants to enhance selenium phytoremediation. Crit Rev Plant Sci 21:567–582

Berti WR, Cunningham SD (2000) Phytostabilization of metals. In: Raskin I, Ensley BD (eds) Phytoremediation of toxic metals—using plants to clean up the environment. Wiley, New York, pp 71–88

Bhattacharyya P, Chakraborty A, Chakrabarti K, Tripathy S, Powell MA (2005) Chromium uptake by rice and accumulation in soil amended with municipal solid waste compost. Chemosphere 60:1481–1486

Bilek F (2004) Prediction of ground water quality affected by acid mine drainage to accompany in situ remediation. Appl Earth Sci 113:B31–B42

Blaylock MJ, Salt DE, Dushenkhov S, Zakharova O, Gussman C, Kapulnik Y, Ensley BD, Raskin I (1997) Enhanced accumulation of Pb in Indian mustard by soil-applied chelating agents. Environ Sci Technol 31:860–865

Bolan NS, Adriano DC, Naidu R (2003) Role of phosphorus in immobilization and bioavailability of heavy metals in the soil-plant system. Rev Environ Contam Toxicol 177:1–44

Boonyapookana B, Parkplan P, Techapinyawat S, DeLaune RD, Jugsujinda A (2005) Phytoaccumulation of lead by sunflower (*Helianthus annus*), tobacco (*Nicotiana tabaccum*), and (*Vetiveria zizanioides*). J Environ Sci Heal A 40:117–137

Borovicka J, Randa Z, Jelinek E, Kotrba P, Dunn CE (2007) Hyperaccumulation of silver by *Amanita strobiliformis* and related species of the section lepidella. Mycol Res 111(11):1339–1344

Boyajian G, Carriera LH (1997) Phytoremediation : a clean transition from laboratory to marketplace. Natur Biotechnol 15:127–128

Bradl H, Xenidis A (2005) Remediation techniques. In: Bradl HB (ed) Heavy metals in environment, pp 165–261. Elsevier Ltd, London

Brooks RR (1977) Copper and cobalt uptake by *Haumanniastrum* species. Plant Soil 48:541–544

Brooks RR (1997) Plants that hyperaccumulate heavy metals. C.A.B. International, Wallingford, pp 88–105

Brown SL, Chaney RL, Angle JS, Baker AJM (1994) Phytoremediation potential of *Thlaspi caerulescens* and *Bladder campion* for zinc and cadmium-contaminated soil. J Environ Qual 23:1151–1157

Burken JG, Schnoor JL (1998) Predictive relationship for uptake of organic contaminants by hybrid poplar trees. Environ Sci Technol 32:3379–3385

Caille N, Swanwick S, Zhao FJ, McGrath SP (2004) Arsenic hyperaccumulation by *Pteris vittata* from arsenic contaminated soils and the effect of liming and phosphate fertilization. Environ Pollut 132:113–120

Cameselle C, Chirakkara RA, Reddy KR (2013) Electrokinetic-enhanced photoremediation of soils: status and opportunities. Chemosphere 93:626–636

Castelo-Grande T, Augusto PA, Monteiro P, Estevej AM, Barbosa D (2010) Remediation of soil contaminated by pesticides: a review. Int J Environ Anal Chem 90(3):438–467

Castelo-Grande T, Augusto PA, Barbosa D (2005) Removal of pesticides from soil by superficial extraction- a preliminary study. Chem Eng J 111:167–171

Chandra Sekhar K, Kamala CT, Chary NS, Balaram V, Garcia G (2005) Potential of *Hemidesmus indicus* for phytoextraction of lead from industrially contaminated soils. Chemosphere 58:507–514

Chaney RL, Angle JS, Baker AJM, Li JM (1998) Method for phytomining of nickel, cobalt, and other metal from soil. U.S. Patent # 5, 711, 784

Chaney RL, Malik M, Li YM, Brown SL, Brewer EP, Angle JS, Baker AJM (1997) Phytoremediation of soil metal. Curr Opin Biotechnol 8:279–284

Chaney RL, Chen KY, Li Y-M, Angle JS, Baker AJM (2008) Effects of calcium on nickel tolerance and accumulation in *Alyssum* species and cabbage grown in nutrient solution. Plant Soil. doi:10.1007/s11104-008-9664-7

Chang P, Kim K-W, Yoshida S, Kim S-Y (2005) Uranium accumulation of crop plants enhanced by citric acid. Environ Geochem Health 27:529–538

Chen BD, Jakobsen I, Roos P, Borggaard OK, Zhu YG (2005) Mycorrhiza and root hairs enhance acquisition of Phosphorus and uranium from phosphate rock but mycorrhiza decreases root to shoot uranium transfer. New Phytol 165:591–598

Chen HM, Zheng CR, Tu C, Shen ZG (2000) Chemical methods and phytoremediation of soil contaminated with heavy metals. Chemosphere 41:229–234

Chu W, Chan KH (2003) The mechanism of the surfactant-aided soil washing system for hydrophobic and partial hydrophobic organics. Sci Total Envir 307(1–3):83–92

Clemente R, Almela C, Bernal PM (2006) A remediation strategy based on active phytoremediation followed by natural attenuation in a soil contaminated by pyrite waste. Environ Pollut 143(3):397–406

Clemente R, Walker DJ, Bernal MP (2005) Uptake of heavy metals and as by *B. juncea* grown in contaminated soil in Aznalcollar (Spain): the effect of soil amendments. Environ Pollut 138:46–58

Clemente R, Walker JD, Roig A, Bernal PM (2003) Heavy metal bioavailability in a soil affected by mineral sulphides contamination following the mine spillage at Aznalcollar (Spain). Biodegradation 14:199–205

Conesa MH, Faz A, Arnaldos R (2006) Initial Studies for the phytostabilization of a mine tailing from the Cartagena–La Union Mining District (SE Spain). Chemosphere 66(1):38–44

Cravotto G, Carlo, S, Tumiatti V, Roggero C, Bremner HD (2005) Degradation of persistent organic pollutants by Fenton's reagent facilitated by microwave or high intensity ultrasound. Environ Technol 26:721–724

Cravotto G, Carlo, S, Tumiatti V, Roggero CM (2007) Decontamination of soil containing POPs by the combined action of solid Fenton-like reagents and microwaves. Chemosphere 69:1326–1329

Cruiz- Jimenez G, Peralta-Videa JR, de la Rosa G, Meitzner G, Parsons JG, Gardea-Torresdey JL (2005) Effect of sulfate on selenium uptake and chemical speciation in *Convolvulus arvensis* L. Environ Chem 2:100–107

Cunningham SD, Anderson TA, Schwab P, Hsu FC (1996) Phytoremediation of soils contaminated with organic pollutants. Adv Agron 56:55–114

Cunningham SD, Berti WR, Huang JW (1995) Phytoremediation of contaminated soils. Trends Biotechnol 13:393–397

Cunningham SD, Ow DW (1996) Promises and prospects of phytoremediation. Plant Physiol 110(3):715–719

Dahmani-Muller H, van Oort F, Gélie B, Balabane M (2000) Strategies of heavy metal uptake by three plant species growing near a metal smelter. Environ Pollut 109:231–238

Davies FT Jr, Puryear JD, Newton RJ, Egilla JN, Saraivag JA (2001) Mycorrhizal fungi enhance accumulation and tolerance of chromium in sunflower (Helianthus annuus). Plant Physiol 158:777–786

Davis MA, Pritchard SG, Boyd RS, Prior SA (2001) Developmental and induced responses of nickel- based and organic defenses of the nickel-hyperaccumulating shrub *Psichotria douarrei*. New Phytol 150:49–58

De la Rosa G, Peralta-Videa JR, Montes M, Parsons JG, Cano-Aguilera I, Gardea-Torresdey JL (2004) Cadmium uptake and translocation in tumbleweed (*Salsola kali*), a potential Cd-hyperaccumulator desert plant species: ICP/OES and XAS studies. Chemosphere 55:1159–1168

Dec J, Bollag JM (1994) Use of plant material for the decontamination of water polluted with phenols. Biotech Bioeng 44:1132–1139

Dermatas D, Meng X (2003) Utilisation of fly ash for stabilisation/solidification of heavy metal contaminated soils. Eng Geo 70(3–4):377–394

Dermont G, Bergeron M, Mercier G, Richer-Lafleche M (2008) Soil washing for metal removal: a review of physical/chemical technologies and field application. J Hazard Mater 152:1–31

Diele F, Notarnicola F, Sgura I (2002) Uniform air velocity field for a bioventing system design: some numerical results. Int J Eng Sci 40(11):1199–1210

Do Nascimento CWA, Xing B (2006) Phytoextraction a review on enhanced metal availability and plant accumulation. Scienta Agricola (Piracicaba, Brazil) 3(3):299–311

Dong J, Wu FB, Huang RG, Zang GP (2007) A chromium-tolerant plant growing in cr-contaminated land. Int J Phytoremediation 9:167–179

Drazic G, Mihalovic N, Lolic M (2006) Cadmium concentration in *Medicago sativa* seedlings treated with salicylic acid. Biol Plant 50:239–244

Dushenkov V, Kumar PBAN, Motto H, Raskin I (1995) Rhizofiltration: the use of plants to remove heavy metals from aqueous streams. Environ Sci Technol 29:1239–1245

Dushenkov D (2003) Trends in phytoremediation of radionuclides. Plant Soil 249:167–175

Ebbs SD, Kochian LV (1997) Toxicity of zinc and copper to *Brassica* species: implications for phytoremediation. J Environ Qual 26:776–781

Ebbs SD, Kochian LV (1998) Phytoextraction of zinc by oat (*Avena sativa*), barley (*Hordeum vulgare*), and Indian mustard (*Brassica juncea*). Environ Sci Technol 32:802–806

Ensley BD (2000) Rational for use of phytoremediation. In: Raskin I, Ensley BD (eds) Phytoremediation of toxic metals: using plants to clean- up the environment. Wiley, New York, pp 3–12

Escarre J, Lefebre C, Gruber W, Leblanc M, Lipart J, Riviere Y, Delay B (2000) Zinc and cadmium hyperaccumulation by *Thlaspi caerulescens* from metalliferous and non metalliferous sites in the miditerranean area: implications for phytoremediation. New Phytol 145:429–437

Evanko CR, Dzombak DA (1997) Remediation of metals-contaminated soils and ground water, technology evaluation report, TE-97-01. Ground Water Remediation Technologies Analysis Center, Pittsburgh, P.A

Evans CS, Asher C, Johnson CM (1968) Isolation of dimethyl diselenide and other volatile selenium compounds from *Astragalus racemosus* (Pursh.). Aust J Biol Sci 21:13–20

Faucon M-P, Shutcha N, Meerts P (2007) Revisiting copper and cobalt concentrations in supposed hyperaccumulators from SC Africa: influence of washing and metal concentrations in soil. Plant Soil 301:29–36

Feng D, Lorenzen L, Aldrich C, Mare PW (2001) Ex- situ diesel contaminated soil washing with mechanical methods. Miner Eng 14(9):1093–1100
Fenus TJ, MacNeil JH (2003) Hyperaccumulation of cadmium by *Helianthus annuus*. In: Abstract of the proceedings of the 225th ACS national meeting, New Orleans, LA, pp 23–27
Ferro AM, Sims RC, Bugbee B (1994) Hycrest crested wheatgrass accelerates the degradation of pentachlorophenol in soil. J Environ Qual 23:272–279
Flathman PE, Lanza GR (1998) Phytoremediation: current views on an emerging green technology. J Soil Contam 7(4):415–432
Fletcher JS, Hegde RS (1995) Release of phenols by perennial plant roots and their potential importance in bioremediation. Chemosphere 31:3009–3016
Flores-Tavizon E, Alarcon-Herrera MT, Gonzalez Elizondo S, Olguin EJ (2003) Arsenic tolerating plants from mine sites and hot springs in the semi arid region of Chihuahua Mexico. Acta Biotechnol 23:113–119
Francesconi K, Visoottiviseth P, Sridokchan W, Goessler W (2002) Arsenic species in an arsenic hyperaccumulating fern, *Pityrogramma calomelanos*: a potential phytoremediater of arsenic-contaminated soils. Sci Total Environ 284:27–35
Frerot H, Lefèbvre C, Gruber W, Collin C, Dos Santos A, Escarre J (2006) Specific interactions between local metallicolous plants improve the phytostabilization of mine soils. Plant Soil 282:53–65
Friberg L, Nordberg GF, Vouk VB (1986) Handbook on the toxicology of metals, 2nd edn. Elsevier, Amsterdam
Gadd GM (2004) Microbial influence on metal mobility and application for bioremediation. Geoderma 122:109–119
Gardea-Torresdey JL, de la Rosa G, Peralta-Videa JR (2004a) Use of phytofiltration technologies in the removal of heavy metals: a review. Pure Appl Chem 76(4):801–813
Gardea-Torresdey JL, Peralta-Videa JR, de La Rosa G, Parsons JG (2005) Phytoremediation of heavy metals and study of the metal coordination by x-ray absorption spectroscopy. Coord Chem Rev 249:1797–1810
Gardea-Torresdey JL, Peralta-Videa JR, Montes M, deLa Rosa G, Corral-Diaz B (2004b) Bioaccumulation of cadmium, chromium and copper by *Convolvulus arvensis* L.: impact on plant growth and uptake of nutritional elements. Bioresour Technol 92(3):229–235
Gavrilescu M, Pavel LV, Cretescu I (2009) Characterization and remediation of soils contaminated with uranium. J Hazard Mater 163:475–510
George CE, Lightsey GR, Jun I, Fan JY (1992) Soil decontamination via microwave and radio-frequency covolatilisation. Environ Prog 11:216–219
Ghosh M, Singh SP (2005) A review on phytoremediation of heavy metals and utilization of it's by product. Appl Ecol Environ Res 3(1):1–18
Gomez-Lahoz C, Rodriguez-Maroto JM, Wilson DJ (1995) Soil clean up by insitu aeration XXII. Impact of natural soil organic matter on clean up rates. Sep Sci Technol 30:659–682
Gonzalez RC, Gonzalez-Chavez MCA (2006) Metal accumulation in wild plants surrounding mining wastes. Environ Pollut 144(1):84–92
Goodson CC, Parker DR, Amrhein C, Zhang Y (2003) Soil selenium uptake and root system development in plant taxa differing in Se- accumulating capability. New Phytol 159:391–401
Haimi J (2000) Decomposer animals and bioremediation of soils. Environ Pollut 107:233–238
Hambuckers A, Dotreppe O, Hornick JL, Istasse L, Dufrasne I (2008) Soil applied selenium effects on tissue selenium concentrations in cultivated and adventitious grassland and pasture plant species. Soil Sci Plant Anal 39:800–811
Han FXX, Sridhar BBM, Monts DL, Su Y (2004) Phytoavailability and toxicity of trivalent and hexavalent chromium to *B. juncea*. New Phytol 162:489–499
Harris AT, Bali R (2008) On the formation and extent of uptake of silver nanoparticles by live plants. J Nanopart Res 10:691–695
Hazardous Waste Consultant (1996) Remediating Soil and sediment contaminated with heavy metals, Nov/Dec. Elsevier science, Netherlands

Heaton ACP, Rugh CL, Wang N, Meagher RB (1998) Phytoremediation of mercury and methyl mercury polluted soils using genetically engineered plants. J Soil Contam 74:497–510
Hoffmann G (1983) Relationships between critical levels of pollutants in soils, fodder, and crops, (In German.). Landwirtsch Forsch Sonderh 39:130–152
Huang JW, Blaylock MJ, Kapulnik Y, Ensley BD (1998) Phytoremediation of uranium contaminated soils: role of organic acids in triggering uranium hyperaccumulation in plants. Environ Sci Technol 32(13):2004–2008
Huang JW, Chen J, Berti WB, Cunningham SD (1997) Phytoremediation of lead-contaminated soils: role of synthetic chelates in lead phytoextraction. Environ Sci Technol 31:800–805
Huang JW, Cunningham SD (1996) Lead phytoextraction: species variation in lead uptake and ttranslocation. New Phytol 134:75–84
Jain SK, Vasudevan P, Jha NK (1989) Removal of some heavy metals from polluted water by aquatic plants: studies on duckweed and water velvet. Biol Wastes 28(2):115
Jiang LY, Shi WY, Yang XE, Fu CX, Chen WG (2002) Hyperaccumulators in mining area. Chinese J Appl Ecol 13(7):906–908
Karavaiko GI, Rossi G, Agates AD, Groudev SN, Avakyan ZA (1988) Biogeotechnology of metals: manual. Center for International Projects GKNT, Moscow
Kavamura VN, Esposito E (2010) Biotechnological strategies applied to the decontamination of soils polluted with heavy metals. Biotech Adv 28:61–69
Keller C, Diallo C, Cosio N, Basic N, Galland N (2006) Cadmium tolerance and hyperaccumulation by *Thlaspi caerulescens* populations grown in hydroponics are related to plant uptake characteristics in the field function. Plant Biol 33:673–684
Khan FI, Husain T, Hejazi R (2004) An overview and analysis of site remediation technologies. J Environ Mgmt 71:95–255
Khan AG (2005) Role of soil microbes in the rhizospheres of plants growing on trace metal contaminated soils in phytoremediation. J Trace Elem Med Biol 18:355–364
Kim SK, Park CB, Koo YM, Yun HS (2003) Biosorption of cadmium and copper ions by *Trichoderma reesei* RUT C30. J Ind Eng Chem 9:403–406
Kinnersely AM (1993) The role of phytochelates in plant growth and productivity. Plant Growth Regul 12:207–217
Kirk I, Klironomos I, Lee H, Trevors JT (2005) The effects of perennial ryegrass and alfalfa on microbial abundance and diversity in petroleum contaminated soil. Environ Pollut 133:455–465
Knabel DB, Vestal JR (1992) Effects of intact rhizosphere microbial communities on the mineralization of surfactants in surface soils. Can J Microbiol 38:643–653
Knasmuller S, Gottmann E, Steinkellner H, Fomin A, Pickl C, Paschke A, God R, Kundi M (1998) Detection of genotoxic effects of heavy metal contaminated soils with plant aioassay. Mutat Res 420:37–48
Kologziej M, Baranowska I, Matyja A (2007) Determination of platinum in plant samples by voltammetric analysis. Electro-analysis 19:1585–1589
Kral'ova K, Masarovicova E (2003) *Hypericum perforatum* L. and *Chamomilla recutita* (L.) rausch.—accumulators of some toxic metals. Pharmazie 58(5):359–359
Kramer U, Cotter-Howells JD, Charnock JM, Baker AJM, Smith JAC (1996) Free histidine as a metal chelator in plants that accumulate nickl. Nature 379:635–638
Krishnaraj S, Dan TV, Saxena PK (2000) A fragment solution to soil remediation. Int J Phytorem 2:117–132
Krishnaraj S, Saxena PK, Perras MR, Michel R (1999) Method of using *Pelargonium* species as hyperaccumulators for remediating contaminated soil. PCT/CA9801027 Int Appl 1–20
Kumar PBAN, Motto H, Raskin I (1995a) Rhizofiltration: the use of plants to remove heavy metals from aqueous streams. Environ Sci Technol 29(5):1239–1245
Kumar PBAN, Dushenkov V, Motto H, Raskin I (1995b) Phytoextraction –the use of plants to remove heavy metals from soils. Environ Sci Technol 29:1232–1238
Lamb AE, Anderson CWN, Haverkamp RG (2001) The induced accumulation of gold in the plants *Brassica juncea, Berkheya codii* and *Chicory*. Chem New Zealand 65(2):34–36

Leblanc M, Robinson BH, Petit D, Deram A, Brooks RR (1999) The phytomining and environmental significance of hyperaccumulation of thallium by *Iberis intermedia* from southern France. Econ Geol 94:109–114

LeDuc DL, Tarun AS, Montes-Bayon M, Meija J, Malit MF, Wu CP, Abdel Samie M, Chiang CY, Tagmount A, DeSouza M, Neuhierl B, Bock A, Caruso J, Terry N (2004) Overexpression of selenocysteine methyltransferase in *Arabidopsis* and Indian mustard increases selenium tolerance and accumulation. Plant Physiol 135:377–383

Lehmann J, Rillig MC, Thies J, Masiello CA, Hockaday WC, Crowley D (2011) Biochar effects on soil biota—a review. Soil Biol Biochem 43:1812–1836

Li Y-M, Chaney RL, Reeves RD, Angle JS, Baker AJM (2006) *Thlaspi caerulescens* sub species for Cd and Zn recovery. US Patent No.7049, 492. Date issued-23 May

Liu D, Jiang W, Liu C, Xin C, How W (2000) Uptake and accumulation of lead by roots, hypocotyls and shoots of Indian mustard (*Brassica juncea* L.). Bioresour Technol 71:273–277

Lombi E, Zhao FJ, Dunham SJ, McGrath SP (2001) Phytoremediation of heavy metal-contaminated soils: natural hyperaccumulation versus chemically enhanced phytoextraction. J Environ Qual 30:1919–1926

Long XX, Yang XE, Ye ZQ, Ni WZ, Shi WY (2002) Differences of uptake of and accumulation of zinc in four species of *Sedum*. Acta Botanica Sinica 44:152–157

Luo L, Lou LP, Cui XY, Wu BB, Hou J, Xun B, Xu XH, Chen YX (2011) Sorption and desorption of pentachlorophenol to black carbon of three different origins. J Hazard Mater 185:639–646

Ma JF, Ryan PR, Delhaize E (2001) Aluminium tolerance in plants and the complexing role of organic acids. Trends Plant Sci 6:273–278

Macnair MR (2002) Within and between population genetic variations for zinc accumulation in *Arabidopsis halleri*. New Phytol 155(1):9–66

Madejon P, Murillo JM, Maranon T, Lepp NW (2007) Factors affecting accumulation of thallium and other trace elements in two wild Brassicaceae spontaneously growing on soils contaminated by tailings dam waste. Chemosphere 67:20–28

Mains D, Craw D, Rufaut CG, Smith CMS (2006a) Phytostabilization of gold mine tailings, New Zealand. Part 1: plant establishment in alkaline saline substrate. Int J Phytorem 8(2):131–147

Mains D, Craw D, Rufaut CG, Smith CMS (2006b) Phytostabilization of gold mine tailings from New Zea-land. Part 2: experimental evaluation of arsenic mobiliza- tion during revegetation. Int J Phytorem 8(2):163–183

Makridis C, Pateras D, Amberger A (1996) Thallium pollution risk to food chain from cement plant. Fresenius Environ Bull 5:643–648

Malaisse F, Gregoire J, Morrison RS, Reeves RD (1979) Copper and cobalt in vegetation of Fungurume, Shaba Province, Zaire. Oikos 33:472–478

Mangabeira PAO, Labejof L, Lamperti A, deAlmeida AAF, Oliveira AH, Escaig F, Severo MIG (2004) Accumulation of chromium in roots tissues of *Eichhornia crassipes* (Mart.) Solms. In Cachoeira river-Brazil. Appl Surf Sci 231(232):497–501

Marris E (2006) Putting the carbon back: black is the New Green. Nature 442:624–626

McGrath SP, Lombi E, Gray CW, Caille N, Dunham SJ, Zhao FJ (2006) Field evaluation of Cd and Zn phytoextraction potential by the hyperaccumulators *Thlaspi caerulescens* and *Arabidopsis halleri*. Environ Pollut 141:115–125

Melendo M, Benítez E, Nogales R (2002) Assessment of the feasibility of endogeneous Mediterranean species for phytoremediation of Pb-contaminated areas. Fresenius Environ Bull 11:1105–1109

Minguzzi C, Vergnano O (1948) Il cotenuto di nichel nelle ceneri di *Alyssum bertolonii*. Atti Soc Tosc Sci Nat 55:49–74

Msuya FA, Brooks RR, Anderson CWN (2000) Chemically-induced uptake of gold by root crops: its significance for phytomining. Gold Bull 33(4):134–137

Mulligan CN, Yong RN, Gibbs BF (1999a) On the use of biosurfactants for the removal of heavy metals from oil-contaminated soil. Environ Prog 18(1):50–54

Mulligan CN, Yong RN, Gibbs BF (2001) Remediation technologies for metal-contaminated soils and groundwater: an evaluation. Eng Geol 60:193–207

Mulligan CN, Galvez-Cloutier R, Renaud N (1999b) Biological leaching of copper mine residues by Aspergillus niger. Presented at AMERICANA 1999, Pan-American Environment Trade Show and Conference, Montreal, Canada, pp 24–26

Nathanail CP, Earl N (2001) Human health risk assessment: guidelines values and magic numbers. In: Hester RE, Harrison RM (eds) Assessment and reclamation of contaminated land. Royal Society of Chemistry, Cambridge, pp 85–102

Neumann PM, DeSouza MP, Pickering IJ, Terry N (2003) Rapid microalgal metabolism of selenate to volatile dimethylselenide. Plant Cell Environ 26:897–905

Newman LA, Reynolds CM (2004) Phytodegradation of organic compounds. Curr Opin Biotechnol 15:225–230

Newman LA, Strand SE, Choe N, Duffy J, Ekuan G (1997) Uptake and biotransformation of trichloroethylene by hybrid poplars. Environ Sci Technol 31:1062–1067

Padmavathiamma PK, Loretta YM (2007) Phytoremediation technology: hyperaccumulation metals in plants. Water Air Soil Pollut 184:105–126

Park G, Shin HS, Ko SO (2005) A laboratory and pilot study of thermally enhanced soil vapor extraction method for the removal of semi-volatile organic contaminants. J Environ Sci Health Part Am 40:881–897

Pazos M, Rosales E, Alcantara T, Gomez J, Sanaroman MA (2010) decontamination of soils containing PAHs by electroremediation, a review. J Hazard Mater 177:1–11

Peng JF, Song YH, Yuan P, Cui XY, Qui GL (2009) The remediation of heavy metals contaminated sediments. J Hazard Mater 161:633–640

Perrier N, Colin F, Jaffre T, Ambrosi JP, Rose J, Bottero JY (2004) Nickel speciation in *Sebertia acuminate,* a plant growing on a lateritic soil of New Caledonia. CR Geosci 336:567–577

Peters RW (1999) Chelant extraction of heavy metals from contaminated Soils. J Hazard Mater 66:151–210

Pickering IJ, Prince RC, George MJ, Smith RD, George GN, Salt DE (2000) Reduction and coordination of arsenic in Indian mustard. Plant Physiol 122:1171–1177

Pilon-Smits E (2005) Phytoremediation. Annu Rev Plant Biol 56:15–39

Pilon-Smits EAH, Hwang S, Lytle CM, Zhu Y, Tai JC, Bravo RC, Chen Y, Leustek T, Terry N (1999) Overexpression of ATP sulfurylase in Indian mustard leads to increased selenate uptake, reduction, and tolerance. Plant Physiol 119:123–132

Pollard AJ, Powell KD, Harper FA, Smith JAC (2002) The genetic basis of metal hyperaccumulation in plants. Crit Rev Plant Sci 21(6):539–566

Pollard J, Cizdziel J, Stave K, Reid M (2007) Selenium concentrations in water and plant tissues of a newly formed arid wetland in Las Vegas. Nevada Env Monit Assess 135:447–457

Prasad VMN, Frietas HMO (2003) Metal hyperaccumulation in plants—biodiversity prospecting for phytoremediation technology. Electron J Biotechnol 6(3):285–321

Rajput VS, Higgins AJ, Singley ME (1994) Cleaning of excavated soil contaminated with hazardous organic compounds by washing. Water Environ Res 66:819–827

Raskin I, Kumar PBAN, Dushenkov S, Salt DE (1994) Bioconcentration of heavy metals by plants. Curr Opin Biotechnol 5:285–290

Raskin I, Smith RD, Salt DE (1997) Phytoremediation of metals: using plants to remove pollutants from environment. Curr Opin Biotechnol 8:221–226

Reed DT, Tasker IR, Cunnane JC, Vandegrift GF (1992) In: Vandgrift GF, Reed DT, Tasker IR (eds) Environmental remediation removing organic and metal ion pollutants. Am Chem Soc, Washington DC, pp 1–9

Reeves RD, Brooks R (1983) Hyperaccumulation of lead and zinc by two metallophytes from a mining area in Central Europe. Environ Pollut 31:277–287

Reeves RD, Baker AJM (1984) Studies on metal uptake by plants from serpentine and non serpentine populations of *Thlaspi goesingense* Halacsy (Crucifera). New Phytol 98:191–204

Reeves RD, Brooks RR, Macfarlane RM (1981) Nickel uptake by Californian *Streptanthus* and *Caulanthus* with particular reference to the hyperaccumulator *S. polygaloides* Gray (Brassicaceae). Am J Bot 68:708–712

Renoux AY, Sarrazin M, Hawari J, Sunahara GI (2000) Transformation of 2,4,6-trinitrotoluene in soil in the presence of the earthworm eisenia andrei. Environ Toxicol Chem 19:1473–1480

Riddle SG, Tran HH, Dewitt JG, Andrews JC (2002) Field, laboratory, and x-ray absorption spectroscopic studies of mercury accumulation by water hyacinths. Environ Sci Technol 36:1965

Rizzi L, Petruzzelli G, Poggio G, Vigna G (2004) Soil physical changes and plant availability of zn and pb in a treatability test of phytostabilization. Chemosphere 57(9):1039–1046

Robinson BH, Brooks RR, Howes AW, Kirkma JH, Gregg PEH (1997a) The potential of the high- biomass nickel hyperaccumulator *Berkheya coddii* for phytoremediation and phyto-mining. J Geochem Explor 60:115–126

Robinson BH, Chiarucci A, Brooks RR, Petit D, Kirkman JH, Gregg PEH, De Dominicis V (1997b) The nickel hyperaccumulator plant *Alyssum bertolonii* as a potential agent for phytoremediation and phytomining of nickel. J Geochem Explor 59:75–86

Roper JC, Dec J, Bollag J (1996) Using minced horseradish roots for the treatment of polluted waters. J Environ Qual 25:1242–1247

Rosenfeld I, Beath OA (1964) Selenium- geobotany, biochemistry, toxicity and nutrition. Academic Press, New York

Rout GR, Samantary S, Das P (1999) Chromium, nickel and zinc tolerance in *Leucaena leucocephala* (K8). Silvae Genet 48:151–157

Rugh CL, Senecoff JF, Meagher RB, Merkle SA (1998) Development of transgenic yellow poplar for mercury phtoremediation. Nature Biotechnol 16:925–928

Rugh CL, Wilde HD, Stack NM, Thompson DM, Summers AO, Meagher RB (1996) Mercuric ion reduction and resistance in transgenic *Arabidopsis thaliana* plants expressing a modified bacterial merA gene. Proc Natl Acad Sci USA 93:3182–3187

Sahi SV, Bryant NL, Sharma NC, Singh SR (2002) Characterization of a lead hyperaccumulator shrub, *Sesbania drummondii*. Environ Sci Technol 36(21):4676–4680

Salt DE, Blaylock M, Kumar PBAN, Dushenkov V, Ensley BD, Chet I, Raskin I (1995a) Phytoremediation: a novel strategy for the removal of toxic metals from the environment using plants. Biotechnology 13:468–475

Salt DE, Prince RC, Pickering IJ, Raskin I (1995b) Mechanism of cadmium mobility and accumulation in Indian mustard. Plant Physiol 109:1426–1433

Sari A, Tuzen M (2009) Kinetic and equilibrium studies of biosorption of Pb(II) and Cd(II) from aqueous solution by macrofungus (Amanita rubescens) biomass. J Hazard Mater 164:1004–1011

Schnoor JL (1997) Phytoremediation: technical and organisatoric issues, key factors. Ground-Water Remediation Technologies Analysis Center, Pittsburgh

Scullion J (2006) Remediating polluted soils. Naturwissenschaften 93:51–65

Scullion J, Malik A (2000) Earthworm effects on aggregate stability, organic matter composition and disposition, and their relationships. Soil Biol Biochem 32:119–126

Sebastiani L, Scebba F, Tognetti R (2004) Heavy metal accumulation and growth responses in poplar clones Eridano (*Populus deltoides x maximowiczii*) and I-214 (*P.* x *euramericana*) exposed to industrial waste. Environ Exp Bot 52:79–88

Sekhar KC, Kamala CT, Chary NS, Sastry ARK, Rao TN, Vairamani M (2004) Removal of lead from aqueous solutions using an immobilized biomaterial derived from a plant biomass. J Hazard Mater 108:111–117

Sheoran V, Sheoran AS, Poonia P (2008) Remediation techniques for contaminated soil. Environ Enging Managt J 7(4):379–387

Sheoran AS, Sheoran V, Choudhary RP (2010) Bioremediation of acid-rock drainage by sulphate-reducing prokaryotes: a review. Miner Eng 23(14):1073–1100

Sheoran V, Sheoran AS, Poonia P (2011) Role of hyperaccumulators in phytoextraction of metals from contaminated mining sites: a review. Crit Rev Environ Sci Technol 41:168–214

Sheoran V, Sheoran AS, Poonia P (2012) Phytoremediation technologies for the reclamation of organic and inorganic polluted soils and water: a review. Environ Res J 6(4/5):1–23

Shu WS, Lan CY, Zhang ZQ, Wong MH (2000) Use of vetiver and other three grasses for revegetation of Pb/Zn Mine tailings at Lechang, Guangdong Province: field experiment. In: 2nd international vetiver conference, Bangkok, Thailand

Sikdar SK, Grosse D, Rogut I (1998) Membrane technologies for remediating contaminated soils: a critical review. J Membrane Sci 151:75–85

Singer AC, van der Gast CJ, Thompson IP (2005) Perspectives and vision for strain selection in bioaugmentation. Trends Biotech 23:74–77

Singh OV, Jain RK (2003) Phytoremediation of toxic aromatic pollutants from soil. Appl Microbiol Biotechnol 63:128–135

Smith RAH, Bradshaw AD (1992) Stabilisation of toxic mine wastes by the use of tolerant plant populations. Trans Inst Min Metall Sect A 81:230–237

Smrkolj P, Osvald M, Osvald J (2007) Selenium uptake and species distribution in *Seliniumaseolus vulgaris* seeds obtained by two different cultivations. Eur Food Res Technol 225:233–237

Song SQ, Zhou X, Wu H, Zhou YZ (2004) Application of municipal garbage compost on revegetation of tin tailings dams. Rural Eco-Environ 20(2):59–61

Srivastava M, Ma LQ, Contruva JA (2005) Uptake and distribution of selenium in different fern species. Int. J Phytorem 7:33–42

Sui H, Li X, Jiang B, Huang G (2007) Simulation of remediation of multiple organic contaminats system by bioventing. Huagong Xuebo (Chinese Edition). 58:1025–1031

Suko T, Fujikawa T, Miyazaki T (2006) Transport phenomena of volatile solute in soil during bioventing technology. J ASTM Int 3:374–379

Susarla S, Medina VF, McCutcheon SC (2002) Phytoremediation: an ecological solution to organic chemical contamination. Ecol Engineer 18:647–658

Suthersan SS (1997) Remediation engineering: design concepts. Lewis Publishers, Boca Raton

Tang J, Zhu W, Kookana R, Arata K (2013) Characteristics of biochar and its applications in remediation of contaminated soils. J Biosci Bioeng 116(6):653–659

Tang SR, Huang CY, Zhu ZX (1997) *Commelina communis* L.: copper hyperaccumulator found in Anhui Province of China. Pedosphere 7(3):207–210

Terry N, Carlson C, Raab TK, Zayed A (1992) Rates of selenium volatilization among crop species. J Environ Qual 21:341–344

Urlings LGCM (1990) In situ cadmium removal-full scale remedial action of contaminated soil. In: International symposium on hazardous waste treatment: treatment of contaminated soil, air, waste association and US.EPA Risk education laboratory, Cincinnati, Ohio, 5–8 Feb 1990

USEPA (1998) Bioventing. Office of the Underground Storage Tank U.S. Environmental Protection Agency, Publication EPA, 510-B-95-007

USEPA (1994) Selection of control technologies for remediation of soil contaminated with arsenic, cadmium, chromium, lead or mercury. Revised Draft Engineering Bulletin, Jan 31

USEPA (1996) Engineering bulletin: technology alternatives for the remediation of soils contaminated with arsenic, cadmium, chromium, mercury and lead. U.S. Environmental Protection Agency, Office Of Emergency And Remedial Response, Cincinnati, OH

Van denhove H (2013) Phytoremediation options for radioactively contaminated sites evaluated. Ann Nucl Energy 62:596–606

Vernay P, Gauthier-Moussard C, Hitmi A (2007) Interaction of bioaccumulation of heavy metal chromium with water relation, mineral nutrition and photosynthesis in developed leaves of *Lolium perenne* L. Chemosphere 68:1563–1575

Vidali M (2001) Bioremediation, An overview. Pure Appl Chem 73:1163–1172

Vinterhalter B, Vinterhalter D (2005) Nickel hyperaccumulation in shoot cultures of *Alyssum narkgrafii*. Biol Plant 49:121–124

Wan QF, Deng DC, Bai Y, Xia CQ (2012) Phytoremediation and electrokinetic remediation of uranium contaminated soils: a review. He-Huaxue yu Fangshe Huaxue. J Nucl Radiochem 34:148–156

Wang J, Feng X, Anderson CWN, Xing Y, Shang L (2012) Remediation of mercury contaminated sites- a review. J Hazard Mater 221–222:1–18

Wang S, Catherine NM (2004) An evaluation of surfactant technology in remediation of ctaminated soil. Chemosphere 57:1079–1089

Wenzel WW, Adrino DC, Salt D, Smith R (1999) Phytoremediation: a plant-microbe-based remediation system. In: Adrino DC, Bollag JM, Frankenberger WT, Sims RC (eds) Bioremediation of contaminated soil, pp 456–508. Agronomy Monograph no. 37, Madison

Whiting SN, Reeves RD, Richards D, Johnson MS, Cooke JA, Malaisse F, Paton A, Smith JAC, Angle JS, Chaney RL, Ginocchio R, Jaffre T, Johns R, Mcintyre T, Wojcik M, Tukiendorf A (2005) Cadmium uptake, localization and detoxification in *Zea mays*. Biol Plant 49:237–245

Williams GM (1988) Integrated studies into ground water pollution by hazardus wastes. In: Gronow JR, Scho-field AN, Jain RK (eds) Land disposal of hazardous waste, engineering and environmental issues. Horwood Ltd. Chichester

Woelders J (1998) Sanirengsmogelijkheden Cadmium: In situ Reiniging Van Cadmium houdede zandground. In Cadmium: Vooromen impact en sanering. Lisec, Genk, Belgium

Wong MH (1982) Metal co-tolerance to copper, lead and zinc in *Festuca rubra*. Environ Res 29:42–47

Wood P (2001) Remediation methods for contaminated land. In: Hester RE, Harrison RM (eds) Assessment and reclamation of contaminated land. Issues in environmental science and technology. pp 115–139. Royal Society f Chemistry, Cambridge

Xue SG, Chen YX, Reeves RD, Lin Q, Fernando DR (2004) Mangenese uptake and accumulation by the hyperaccumulator plant *Phytolacca acinosa Roxb*. (Phytolaccaeae). Environ Pollut 131:393–399

Yang X, Baligar DC, Martens DC, Clark RB (1996) Plant tolerance to nickel toxicity: I. Influx, transport, and accumulation of nickel in four species. J Plant Nutr 19:73–85

Yang XE, Long XX, Ni WZ (2002) Physiological and molecular mechanisms of heavy metal uptake by hyperaccumulting plants. Plant Nutr Fertilizer Sci 8(1):8–15

Yang ZY, Yuan JG, Xin GR, Chang HT, Wong MH (1997) Germination, growth and nodulation of *Sesbania rostrata* grown in Pb/Zn mine tailings. Environ Manage 21:617–622

Ye ZH, Baker AJM, Wong MH, Willis AJ (1997) Zinc, lead and cadmium tolerance, uptake and accumulation by *Typha latifolia*. New Phytol 136:469–480

Ye ZH, Wong MH, Baker AJM, Willis AJ (1998) Comparison of biomass and metal uptake between two populations of *Phragmites australis* grown in flooded and dry conditions. Ann Bot 80:363–370

Yong-pisanphop J, Kruatrachue M, Pokethitiyook P (2005) Toxicity and accumulation of lead and chromium in *Hydrocotyle umbellate*. J Environ Biol 26:79–89

Yu X-Z, Gu J-D (2008) The role of EDTA in Phytoextraction of hexavalent and trivalent chromium by two willow trees. Ecotoxicology 17:143–152

Zhao FJ, Dunham SJ, McGrath SP (2002) Arsenic hyperaccumulation by different fern species. New Phytol 156:27–31

Zhu YL, Zayed AM, Quian JH, De Souza M, Terry N (1999) Phytoaccumulation of trace elements by wetland plants: II. Water hyacinth. J Environ Qual 28:339–344

Zou JH, Wang M, Jiang WS, Liu DH (2006) Chromium accumulation and its effect on other mineral elements in amaranthus viridis L. Acta Biol Crac Ser Bot 48:7–12

Chapter 18
Microbial Fuel Cells for Wastewater Treatment

Cuijie Feng, Subed Chandra Dev Sharma and Chang-Ping Yu

Abstract Microbial fuel cells (MFCs), which are the bioelectrochemical systems, have been developed rapidly over the past few decades and are considered as a promising technique to obtain renewable resources from wastewater. MFCs can be used to harness electricity from microorganisms during wastewater treatment. This chapter reviews recent literature on MFCs for wastewater treatment. We first introduce the concept of MFCs and summarize the materials and design of MFCs afterward. It shows that through innovative materials and design, the current density of MFCs has been greatly improved during the last decade. Microorganisms play a major role in the electricity production of MFCs and therefore, an in-depth discussion of the microbiology of MFCs was also included in this chapter. Extensive studies on exoelectrogenic bacteria and consortia are beginning to expose the mechanistic and ecological complexities of MFC biofilm communities. Yet, our understanding of electrochemically active microbes is still in its infancy, as the diverse communities have a multitude of undiscovered populations in different MFC applications. Further study is warranted to optimize design, materials, and microbiology to improve electricity recovery from MFCs.

18.1 Introduction

Wastewater treatment currently consumes substantial energy about 15 GW (McCarty et al. 2011), or accounts for approximately 3 % of the U.S. electrical energy load (EPA Office of Water 2006), and has similar level to that in other developed countries (Curtis 2010). However, there is abundant potential energy of approximately 17 GW of power (1.5 × 1011 kWh) contained in domestic,

C. Feng · S.C.D. Sharma · C.-P. Yu (✉)
Institute of Urban Environment, Chinese Academy of Sciences,
361021 Xiamen, People's Republic of China
e-mail: cpyu@iue.ac.cn

F. Pacheco Torgal et al. (eds.), *Biotechnologies and Biomimetics for Civil Engineering*,
DOI 10.1007/978-3-319-09287-4_18

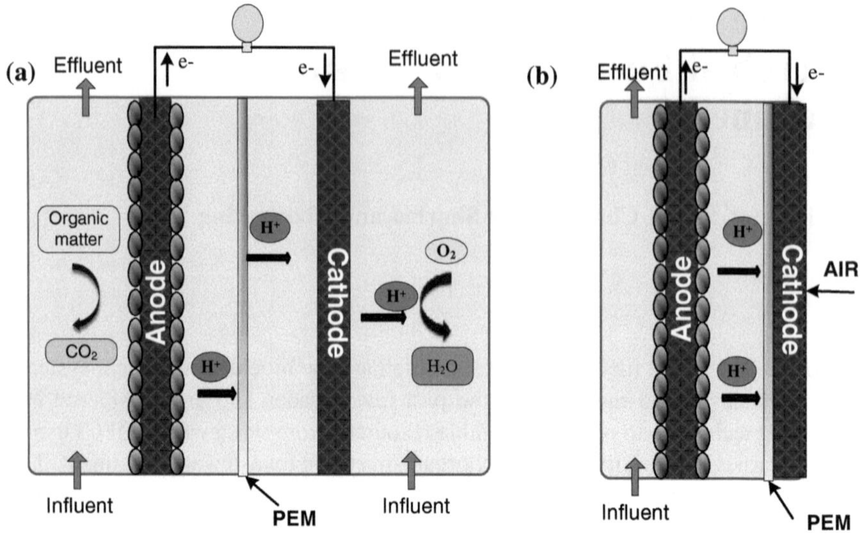

Fig. 18.1 Schematic diagrams of MFCs: **a** a two-chamber MFC; **b** a single-chamber MFC with open air cathode

industrial, and animal wastewater together (Logan 2004). Thus, capturing part of this energy would provide a new source of electrical power and would also compensate the consumption of energy for wastewater treatment.

Recently, microbial fuel cells (MFCs) (Allen and Bennetto 1993; Logan et al. 2006; Lovley 2006), which are the bioelectrochemical systems, are generally regarded as a promising and sustainable technology for their direct electrical power production from wastewaters (Rabaey and Verstraete 2005). A conventional MFC consists of a biological anode and a cathode (Fig. 18.1a), where exoelectrogenic microorganisms could catalyze electrochemical reactions through interaction with the electrodes (Logan et al. 2006; Rabaey et al. 2007; Clauwaert et al. 2008). The electrons available through the metabolism of the electron donors by microorganisms are transferred to the anode and then to the cathode through the circuit; in the cathode, the oxidant is reduced with the consumption of protons available through the membrane from the anode (Allen and Bennetto 1993).

In terms of potential applications, MFCs and related bioelectrochemical systems can be utilized for renewable energy generation and wastewater treatment, i.e., organic matter elimination and nitrogen removal (Logan and Regan 2006b; Clauwaert et al. 2007; Rozendal et al. 2008; Yu et al. 2011), for the potential production of valuable products, such as hydrogen, methane or hydrogen peroxide (Liu et al. 2005c; Rozendal et al. 2006, 2009), for bioremediation of recalcitrant compounds (Catal et al. 2008; Morris and Jin 2008), for desalination (Forrestal et al. 2012; Yuan et al. 2012; Feng et al. 2013a), and as biosensors for on-line monitoring of treatment processes (Kim et al. 2007b) and biological oxygen demand or toxic contaminants in wastewater (Kim et al. 2003; Chang et al. 2004; Kim et al. 2007b).

Recent investigations have shown that during the last 10 years, the current density of MFCs has been improved by 10,000-fold (Debabov 2008). Power densities of MFCs have increased from less than 1 W/m^3 to over 4000 W/m^3, which is the highest MFC power density reported up to date (Logan 2008; Biffinger et al. 2009). Despite their potential applications and continuously improved power, limited maximum power production by these systems impedes commercial applications of bioelectrochemical wastewater treatment, primarily because of high internal resistance including anode limitations and electrochemical losses. Improvements of power generation are also dependent on the materials and design of MFCs and capabilities of the microorganisms. Analysis of the community profiles of exoelectrogenic microbial consortia shows great diversity, ranging from primarily δ-Proteobacteria that dominate in sediment MFCs to communities composed of α-, β-, γ- or δ-Proteobacteria, Firmicutes, and uncharacterized clones in other types of MFCs. Much remains to be discovered about the physiology of these bacteria (collectively referred to as exoelectrogens) capable of exocellular electron transfer.

This chapter is intended to provide an overview of recent development and challenges in MFCs with a special focus on the materials, design, and microbiology of MFC research. Since microorganisms play a crucial role in the MFCs, comprehensive reviews focused on isolated exoelectrogens that have been identified to produce electricity, their mechanisms of exocellular electron transfer, and the microbial communities found in MFCs. In the end, the prospects for this emerging bioelectrochemical technology were discussed.

18.2 History of MFCs

Currently, MFCs have been recognized as a promising green technology for the generation of electricity through the microbial oxidation of biodegradable organic matters. The concept of generating electricity by bacteria was introduced more than 100 years ago. The electricity generated by microorganisms was firstly demonstrated in 1911 by Potter, a Professor of Botany Department at the University of Durham (Potter 1911). To examine the electricity producing capability of microorganism, he conducted his experiment using yeast and certain other bacteria in an apparatus consisted of a glass jar containing a porous cylinder. He observed that *Saccharomyces cerevisiae* and *Bacillus coli communis* (now called *Escherichia coli*) produced electric current when glucose was used as substrate. After that there was no important research on MFCs up to 1966 (Lewis 1966) and most studies on MFCs did not appear until the late twentieth century. However, experiments carried out by researchers used artificial electrochemical mediators to facilitate electron transfer between microbes and electrodes. Thurston and his colleagues used thionine as a redox mediator and *Proteus vulgaris* culture as catalyst in a two-chamber MFC to evaluate coulombic yield from glucose oxidation (Thurston et al. 1985). These chemicals were considered important for obtaining a higher electron transfer rate and electron recovery between microbial cells and electrodes. In 1999, a breakthrough in MFCs was published by Kim and his colleagues, who showed that exogenous mediators were not necessary to be added to transfer electrons from bacterial cells to electrodes and they developed the first mediator-less MFC using a Fe(III)-reducing bacterium, *Shewanella putrefaciens* IR-1 (Kim et al. 1999). The cell suspension of *Shewanella putrefaciens* IR-1 was able to generate current without redox mediator in the presence of lactate as the main carbon source. Another important bacterium *Geobacter sulfurreducens* can transfer electrons to electrode in the absence of the mediators with high current generation (Bond and Lovley 2003) and has become an important issue on MFC research. After the discovery of mediator-less MFCs, scientists have become more interested to do research on MFCs, especially in wastewater treatment because mediator-less MFCs provide a more practical and promising approach to recover electricity from organic waste and wastewater through microbial systems (Liu and Logan 2004; Min and Logan 2004). Presently many research laboratories have been engaged in improving MFC technologies to enhance the electricity production and efficient removal of wastewater by designing different configurations of MFCs such as single chamber MFC, tubular MFC (Rabaey et al. 2005b), stacked MFC (Aelterman et al. 2006) and also membrane-less MFC (Feng et al. 2013b). The advancement of research on MFCs in the future may be the solution to energy scarcity and the clean-up of wastewater. Thus, MFCs have received a great deal of attention as a novel green technology for alternative energy generation and wastewater treatment.

18.3 Design and Operations of MFCs

An appropriate design and architecture is of great significance for improving performance in MFC systems (Du et al. 2007; Pant et al. 2010). The mode of operation and components of a typical two-chamber and a single-chamber MFC are shown in Fig. 18.1.

18.3.1 Two-Chamber MFC Systems

Traditional two-chamber MFCs consist of an anaerobic anode chamber and an aerobic cathode chamber separated by a proton exchange membrane (PEM) or sometimes a salt bridge, allowing proton transfer from anode to cathode and preventing oxygen diffusion to the anode chamber, as shown in Fig. 18.1a. Regardless of the problems in scale-up, the dual-chamber MFCs have remained the most popular devices for testing microbial activity and optimizing materials. There are a variety of designs and structures occurred based on the principles of two chamber MFC systems, e.g., the widely used and inexpensive H-type MFCs

(Min et al. 2005) and U-shaped MFCs (Milliken and May 2007) (Fig. 18.2a, b). In the H-configuration, the membrane is clamped in the middle of the tubes connecting the bottle. Although H-shape systems are usually available for basic parameter research, they generate low power densities. This may attribute to high internal resistance and electrode-based losses. Oh and colleagues demonstrated that the power densities had a close relationship with the relative sizes (cross sections) of the cathode to that of the anode and the membrane (Oh et al. 2004; Oh and Logan 2006).

Ringeisen and colleagues provided a miniature configuration of MFC (Mini-MFC) with a total volume of 1.2 cm^3 (Fig. 18.2c) (Ringeisen et al. 2006). As the result of its specific structure, the mini-MFC maintains a large surface area to volume ratio when graphite felt electrodes were used, enabling high power densities to be attained. Min and Logan (2004) designed a Flat Plate MFC (FPMFC) to treat domestic wastewater. The FPMFC was comprised of a single channel formed between two nonconductive (polycarbonate) plates that were separated into two halves by the electrode/PEM assembly (Fig. 18.2d). The anodc electrode was a plain porous carbon paper (10×10 cm^2), while a carbon cloth combining a platinum catalyst (0.5 mg/cm^2 catalyst containing 10 % Pt) servers as cathode electrode. The wastewater was fed into the anode chamber and dry air could pass through the cathode chamber without any catholyte, both in a continuous flow mode. Average power density was obtained at 72 mW/m^2 (Min and Logan 2004).

Another reactor design, named upflow MFC (UMFC) working in continuous flow mode, was first tested by He et al. (2005) (Fig. 18.2e). Its configuration was improved by combining the advantages of upflow anaerobic sludge blanket system, which were operated in continuous mode. Another UMFC with a U-shaped cathode installed inside the anode chamber was developed based on the above configuration (He et al. 2006) (Fig. 18.2f). A U-shaped cathode compartment with a 2 cm diameter was constructed by gluing two tubes made from PEM into a plastic base connector. In addition to a practical configuration, UMFC achieved promising power outputs with a maximum volumetric power density of 29.2 W/m^3 with an overall internal resistance of 17.3 Ω (He et al. 2006). They suggested that the main limitation to power generation was the internal resistance. Overall, these systems seem to be more available for practical implementation as they are relatively easy to scale up.

18.3.2 Single-Chamber MFC Systems

In the single-chamber MFC (SCMFC), the cathode is exposed directly to the air by eliminating the cathodic compartment containing air-sparged solution (Park and Zeikus 2003; Liu and Logan 2004; Liu et al. 2004) (Fig. 18.1b). They typically possess only an anode chamber without the requirement of aeration in the cathode chamber. In comparison with dual chamber system, a SCMFC provides the simplified design, increased mass transfer to the cathode, cost savings and an overall decrease in reactor volume.

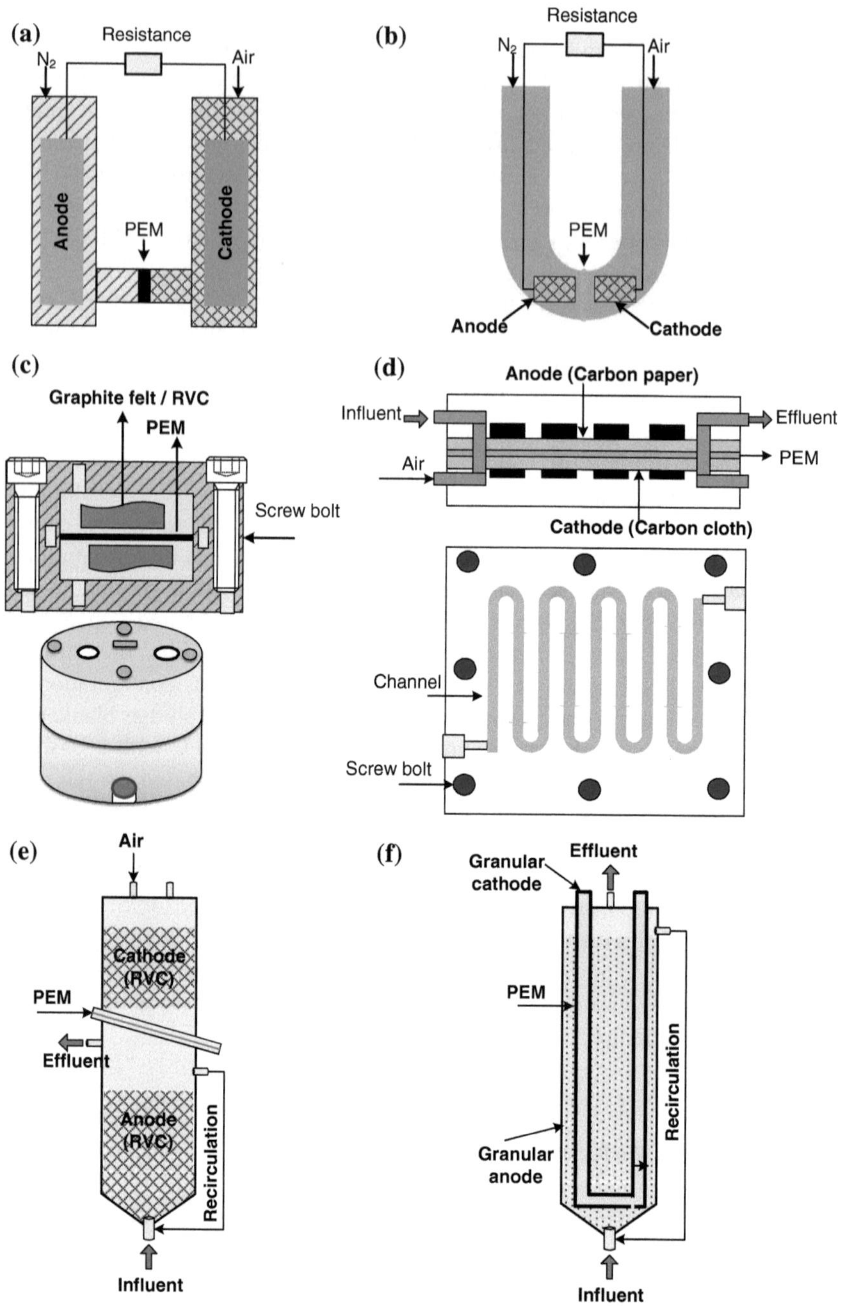

Fig. 18.2 Schematics of typical two-chamber MFCs: **a** H-type MFC with PEM or salt bridge; **b** U-shaped MFC; **c** Mini-MFC; **d** Flat plate MFC; **e, f** Up flow MFC with cylindrical shape

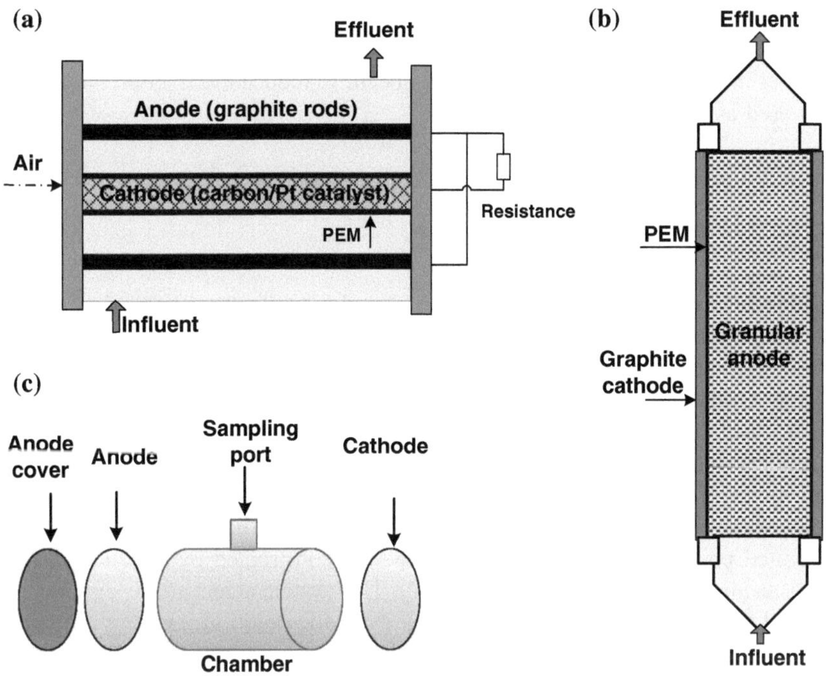

Fig. 18.3 Schematics of typical single-chamber MFCs: **a** The first SCMFC for domestic wastewater treatment; **b** Tubular MFC; **c** a lab-scale single-chamber MFC

Liu et al. (2004) first demonstrated that domestic wastewater could be used as the substrate in MFCs without actively feeding air into a cathode chamber. Their MFC consisted of a single chamber with eight graphite electrodes (anodes) and a single air cathode as shown in Fig. 18.3a. Most importantly, the promising idea of using MFC technology to reduce energy costs in wastewater treatment was initiated. A tubular MFC (TMFC), designed by Rebeay and colleagues (Rabaey et al. 2005b) was shown in Fig. 18.3b. The TMFC had a wet anode volume of 210 mL and generated a maximum volumetric power of 90 W/m^3 using graphite granules as the anode and a ferricyanide solution in the cathode chamber. A relatively low internal resistance of 4 Ω was achieved by sustaining a short distance between the anode and cathode electrodes and a large PEM surface area. Rabaey et al. (2005b) believed that the use of sustainable open air cathodes was a promising design for practical implementation.

It has been demonstrated that power output can further be increased in a single-chamber MFC by removing the PEM. Liu et al. (2004) found that there was a significant rise in power density by a factor of approximately 1.9 for glucose and 5.2 for wastewater through removing the PEM from a single chamber MFC (Fig. 18.3c). This increase was partly attributed to an enhancement of the proton flux from the anode to the cathode. The lack of a PEM substantially reduce the expenditure on the materials needed to make a MFC and eliminated the disturbing biofouling of membrane. However, substantial oxygen diffusion into the anode chamber in the absence of the PEM could occur to reduce the fraction of electrons recovered as current.

18.3.3 Other MFC Configuration

Besides the above configurations, a series of variations on these basic designs have emerged in order to achieve different purposes, such as increase of power density, achieving continuous flow or nutrient removal. For example, to increase the overall system voltage, MFCs can be stacked or linked together in series (Aelterman et al. 2006). Another type of MFC, nitrifying and denitrifying MFC for decentralized wastewater treatment was reported by Feng et al. (2013b). Their MFC system was built on the basis of conventional anoxic/oxic wastewater treatment system and achieved the continuous flow mode by using a baffle with holes instead of PEM. An integrated photobioelectrochemical system was constructed by installing a MFC inside an algal bioreactor (Xiao et al. 2012). This system achieves the simultaneous removal of organics and nutrients from a synthetic solution, and the production of bioenergy in electricity and algal biomass through bioelectrochemical and microbiological processes.

18.4 Materials

MFCs are generally made of three major parts: anode, cathode, and PEM (if present). There are a variety of materials for their construction. Electrode materials play an important role both in the performance and cost of MFCs. A good anode material should have the following properties (Logan et al. 2006; Zhou et al. 2011): large surface area; excellent electrical conductivity, strong biocompatibility, chemical stability, appropriate mechanical strength and toughness. Up to now, various materials are used for electrodes including carbon materials, e.g., carbon paper (Liu et al. 2005a), carbon cloth, carbon felt (Chaudhuri and Lovley 2003) and reticulated vitreous carbon (RVC) (He et al. 2005; Rabaey et al. 2005b), graphite materials, e.g., graphite granules and graphite fiber brushes (Aelterman et al. 2006; He et al. 2006; Rinaldi et al. 2008), etc. Since different electrode materials vary obviously in their physical and chemical characteristics, they have impacts on microbial attachment, electron transfer, electrode resistance and the rate of electrode surface reaction. Thus, some strategies could be applied to boost the performance in terms of increasing the surface area and the biocompatibility. The anode materials could be fabricated with C/polyaniline (PANI) composites, carbon nanofibers, or nitric acid carbon activation (Scott et al. 2007), or integration of carbon nanotubes to PNAI (Qiao et al. 2007), etc.

Cathodes are made from the same materials as anodes, and catalysts are usually contained but not necessary. Because oxygen is the terminal electron acceptor in most cases, the high overpotential arising from oxygen reduction reaction causes the noncatalyst cathodes to be inefficient. Thus, catalysts and/or artificial mediators are generally required to improve performance. They are generally mounted on the cathodes with a binder such as Nafion (perfluorosulfonic acid) or polytetrafluoroethylene. Pt has become the most popular catalyst (Thurston et al. 1985), but its high cost and reduced activities due to formation of a PtO layer on the electrode surface restrict its practical application. For this reason non-Pt catalysts including nonabundant metals, e.g., Pd or Ru (Vante and Tributsch 1986; Fernández et al. 2005; Raghuveer et al. 2005) and nonprecious materials, e.g., Fe, Mn and Co (Park and Zeikus 2003) tend to be more appealing. They could exhibit essentially equal or slightly better performance than the more expensive Pt. Among the non-Pt catalysts, the most promising CoTMPP and iron (II) phthalocyanine (FePc) (Zhao et al. 2005) were proved to be inexpensive and efficient alternatives for MFC applications. Integration of noncorrosive metals (titanium and nickel) and carbon fibers can be used as cathode materials as well (Hasvold et al. 1997; Zhao et al. 2009). Additionally, catalysts are not required for catholyte cathodes, which use the redox mediators such as ferricyanide (Oh et al. 2004; Venkata Mohan et al. 2008) or permanganate (You et al. 2006). Using them as terminal electron acceptors could result in alternative cathodic reactions and further improve power output to 258 W/m^3 (Aelterman et al. 2006). These catholytes seemed to be impractical and unsustainable for practical application owning to the requirement of regeneration of the chemicals. On the basis of the above introduction, a large number of materials have been investigated to improve cathode performance. However, their long-term stability on the cathode should be further evaluated for future application.

The PEM is also an important component in the PEM-MFC configuration. It provides a separation between the anode and cathode chambers and allows for transport of positive charges to compensate the electron transport. Currently, the most widely used membrane material is Nafion^{TM} (Park and Zeikus 2000; Bond and Lovley 2003), which has set the industry standard for PEM. Its properties have been extensively reviewed (Mauritz and Moore 2004). Obviously, Nafion^{TM} was the predominant choices for current MFCs. Nevertheless, it has been recently found that the use of Nafion^{TM} leads to some side effects such as pH imbalance and power reduction (Gil et al. 2003; Kim et al. 2004). In addition to Nafion^{TM}, polyether ether ketone (PEEK) is a promising polymer being actively studied by the MFC researchers to overcome the drawbacks of Nafion^{TM} (Roziere and Jones 2003; Mecheri et al. 2006). In fact, membranes can be omitted from the bioelectrochemical configuration. The lack of a PEM could decrease the cost of the materials for a MFC, but substantial oxygen diffusion into the anode chamber in the absence of the PEM could reduce the fraction of electrons recovered as current.

18.5 Exoelectrogens

In nature, there are many microorganisms possessing the ability to transfer electrons derived from the metabolism of organic matters to the anode. Microorganisms capable of extracellular electron transfer are generally called "exoelectrogens". These microorganisms attain their required energy by oxidizing organic matter with the release of protons and electrons that are used in MFC to produce electricity. Marine sediment, soil, wastewater, fresh water sediment and activated sludge are rich sources for these microorganisms (Niessen et al. 2006; Zhang et al. 2006). In the beginning, it was considered that only a few types of bacteria were capable of producing electricity and most of them were gram negative Proteobacteria such as *Shewanella putrefaciens* (Park and Zeikus 2002), *Geobacter sulfurreducens* (Bond and Lovley 2003), etc. However, now gram positive bacteria also have been discovered to produce electricity, including *Clostridium butyricum* within the Firmicutes (Park et al. 2001). The capability to produce electricity generally depends on the nature of bacterial species and their ability to utilize different substrates. Power generation also depends on the optimal growth condition of bacteria, e.g., pH and temperature.

Up to now, most of isolated exoelectrogens are bacteria (Table 18.1) and were isolated from different MFCs using large varieties of substrates. Scientists are trying to discover new exoelectrogenic bacteria, which will have the capacity to achieve high power density. The pure strain *Geobacter sulfurreducens* operated in a two-chamber MFC with PEM and graphite electrode produced an electric current density of 65 mA/m^2 using acetate as the substrate (Bond and Lovley 2003). Other pure strains such as *Comamonas denitrificans* DX-4 and *Citrobacter sp.* SX-1 produced the highest power and current density of 35 mW/m^2 and 205 mA/m^2 using acetate and citrate as electron donors in MFC respectively (Xing et al. 2010; Xu and Liu 2011). One scientific report showed that power output in a MFC inoculated with a pure culture (*Geobacter metallireducens*) or a mixed culture (wastewater inoculums) was similar, with 40 ± 1 mW/m^2 for *Geobacter metallireducens* and 38 ± 1 mW/m^2 for the wastewater inocula (Min et al. 2005). However, *Rhodopseudomonas palustris* DX-1, isolated from an air cathode MFC, produced electricity at higher power densities (2720 ± 60 mW/m^2) than mixed culture in the same device using complex substrates including volatile acids, yeast extract and thiosulfate (Xing et al. 2008). In addition, some bacteria require exogenous redox compound to increase maximum power production. For example, *Shewanella putrefaciens* generated the maximum power density of 10.2 mW/m^2 when operated in the absence of exogenous electron acceptors in a single chambered MFC, but current production by *Shewanella putrefaciens* was enhanced 10-folds when an electron mediator, i.e., Mn^{4+} or neutral red was incorporated into the graphite anode (Park and Zeikus 2002).

Table 18.1 Some examples of isolated exoelectrogens

Class	Microorganism	Substrate	Mediator	Power or current density	References
α-Proteobacteria	*Rhodopseudomonas palustris* DX-1	Volatile acids, yeast extract	Mediator-less	2720 ± 60 mW/m^2	(Xing et al. 2008)
	Gluconobacter oxydans	Glucose	2-hydroxy-1,4-naphthoquinone	–	(Lee et al. 2002)
β-Proteobacteria	*Comamonas denitrificans* DX-4	Acetate	Mediator-less	35 mW/m^2	(Xing et al. 2010)
γ-Proteobacteria	*Shewanella oneidensis*	Lactate	Mediator-less	24 mW/m^2	(Ringeisen et al. 2006)
	Citrobacter sp. SX-1	Citrate	Mediator-less	205 mA/m^2	(Xu and Liu 2011)
	Shewanella putrefaciens	Sodium lactate	Mn^{4+} or neutral red	10.2 mW/m^2	(Park and Zeikus 2002)
	Klebsiella pneumoniae	Glucose	2-hydroxy-1,4-naphthoquinone	126.7 ± 31.5 mW/m^2	(Rhoads et al. 2005)
	Actinobacillus succinogenes	Glucose	Neutral red	–	(Park and Zeikus 1999)
	Proteus mirabilis	Glucose	Thionin	–	(Thurston et al. 1985)
	Proteus vulgaris	Sucrose	Thionine	–	(Bennetto et al. 1985)
	Escherichia coli	Glucose	Neutral red	–	(Park and Zeikus 2000)
	Aeromonas hydrophila	Acetate	Mediator-less	–	(Pham et al. 2003)
	Pseudomonas aeruginosa KRP1	Glucose	Pyocyanin and phena-zine-1-carboxamide	–	(Rabaey et al. 2005a)
δ-Proteobacteria	*Geobacter metallireducens*	Acetate	Mediator-less	40 ± 1 mW/m^2	(Min et al. 2005)
	Geobacter sulfurreducens	Acetate	Mediator-less	65 mA/m^2	(Bond and Lovley 2003)
Clostridia	*Clostridium butyricum*	Glucose	Mediator-less	–	(Park et al. 2001)

18.6 Electron Transfer Mechanism of Exoelectrogens

The electron transfer mechanism is a key issue to understand the theory of how MFCs work. Numerous investigations were conducted to study how electrons were transferred from microbial cells to anode surface in the MFCs. There are generally two main mechanisms that are direct or mediator-less and indirect or mediated electron transfer (MET).

18.6.1 Direct or Mediator-Less Electron Transfer

Direct electron transfer (DET) requires a physical contact between the microbial cell membrane or a membrane organelle and the electrode surface. *Shewanella putrefaciens* (Kim et al. 2002), *Geobacter sulferreducens* (Bond and Lovley 2003), and *Geobacter metallireducens* (Min et al. 2005) can effectively transfer electrons directly to an electrode across the membrane. Some of DET bacteria transfer electrons through direct attachment of cell membrane to anode (Fig. 18.4a), while the rest use their pili or nanowires to transfer electrons to anode (Fig. 18.4b). Generally c-type cytochromes associated with bacterial outer membrane and conductive nanowires or pili can be used for DET (Peng et al. 2010).

18.6.2 Indirect or Mediated Electron Transfer

Although some bacteria can transfer electrons directly, many other microbes need redox-active chemical species or mediators to carry out electron transfer to anode; this type of mechanism is known as indirect or MET. In MET, direct contact between the bacterial cell membrane and the electrode surface is not required, but a redox mediator is essential. An electron mediator is a molecule that functions as an electron shuttle between microbes and an electrode. Mediators in the oxidized state are easily reduced by capturing electrons from within the bacterial cell membrane or the cytoplasm (Fig. 18.4c). The reduced mediators after passing across the membrane release their electrons to the electrode and become oxidized again in anode chamber and thus are reutilized. Generally chemical mediators are supplied from outside into the anode chamber of a MFC. Apart from externally provided mediators, some microorganisms are able to excrete their own mediators such as phenazine, 2-amino-3-carboxy-1,4-naphthoquinone and 2,6-di-tert-butyl-p-benzoquinone (Rabaey et al. 2005a; Freguia et al. 2009; Deng et al. 2010) that are used to transfer electron from cytoplasm to anode (Fig. 18.4d). In addition, there is another way by which some bacteria, especially fermentative bacteria, produce energy rich reduced metabolites such as H_2, ethanol or formate, which can be subsequently oxidized to provide electron to anode (Schroder 2007) (Fig. 18.4e). Furthermore, in

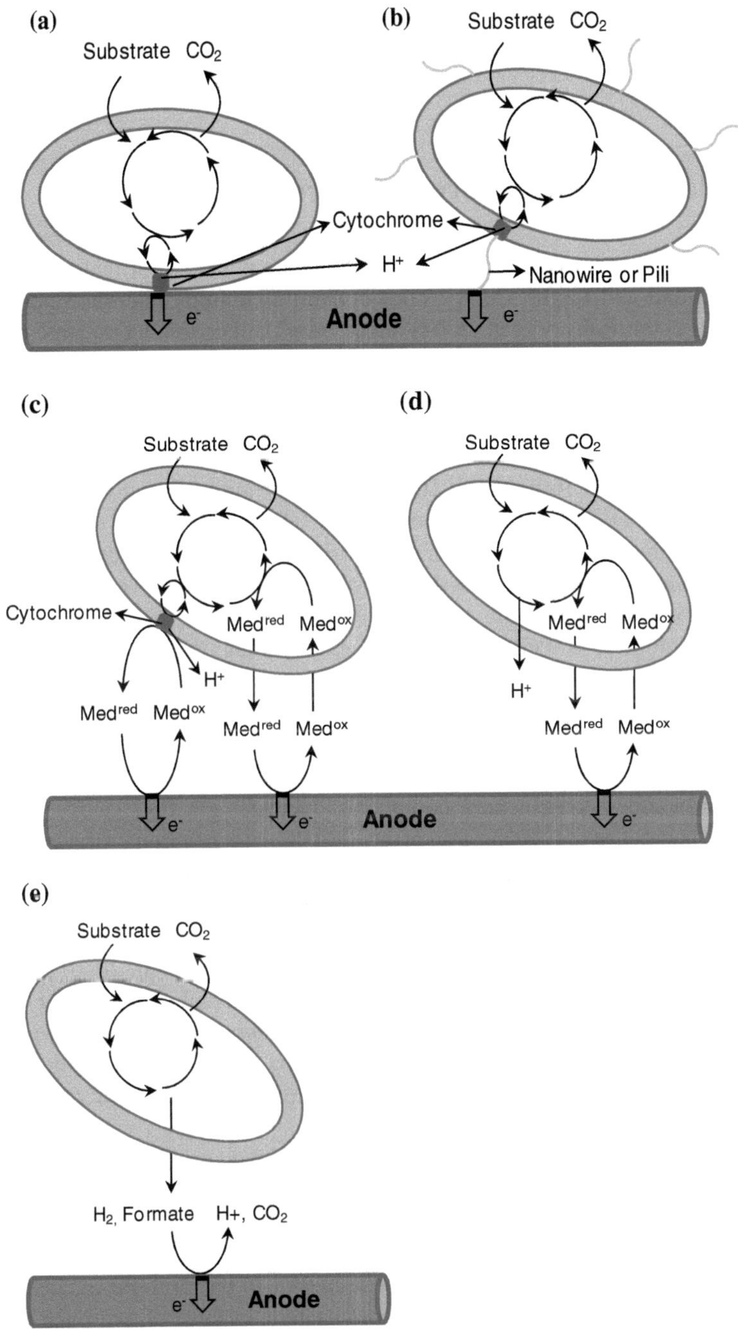

Fig. 18.4 Electron transfer mechanism of exoelectrogens: Direct electron transfer by **a** attachment of cell membrane or **b** nanowire; Indirect electron transfer by **c** exogenous mediators, **d** endogenous secondary metabolites or **e** primary metabolites

a synergistic biofilm consortium, it is likely that a nonelectrogenic microbe may secrete mediators that may help the electrogenic microbe to perform better electron transfer.

18.7 Microbial Community of Electroactive Biofilms

Biofilms more than ten micrometers in thickness are typically formed on the anode surfaces (Bond and Lovley 2003). They contain a complex microbial population (Kim et al. 2004; Rabaey et al. 2004), apart from the known electrogenic bacteria (*Geobacter, Shewanella*). Identifying members of the microbial community will be a valuable aid in terms of improving the performance of MFCs and a more comprehensive understanding of the key microbes required for exoelectrogenesis. Up to now, there are many publications associated with microbial communities in MFCs by means of PCR-amplified 16S rRNA gene fragments and sequencing such as denaturing gradient gel electrophoresis (DGGE) (Table 18.2). Analysis of the populations inhabiting such systems demonstrates that microbial communities are phylogenetically diverse in most MFCs. Microbial populations are affected by numerous factors, such as the substrate, cultivation mode, system architectures, anaerobiosis degree, as well as the conditions within the cathode chamber (Logan and Regan 2006a).

The composition of substrates has a close relationship with the microbial populations within the anode biofilms and MFC performance, as they serve as the carbon (nutrient) and energy source for the microbiological process. Commonly, the carbon sources contain pure compounds (acetate, glucose, lactic acid, etc.) (Chaudhuri and Lovley 2003; Liu et al. 2005b) and a variety of wastewaters (brewery, chocolate, meat packing and paper recycling wastewaters, etc.) (Feng et al. 2008; Huang and Logan 2008). The pure substrate inoculated systems are found to produce more power than those fed with wastewater perhaps as the result of different solution conductivity and buffer capacity (Pant et al. 2010). Based on 16S rRNA gene sequences, the dominant community members in the MFCs with pure substrate are more known exoelectrogens (*Geobacter* sp., *Desulfuromonas* sp., *Rhodopseudomonas* sp., etc.) and other bacteria with special function, such as *Clostridium* sp., which is useful for lignocellulose degradation in cellulose-fed MFCs (Cheng et al. 2011) (Table 18.2).

The highest power density of 4.31 W/m^2 was achieved using a mixed culture in a fed-batch MFC and glucose as the substrate in the reactor with a Coulombic efficiency (defined as the fraction of electrons recovered as current versus the maximum possible recovery) of 81 %. The analysis of the population using DGGE showed great phylogenetic diversity, with a complex mixture of bacteria (Firmicutes, γ-, β-, and α-Proteobacteria). Facultative anaerobic bacteria capable of hydrogen production (*Alcaligenes faecalis, Enterococcus gallinarum*) were predominant (Rabaey et al. 2004), probably owning to using a fermentable substrate with a mixed culture inocula (Debabov 2008). It was deduced that mediator production accounted for the

Table 18.2 Summary of dominant microbes present in bacterial community of the anode biofilm

Substrates	MFC	Technique	Dominant community members	Power density (mW/m^2)	Coulombic efficiency (%)	References
Acetate	Two-chamber	DGGE	*Geobacter sulfurreducens*	N/A	72	(Jung and Regan 2007)
	Single-chamber	Clone library	*Pelobacter propionicus*	835 ± 21	20	(Kiely et al. 2011b)
	Single-chamber	DGGE	*Rhodopseudomonas palustris*, *Geobacter sulfurreducens*, *Pseudomonas alcaligenes*	1797 ± 10	N/A	(Xing et al. 2009)
	Two-chamber	Clone library	*Thauera aromatica*, *Geobacter sulfurreducens*	64.3	72.3	(Chae et al. 2009)
	Two-chamber	Clone library	*Geobacter sulfurreducens*, *Pelobacter propionicus*	N/A	50	(Chae et al. 2008)
Ethanol	Single-chamber	Clone library	*Geobacter sulfurreducens*, *Pelobacter propionicus*	820 ± 24	11	(Kiely et al. 2011b)
	Two-chamber	RFLP	*Azoarcus* sp., *Desulfuromonas* sp.	40 ± 2	10	(Kim et al. 2007a)
Lactate	Single–chamber	Clone library	*Pelobacter propionicus*, *Desulfuromonas* sp.	739 ± 32.2	20	(Kiely et al. 2011b)
Propionate	Two-chamber	Clone library	*Bacillus* sp.	58	36	(Chae et al. 2009)
Butyrate	Two-chamber	DGGE	*Bacillus* sp.	N/A	46–67	(Freguia et al. 2010)
	Two-chamber	Clone library	*Dechloromonas* sp., *Geobacter* sp.	51.4	43	(Chae et al. 2009)
Formate	Two-chamber	Clone library	*Paracoccus* sp., *Geobacter* sp.	10	3–11	(Kiely et al. 2010)
	Two-chamber	DGGE	*Geobacter* sp.	N/A	5–6	(Ha et al. 2007)

(continued)

Table 18.2 (continued)

Substrates	MFC	Technique	Dominant community members	Power density (mW/m^2)	Coulombic efficiency (%)	References
Succinate	Single-chamber	Clone library	*Geobacter sulfurreducens*, *Pelobacter propionicus*	444 ± 12.5	16	(Kiely et al. 2011b)
Glucose	Two-chamber	Clone library	*Geobacter sulfurreducens*	156	15	(Chae et al. 2009)
	Single-chamber	DGGE	*Rhodopseudomonas palustris*, *Geobacter sulfurreducens*, *Clostridium* sp.	1000 ± 19	N/A	(Xing et al. 2009)
Cellulose	Single-chamber	Clone library	*Geobacter sulfurreducens*, *Clostridium* sp.	1070	25–50	(Xing et al. 2009)
	Two-chamber	Clone library	Clostridiales, Chloroflexi, Rhizobiales, Methanobacterium	10	N/A	(Chae et al. 2009)
	Single-chamber	DGGE	*Clostridium* sp.	331	4	(Wang et al. 2009)
Cysteine	Two-chamber	DGGE	*Shewanella* sp.	39	14	(Logan et al. 2005)
Organic wastewater	Two-chamber	DGGE	*Azoarcus* sp, *Thauera* sp.	N/A	N/A	(Kim et al. 2004)
Dairy manure wastewater	Single-chamber	Clone library	*Thauera aromatica*, *Clostridium* sp., *Geobacter* sp.	N/A	12	(Kiely et al. 2011a)
Potato wastewater	Single-chamber	Clone library	*Geobacter* sp., *Pelobacter propionicus*	N/A	21	(Kiely et al. 2011a)

excellent power generation, as large concentrations of highly colored mediators from this reactor were detected (Logan and Regan 2006a).

However, there are complicated organic matters in wastewater and complex metabolisms such as fermentation could get involved in MFCs. Molecular characterizations of anodic communities with complex wastewater sources revealed a high diversity of microbial species, dominant with α- (Phung et al. 2004), β- (Kim et al. 2004; Phung et al. 2004), and γ-Proteobacteria (Logan et al. 2005). For example, the characterization of anodic communities present in a two-chamber MFC treating chocolate wastewater showed a high percentage of β-Proteobacteria (51 %) (Patil et al. 2009). Whereas, microbial communities that developed by MFCs supplied with winery or potato wastewater, were a mixed consortia predominated by *Geobacter sulfurreducens*, representing 44 % and 60 % of 16S rRNA gene clones, respectively (Cusick et al. 2010; Kiely et al. 2011a). Most importantly, a large proportion of clones is uncharacterized in these mixed-culture systems, especially with complex wastewater sources. The lower frequency to detect known exoelectrogens implies a greater diversity of this phenotype than presently realized. The significance of the potential function of these dominant community members is still unknown.

Cultivation mode including fed-batch and continuous flow could affect microbial communities as well. In a continuous flow mode MFC supplied with acetate, the composition of anodic community revealed that the most dominant phyla were Proterobacteria (23–33 %), Bacteroidetes (17–40 %) and Chloroflexi (21–30%) on the basis of 454 pyrosequencing technique (Feng et al. 2013b). In an upflow system, a large number of methanogenic archaea in the mixed biomass appeared on the anode based on fluorescence in situ hybridization (He et al. 2005). Literature studies have demonstrated that δ-Proteobacteria (50–90 %) were dominant in the anode community of sediment MFC (Bond et al. 2002; Bond and Lovley 2003), while Cytophagales (up to 33 %), Firmicutes (11.6 %), and γ-Proteobacteria (9–10 %) were the minor components in the anodophilic consortia (Tender et al. 2002; Holmes et al. 2004).

18.8 The MFC's Full-Scale Applications

The development of MFC's practical application is still in the early stage. To date, most MFCs have been investigated in the bench-scale, generally less than 1 L and produced a maximum potential approximately 0.8 V. Apparently, the power density and MFC configuration have not yet reached a widely applicable level, remaining the challenging obstacle.

Sediment MFCs have been demonstrated at scales effective to be an alternative renewable power source in seawater applications (Bond et al. 2002; Lowy et al. 2006; Dewan et al. 2014). According to Fig. 18.5, in principle sediment MFCs consist of two electrodes made of conductive material. The anode is buried under

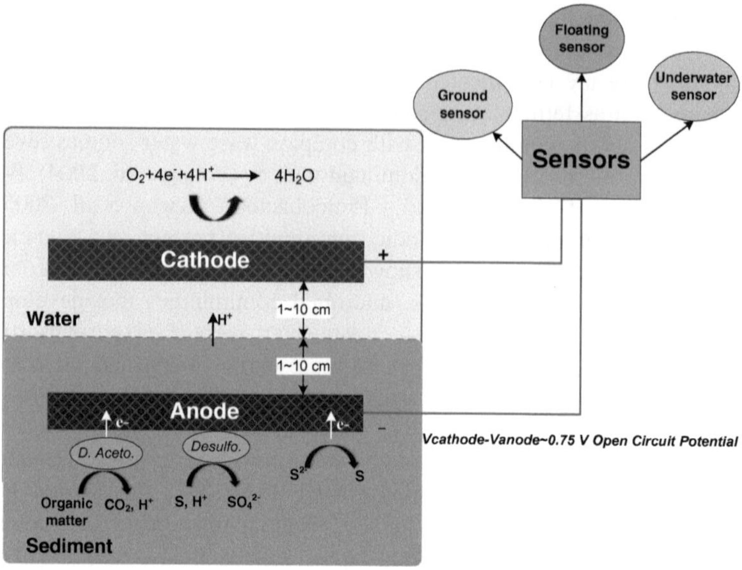

Fig. 18.5 Schematic representation of fundamental principle of the mediator less sediment MFCs used to provide energy for on-site sensors. Microorganisms colonizing the anode are most similar to *Desulfuromonas acetoxidans* (*D. Aceto.*), which could oxidize acetate in sediment and transfer electrons to the anode. *Desulfo.* represents the species in the *Desulfobulbus* or *Desulfucapsa* genera, which could oxidize anode generated S_0 to SO_4^{2-}

surface water or marine sediment and cathode is placed in the water above the sediment (Tender et al. 2002; Logan and Regan 2006b; Rezaei et al. 2007). The sedimentary organic carbon (Aller 1994) or sulfate compounds (Rabaey et al. 2006) present in the sediment are oxidized by microorganisms growing on the anode surface for production of electricity. There are several attempts to demonstrate the availability of sediment MFCs as power source for underwater (Donovan et al. 2013), ground (Donovan et al. 2008), and floating sensors (Nielsen et al. 2007; Tender et al. 2008; Donovan et al. 2011). The first demonstration of scale-up of MFC was used to power a weather buoy embedded with temperature and humidity sensors using two sediment MFCs that generated 24 mW and 36 mW (Tender et al. 2008). The sediment MFCs were deployed in the Potomac River, at Washington, DC and Tukerton, NJ, USA. Donovan et al. used sediment MFCs to operate a low-power (11 mW) and a high-power (2500 mW) wireless temperature sensors in a creek at Palouse, WA, USA. (Donovan et al. 2008; 2011). The average power generation to power a remote device via a sediment MFC ranges from 3.4 to 36 mW (Dewan et al. 2014). These studies illustrate that MFCs deployed in natural aquatic environment (i.e., rivers, lakes, or oceans) can produce enough energy to operate sensors requiring low power.

However, MFCs for wastewater treatment have faced a variety of restrictions in terms of practical implementation. First, the real wastewater contains complex organics and diverse microorganisms such as methanogens. This may lead to an inferiority of electroactive biofilm due to methanogenic competition or metabolic diversity. The low ionic strength of real wastewater can limit the power output of MFCs as well (Rozendal et al. 2008). In addition, there are physical constraints with regard to linearly scaling up MFCs. Excessive pressure because of hydrostatic head could require variable permeability to regulate water loss and cathode hydration in the case of permeable membrane. Most importantly, the greatest hindrance lies in the increasing electrical losses and overpotentials with enlarged size (Oh et al. 2010). All of this means that innovative reactor designs are required for practically useful MFCs. As a consequence, after more than two decades of development, in which numerous studies have focused on MFC's application for wastewater treatment (Habermann and Pommer 1991), successful full-scale application is still relatively rare.

In view of the concept of MFCs with current wastewater treatment system, several types of MFCs have been proposed. In order to enhance the quality of effluent, Logan (2008) proposed an integrated bioprocess, which combined the post-treatment process, e.g., solids contact (SC) process or membrane bioreactor (MBR) with MFC system (Fig. 18.6a and b). However, performance of post bioreactor can be inhibited due to consumption of most organic matter in the preceding MFC. The MFC can be combined into the existing wastewater treatment facilities as well. Min and Angelidaki (2008) developed a submersible MFC by immersing an air-cathode MFC in an anaerobic reactor. Similarly, Cha et al. (2010) submerged a single chamber MFC into the aeration tank of the activated sludge process to optimize the cell configuration and electrode materials. The submersible MFC can be applied to the anaerobic (or aerobic) facility as an anode (or cathode) chamber without additional constructions (Min and Angelidaki 2008; Cha et al. 2010) (Fig. 18.6c, d). Yu et al. (2011 and Feng et al. (2013b) designed another configuration for decentralized wastewater treatment through immersing the anode into an anaerobic tank and the cathode into an aerobic tank of the A/O system, respectively (Fig. 18.6e). These types of configuration enable MFCs to be applied to existing wastewater treatment systems.

Meanwhile, the work on scaling up MFCs for wastewater treatment is moving forward. According to some information on the Internet or public literatures, there are at least two pilot-scale MFCs for wastewater treatment available for practical implementation. The first large-scale test of tubular MFCs was located at Foster's brewery in Yatala, Queensland (Australia) (http://www.microbialfuelcell.org). This system was constructed by the Advanced Water Management Center of the University of Queensland, led by Jurg Keller and Korneel Rabaey. MFCs consisted of 12 modules with an entire volume of 1 m^3. The anodes and cathodes are made of carbon fiber based on a brush design. Another pilot-scale multi-anode/cathode MFC (MAC MFC) was developed by researchers of University of Connecticut and their collaborators (Fuss and O'Neill, and Hydroqual Inc.) in the USA (Jiang et al. 2011). The MAC MFC contained 12 anodes/cathodes with a total volume of 20 L. The reactors contain graphite rods as the anode, with Cu-MnO_2 or Co-MnO_2 catalyzed carbon cloth cathodes. The systems are treating wastewater, achieving

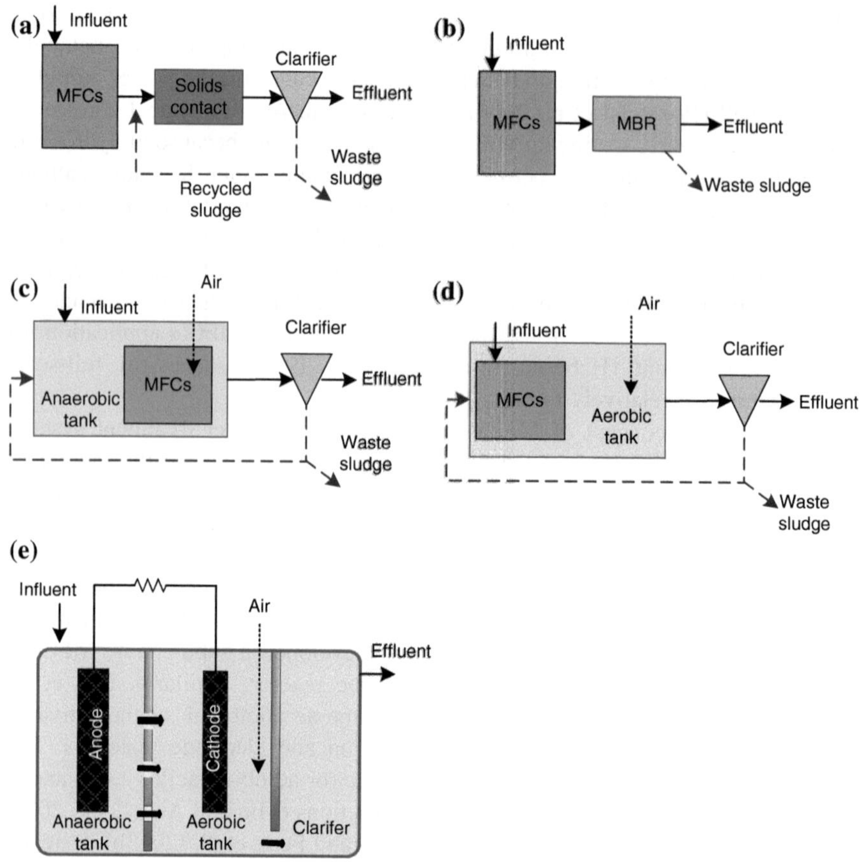

Fig. 18.6 Schematic diagram of MFCs combined wastewater treatment process: **a, b** MFC combined with a solids contact tank or a MBR; **c, d** MFC submerged into an anaerobic or aerobic tank of existing wastewater treatment process; **e** a decentralized wastewater treatment based on A/O system

80 % of contaminant removal at different organic loading rates (0.19–0.66 kg/m^3/d). The power density of MAC MFC reached 380 mW/m^2. In addition to the pilot scale MFCs, Ieropoulos et al. (2013) originally exploited a stack of small ceramic MFCs (6.25 mL) fed with real urine to power a mobile phone, which was previously considered impossible.

Therefore, tremendous efforts should be dedicated in terms of utilizing the voltage from MFCs in the near future. Dewan et al. (2014) pointed out that renewable energy sources tend to be applied to power remote sensors, due to the potential environmental risks and operational cost associated with batteries. More research is also required to focus on assessment of lifetime, reliability and renewability, which are of great significance in the process of promoting the MFCs widespread application.

18.9 The Conclusion and Perspective

Substantial efforts have been devoted to the development and improvement of MFC technology to reduce its operating cost, and to increase power output although MFC technology has not been widely scaled up for commercial application. MFC technology covers many distinct scientific disciplines, including material sciences, microbial ecology, and engineering design. Previous studies have proposed innovative designs of MFC reactors to improve the performance together with reduced capital costs. It has been demonstrated that different electrode materials exhibited different behaviors and electrode modification offers a good and effective approach for enhancing the performance. Development of the electrode with excellent proprieties and the reasonable price could be crucial for the practical application. Furthermore, appropriate integration or combination of MFCs with present wastewater treatment technologies should be taken into consideration.

MFCs provide us with a model system to study the different microbial populations present in the exoelectrogenic biofilms, and it would be an important research area in understanding how the microbial ecology of electricity producing communities develops and shifts over time. Extensive studies on exoelectrogenic bacteria and consortia begin to expose the mechanistic and ecological complexities of MFC biofilm communities. Yet, our understanding of electrochemically active microbes is still in its infancy, as the diverse communities have a multitude of undiscovered electrochemical capabilities that can be exploited in different MFC applications. Discovery of the potential exoelectrogenic bacteria is important in understanding the function of anodic microbial communities and to improve the electron transfer efficiency of MFCs.

References

Aelterman P, Rabaey K, Pham HT, Boon N, Verstraete W (2006) Continuous electricity generation at high voltages and currents using stacked microbial fuel cells. Environ Sci Technol 40(10):3388–3394

Allen RM, Bennetto HP (1993) Microbial fuel-cells: electricity production from carbohydrates. Appl Biochem Biotechnol 39(1):27–40

Aller RC (1994) The sedimentary Mn cycle in long island sound: its role as intermediate oxidant and the influence of bioturbation, O_2, and Corg flux on diagenetic reaction balances. J Mar Res 52(2):259–295

Bennetto H, Delaney G, Mason J, Roller S, Stirling J, Thurston C (1985) The sucrose fuel cell: efficient biomass conversion using a microbial catalyst. Biotechnol Lett 7(10):699–704

Biffinger J, Ribbens M, Ringeisen B, Pietron J, Finkel S, Nealson K (2009) Characterization of electrochemically active bacteria utilizing a high-throughput voltage-based screening assay. Biotechnol Bioeng 102(2):436–444

Bond DR, Holmes DE, Tender LM, Lovley DR (2002) Electrode-reducing microorganisms that harvest energy from marine sediments. Science 295(5554):483–485

Bond DR, Lovley DR (2003) Electricity production by *Geobacter sulfurreducens* attached to electrodes. Appl Environ Microbiol 69(3):1548–1555

Catal T, Fan Y, Li K, Bermek H, Liu H (2008) Effects of furan derivatives and phenolic compounds on electricity generation in microbial fuel cells. J Power Sources 180(1):162–166

Cha J, Choi S, Yu H, Kim H, Kim C (2010) Directly applicable microbial fuel cells in aeration tank for wastewater treatment. Bioelectrochemistry 78(1):72–79

Chae KJ, Choi MJ, Lee J, Ajayi F, Kim IS (2008) Biohydrogen production via biocatalyzed electrolysis in acetate-fed bioelectrochemical cells and microbial community analysis. Int J Hydrogen Energy 33(19):5184–5192

Chae KJ, Choi MJ, Lee JW, Kim KY, Kim IS (2009) Effect of different substrates on the performance, bacterial diversity, and bacterial viability in microbial fuel cells. Bioresour Technol 100(14):3518–3525

Chang IS, Jang JK, Gil GC, Kim M, Kim HJ, Cho BW, Kim BH (2004) Continuous determination of biochemical oxygen demand using microbial fuel cell type biosensor. Biosens Bioelectron 19(6):607–613

Chaudhuri SK, Lovley DR (2003) Electricity generation by direct oxidation of glucose in mediatorless microbial fuel cells. Nat Biotechnol 21(10):1229–1232

Cheng S, Kiely P, Logan BE (2011) Pre-acclimation of a wastewater inoculum to cellulose in an aqueous-cathode MEC improves power generation in air-cathode MFCs. Bioresour Technol 102(1):367–371

Clauwaert P, Aelterman P, de Schamphelaire L, Carballa M, Rabaey K, Verstraete W (2008) Minimizing losses in bio-electrochemical systems: the road to applications. Appl Microbiol Biotechnol 79(6):901–913

Clauwaert P, Rabaey K, Aelterman P, de Schamphelaire L, Ham TH, Boeckx P, Boon N, Verstraete W (2007) Biological denitrification in microbial fuel cells. Environ Sci Technol 41(9):3354–3360

Curtis TP (2010) Low-energy wastewater treatment: strategies and technologies. In: Mitchell R, Gu J-D (eds) Environmental microbiology, 2nd edn. Wiley, Hoboken, New Jersey

Cusick RD, Kiely PD, Logan BE (2010) A monetary comparison of energy recovered from microbial fuel cells and microbial electrolysis cells fed winery or domestic wastewaters. Int J Hydrogen Energy 35(17):8855–8861

Debabov V (2008) Electricity from microorganisms. Microbiology 77(2):123–131

Deng L, Li F, Zhou S, Huang D, Ni J (2010) A study of electron-shuttle mechanism in *Klebsiella pneumoniae* based-microbial fuel cells. Chin Sci Bull 55(1):99–104

Dewan A, Ay SU, Karim MN, Beyenal H (2014) Alternative power sources for remote sensors: a review. J Power Sources 245:129–143

Donovan C, Dewan A, Heo D, Beyenal H (2008) Batteryless, wireless sensor powered by a sediment microbial fuel cell. Environ Sci Technol 42(22):8591–8596

Donovan C, Dewan A, Heo D, Lewandowski Z, Beyenal H (2013) Sediment microbial fuel cell powering a submersible ultrasonic receiver: new approach to remote monitoring. J Power Sources 233:79–85

Donovan C, Dewan A, Peng H, Heo D, Beyenal H (2011) Power management system for a 2.5 W remote sensor powered by a sediment microbial fuel cell. J Power Sources 196(3):1171–1177

Du ZW, Li HR, Gu TY (2007) A state of the art review on microbial fuel cells: a promising technology for wastewater treatment and bioenergy. Biotechnol Adv 25(5):464–482

EPA Office of Water (2006) Wastewater management fact sheet, energy conservation, EPA 832-F-06-024, Washington DC

Feng C, Hou CH, Chen S, Yu CP (2013a) A microbial fuel cell driven capacitive deionization technology for removal of low level dissolved ions. Chemosphere 91(5):623–628

Feng C, Hu A, Chen S, Yu CP (2013b) A decentralized wastewater treatment system using microbial fuel cell techniques and its response to a copper shock load. Bioresour Technol 143:76–82

Feng Y, Wang X, Logan BE, Lee H (2008) Brewery wastewater treatment using air-cathode microbial fuel cells. Appl Microbiol Biotechnol 78(5):873–880

Fernández JL, Raghuveer V, Manthiram A, Bard AJ (2005) Pd-Ti and Pd-Co-Au electrocatalysts as a replacement for platinum for oxygen reduction in proton exchange membrane fuel cells. J Am Chem Soc 127(38):13100–13101

Forrestal C, Xu P, Ren Z (2012) Sustainable desalination using a microbial capacitive desalination cell. Energy Environ Sci 5:7161–7167

Freguia S, Masuda M, Tsujimura S, Kano K (2009) *Lactococcus lactis* catalyses electricity generation at microbial fuel cell anodes via excretion of a soluble quinone. Bioelectrochemistry 76(1):14–18

Freguia S, Teh EH, Boon N, Leung KM, Keller J, Rabaey K (2010) Microbial fuel cells operating on mixed fatty acids. Bioresour Technol 101(4):1233–1238

Gil G-C, Chang I-S, Kim BH, Kim M, Jang J-K, Park HS, Kim HJ (2003) Operational parameters affecting the performannce of a mediator-less microbial fuel cell. Biosens Bioelectron 18(4):327–334

Ha PT, Tae B, Chang IS (2007) Performance and bacterial consortium of microbial fuel cell fed with formate. Energy Fuels 22(1):164–168

Habermann W, Pommer E (1991) Biological fuel cells with sulphide storage capacity. Appl Microbiol Biotechnol 35(1):128–133

Hasvold Ø, Henriksen H, Citi G, Johansen BØ, Kjønigsen T, Galetti R (1997) Sea-water battery for subsea control systems. J Power Sources 65(1):253–261

He Z, Minteer SD, Angenent LT (2005) Electricity generation from artificial wastewater using an upflow microbial fuel cell. Environ Sci Technol 39(14):5262–5267

He Z, Wagner N, Minteer SD, Angenent LT (2006) An upflow microbial fuel cell with an interior cathode: assessment of the internal resistance by impedance spectroscopy. Environ Sci Technol 40(17):5212–5217

Holmes D, Bond D, O'neil R, Reimers C, Tender L, Lovley D (2004) Microbial communities associated with electrodes harvesting electricity from a variety of aquatic sediments. Microb Ecol 48(2):178–190

Huang L, Logan BE (2008) Electricity generation and treatment of paper recycling wastewater using a microbial fuel cell. Appl Microbiol Biotechnol 80(2):349–355

Ieropoulos IA, Ledezma P, Stinchcombe A, Papaharalabos G, Melhuish C, Greenman J (2013) Waste to real energy: the first MFC powered mobile phone. Phys Chem Chem Phys 15(37):15312–15316

Jiang D, Curtis M, Troop E, Scheible K, McGrath J, Hu B, Suib S, Raymond D, Li B (2011) A pilot-scale study on utilizing multi-anode/cathode microbial fuel cells (MAC MFCs) to enhance the power production in wastewater treatment. Int J Hydrogen Energy 36(1):876–884

Jung S, Regan JM (2007) Comparison of anode bacterial communities and performance in microbial fuel cells with different electron donors. Appl Microbiol Biotechnol 77(2):393–402

Kiely PD, Call DF, Yates MD, Regan JM, Logan BE (2010) Anodic biofilms in microbial fuel cells harbor low numbers of higher-power-producing bacteria than abundant genera. Appl Microbiol Biotechnol 88(1):371–380

Kiely PD, Cusick R, Call DF, Selembo PA, Regan JM, Logan BE (2011a) Anode microbial communities produced by changing from microbial fuel cell to microbial electrolysis cell operation using two different wastewaters. Bioresour Technol 102(1):388–394

Kiely PD, Rader G, Regan JM, Logan BE (2011b) Long-term cathode performance and the microbial communities that develop in microbial fuel cells fed different fermentation endproducts. Bioresour Technol 102(1):361–366

Kim B, Park H, Kim H, Kim G, Chang I, Lee J, Phung N (2004) Enrichment of microbial community generating electricity using a fuel-cell-type electrochemical cell. Appl Microbiol Biotechnol 63(6):672–681

Kim BH, Chang IS, Gil GC, Park HS, Kim HJ (2003) Novel BOD (biological oxygen demand) sensor using mediator-less microbial fuel cell. Biotechnol Lett 25(7):541–545

Kim H, Hyun M, Chang I, Kim BH (1999) A microbial fuel cell type lactate biosensor using a metal-reducing bacterium, *Shewanella putrefaciens*. J Microbiol Biotechnol 9(3):365–367

Kim HJ, Park HS, Hyun MS, Chang IS, Kim M, Kim BH (2002) A mediator-less microbial fuel cell using a metal reducing bacterium, Shewanella putrefaciens. Enzyme Microb Technol 30(2):145–152

Kim JR, Jung SH, Regan JM, Logan BE (2007a) Electricity generation and microbial community analysis of alcohol powered microbial fuel cells. Bioresour Technol 98(13):2568–2577

Kim M, Hyun MS, Gadd GM, Kim HJ (2007b) A novel biomonitoring system using microbial fuel cells. J Environ Monit 9(12):1323–1328

Lee S, Choi Y, Jung S, Kim S (2002) Effect of initial carbon sources on the electrochemical detection of glucose by *Gluconobacter oxydans*. Bioelectrochemistry 57(2):173–178

Lewis K (1966) Symposium on bioelectrochemistry of microorganisms: IV. Biochemical fuel cells. Bacteriol Rev 30(1):101–113

Liu H, Cheng S, Logan BE (2005a) Power generation in fed-batch microbial fuel cells as a function of ionic strength, temperature, and reactor configuration. Environ Sci Technol 39(14):5488–5493

Liu H, Cheng S, Logan BE (2005b) Production of electricity from acetate or butyrate using a single-chamber microbial fuel cell. Environ Sci Technol 39(2):658–662

Liu H, Grot S, Logan BE (2005c) Electrochemically assisted microbial production of hydrogen from acetate. Environ Sci Technol 39(11):4317–4320

Liu H, Logan BE (2004) Electricity generation using an air-cathode single chamber microbial fuel cell in the presence and absence of a proton exchange membrane. Environ Sci Technol 38(14):4040–4046

Liu H, Ramnarayanan R, Logan BE (2004) Production of electricity during wastewater treatment using a single chamber microbial fuel cell. Environ Sci Technol 38(7):2281–2285

Logan BE (2004) Feature article: biologically extracting energy from wastewater: biohydrogen production and microbial fuel cells. Environ Sci Technol 38:160A–167A

Logan BE (2008) Microbial fuel cells. Wiley, Hoboken, New Jersey

Logan BE, Hamelers B, Rozendal R, Schröder U, Keller J, Freguia S, Aelterman P, Verstraete W, Rabaey K (2006) Microbial fuel cells: methodology and technology. Environ Sci Technol 40(17):5181–5192

Logan BE, Murano C, Scott K, Gray ND, Head IM (2005) Electricity generation from cysteine in a microbial fuel cell. Water Res 39(5):942–952

Logan BE, Regan JM (2006a) Electricity-producing bacterial communities in microbial fuel cells. Trends Microbiol 14(12):512–518

Logan BE, Regan JM (2006b) Microbial challenges and applications. Environ Sci Technol 40(17):5172–5180

Lovley DR (2006) Microbial fuel cells: novel microbial physiologies and engineering approaches. Curr Opin Biotechnol 17(3):327–332

Lowy DA, Tender LM, Zeikus JG, Park DH, Lovley DR (2006) Harvesting energy from the marine sediment–water interface II: kinetic activity of anode materials. Biosens Bioelectron 21(11):2058–2063

Mauritz KA, Moore RB (2004) State of understanding of Nafion. Chem Rev 104(10):4535–4586

McCarty PL, Bae J, Kim J (2011) Domestic wastewater treatment as a net energy producer—can this be achieved? Environ Sci Technol 45(17):7100–7106

Mecheri B, D'Epifanio A, di Vona ML, Traversa E, Licoccia S, Miyayama M (2006) Sulfonated polyether ether ketone-based composite membranes doped with a tungsten-based inorganic proton conductor for fuel cell applications. J Electrochem Soc 153(3):A463–A467

Milliken C, May H (2007) Sustained generation of electricity by the spore-forming, gram-positive, *Desulfitobacterium hafniense* strain DCB2. Appl Microbiol Biotechnol 73(5):1180–1189

Min B, Angelidaki I (2008) Innovative microbial fuel cell for electricity production from anaerobic reactors. J Power Sources 180(1):641–647

Min B, Cheng S, Logan BE (2005) Electricity generation using membrane and salt bridge microbial fuel cells. Water Res 39(9):1675–1686

Min B, Logan BE (2004) Continuous electricity generation from domestic wastewater and organic substrates in a flat plate microbial fuel cell. Environ Sci Technol 38(21):5809–5814
Morris JM, Jin S (2008) Feasibility of using microbial fuel cell technology for bioremediation of hydrocarbons in groundwater. J Environ Sci Health Part A Toxic/Hazard Subst Environ Eng 43(1):18–23
Nielsen ME, Reimers CE, Stecher HA (2007) Enhanced power from chambered benthic microbial fuel cells. Environ Sci Technol 41(22):7895–7900
Niessen J, Harnisch F, Rosenbaum M, Schroder U, Scholz F (2006) Heat treated soil as convenient and versatile source of bacterial communities for microbial electricity generation. Electrochem Commun 8(5):869–873
Oh S-E, Logan BE (2006) Proton exchange membrane and electrode surface areas as factors that affect power generation in microbial fuel cells. Appl Microbiol Biotechnol 70(2):162–169
Oh S, Min B, Logan BE (2004) Cathode performance as a factor in electricity generation in microbial fuel cells. Environ Sci Technol 38(18):4900–4904
Oh ST, Kim JR, Premier GC, Lee TH, Kim C, Sloan WT (2010) Sustainable wastewater treatment: how might microbial fuel cells contribute. Biotechnol Adv 28(6):871–881
Pant D, van Bogaert G, Diels L, Vanbroekhoven K (2010) A review of the substrates used in microbial fuel cells (MFCs) for sustainable energy production. Bioresour Technol 101(6):1533–1543
Park D, Zeikus J (2002) Impact of electrode composition on electricity generation in a single-compartment fuel cell using *Shewanella putrefaciens*. Appl Microbiol Biotechnol 59(1):58–61
Park D, Zeikus J (1999) Utilization of electrically reduced neutral red by *Actinobacillus succinogenes*: physiological function of neutral red in membrane-driven fumarate reduction and energy conservation. J Bacteriol 181(8):2403–2410
Park DH, Zeikus JG (2000) Electricity generation in microbial fuel cells using neutral red as an electronophore. Appl Environ Microbiol 66(4):1292–1297
Park DH, Zeikus JG (2003) Improved fuel cell and electrode designs for producing electricity from microbial degradation. Biotechnol Bioeng 81(3):348–355
Park HS, Kim BH, Kim HS, Kim HJ, Kim GT, Kim M, Chang IS, Park YK, Chang HI (2001) A novel electrochemically active and Fe (III)-reducing bacterium phylogenetically related to *Clostridium butyricum* isolated from a microbial fuel cell. Anaerobe 7(6):297–306
Patil SA, Surakasi VP, Koul S, Ijmulwar S, Vivek A, Shouche Y, Kapadnis B (2009) Electricity generation using chocolate industry wastewater and its treatment in activated sludge based microbial fuel cell and analysis of developed microbial community in the anode chamber. Bioresour Technol 100(21):5132–5139
Peng L, You S-J, Wang J-Y (2010) Carbon nanotubes as electrode modifier promoting direct electron transfer from *Shewanella oneidensis*. Biosens Bioelectron 25(5):1248–1251
Pham CA, Jung SJ, Phung NT, Lee J, Chang IS, Kim BH, Yi H, Chun J (2003) A novel electrochemically active and Fe (III)-reducing bacterium phylogenetically related to *Aeromonas hydrophila*, isolated from a microbial fuel cell. FEMS Microbiol Lett 223(1):129–134
Phung NT, Lee J, Kang KH, Chang IS, Gadd GM, Kim BH (2004) Analysis of microbial diversity in oligotrophic microbial fuel cells using 16S rDNA sequences. FEMS Microbiol Lett 233(1):77–82
Potter MC (1911) Electrical effects accompanying the decomposition of organic compounds. Proc R Soc B Biol Sci 84(571):260–276
Qiao Y, Li CM, Bao S-J, Bao Q-L (2007) Carbon nanotube/polyaniline composite as anode material for microbial fuel cells. J Power Sources 170(1):79–84
Rabaey K, Boon N, Höfte M, Verstraete W (2005a) Microbial phenazine production enhances electron transfer in biofuel cells. Environ Sci Technol 39(9):3401–3408
Rabaey K, Boon N, Siciliano SD, Verhaege M, Verstraete W (2004) Biofuel cells select for microbial consortia that self-mediate electron transfer. Appl Environ Microbiol 70(9): 5373–5382

Rabaey K, Clauwaert P, Aelterman P, Verstraete W (2005b) Tubular microbial fuel cells for efficient electricity generation. Environ Sci Technol 39(20):8077–8082

Rabaey K, Rodriguez J, Blackall LL, Keller J, Gross P, Batstone D, Verstraete W, Nealson KH (2007) Microbial ecology meets electrochemistry: electricity-driven and driving communities. ISME J 1(1):9–18

Rabaey K, van de Sompel K, Maignien L, Boon N, Aelterman P, Clauwaert P, de Schamphelaire L, Pham HT, Vermeulen J, Verhaege M (2006) Microbial fuel cells for sulfide removal. Environ Sci Technol 40(17):5218–5224

Rabaey K, Verstraete W (2005) Microbial fuel cells: novel biotechnology for energy generation. Trends Biotechnol 23(6):291–298

Raghuveer V, Manthiram A, Bard AJ (2005) Pd-Co-Mo electrocatalyst for the oxygen reduction reaction in proton exchange membrane fuel cells. J Phys Chem B 109(48):22909–22912

Rezaei F, Richard TL, Brennan RA, Logan BE (2007) Substrate-enhanced microbial fuel cells for improved remote power generation from sediment-based systems. Environ Sci Technol 41(11):4053–4058

Rhoads A, Beyenal H, Lewandowski Z (2005) Microbial fuel cell using anaerobic respiration as an anodic reaction and biomineralized manganese as a cathodic reactant. Environ Sci Technol 39:4666–4671

Rinaldi A, Mecheri B, Garavaglia V, Licoccia S, di Nardo P, Traversa E (2008) Engineering materials and biology to boost performance of microbial fuel cells: a critical review. Energy Environ Sci 1(4):417–429

Ringeisen BR, Henderson E, Wu PK, Pietron J, Ray R, Little B, Biffinger JC, Jones-Meehan JM (2006) High power density from a miniature microbial fuel cell using *Shewanella oneidensis* DSP10. Environ Sci Technol 40(8):2629–2634

Rozendal RA, Hamelers HV, Euverink GJ, Metz SJ, Buisman CJ (2006) Principle and perspectives of hydrogen production through biocatalyzed electrolysis. Int J Hydrogen Energy 31(12):1632–1640

Rozendal RA, Hamelers HV, Rabaey K, Keller J, Buisman CJ (2008) Towards practical implementation of bioelectrochemical wastewater treatment. Trends Biotechnol 26(8):450–459

Rozendal RA, Leone E, Keller J, Rabaey K (2009) Efficient hydrogen peroxide generation from organic matter in a bioelectrochemical system. Electrochem Commun 11(9):1752–1755

Roziere J, Jones DJ (2003) Non-fluorinated polymer materials for proton exchange membrane fuel cells. Ann Rev Mater Res 33(1):503–555

Schroder U (2007) Anodic electron transfer mechanisms in microbial fuel cells and their energy efficiency. Phys Chem Chem Phys 9(21):2619–2629

Scott K, Rimbu G, Katuri K, Prasad K, Head I (2007) Application of modified carbon anodes in microbial fuel cells. Process Saf Environ Prot 85(5):481–488

Tender LM, Gray SA, Groveman E, Lowy DA, Kauffman P, Melhado J, Tyce RC, Flynn D, Petrecca R, Dobarro J (2008) The first demonstration of a microbial fuel cell as a viable power supply: powering a meteorological buoy. J Power Sources 179(2):571–575

Tender LM, Reimers CE, Stecher HA, Holmes DE, Bond DR, Lowy DA, Pilobello K, Fertig SJ, Lovley DR (2002) Harnessing microbially generated power on the seafloor. Nat Biotechnol 20(8):821–825

Thurston C, Bennetto H, Delaney G, Mason J, Roller S, Stirling J (1985) Glucose metabolism in a microbial fuel cell. Stoichiometry of product formation in a thionine-mediated Proteus vulgaris fuel cell and its relation to coulombic yields. J Gen Microbiol 131(6):1393–1401

Vante NA, Tributsch H (1986) Energy conversion catalysis using semiconducting transition metal cluster compounds. Nature 323(6087):431–432

Venkata Mohan S, Saravanan R, Raghavulu SV, Mohanakrishna G, Sarma P (2008) Bioelectricity production from wastewater treatment in dual chambered microbial fuel cell (MFC) using selectively enriched mixed microflora: effect of catholyte. Bioresour Technol 99(3):596–603

Wang X, Feng Y, Wang H, Qu Y, Yu Y, Ren N, Li N, Wang E, Lee H, Logan BE (2009) Bioaugmentation for electricity generation from corn stover biomass using microbial fuel cells. Environ Sci Technol 43(15):6088–6093

Xiao L, Young EB, Berges JA, He Z (2012) Integrated photo-bioelectrochemical system for contaminants removal and bioenergy production. Environ Sci Technol 46(20):11459–11466

Xing D, Cheng S, Logan BE, Regan JM (2010) Isolation of the exoelectrogenic denitrifying bacterium *Comamonas denitrificans* based on dilution to extinction. Appl Microbiol Biotechnol 85(5):1575–1587

Xing D, Cheng S, Regan JM, Logan BE (2009) Change in microbial communities in acetate-and glucose-fed microbial fuel cells in the presence of light. Biosens Bioelectron 25(1):105–111

Xing D, Zuo Y, Cheng S, Regan JM, Logan BE (2008) Electricity generation by *Rhodopseudomonas palustris* DX-1. Environ Sci Technol 42(11):4146–4151

Xu S, Liu H (2011) New exoelectrogen Citrobacter sp. SX-1 isolated from a microbial fuel cell. J Appl Microbiol 111(5):1108–1115

You S, Zhao Q, Zhang J, Jiang J, Zhao S (2006) A microbial fuel cell using permanganate as the cathodic electron acceptor. J Power Sources 162(2):1409–1415

Yu CP, Liang ZH, Das A, Hu ZQ (2011) Nitrogen removal from wastewater using membrane aerated microbial fuel cell techniques. Water Res 45(3):1157–1164

Yuan L, Yang X, Liang P, Wang L, Huang ZH, Wei J, Huang X (2012) Capacitive deionization coupled with microbial fuel cells to desalinate low-concentration salt water. Bioresour Technol 110:735–738

Zhang E, Xu W, Diao G, Shuang C (2006) Electricity generation from acetate and glucose by sedimentary bacterium attached to electrode in microbial-anode fuel cells. J Power Sources 161(2):820–825

Zhao F, Harnisch F, Schroder U, Scholz F, Bogdanoff P, Herrmann I (2005) Application of pyrolysed iron(II) phthalocyanine and CoTMPP based oxygen reduction catalysts as cathode materials in microbial fuel cells. Electrochem Commun 7(12):1405–1410

Zhao H, Bian P, Ju D (2009) Electrochemical performance of magnesium alloy and its application on the sea water battery. J Environ Sci 21:S88–S91

Zhou M, Chi M, Luo J, He H, Jin T (2011) An overview of electrode materials in microbial fuel cells. J Power Sources 196(10):4427–4435 [illegible] Feng Y, Wang [illegible] (2009) [illegible]

[illegible] for [illegible] bacteria [illegible]

[illegible]

Xiao L, Young EB, [illegible] (2014) Integrated photo-bioelectrochemical system for contaminants removal and bioenergy production. Environ Sci Technol 46(20):11459–11466

Xing D, Cheng S, Logan BE, Regan JM (2010) Isolation of the exoelectrogenic denitrifying bacterium [illegible] based on [illegible] selection. Appl Microbiol Biotechnol 85:1575–1587

Xing D, [illegible] Logan BE [illegible]

[illegible]

Xing D, [illegible]

[illegible]

[illegible]

[illegible]

MIX
Papier aus verantwortungsvollen Quellen
Paper from responsible sources
FSC® C105338

If you have any concerns about our products,
you can contact us on
ProductSafety@springernature.com

In case Publisher is established outside the EU,
the EU authorized representative is:
Springer Nature Customer Service Center GmbH
Europaplatz 3, 69115 Heidelberg, Germany

Printed by Libri Plureos GmbH
in Hamburg, Germany